Practical Electrical Wiring

Practical Electrical Wiring

RESIDENTIAL, FARM, AND INDUSTRIAL

THIRTEENTH EDITION

Based on the 1984 National Electrical Code®

H. P. Richter

W. Creighton Schwan

Electrician, Inspector, Professional Engineer

MCGRAW-HILL BOOK COMPANY

**New York St. Louis San Francisco Auckland Bogotá
Hamburg Johannesburg London Madrid
Mexico Montreal New Delhi Panama
Paris São Paulo Singapore
Sydney Tokyo Toronto**

Library of Congress Cataloging in Publication Data

Richter, H. P. (Herbert P.), 1900–1978.
Practical electrical wiring.

"Based on the 1984 National Electrical Code.®"
Bibliography: p.
Includes index.
1. Electric wiring, Interior. I. Schwan, W. Creighton.
II. National Electrical Code.® III. Title.
TK3271.R48 1984 621.319'24 83-26734

ISBN 0-07-052390-8

The editors for this book were Harold B. Crawford and Beatrice E. Eckes, the designer was Naomi Auerbach, and the production supervisor was Sally Fliess. It was set in Times Roman by Progressive Typographers, Inc.

Contents

Preface to the Thirteenth Edition

For over forty years *Practical Electrical Wiring* has been serving the needs of both students and practitioners of the art of electrical wiring. The first 10 editions were the work of H. P. Richter, whose ability to translate a complex subject into clear, simple, everyday language was outstanding. I have tried to follow the example he set.

New editions are issued as soon as possible after the appearance of the latest National Electrical Code®, which is revised on a three-year schedule. This thirteenth edition incorporates the applicable requirements found in the 1984 National Electrical Code®.

Electrical construction is a dynamic activity, with new products and new methods appearing continually, and I have tried to reflect herein those falling within the scope of this volume.

Suggestions for improvement of the book from readers will be most welcome.

W. CREIGHTON SCHWAN

Preface to the Ninth Edition

In preparing this book it has been the author's aim to make it simple enough for beginners, yet complete enough so that it will be of value also to those already engaged in electrical work. It is intended to be, not a manual that merely recites the methods used in wiring buildings for the use of electricity, but rather a book that explains the subject in such fashion that readers will learn both the *way* things are done and *why* they are done in that particular way. Only in this manner can students master the subject so that they can solve their own problems as they arise in actual practice, for no book can possibly cover all the different problems that are likely to arise.

Since this book is not intended to include the subject of electrical engineering, only so many basic engineering data as are essential have been included, and these so far as possible have been boiled down to ABC proportions.

All methods shown are in strict accordance with the National Electrical Code®, but no attempt has been made to include a detailed explanation of *all* subjects covered by the Code. The Code is written to include any and all cases that might arise in wiring every type of structure from the smallest cottage to the largest skyscraper; it covers ordinary wiring as well as those problems that come up only very rarely. The scope of this book has been limited to the wiring of structures of limited size and at ordinary voltages, at 600 volts or less. Skyscrapers and steel mills and projects of similar size involve problems that the student will not meet until long after he has mastered the contents of this book.

The book consists of three parts:

Part 1 presents the fundamentals of electrical work; terminology; basic principles; the theory behind general practices. Part 2 deals with the actual wiring of residential buildings and farms. Part 3 covers the actual wiring of nonresidential buildings, such as stores, factories, schools, and similar structures.

The science or principles of electricity do not change; the art or method of application does change. That portion of this book having to do with *principles* has in this ninth edition been revised and amplified to present such principles more clearly. That portion concerned with *methods* has been revised or rewritten as required by Code changes, to describe new materials and methods, and to outline the methods more clearly.

H. P. RICHTER

About the Authors

The late Herbert P. Richter, who died in 1978, was long regarded as both the electrician's electrician and the do-it-yourselfer's electrician. With the publication of the first edition of Practical Electrical Wiring *in 1939 and with the publication of each succeeding edition, he demonstrated again and again a rare ability to translate the technical and complex language of the National Electrical Code® into simple, everyday language. A member of the International Association of Electrical Inspectors and the National Fire Protection Association, he was also coauthor of the homeowners' guide* Wiring Simplified.

W. Creighton Schwan, a registered professional engineer, served as supervising electrical inspector of Alameda County, California, as electrical safety engineer for the California Division of Industrial Safety, and as senior field engineer for 10 western states for the National Electrical Manufacturers Association. In addition to revising Practical Electrical Wiring, *he is coauthor of* Wiring Simplified, *and he contributes a column, "Code Comments,"* to Electrical Contractor, *the monthly magazine of the National Electrical Contractors Association. He has taught National Electrical Code classes at several California schools and colleges and is a member of the National Fire Protection Association Electrical Section as well as of the International Association of Electrical Inspectors.*

Practical Electrical Wiring

PART 1

Theory and Basic Principles

Part 1 of this book is the introduction to practical electrical work, the ABC of the science and the art of electrical wiring. In order to master the art, naturally you must clearly understand the science or the principles involved. The terms used in the measurement of electricity, as well as the names of the devices used in wiring, must be at your finger tips.

For that reason a considerable portion of the material presented in Part 1 of this book emphasizes the "why" more than the "how." If you understand the "why," the "how" becomes obvious. Master both, and Parts 2 and 3 of this book will be relatively simple.

1

Codes, Inspections, and Product Approvals

The purpose of this book is to explain the types of electrical wiring, wiring materials, and equipment that are safe to use; how to install them in a safe manner; and why they are installed in certain ways so that electricity may be safely used as needed inside and outside of buildings.

Electricity Is Powerful. Under control, electricity safely performs an endless variety of work. But uncontrolled, it can be destructive. Electricity can be controlled if the right kinds of wiring and equipment are installed the right way. It can be dangerous if the wrong kinds of wiring or equipment are used, or even if the right kinds are improperly installed. And properly installed wiring and equipment can become dangerous if they are not properly maintained.

At best, improper electrical installations can be uneconomical and inconvenient. Moreover, they can result in higher insurance rates. But worst of all, unsafe installations can burn, maim, and kill.

The purpose of this book, then, is to explain how to properly install electrical wiring and equipment so that the completed installations will be safe, economical, and convenient—in short, so that instead of destroying, electricity will serve our needs.

National Electrical Code. The National Electrical Code (NEC)®* is a set of rules and regulations that sets forth the proper methods for installing electrical materials so that the finished job will be *safe*. It is published by the National Fire Protection Association (NFPA), Batterymarch Park, Quincy, MA 02269. Hereafter, the National Electrical Code will be referred to in this book either as the Code or as the NEC. This book will contain excerpts, tables, examples, etc., from the Code, all of which are reproduced by permission of the NFPA, the copyright holder. Any further reproduction of such excerpted material is not authorized without the express permission of the National Fire Protection Association.

NEC Committee. The National Electrical Code Committee of NFPA formulates the NEC. The NEC Committee has members representing all facets of the electrical industry. The NEC Committee consists of a Correlating Committee and 20 technical committees, which are called Code-Making Panels. Each panel consists of nationally recognized experts in one or more fields. These experts revise and update the NEC at three-year intervals, in keeping with the state of the art, at which time new materials and methods are covered and any obsolete rules are deleted. Hence, some *section numbers* of the Code are changed from time to time even when the text itself remains unchanged.

After general agreement has been reached by the members of the various Code-Making Panels in regard to proposed additions, deletions, or revisions, the Correlating Committee determines that there are no conflicts among the requirements adopted by the several panels. After that has been done, the proposed new edition of the Code is voted on by the NFPA membership. If adopted, the new edition is submitted to the American National Standards Institute (ANSI). After adoption by ANSI, the new NEC becomes an American National Standard. Hence, the NEC is "NFPA 70-1984" and also "ANSI/NFPA 70."

Clearly, then, the NEC is an authoritative code, and compliance with it will result in a safe installation if the installation is also maintained so as to *remain* in compliance. The word "authoritative" as used here means "an accepted source of expert information" rather than "the power to enforce," since the Code does not become law unless legally adopted as a law or code in a particular locality.

Local Codes. Almost every locality has an electrical code or ordinance. Some of these "local" codes are state codes. Others are municipal, town,

*National Electrical Code® and NEC® are Registered Trademarks of the National Fire Protection Association, Inc., Quincy, MA.

or county codes. Local codes are based primarily on the National Electrical Code. Electrical installations in any locality must meet the requirements of the electrical code in that locality.

Some localities have local codes that are more restrictive than the National Electrical Code. For instance, some wiring methods that are permitted by the National Electrical Code either are not permitted at all by some local codes or else have additional restrictions imposed by some local codes. Other localities adopt the National Electrical Code without any changes.

Enforcement of the NEC. When the NEC is legally adopted by a city, town, state, or other governmental body as an official code of that governmental body, it becomes law; and compliance with the Code in that locality then becomes mandatory on the date it is officially adopted.

Author's Interpretation of the Code. The Code is concerned with safety only, as is proper for a safety code and as is made clear by the following statements excerpted from Sec. 90-1 of the Code:

> This Code contains provisions considered necessary for safety. Compliance therewith and proper maintenance will result in an installation essentially free from hazard, but not necessarily efficient, convenient, or adequate for good service or future expansion of electrical use.
>
> This Code is not intended as a design specification nor an instruction manual for untrained persons.*

The safety measures advocated in this book are based on Code requirements as the author interprets the 1984 Code. But since the Code is not intended as a design or instruction manual, this book will also cover practical design and installation methods and the economic and convenience aspects of the art as well as safety.

Anyone using this book must recognize that the author does not and cannot accept any liability from its use. However, much thought and care have gone into the development of the material covered in an effort to make the book useful to all who wish to use it.

It is essential for every person who installs electrical wiring and equipment to have and use a copy of the Code. In addition to other objectives of this book, one of the objectives is to make it easier for you to under-

*Reprinted with permission from NFPA 70-1984, National Electrical Code®, Copyright© 1983, National Fire Protection Association, Quincy, Massachusetts 02269. This reprinted material is not the complete and official position of the NFPA on the referenced subject, which is represented only by the standard in its entirety.

stand and use the Code. So get a copy of the 1984 Code, which is the current edition.[1]

Format of the Code. The Code consists of an introduction and nine chapters. Each chapter contains several articles, except Chap. 9, which contains general tables and examples. Some articles are divided into parts, such as Parts A, B, etc. All articles are divided into sections. Many sections have subsections, and some have sub-subsections. Some sections or their subdivisions have numbered paragraphs. Some sections, subsections, etc., are followed by fine-print notes. All mandatory rules of the Code are in full-size print. The fine-print notes contain explanatory or informational material and are not mandatory rules.

In the Code, vertical lines in the margin of various pages show where changes, such as an added or a revised rule or an editorial change, have been made from the preceding edition; they show at a glance which material has been changed or is new.

History of the Code. The first electrical code was the "Standard for Electric Light Wires, Lamps, Etc." that was adopted by the New York Board of Fire Underwriters on Oct. 19, 1881. It contained seven rules, plus instructions for applying for permission to use electric lights. That was the ancestor of the NEC.

In May 1882, the National Board of Fire Underwriters adopted the rules drawn up by the New York Board of Fire Underwriters. Later, similar rules were adopted by various other groups. Finally, a committee consisting of representatives from the fire-insurance groups, the electric utility company association, and a number of other national organizations produced the first National Electrical Code in 1897. That was the first electrical code that represented *a nationwide consensus of the entire electrical industry without any single group having a dominant position.*

What Is the NFPA? The National Fire Protection Association is a nonprofit voluntary membership organization. It has members not only in the United States but also in Canada and some 80 other foreign countries. No single group has a dominant role. Its membership is representative not only of the fire services, but also of architects, engineers, and other professions; of industry and commerce; of government agencies at local, state, regional, and national levels, including the military branches; of hospital and school administrators; and of any others that have voca-

[1] The 1984 Code can be obtained by sending $15 per copy to the National Fire Protection Association, Batterymarch Park, Quincy, MA 02269. Ask for "NFPA 70-1984."

tional or even avocational interests in achieving a fire-safe environment, which has been the primary objective of the NFPA since its inception in 1896. (As used here, the term "fire-safe" is used in its broadest sense and includes *all* facets of fire safety, including—but not limited to—electrical safety.)

The many codes, standards, recommendations, and manuals that have been developed by NFPA technical committees, which are composed of nationally recognized experts in their respective fields, have been widely adopted by national and local governments and by private industry. One of these, the NEC, is the most widely adopted and used in the world.

Interpretations. Unfortunately many parts of the Code are interpreted differently by different people; the words of the Code mean different things to different people. The interpretations in this book are the opinions of the author, based on his experience and the opinions of others. There is no "official interpretation"[2] of the Code, and the Code in Sec. 90-4 makes it entirely clear that the local electrical inspector has the final word in any situation.

Permits. In many places it is necessary to get a permit from city, county, or state authorities before a wiring job can be started. The fees charged for permits generally are used to pay the expenses of electrical inspectors, whose work leads to safe, properly installed jobs. Power suppliers usually will not furnish power until an inspection certificate has been turned in.

Licenses. In some localities, electrical installations can be legally made only by licensed electricians or by persons who are supervised by a licensed electrician. Many localities in which licensing is required make an exception to the licensing law, allowing you to install wiring in your own home or farm. Other localities do not have such an exception. Ask the inspector in your locality about local rules, and *get a copy of the local code.*

Testing Laboratories. Although approval of all installations is the responsibility of the authority having jurisdiction (the inspector or the inspector's boss), that person may (and normally does) base approval of wiring materials and equipment on prior listing of the materials and equipment by a qualified testing laboratory, as provided in Sec. 90-6 of

[2] There are indeed "Formal Interpretations" of specific parts of the Code, but not of the Code as a whole. The method for requesting a Formal Interpretation of a specific provision is given in the "NFPA Regulations Governing Committee Projects." Very few such interpretations are handed down.

the Code, since the inspector has no means of making suitable tests of most types of material and equipment. Such laboratory "listing" (after testing) of products indicates the suitability of the products for their intended purpose. Then after the wiring and materials have been installed, it is the inspector who must determine whether acceptable materials have been used and properly installed. The inspector then approves the completed installation or turns it down until any deficiencies have been corrected.

There are several qualified testing laboratories, such as the Canadian Standards Association (CSA) and the Underwriters Laboratories of Canada, which operate in Canada; and the Factory Mutual Research Corporation (FM), Electrical Testing Laboratories, Inc. (ETL), and United States Testing Co., Inc., which test a limited range of equipment. But the testing laboratory that is most widely used in the United States is Underwriters Laboratories Inc. (UL). Hence, for the sake of simplicity and brevity, the term "UL" will be used in this book instead of the term "qualified testing laboratory." Similarly, the term "inspector" will be used instead of the term "authority having jurisdiction."

But to amplify and clarify this discussion, note again that the UL does not "approve" anything because approval would connote something beyond its service. UL investigates, studies, experiments, and tests products, materials, and systems, and if they are found to meet the UL safety requirements, UL will list them, and the items are then described as "Listed by UL" or "UL Listed." UL does not *approve* the item; it "lists" it. It is therefore wrong to say "approved by UL." The Code in Art. 100 defines "approved" as "acceptable to the authority having jurisdiction."* The inspector *approves* an item after UL has *listed* it and it has been used or installed in accordance with any instructions included in the listing. See Sec. 110-3(b) of the Code.

Requirement for UL Listed Equipment. Some local ordinances require that only UL Listed materials and equipment be used, if the material and equipment is of the type that is normally Listed by UL. [Some types of material and equipment, such as some fastening devices, lamps (light bulbs),[3] etc., are not Listed by UL.] Whether the ordinance requires such

*Reprinted with permission from NFPA 70-1984, **National Electrical Code**®, Copyright© 1983, National Fire Protection Association, Quincy, Massachusetts 02269. This reprinted material is not the complete and official position of the NFPA on the referenced subject, which is represented only by the standard in its entirety.

[3] A lamp consists of a bulb—the glass enclosure—and various other parts. Sometimes a lamp is called a bulb. See Chap. 14 for a discussion of an incandes-

Listing or not, the inspector usually will not approve an installation where unlisted material or equipment has been used, if it is the type that could be Listed. This applies to material and equipment in general, such as wires, cords, cables, boxes, fuses, circuit breakers, switches, receptacles, panelboards, etc.[4]

Unsafe Junk. Some manufacturers do not submit their products for Listing for a good reason—they could not pass the required examination and tests. These products are junk—shoddy and unsafe to use. Even if slightly lower in initial cost, such unlisted products are by no means inexpensive. They are "cheap" but not inexpensive, for they may cost as much as the building in which they are installed—plus the contents of the building. The ultimate cost might be one or more human lives. Figure 1-1 shows the results of UL tests of two plug fuses tested in accordance with standard testing procedures. The upper one failed; the lower one passed, as will be explained under the next heading.

UL Tests Are Meaningful. In order for a test to be meaningful, the product must be tested under the most severe conditions that would be encountered when the product is being used as intended. There must also be a safety-factor allowance. A fuse, as one example, must be able to operate (blow) and open the circuit (turn the circuit off as a switch does) within a predetermined time, such as when a wire is overloaded beyond its safe capacity. However, it must also blow if there is a short circuit, in which case the current might be a hundred or more times as great as is the case in just an overloaded wire. In other words it must be able to blow when the current is the *maximum* amount available, and it must do that without exploding or disintegrating, which could start a fire.

As an example, the familiar plug fuse used in homes (and in other occupancies) is made only in ratings up to (but not exceeding) 30 amp. But UL tests plug fuses on an alternating-current (ac) circuit that can deliver 10 000 amp at 0.85–0.95 power factor, which was the amount of alternating current UL used to test the two fuses shown in Fig. 1-1. Each fuse is entirely surrounded by surgical cotton. The cotton on the lower fuse did not ignite when the fuse operated (blew). The fuse shown at the top not only burned up the cotton but severely damaged the fuseholder as well as itself. Some fuses explode while undergoing this test, sending

---

cent lamp. The term "lamp" has been extended to include floor or table "lamps," which are actually portable lighting fixtures.

[4] In addition to electrical products, UL also tests materials related to Burglary Protection and Signaling, Casualty and Chemical Hazards, Fire Protection, Heating, Air Conditioning, Refrigeration, and Marine.

Fig. 1-1 The UL tested both fuses. It Listed the one at the bottom. The fuse at the top had no chance of becoming a Listed item. (*Underwriters Laboratories Inc.*)

molten particles flying against the walls and ceiling of the test enclosure. The fuse at the top was an unlisted fuse purchased on the open market. (In some localities, the law prohibits the sale of unlisted fuses and similar products; but in many localities, both Listed and unlisted products of this type are sold in various stores.) The fuse at the bottom in Fig. 1-1 was one of a group that had been submitted for testing by a reliable manufacturer. Similar meaningful tests are applied to all products submitted for UL Listing, such as wires, cables, and appliances.

Underwriters Laboratories Inc. Locations. UL has testing laboratories in its headquarters at 333 Pfingsten Road, Northbrook, IL 60062; as well as in Melville, N.Y.: Tampa, Fla.; and Santa Clara, Calif. It also has inspection centers in 134 cities in the United States and in over 66 foreign countries. Therefore foreign-made products intended for use in the United States can obtain UL Listing. Thus an inspector can reject foreign-made (as well as United States–made) products that are *not* Listed.

Identifying Listed Products. Tested products that meet UL standards are Listed under various categories. The Listed products that are of interest to readers of this book are contained in UL's "Electrical Construction Materials Directory," "Electrical Appliance and Utilization Equipment Directory," and "Hazardous Location Equipment Directory." These directories are usually available for checking Listings in the office of the electrical inspector as well as the local UL field office. You can inquire about purchasing a set of these directories from the UL office in Northbrook.

The various UL product directories and semiannual supplements should be used to find out which manufacturers have UL Listed products of the kind in question. Do not depend on such directories entirely, however, because a manufacturer may produce both Listed and unlisted products. Therefore, it is essential that you look for the UL Listing Mark (or label) itself, which indicates that a product is produced under the UL Follow-Up Service. (This service is explained under the heading "UL Follow-Up Service.")

For some small items, such as wire connectors, lampholders (sockets), and outlet boxes, the UL Listing Mark or symbol as shown in Fig. 1-2 is stamped or molded on the surface of each item. In these cases, the com-

Fig. 1-2 If because of size, shape, material, or surface texture, the product bears only the symbol at left, a complete Listing Marking with the four elements shown above will be found on the package. (*Underwriters Laboratories Inc.*)

plete Listing Mark will appear on the smallest package containing the devices. For larger items, such as raceways, lighting fixtures, panelboards, and motor controllers, a UL Listing Mark of one of the types shown in Fig. 1-3 is attached to the surface of each item. The Listing Mark may be on an interior or exterior surface, such as the interior surface of a panelboard enclosure and the exterior surface of a length of conduit or a fuse. Some products have a Listing Mark on the carton or reel.

Flexible cords sold by the foot have a Listing Mark on the spool. A

Fig. 1-3 Many items have an individual Listing Mark on each piece. A few samples are shown above. (*Underwriters Laboratories Inc.*)

"replacement power supply cord" is a length of flexible cord having an attachment plug on one end and having the other end prepared for permanent attachment to an appliance, such as a vacuum cleaner. Power supply cords intended for new appliances do not bear a Listing Mark, but the carton is so identified. Thus a consumer will not be duped into thinking that the entire appliance is Listed when only the power supply cord is covered. A replacement power supply cord does bear a Listing Mark and is intended to replace a cord that has deteriorated or been damaged. UL Listed cords of this type have a tan flag-type label as shown in Fig. 1-4.

A general-use "cord set" is a length of cord at least 6 ft long with an attachment plug on one end and the female member of a separable connector on the other end. This type of cord, commonly called an "extension cord," is for temporary use only and should be disconnected when the tool or appliance with which it is used has been turned off. An ex-

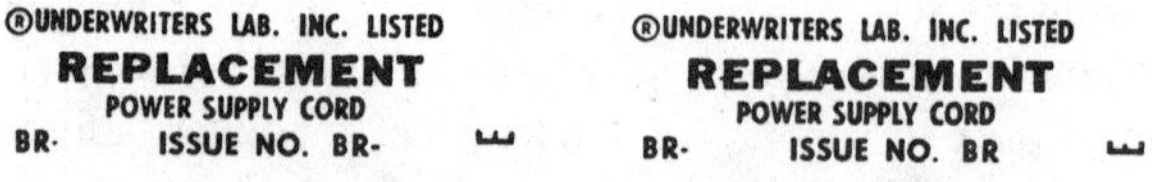

®UNDERWRITERS LAB. INC. LISTED
REPLACEMENT
POWER SUPPLY CORD
BR- ISSUE NO. BR-

®UNDERWRITERS LAB. INC. LISTED
REPLACEMENT
POWER SUPPLY CORD
BR- ISSUE NO. BR

Fig. 1-4 Listed replacement cords for appliances bear the tan label shown above. (*Underwriters Laboratories Inc.*)

tension cord should never be used as a permanent extension of a branch circuit. UL Listed cords of this type have either a blue flag label or a doughnut-type label, both of which are shown in Fig. 1-5.

Fig. 1-5 Cord assemblies, if Listed, bear one of the two blue labels above. (*Underwriters Laboratories Inc.*)

UL Listed appliances bear the UL Listing Mark shown in Fig. 1-2 combined with the word "Listed," a control number, and the manufacturer's name, trademark, or similar identifying symbol. The product name is also required unless it is obvious what the product is. The product category name may then be found in the UL product directories which provide information on the intended use and restrictions for the product.

UL Follow-Up Service. Each manufacturer that obtains UL Listing of its products signs a Listing and Follow-Up Services Agreement enabling a UL inspector to visit the factory at random to conduct inspections at the factory and to determine that the products are manufactured and tested in accordance with the UL standard and test procedures by which the original Listing was obtained. UL inspectors witness follow-up tests at the plant and sometimes send randomly selected samples to a UL testing laboratory for follow-up testing. Occasionally, UL inspectors purchase Listed products on the open market for follow-up testing. If the safety level of the product is not maintained, the UL Listing is withdrawn, and the manufacturer is prohibited further use of the UL Listing Mark until the safety requirements are again met and maintained.

What the UL Listing Means. Like the NEC, UL is concerned with *safety* only, and not with efficiency or convenience. Hence, a UL Listed product of one manufacturer may be superior to the same type of product of another manufacturer that is also UL Listed. Yet both products meet the *minimum* standards of safety. It is always prudent, therefore, to use good judgment when buying merchandise. In other words, a UL Listing is not the only factor to be considered, even though it is an essential factor.

Another important factor to remember is that a product is Listed for a specific purpose or type of use. As an example, the fact that a flexible cord is UL Listed does not mean that the cord is safe to use in a gasoline-dispensing pump, or in a location where it will be subject to physical damage, or used in any other way that was not intended. UL does not List any product for any use that would be in violation of any requirement of the NEC.

And finally, it is important to remember that a UL Listed cord attached to an appliance does not mean that the appliance itself is UL Listed. UL tests involve other features besides electrical insulation, clearances, etc. Hence, if an appliance is unsafe to use for its intended purpose in the intended manner, it will not pass UL tests even if the electrical parts are safe, because UL is interested in *total* safety, which includes (but is not limited to) electrical safety. So be sure that an entire assembly is UL Listed.

2

Numbers, Measurements, and Electricity: Basic Principles

The hows and whys of practical electrical wiring begin in Chaps. 3 and 4. But just as Chap. 1 contained basic information that you will need before starting to plan and install electrical wiring and equipment, this chapter contains much background information that you will also need about numbers and basic principles. The information about numbers (including the metric system) will help you to understand what you read. In other words, after studying this chapter, if you see a prefix such as "milli-" or "kilo-" as part of a word, you will know what it means. If you forget, you can refer to Table 2-1 from time to time. In addition, the basic principles of electricity and magnetism will be very briefly reviewed.

Numbers. The manner of writing numbers in all branches of science and technology has changed in recent years, so the new method will be used in this book. Some people are not entirely happy with every facet of the new method, but to attain international standardization, each country, including the United States, had to agree to a few changes—had to give a little in order to keep or gain a little.

For instance, in the United States and some other countries, numbers were separated at intervals of three digits on each side of a decimal point

TABLE 2-1

Unit	Multiple	Prefix	Symbol
1	(one)	None	None
$10 = 10^1$	(ten)	deka	da
$0.1 = 10^{-1}$	(one-tenth)	deci	d
$100 = 10^2$	(one hundred)	hecto	h
$0.01 = 10^{-2}$	(one-hundredth)	centi	c
$1000 = 10^3$	(one thousand)	kilo	k
$0.001 = 10^{-3}$	(one-thousandth)	milli	m
$1\,000\,000 = 10^6$	(one million)	mega	M
$0.000\,001 = 10^{-6}$	(one-millionth)	micro	μ*
$1\,000\,000\,000 = 10^9$	(one billion)	giga	G
$0.000\,000\,001 = 10^{-9}$	(one-billionth)	nano	n
$1\,000\,000\,000\,000 = 10^{12}$	(one trillion)	tera	T
$0.000\,000\,000\,001 = 10^{-12}$	(one-trillionth)	pico	p

*This is the lowercase Greek letter mu.

by a comma. In some other countries, a period was used instead of a comma. In the United States and some other countries, a period was used as a decimal point. In some other countries, a comma was used instead of a period. In the United States and some other countries, a billion was (and still is) a thousand millions. In some other countries, a billion was (and still is) a million millions.

Such variations could be tolerated for some things, but with international trade growing in an ever-shrinking world, it became desirable—essential in science and technology—to have international standardization. Like other branches of science and technology, electricity knows no political boundaries. Its laws are the laws of physics and chemistry, which are the same everywhere.

So, by international agreement, a space is now used instead of a comma, period, or any other mark between every third digit, except where the number consists of four or fewer digits, in which case the numbers are grouped together without spacings. And a period is used only for a decimal point. Examples: 7; 21; 356; 1245; 49 681; 0.496 81; 6 348 907.301 25; 0.634 890 7; 75 372.860 4. Always use a zero before a decimal point if there is no other number preceding it, thus: 0.34. Never this way: .34.

Numbers are based on the unit 1 and multiples of 10. Each multiple of 10 has a prefix, as shown in Table 2-1. You can readily see from Table 2-1 that if the unit is 1 ampere (as an example) then $^{1}/_{1000}$ (0.001 or one-thousandth) of an ampere is 1 milliampere. Similarly, if the unit is 1 watt, then 1000 watts is 1 kilowatt, etc.[1]

Abbreviations. The international agreement also requires that most abbreviations be in capital letters for terms bearing or derived from the name of a person, and in lowercase letters for other terms. (The terms used here in explaining abbreviations will themselves be explained later as they are used.) For example, "watt" is the name of a person, so it is abbreviated with a capital W. Similarly, such terms as "ampere," "volt," and "ohm" (each of which bears the name of a person) are abbreviated with a capital A, capital V, and capital Ω (which is the Greek capital letter omega). But such terms as "kilo" and "hour" are not derived from a person's name and are abbreviated with lowercase letters (k and h, respectively, for the examples given). So "kilowatthour" is abbreviated kWh, and "kilovolt-ampere" is abbreviated kVA.

However, this is not always true. As an example, the abbreviation for "mega" is a capital M, even though mega is not derived from a person's name. See Table 2-1. This is necessary in some cases to avoid confusion. In this case, capital M is used for mega, because lowercase m was already in use for the more commonly used term "milli." Another example is the term "kilovar," which is abbreviated kvar, even though the term (which stands for kilovolt-ampere-reactive) contains the words "volt" and "ampere," both of which are derived from a person's name. But again, this and other seeming contradictions are necessary to avoid confusion where infrequently used terms have the same letter abbreviations as those used in more frequently used terms. *When terms are spelled out, they are*

[1] For a complete treatise on the material shown in brief in Table 2-1, see "Letter Symbols for SI Units and Certain Other Units of Measurement" (ANSI/IEEE 260-1978). The abbreviation IEEE stands for Institute of Electrical and Electronics Engineers. The abbreviation ANSI stands for American National Standards Institute. The above standard is sponsored by the IEEE and has been adopted as an ANSI standard. It is also published by the American Society of Mechanical Engineers (ASME). The standard can be purchased from IEEE or from ASME; and all ANSI standards can be purchased from ANSI. The address of both IEEE and ASME is 345 East 47th St., New York, NY 10017. The address of ANSI is 1430 Broadway, New York, NY 10018. Any standards referenced hereafter as IEEE or ANSI standards may be purchased from either organization referenced.

never capitalized (unless the term is the first word of a sentence) *even if derived from a person's name*. In addition to the abbreviations shown in Table 2-1, some of the more common terms and their abbreviations, together with a brief explanation of their meanings, are listed in Table A-1 in the Appendix. For a complete list, see "Letter Symbols for SI Units and Certain Other Units of Measurement" (see footnote on page 17).

Metric System. In keeping with the worldwide process of abandoning the customary English units of measurement, the 1984 Code includes approximate values in SI units for specific values where they appear in the text and tables. "SI" is the abbreviation for the International System of Measuring Units, a modernized version of the metric system, adopted in 1960 at the Eleventh General Conference on Weights and Measures. Table 2-2 shows some commonly used English-metric equivalents for your convenience. You can refer to it from time to time if you need to.

Temperature. It will be helpful for you to know how to convert a given temperature from the Fahrenheit (°F) to the Celsius (°C) scale, and vice versa. The temperature scale that was formerly called the "centigrade" scale is now correctly called the "Celsius" scale in honor of the scientist who originated it. So there is no difference between the Celsius scale and the scale that was formerly (and often still is) called the centigrade scale.[2]

Temperature Conversion. The formulas for converting the temperatures from one scale to the other are as follows:

$$(\text{Degrees F} - 32)(5/9) = \text{degrees Celsius}$$
$$(\text{Degrees C})(9/5) + 32 = \text{degrees Fahrenheit}$$

In other words, to convert degrees F to degrees C, first subtract 32 from degrees F, then multiply the remainder by $5/9$. The answer is the same temperature on the Celsius scale. Example: Convert 86°F to degrees C: 86° − 32 = 54; then

$$(54)(5/9) = \frac{(54)(5)}{9} = \frac{270}{9} = 30°\text{C}$$

To convert degrees C to degrees F, first multiply the degrees C by $9/5$,

[2] "Centigrade" is a word of French and Latin derivation meaning "hundred grades" or "hundred marks," because in that scale water freezes at 0° and boils at 100°. (In the Fahrenheit scale water freezes at 32° and boils at 212°.)

TABLE 2-2 Commonly Used English-Metric Equivalents

English to Metric	Metric to English
LENGTH	
1 in = 25.4 mm (millimeters)	1 mm = 0.0394 in
1 in = 2.54 cm (centimeters)	1 cm = 0.394 in
1 ft = 304.8 mm	1 cm = 0.033 ft
1 ft = 30.48 cm	1 m = 39.37 in
1 ft = 0.305 m (meter)	1 m = 3.28 ft
1 yd = 0.915 m	1 km = 3280.83 ft
1 mi = 1609.34 m	1 km = 0.621 mi (mile)
1 mi = 1.609 km (kilometers)	
AREA	
1 sq in = 645.16 sq mm (mm^2)	1 sq cm = 0.155 sq in
1 sq in = 6.45 sq cm (cm^2)	1 sq cm = 0.0011 sq ft
1 sq ft = 929.03 sq cm (cm^2)	1 sq m = 10.764 sq ft
1 sq ft = 0.093 sq m (m^2)	1 sq m = 1.2 sq yd (yards)
VOLUME	
1 cu in = 16.38 cc (cm^3)	1 cc = 0.061 cu in
1 cu in = 0.016 liter (L or l)	1 L = 61.02 cu in
1 cu ft = 28.32 (L or l)	1 L = 0.035 cu ft
1 liquid qt = 0.9475 (L or l)	1 L = 1.056 liquid qt
1 liquid gal = 3.79 (L or l)	1 L = 0.264 liquid gal
MISCELLANEOUS CONVERSIONS	
946 cubic centimeters (cm^3) = 1 liquid qt (quart)	
1000 cubic centimeters (1000 cm^3) = 1 cubic decimeter = 1 L or l	
1 lb = 0.454 kg (kilogram) 1 kg = 2.2 lb (pounds)	
1 oz (ounce avoirdupois) = 28.349 g (grams)	

then add 32. The answer is the same temperature on the Fahrenheit scale. Example: Convert 30°C to degrees F:

$$(30°\text{C})(9/5) = \frac{(30)(9)}{5} = \frac{270}{5} = 54$$

Then add 32:

$$54 + 32 = 86°\text{F}$$

But be careful if you are trying to convert the *change in number of degrees* on a Celsius thermometer into the *change in number of degrees* on a Fahrenheit thermometer (or vice versa), rather than the actual reading in degrees on the two different thermometers. This is often confusing to some people, but note:

	Celsius	*Fahrenheit*
Water boils at	100°	212°
Water freezes at	0°	32°
Difference	100°	180°

From this you can see that each degree of *change* on the Celsius scale is equivalent to $^{180}/_{100}$ or $^{9}/_{5}$ or 1.8° change on the Fahrenheit scale; a degree of change on the Fahrenheit scale is equivalent to $^{100}/_{180}$ or $^{5}/_{9}$ or 0.555° change on the Celsius scale.

Arithmetic Refresher. When two or more numbers, each in parentheses, follow each other, it means that they are to be multiplied. Thus(5)(11)(13) means 5 × 11 × 13, or 715; and $2X$ means $(2)(X)$, which means $2 \times X$.

But when *letters* that are abbreviations of terms (such as V for volts and A for amperes) are to be multiplied, they need not be (but may be) placed in parentheses. Thus VA means (V)(A) or V × A, all indicating that the volts must be multiplied by the amperes.

Where multiplication, division, addition, and subtraction are shown in a formula, you must perform all multiplications and divisions first, and then perform the remaining operations. This is true *except* where additions or subtractions are enclosed in parentheses ( ) or brackets [ ], in which case you must perform the enclosed operations first. Examples:

$$(4)(6) - 2 = 24 - 2 = 22$$

$$(4)(6 - 2) = (4)(4) = 16$$

$$(^{4}/_{5})(^{5}/_{8})(3) + 3 - 1 = \frac{(4)(5)}{(5)(8)}(3) + 3 - 1$$

$$= \frac{20}{40}(3) + 3 - 1$$

$$= \frac{60}{40} + 3 - 1 = 1.5 + 3 - 1$$

$$= 3.5$$

$$(^4/_5)(^5/_8)(3 + 3 - 1) = \frac{(4)(5)}{(5)(8)}(3 + 3 - 1)$$

$$= \frac{20}{40}(5) = \frac{100}{40} = \frac{5}{2}$$

$$= 2.5$$

$$[(4) + (6 - 2)](3) = (4 + 4)(3) = (8)(3) = 24$$

When a square root is to be extracted, as indicated by the radical sign $\sqrt{\ }$, everything under the sign must be done first, then the square root is extracted, and finally any other indicated operations are performed. Examples: $\sqrt{4} = 2$ (because 2 multiplied by itself equals 4).

$$\sqrt{(2)(8)} = \sqrt{16} = 4$$

$$2\sqrt{(4)(4)} = 2\sqrt{16} = (2)(4) = 8$$

Where a number is followed by a second number that is placed just above and to the right of the first number, the second number is called an exponent. It means that the first number is to be used as a factor in multiplication, as often as indicated by the exponent. Thus 2^2 means 2 "to the second power" or 2 "squared," which means $2 \times 2 = 4$; 3^4 means 3 raised to the fourth power, or $3 \times 3 \times 3 \times 3 = 81$; 10^5 means 10 raised to the fifth power, or $10 \times 10 \times 10 \times 10 \times 10 = 100\,000$. It is common practice in speech to omit the word "power," by saying, for example, "10 to the fifth," meaning 10 raised to the fifth power, or 10^5.

Units of Electrical Measurement. In the study of electricity you will meet many terms that have to do with measurement: "volts," "amperes," "watts," and others. It will be much easier to learn how one is related to the other than to get an idea of the absolute value of each. That is because electric power is measured in units that cannot be compared directly with feet, pounds, quarts, or any other measure familiar to you. If you consult your dictionary, you will find definitions like these:

Volt: the pressure required to force one ampere through a resistance of one ohm.

Ampere: the electric current that will flow through one ohm under a pressure of one volt.

Ohm: the resistance through which one volt will force one ampere.

These definitions show a clear interrelationship between the three items, but unfortunately define each in terms of the other two. How "big" is each unit? It is as confusing as the beginner's first encounter with the metric system: 10 millimeters make a centimeter, 100 centimeters make a meter, 1000 meters make a kilometer; further, 1000 cubic centimeters make a liter; a liter of water weighs a kilogram. Now pretend you have neither seen nor studied Table 2-2 in this chapter. These metric terms are meaningless to you unless you can translate them into more familiar terms such as inches, miles, pounds, quarts.

After using these metric terms for a while, you will indeed begin to see some relationship between them and the more familiar terms you are accustomed to using. So also in electrical work; after a while you will begin to see some meaningful relationships among the various terms such as volts, amperes, ohms, and others.

Unfortunately in electrical work, it is difficult to translate an ampere, a volt, or an ohm into something that is familiar to you. But we can compare these terms with other measures that behave in a similar fashion. The best of comparisons or analogies are not very good, but they are better than none.

Gallons of Water. We can measure water in pounds or cubic feet or acre-feet or in many other ways. The measure known to most people is no doubt the gallon. A gallon of water is a specific quantity of water.

Coulombs of Electricity. When we come to measure electric current, the term that corresponds directly to the gallon in the case of water is the "coulomb." Ask all the people you know in the electrical business, "How much is a coulomb of electric power?" and 99% or more will answer, "Coulomb? I vaguely remember the term from way back when, but I don't know what a coulomb is." That being the case, why should we talk about it here? The only answer is that, while very few people indeed remember the definition of the term, it is nevertheless very helpful in getting to understand other electrical terms. We can use it as a temporary tool, just as a child learns how to ride a tricycle before learning how to ride a bicycle. Just accept it as a fact that a coulomb is a very definite quantity of electric current; do not try to understand how big that quantity is.

Water in Motion. Gallons of water standing in a tank are just quantities of water. But if there is a small hose connected to the tank, water will flow out of it. If there is a big hose, water will flow out of it faster than out of the small hose. If you want to talk about how much faster it flows out of the big hose than out of the small one, you must use some measure to denote the rate of flow, so usually you talk about "gallons per minute." This phrase indicates the quantity (gallons) and the time (per minute). Together, they indicate the rate of flow.

Electricity in Motion. Instead of a tank of water, consider a battery, a generator, or other source of electric power. Instead of a hose connected to the tank, think of a wire through which electric current will flow. Coulombs of electricity will flow through that wire, just as gallons of water flowed through the hose. But instead of gallons of water per minute, we have coulombs of electric current per minute. Only in the case of electric current, we say coulombs per *second*. But instead of using such an awkward phrase as "10 coulombs per second," we say "10 amperes," because an ampere[3] is defined as a flow of one coulomb per second.

Don't say that current is flowing at 10 amperes per *second*, though, for that would be the same as saying that it is flowing at the rate of 10 coulombs per second per *second*. Just remember that a coulomb is a quantity; an ampere, a rate of flow of that quantity. Once you clearly understand that, you can forget the coulomb and think in terms of amperes.[4]

Water under Pressure. A gallon of water standing in a tank is an inert, static quantity. Water dribbling out of your not-quite-shut-off garden hose at a gallon per minute is just a nuisance. Water coming out of your sprinkler on the lawn at a few gallons per minute waters your lawn and can be a delightful shower for a child playing in it. The same gallons-per-minute flow coming out of a hose in the form of one tiny stream but at a higher pressure will be painful when directed at the same child who, a moment earlier, thought it was fun.

Suppose there are three water tanks located 10, 20, and 100 ft above

[3] The ampere is named after André Marie Ampère (1775–1836), one of the great scientists of the early nineteenth century, who discovered many of the fundamental laws concerning the flow of electric current.

[4] In these days of electronics and atomic and nuclear bombs, a word may be in order to those who are interested in the purely scientific aspect of the subject. A flow of one ampere is equivalent to a flow of 6 280 000 000 000 000 000 electrons per second past a given point. However, it is entirely safe to forget all about coulombs and electrons per second, except for those who intend to delve into electrical engineering, and even for those the exact figure is of more academic than practical interest. For those who are interested in comparing numbers as such, it may be interesting to note that the big number shown above is over 100 million times greater than the number that represents the number of seconds that have elapsed in the more than 1900 years since the beginning of the Christian era.

ground level. A pipe runs from each tank to the ground level, and a pressure gauge is connected at the ground level. The gauges on the three tanks will show pressures of 4.3, 8.6, and 43 lb per sq in. (This disregards such things as, for example, pipe friction; let us not complicate the problem by going into details that are for the finished engineer, details that theoretically should be taken into consideration but which may be overlooked for the sake of simplification.) If a tank were located 1000 ft above ground level, the pressure would be 430 lb per sq in.

One gallon per minute running out of the first tank would be a nuisance; a gallon per minute out of the last tank at 430 lb per sq in could do a lot of damage. The difference lies in the difference in pressure between the two. The difference is measured in pounds per square inch.

Electric Power under Pressure. Electric power is also under pressure, but instead of being measured in pounds per square inch, it is measured in volts. One volt[5] is a very low pressure; in commercial work higher pressures or voltages are used.

An ordinary flashlight dry-cell battery (regardless of size) will, if fresh, develop approximately 1½ volts. A single cell of a storage battery develops about 2 volts when fully charged; the six cells of an automobile battery develop 6 × 2, or 12 volts. An ordinary house-lighting circuit operates at about 120 volts. The high-voltage lines feeding the transformers in alleys or city streets usually operate at 4160 to 13 800 volts.

The amount of electric current in amperes depends on the number of electrons flowing past a given point in one second. The pressure forcing the electrons or electric current to flow is called the voltage. Compare the electrons with rifle bullets. If I toss a bullet at you, it will hit you at a low pressure, and no harm is done. If I fire one at you from a rifle, it will hit you at a high pressure and kill you.

However, voltage does not affect the speed at which electric power flows. That is a constant speed, the same as the speed of light, or 186 000 miles per second. It might be interesting to note that this is a speed of just under 1 billion feet per second, which would seem fast enough for any purpose. But, many modern computers perform operations in billionths of a second, so if in the circuits within a computer the current has to

[5] The volt is named after Count Alessandro Volta (1745–1827), one of the great pioneer scientists who had much to do with the early research in electricity. For example, he discovered that when two dissimilar metals are immersed in an acid, an electric current will flow through a wire connecting the two metals. In other words, he discovered the principle of batteries.

travel over long wires, every foot of extra length slows an operation down by a billionth of a second. That is enough to significantly slow down the speed at which some computers operate.

Water Once More. Suppose that each of the water tanks we have talked about has a capacity of 100 gal and that each tank is empty. You must fill each tank using a hand pump (pumping out of a source of water at pump level, just to simplify the problem). You will have to work quite hard to fill the tank located 10 ft above ground level in, say, 5 min. You will have to work twice as hard to fill the one located 20 ft above ground level in the same time. To fill the one located 100 ft above ground level in the same 5 min, you would have to be "Superman"; to fill the one located 1000 ft above ground level would be a fantastic job to do by hand.

It takes you 5 min to fill the 100-gal tank located 10 ft above ground level. Then to fill a tank ten times as large (1000 gal) located at the same level should take you ten times as long, or 50 min. By the same token, to fill the original 100-gal tank but now located ten times as high (100 ft) should take you ten times as long, also 50 min. In other words, it should take you 50 min to fill a 1000-gal tank located 10 ft above ground level or a 100-gal tank located 100 ft above ground level. All this is probably easier to grasp in tabular form.

Capacity of tank, gal	*Height above ground, ft*	*Approx. pressure, lb per sq in*	*Time to fill, min*
100	10	4.3	5
1000	10	4.3	50
100	100	43.0	50

From this discussion it is easy to see that gallons per minute alone do not determine the amount of power involved in filling the tank; and pressure alone does not determine it; but the two in combination do. In other words, gallons per minute times pressure per square inch equals the amount of power being consumed or used at a specific moment in pumping.

Watts. In measuring electric power, neither the current in amperes nor the pressure in volts tells us the amount of power in a circuit at any moment. A combination of the two does tell us the answer very simply, for volts × amperes = watts. Watts equal the total power of a circuit at any given moment just as horsepower equals the power developed by an engine at any given moment. Indeed, horsepower and watts are merely two different ways of measuring or expressing the rate of work, or the power; 746 watts equals 1 hp.

If a lamp consumes 746 watts, you would be correct if you called it a 1-hp lamp, although that method of designating lamps is never used. Similarly you would be correct in saying that a 1-hp motor is a 746-watt[6] motor, although that method of designating power of motors is not used (except in the case of "flea-power" motors, which are sometimes rated, for example, "approximately 1½ watts output," instead of being rated as "1/500 hp").

Again note that the electrical term is just "watts," not "watts per hour." You would not say that the engine in your automobile delivers 150 hp per h; but at any given moment it delivers 150 hp. Both watts and horsepower denote a rate at which work is being done at a particular moment, not a quantity of work being done during a given time.

From the above you can see that a given amount of power in watts may be obtained from a combination of various voltage and current values. For example:

$$
\begin{aligned}
3 \text{ volts} \times 120 \text{ amp} &= 360 \text{ watts} \\
6 \text{ volts} \times 60 \text{ amp} &= 360 \text{ watts} \\
12 \text{ volts} \times 30 \text{ amp} &= 360 \text{ watts} \\
60 \text{ volts} \times 6 \text{ amp} &= 360 \text{ watts} \\
120 \text{ volts} \times 3 \text{ amp} &= 360 \text{ watts} \\
360 \text{ volts} \times 1 \text{ amp} &= 360 \text{ watts}
\end{aligned}
$$

Carrying the illustration further, a lamp in an automobile headlight consuming 5 amp from a 12-volt battery consumes a total of 5 × 12, or 60 watts; a lamp consuming 1/2 amp from a 120-volt lighting circuit in a home consumes a total of 1/2 × 120—also 60 watts. The voltage and the current values differ widely, but the power in watts of the two lamps is the same.

This simple formula is not correct under all circumstances; the exceptions will be covered in Chap. 3.

Kilo-, Mega-, Milli-, Micro-. A watt is a very small amount of power; it is only 1/746 hp. As you learned from Table 2-1, the term "kilo-" means

[6] Note, however, that a motor that *delivers* 1 hp or 746 watts of power actually consumes more nearly 1100 watts from the power line. The difference between the 1100 watts consumed and the 746 watts delivered as useful power is consumed as heat in the motor, to overcome bearing friction, air resistance of the moving parts, and similar factors.

thousand, so when we say "1 kilowatt," it is just another way of saying "1000 watts." The abbreviation is kW, and 25 kW then means 25 000 watts. And 20 kilovolts (kV) means 20 000 volts, and so on.

Practical examples of the use of other prefixes that you learned from Table 2-1 are:

Mega-: a million. Examples: megawatts, megacycles. Thus 400 megawatts means 400 000 kW or 400 000 000 watts. The abbreviation for mega- is the capital letter M.

Milli-: one-thousandth. Examples: milliamperes, milliwatts. Thus 25 milliamperes means $^{25}/_{1000}$ amp. The abbreviation for milli- is the lowercase letter m.

Micro-: one-millionth. Examples: microvolts, microamperes, microseconds. Thus 25 microamperes means $^{25}/_{1\,000\,000}$ amp. The abbreviation for micro- is the lowercase Greek letter mu (μ).

Watthours. The watt merely indicates the total amount of electric power that is being delivered *at a given moment;* it tells us nothing about the total quantity of electric power that was delivered during a period of time. The fact that a man earns $5 per h tells us nothing about his earnings per year unless we know how many hours he works. Multiplying the watts by the number of hours during which this number of watts was being delivered gives us watthours, abbreviated Wh, which equals the total amount of electric power consumed *during a given time*. For example:

$$
\begin{aligned}
10 \text{ watts} \times 1000 \text{ h} &= 10\,000 \text{ Wh} \\
100 \text{ watts} \times 100 \text{ h} &= 10\,000 \text{ Wh} \\
1\,000 \text{ watts} \times 10 \text{ h} &= 10\,000 \text{ Wh} \\
5\,000 \text{ watts} \times 2 \text{ h} &= 10\,000 \text{ Wh} \\
20\,000 \text{ watts} \times \tfrac{1}{2} \text{ h} &= 10\,000 \text{ Wh}
\end{aligned}
$$

Kilowatthours. One kilowatthour (kWh) is 1000 Wh, 20 kWh is 20 000 Wh, etc. Power is paid for by the kilowatthour. (The rate schedule of a power supplier may also include a "demand factor" that allows an added charge for power under certain conditions, but the basic rate is based on the kilowatthours consumed. A demand factor is rarely involved in residential or farm installations.)

Amperage and Wattage. The terms "amperage" for the amount of current in amperes and "wattage" for the amount of power in watts are quite regularly used informally. They are not formal terms, but are generally considered acceptable for informal usage. Similarly, the abbreviations "amp" and "amps" are not formal abbreviations, but they are also regularly used. Of the two, "amp" is preferable to "amps" even where the plural is intended.

Reading Meters. A meter that has a simple scale is easy to read and needs no explanation. Such a meter is shown in Fig. 2-1. It is a voltmeter,

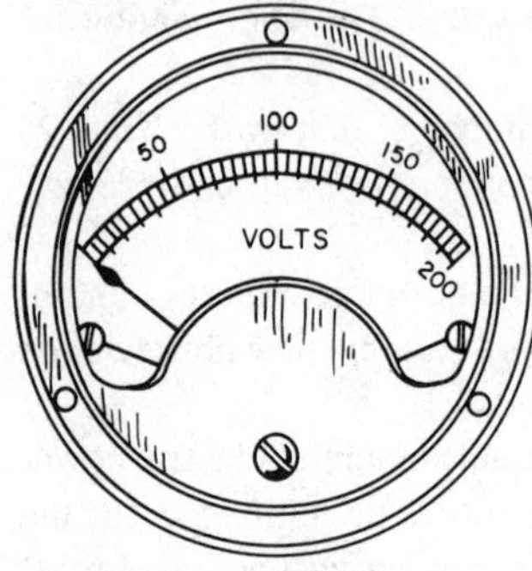

Fig. 2-1 Meters of this type are easy to read.

which is used to measure the potential difference in volts between two conductors or between two points. Other meters also have simple scales, such as an ammeter, which is used to measure the amount of current in amperes; and an ohmmeter, which is used to measure the resistance in ohms of a conductor, resistor, etc. There are combination meters that have more than one scale, such as a combination voltmeter, ammeter, and ohmmeter, that are more difficult to read than a meter with a single scale. Yet, a careful examination of the instrument, and care in following the manufacturer's instructions, are all that is necessary to enable you to read each of the scales properly. You must always be careful when using any meter to avoid damaging it by an incorrect connection. Never use a dc meter on an ac circuit, or vice versa. However, some combination meters permit using at least some of the scales on either ac or dc circuits. DC meters have one terminal marked "pos." or "+," the other marked "neg." or "−." If you connect the terminals of a dc meter the wrong way, the needle of the meter will press against the zero stop peg instead of indicating a reading; in such a case, just reverse the connections.

A *voltmeter* is always connected "across" a circuit, or points on a circuit, i.e., from one conductor to another or from one conductor to ground, to determine the circuit voltage, or from one point on a conductor or point of a circuit to another point to determine the voltage drop between the two points. This must be done while the circuit is energized or "hot."

An ordinary *ammeter* is always connected in series with the load so that the entire current flows through the meter. Always turn off the circuit while making such a connection.

A special kind of ac ammeter is the clamp-on type; two are shown in Fig. 2-2. Each has a movable member that is clamped around any con-

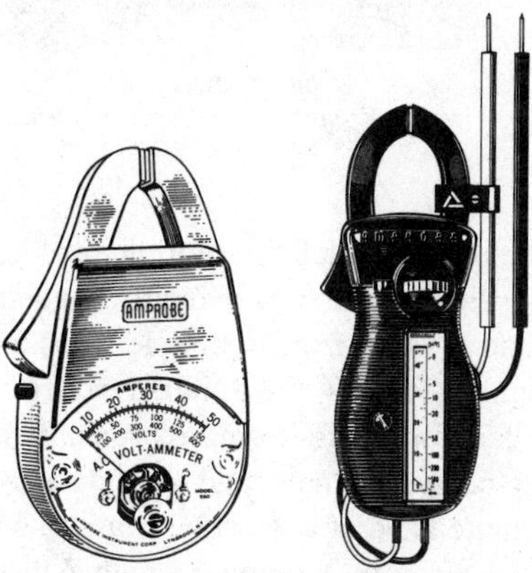

Fig. 2-2 Clamp-on ammeters. They also have voltage scales. (*Amprobe Instrument Division, Core Industries Inc.*)

tinuous wire carrying ac current; the meter then shows the number of amperes flowing in that wire. Each meter is also supplied with a pair of wire leads, permitting the meter to be used also as an ac voltmeter. Although usually not quite as accurate as an ordinary ac ammeter (which requires you to disconnect a wire and connect the meter in series with it), the clamp-on meter is very useful, especially in troubleshooting. Such a meter cannot be used on dc circuits.

Always be alert and careful when using any meter. It must be connected only to a circuit within the voltage and current range for which it was designed. The proper connections and proper setting of the selector switch must always be made on any combination meter.

Reading Kilowatthour Meters. Some kilowatthour meters are more difficult to read than others. Some are equipped with a cyclometer-type dial, as shown in Fig. 2-3, which is easy to read. Others have register-type dials, as shown in Fig. 2-4. The dials of this meter are shown enlarged to approximately full size in Figs. 2-5 and 2-6. The dials are numbered, alternately, clockwise and counterclockwise. In order to determine how much power has been consumed during any given period of time, such as a month, the dials must be read at the beginning and ending of that time span. The reading obtained at the beginning is then subtracted from the later reading to obtain the kilowatthours used during that time span.

Fig. 2-3 This type of register is found on some kilowatthour meters. (*General Electric Co.*)

Fig. 2-4 Ordinary meters have this type of register. (*General Electric Co.*)

How to Read the Dials. Although the number resulting from the reading of the dials in Figs. 2-5 and 2-6 is an ordinary number with digits from left to right, it is easier to read the dials one at a time from right to left, just as a column of figures is added from right to left, but the total is read from left to right. This is true with the dials for the following reasons: *Any three adjacent dials* can be likened to the hour, minute, and second hands of a watch. In other words, if the minute hand of a watch appears to be pointing directly to a number, the only way to determine whether that hand is actually on the number, has just passed the number, or has not quite reached the number is to see whether the second hand has passed the 12 o'clock mark. The same thing is true with the hour hand and the minute hand of a watch. And the same thing is true with the meter dials.

For example, in Fig. 2-5, the dials can be easily read from left to right, but the dials shown in Fig. 2-6 cannot, because it is impossible to determine whether the pointer on the *third* dial from the right is on 4, has passed 4, or has not quite reached 4 until the pointer on the *second* dial from the right has been read to determine its position in respect to the zero (12 o'clock) position. Therefore, it is easier to write down the numbers from right to left (as when adding a column of figures) and then read

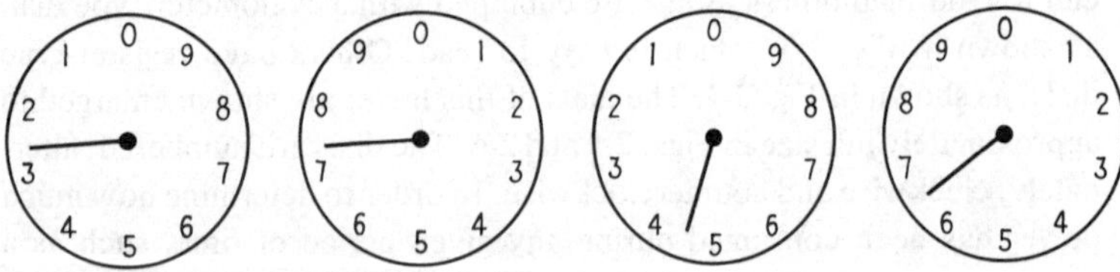

Fig. 2-5 Enlarged register of a kilowatthour meter. Read the figure that the pointer has passed. The reading above is 2746 kWh.

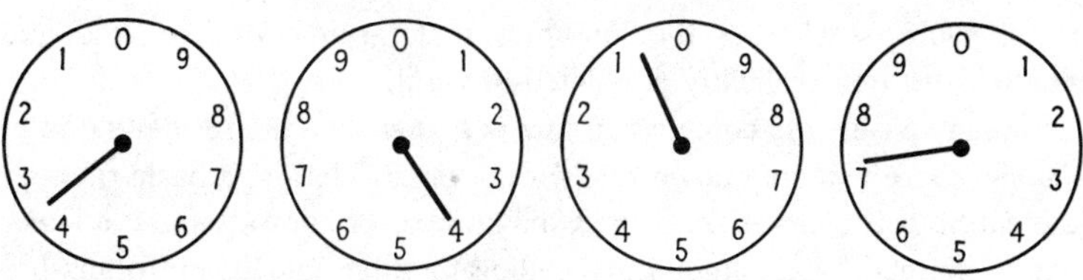

Fig. 2-6 The meter now reads 3407 kWh.

the results from left to right. The meter shown in Fig. 2-5 was the position of the dial pointers at the beginning of the month (or other reading period). The same meter is shown in Fig. 2-6 at the end of the month. So the readings are taken as follows:

For the reading at the beginning of the month in Fig. 2-5, the pointer of the right-hand dial is between 6 and 7, so write down 6. (Always write down the number that the pointer has *passed*. If the pointer has not *passed* a number, write down the next *lower* number.) On the second dial from the right, write down 4 to the left of 6. Similarly, on the third and fourth dials from the right, write down, respectively, 7 and 2 to the left of 46. The final reading from left to right is 2746.

For the reading at the end of the month in Fig. 2-6, the pointer of the right-hand dial has passed 7, so write down 7. The pointer on the second dial from the right is between 0 and 1, so write down 0. Now, on the third dial from the right, the pointer appears to be right on 4, but you have just seen that the pointer on the dial to its right has *passed* 0, so you know that the pointer on the *third* dial from the right has also just *passed* 4, so write down 4. The pointer on the fourth dial from the right is clearly between 3 and 4, so write down 3. The reading at the end of the month, then, is 3407. Subtract the first reading, 2746, from the second, 3407, to obtain 661 kWh consumed during the month.

Where a large amount of power is consumed, the kilowatthour meter will have a multiplier marked on the face of the meter, such as "× 10" or "Multiply by 10" or "× 100." In such cases the reading obtained from the dials must be multiplied by the specified multiplier to determine the actual kilowatthours consumed.

Figuring Electric Bills. The cost of electricity varies from one locality to another, depending on production and delivery costs, such as on the type of fuel available; the cost of obtaining and transporting the fuel; the type of generating plant, such as whether hydroelectric or fuel type; transmission distances and type of terrain on transmission routes; distribution costs, such as whether the lines are underground or overhead and whether the area is thickly populated or rural.

In many cases, the commercial rate is higher than the residential rate. Usually there is a step-down rate for all users. That is, a basic rate per kilowatthour is charged for a specified number of kilowatthours; a lower rate is charged for an amount exceeding the first amount up to another specified amount, with another reduction for amounts beyond that, etc., until the minimum rate is reached for all power over a specified amount. Table 2-3 is an example of a typical rate schedule, though the step-down points and the rate shown at each level will vary from one locality to another.

TABLE 2-3 Typical* Electric Rates

First 100 kWh used per month	7.50¢ per kWh
Next 100 kWh used per month	6.75¢ per kWh
All over 200 kWh used per month	5.50¢ per kWh
Assuming you used 661 kWh during a month, your bill would be figured this way:	
100 kWh at 7.50¢	$ 7.50
100 kWh at 6.75¢	6.75
461 kWh at 5.50¢	25.35
661 kWh total	$39.60
Average per kilowatthour (approx.)	5.84 cents

* The figures above are typical of rates as this is being written. In view of the rate of inflation now in progress, it is probable that rates will be considerably or even drastically increased in coming years. Fuel-cost adjustments authorized by some rate-setting bodies and basic monthly service charges independent of usage further add to the cost of electric power.

To encourage conservation of energy, many rate schedules are now *inverted:* the more power you use, the *higher* the average cost per kilowatthour.

Energy Conservation. Since the early 1970s, power suppliers have been faced with rising fuel costs, environmental restraints on the construction of new generating facilities, and safety restraints on the construction and operation of nuclear plants. Load growth continues, making it necessary that the existing capacity of each power supply system be used effectively. One way to do this is to reduce the load on the system at peak demand periods, and increase the load at other times. Some of the methods used, or being considered, to accomplish load management include:

1. Peak-load pricing (already in effect for some large *individual*

users, and discussed later under "Demand Metering"), by which a higher rate per kWh would be charged for consumption during peak periods of system demand. The assumption is that users would defer some consumption to the off-peak periods, to avoid the higher rates. On a daily basis this would require special metering equipment for each customer. On a seasonal basis this is already being done in some localities, with a higher rate per kWh being charged during, for example, the air-conditioning (summer) season.

2. Direct control, by the power supplier (by radio, over telephone lines, or by signals carried on the power distribution lines), of selected loads, such as water heaters or air-conditioner compressors, whereby the individual loads could be cut off for short intervals. Again, special equipment would be required at each service location.

Determining Electrical Operating Costs. Two methods can be used to determine the cost per hour of operating any specific load—the conventional method and a short-cut method. To illustrate both methods, a 1000-watt iron and a 6-cent-per-kWh electric rate will be used, as follows:

Conventional method: 1000 watts divided by 1000 equals 1 kW; 1 kW times 1 h equals 1 kWh; 1 kWh times 6 cents equals 6 cents per h to operate the iron.

Short-cut method: Note that 1000 watts was changed to 1 kW by dividing by 1000, which is equivalent to moving the decimal point to the left three places, changing 1000 to 1.000 or 1. Note also that by moving the decimal point to the left two places, 6 becomes 0.06. Hence, the short-cut method simply consists of multiplying the load in watts by the number representing the electric rate in cents per kilowatthour and then moving the decimal point to the left five places, which will show the cost per hour in cents or in dollars and cents, thus: 1000 watts times 6 equals 6000; moving the decimal to the left five places changes the figure to 0.060 00 or $0.06 or 6 cents.

Similarly, the cost per hour of a 40-watt lamp at a 6-cent rate or a 5000-watt oven at a 5-cent rate can be quickly determined. Lamp: (40)(6) = 240 or $0.002 40 or $^{24}/_{1000}$ dollars or $2.^{4}/_{10}$ cents per hour. Oven: (5000)(5) = 25 000 or $0.250 00 or 25 cents per hour.

To determine the number of hours any load can be operated to consume exactly 1 kWh, simply divide 1000 by the number of watts stamped on the load. This will show that a 1000-watt load can be operated for 1 h, a 50-watt load for 20 h, a 2000-watt load for $^{1}/_{2}$ h, etc.

Demand Metering. Large commercial and industrial users having widely fluctuating loads may be subject to an added charge by the power supplier, based on the demand factor. (Demand is the average kW consumed over a specific time interval, usually 15, 30, or 60 min. Demand factor is the ratio of the maximum demand to the connected load.) If large amounts of electricity are used for short periods of time, the power supplier may not be receiving adequate revenue to compensate for the investment in the generation and distribution system necessary to make available the peak power required by the customer. Demand meters are installed which record the peak demand. The rate per kWh goes up as the demand increases. Thus it is sometimes to the advantage of the user to control consumption, to lessen the peaks which would otherwise occur. Sophisticated equipment is available which anticipates load increases and sheds nonessential loads automatically, tending to "level out" the load curve. This is beneficial to the power supplier as well, especially during periods when construction of new generating facilities is curtailed due to environmental considerations, or maximum use of existing facilities cannot be realized because of fuel shortages.

Power Consumed by Various Pieces of Equipment. The watt rating of a particular appliance, such as a toaster, may vary depending on who the manufacturer is: sometimes one manufacturer may make two sizes of the same appliance. However, it is helpful to have a general idea of ratings of different kinds of appliances; see Table 2-4.

Conductor Insulation. If two wires are used to connect a 1½-volt lamp to the terminals of a dry-cell battery, as shown in Fig. 2-7, the lamp will

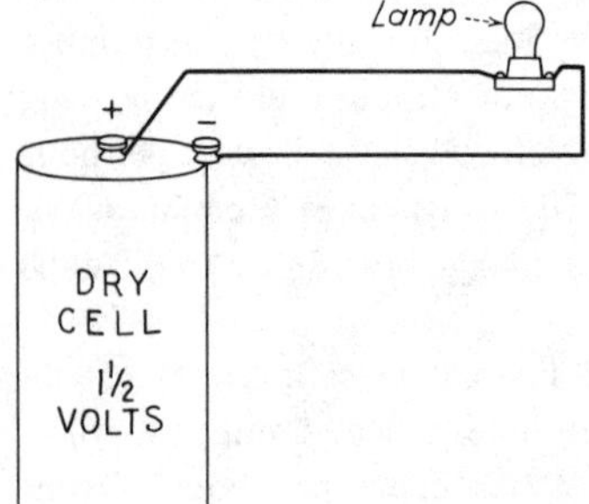

Fig. 2-7 An ordinary dry-cell battery, no matter how large or small, develops 1½ volts when new.

light, thus indicating that electric current is flowing through the lamp filament. Since the current can reach the lamp only through the wires, there is also current flowing in the wires. Yet there is no current through the wax or other material that was poured around the battery terminals at the top of the cell, nor is there a current through the paper or other material wrapped around the cell.

TABLE 2-4 Typical Appliance Ratings

	Watts
FOOD PREPARATION	
Blender	500–1 000
Coffee maker	500–1 000
Convection oven	1 500–1 700
Dehydrator	250–875
Dishwasher	1 000–1 500
Food processor	500–750
Frying pan	1 000–1 200
Hot plate, per burner	600–1 000
Knife	100
Microwave oven	1 000–1 500
Mixer	120–250
Oven, separate	4 000–5 000
Range	8 000–14 000
Range top, separate	4 000–8 000
Roaster	1 200–1 650
Rotisserie (broiler)	1 200–1 650
Toaster	500–1 200
Waste disposer	500–900
FOOD STORAGE	
Freezer, household	300–500
Refrigerator, household	150–300
Refrigerator, frostless	400–600
ENVIRONMENTAL COMFORT	
Air conditioner, central	2 500–6 000
Air conditioner, room	800–2 500
Blanket	150–200
Dehumidifier	250
Fan, portable	50–200
Heat lamp (infrared)	250
Heater, portable	1 000–1 500
Heater, wall-mounted	1 000–4 500
Heater, water-bed	800
Humidifier	450
LAUNDRY	
Dryer	4 000–6 000
Iron, hand (steam or dry)	600–1 200
Washing machine	300–550
Washing machine, automatic	500–800
Water heater	2 000–5 000
PERSONAL CARE	
Hair dryer	350–1 200
Heating pad	50–75
Shaver	8–12
ENTERTAINMENT	
Projector, slide or movie	300–500
Radio (tube type)	40–150
Stereo (solid state)	30–100
TV, black-and-white (tube type)	150–325
TV, black-and-white (solid state)	50–100
TV, color (tube type)	300–450
TV, color (solid state)	150–250
Video tape player/ recorder	40–70
MISCELLANEOUS	
Clock	2–3
Lamps, fluorescent	15–60
Lamps, incandescent	10 upward
Sewing machine	60–90
Vacuum cleaner	250–1 200
MOTORS	
1/4 hp	300–400
1/2 hp	450–600
Over 1/2 hp, per hp	950–1 000

If a material will conduct an electric current, it is called a "conductor." If a material will not conduct an electric current, it is called an "insulator." Insulating materials are therefore used as insulation for conductors by enclosing conductors, such as wires, in a layer of the insulation. (The thickness of the layer depends on the kind of insulation, the purpose of the wire, and the voltage for which the wire was designed.) The insulation thus confines the electric current to the conductor, and prevents it from spreading to unwanted areas. Although there are neither perfect conductors nor perfect insulators, insulating materials are available that will effectively insulate conductors.

Resistance of Wires. If the ends of a 1000-ft length of No. 10 copper wire are connected to a 1-volt source of electricity, 1 amp will flow through the wire, because 1000 ft of No. 10 copper wire has a resistance of 1 ohm (abbreviated Ω, the Greek capital letter omega). If a 1000-ft length of No. 10 aluminum wire is connected across the same 1-volt source of supply, only 6/10 amp will flow, because aluminum wire of the same length and diameter has a higher resistance than copper wire. If the aluminum wire is shortened to 600 ft, 1 amp will flow again because a 600-ft length of it has a resistance of 1 ohm. Only about 1/6 amp will flow through a 1000-ft length of iron wire of the same size, but 1 amp will flow again if the iron wire is shortened to about 167 ft; a 167-ft length of it has a resistance of 1 ohm. (The exact resistance of all types of wire depends on the chemical purity of the metal, the temperature, etc. Hence the figures above will not be precise in all cases.) This experiment shows that some electric conductors have lower resistance than others, and that the ones with the lowest resistance are the best conductors. Aluminum, for example, has a current-carrying capacity about 78% of that of copper; so if aluminum wires are used, they must be larger than copper wires for the same amount of current. Of course, the load supplied—lights, motors, etc.—imposes far greater resistance in the circuit than the conductors supplying the load. Nevertheless, it is important that the resistance of the conductors be held to a minimum.

The safe current-carrying capacity (ampacity) of different sizes and kinds of wire will be further discussed in Chaps. 4 and 7. All references to wire in this book will be to *copper* wire, unless otherwise indicated.

Ohm's Law. As already explained, 1 volt will force 1 amp through a conductor having a resistance of 1 ohm. Additional experiments will show that if the voltage is increased to 2 volts, the current will be 2 amp through 1 ohm. If the voltage is increased to 5 volts, the current will be 5 amp through 1 ohm. Additional experiments will also show that if the resistance is reduced by one-half, to ½ ohm, the current will be doubled to 2 amp if the voltage remains unchanged at 1 volt.

Doubling the cross-sectional area of a conductor of any given length will reduce the resistance of the conductor by one-half. An increase in conductor size (which reduces its resistance) will not reduce the resistance of the connected load. But the point to remember is that *if all other factors remain constant, an increase in current will be directly proportional to an increase in voltage and inversely proportional to an increase in resistance*. This is the principle of Ohm's law, which is one of the most basic laws of electricity (The term "directly proportional" means that one factor will be *increased* in proportion to an *increase* in another factor. The term "inversely proportional" means that one factor will be *decreased* in proportion to an *increase* in another factor, and *increased* in proportion to a *decrease* in another factor. Thus the current will *increase* in proportion to an *increase* in voltage, but the current will *decrease* in proportion to an *increase* in resistance, and vice versa.)

Ohm's-Law Formulas. Ohm's law can be expressed in simple formulas, but first you must understand some symbols: I, E, R.

I = *current*. I is a sort of shorthand symbol meaning that we don't know at the moment how many amperes are flowing in the circuit, and until we know, we'll just call it I. But when we do know what the current is, we say 10 amp, 200 amp, etc.

E = *voltage*. E is used for voltage until we know the specific number of volts involved; then we simply say 12 volts, 120 volts, etc.

R = *resistance*. R is used for resistance until we know the specific number of ohms involved; then we say 15 ohms, 3.5 ohms, etc.

The basic formula for Ohm's law is

$$\text{Number of amperes in circuit} = \frac{\text{voltage of circuit}}{\text{number of ohms in circuit}}$$

Now that is a very long, clumsy formula; therefore substitute the basic symbols discussed, and it becomes the much simpler formula $I = E/R$.

In the above formula, the current is the unknown factor that can be determined by dividing the known voltage by the known resistance. If the voltage and the current are known, the resistance can be found by changing the formula to $R = E/I$. If the resistance and current are known, the voltage can be found by changing the formula to $E = IR$.

There are also several formulas for determining the power in watts (abbreviated W), as follows: $W = EI$, or $W = I^2R$, or $W = E^2/R$. These and other related formulas are shown in Table 2-5.

Corrections for AC Circuits. The resistance of conductors larger than about 4/0 is slightly higher when carrying alternating current than when carrying direct current owing to the "skin effect": the current tends to

TABLE 2-5 Ohm's Law and Other Formulas (for 2-Wire Circuits)

If circuit is dc or is ac having ohmic resistance only	If circuit is ac and has both ohmic resistance and reactance
$E = IR$ or $\frac{W}{I}$ or $\sqrt{WR}$	$E = IZ$
$I = \frac{E}{R}$ or $\frac{W}{E}$ or $\sqrt{\frac{W}{R}}$	$I = \frac{E}{Z}$
$R = \frac{E}{I}$ or $\frac{W}{I^2}$ or $\frac{E^2}{W}$	$Z = \sqrt{R^2 + X^2}$
$W = EI$ or I^2R or $\frac{E^2}{R}$	

NOTE:
E = voltage in volts
I = current in amperes
R = resistance in ohms
W = power in watts
Z = impedance in ohms
X = reactance in ohms

travel on the outside of the conductor. The resistances in Table 8 of the Appendix are for direct current. Use Table 9 of the Code for ac resistance. In sizes 4/0 and smaller the skin effect can be ignored for all practical purposes.

The formulas shown for Ohm's law are always correct for dc circuits and also for ac circuits containing only what is known as "ohmic resistance," which is the kind of resistance imposed by incandescent lamps, resistance-type heating elements such as toasters, ranges, water heaters, etc., or a wire. But in most ac circuits there is an additional deterrent known as *reactance* (abbreviated X), also measured in ohms. While there is some degree of reactance in all ac circuits, it becomes significant mostly in equipment involving windings of wire on a steel core (motors, transformers, fluorescent lighting ballasts, electromagnets, etc.). In any piece of equipment the reactance must be combined with the ohmic resistance to determine the *impedance* (abbreviated Z), also measured in ohms, thus: $Z = \sqrt{R^2 + X^2}$. When using Ohm's-law formulas, for example, to determine voltage drop, if impedance is involved, simply substitute Z for R in the formula. A complete explanation of reactance and impedance is beyond the scope of this book; study any good book on electrical *engineering*.

Factors Apply to Same Part of Circuit. In using these or other formulas, as shown in Table 2-5, it is essential that you remember this: *All* factors must be applied to *all* of the circuit, or to the *same part* of the circuit. In other words, you cannot divide the voltage of an entire circuit by the resistance of one part of the circuit and find the current in some other part of the circuit.

Volt-Amperes. A very important fact about ac circuits concerns power. Although *EI* (voltage times current) or VA (volts times amperes, as usually shown in power formulas) equals watts in dc circuits, this is not true with ac circuits if reactance is involved. Just about all ac circuits have some reactance, but in residential and farm wiring it is usually so low that it can be ignored. But in commercial and industrial installations, often the reactance is high enough that it must be considered. In ac circuits, volts times amperes simply equals volt-amperes (VA), which is also known as "apparent power." To determine watts (actual power) in ac circuits, the volt-amperes (VA) must be multiplied by the power factor (pf) of the circuit. The pf may be low (60 to 70% or lower), average (80 to 90%), or high (above 90%). If the pf is 100%, it is equivalent to a dc circuit.

Power Factor. The power factor (pf) can be found by dividing the actual power in watts (as measured by a wattmeter) by the apparent power in volt-amperes (VA; as measured by a voltmeter and an ammeter). The formula is pf $= W/VA$ or pf $= kW/kVA$. For 3-phase circuits, the phase-to-phase voltage in formulas must be multiplied by 1.732 (which is the square root of 3). (These terms will be explained further in Chap. 3.)

Percent Efficiency (% EFF). The percent efficiency is usually marked on the name-plate of motors, transformers, etc. The formula for finding the percent efficiency is

$$\%\ \text{eff} = \frac{\text{output power}}{\text{input power}}$$

Voltage Drop. All conductors have resistance, and it takes power to force current through them. An example is two very long wires connecting a motor to a source of electricity. If you connect a voltmeter directly to the two terminals at the circuit breaker or fuse location, as shown in Fig. 2-8, the meter will probably read 120 volts (if it is a 120-volt circuit).

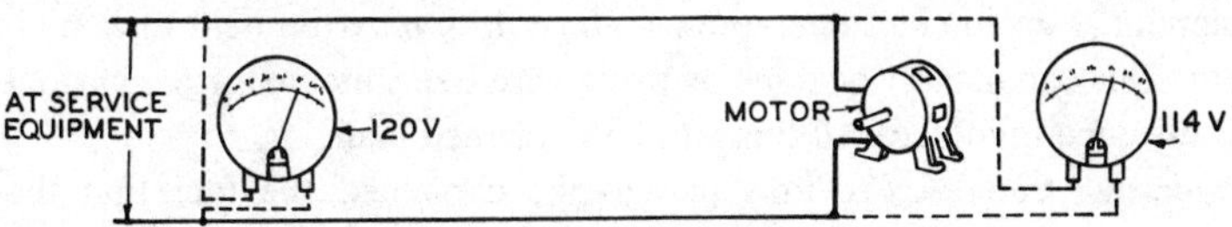

Fig. 2-8 This circuit illustrates voltage drop.

But if you connect the same meter to the motor terminals, the meter will indicate a lower voltage, possibly 114 volts, depending on the length and size of the wires. The difference between 120 and 114 volts indicates a 6-volt or 5% drop between the two points, which is a result of the power consumed in heating the wires because of the resistance.

This means that less power is available to operate the motor; when operated at 95% of its rated voltage, it will deliver only about 90% as much power as it would if operated at its rated voltage; if operated at 90% of its rated voltage, it will deliver only about 81% of its normal power. A lamp burned at 95% of its rated voltage will produce only about 83% of its normal light; if burned at 90% of its rated voltage, it will produce only about 68% of its normal light. Voltage drop therefore not only leads to power wasted in heating the wires, but also seriously affects the operation of electrical equipment.

Voltage drop and power loss can be computed if the resistance of the conductors in ohms and the current in amperes are known. As shown in Table 2-5, the power in watts equals I^2R. Since reactive ac power heats as much as actual power (watts), the heating effect for both ac and dc circuits equals I^2R without multiplying by the power factor for ac circuits. As an example, if the current in the wires is 7½ amp and the resistance of the wires is 2 ohms, the power consumed in heating the wires will be I^2R, or 7½ (amp) × 7½ (amp) × 2 (ohms), or 112.5 VA (or watts for dc circuits). By Ohm's law, $E = IR$, so the voltage drop in the wires equals 7½ (amp) × 2 (ohms), or 15 volts.

But suppose the current should be doubled to 15 amp, using the same size wire at the same voltage. The voltage drop would then be $IR = 15 \times 2$, or 30 volts drop. *Doubling the amperes doubled the voltage drop.* But will the *power loss* also be doubled? Let's find out. $I^2R = 15 \times 15 \times 2$, or 450 VA. The power loss is therefore four times greater! Hence, the power loss for both ac and dc circuits is directly proportional to the *square* of the current in amperes. So if you double the ampere load on the same conductors at the same voltage, your power loss will be four times greater; and if you triple the ampere load, the power loss will be nine times greater, etc. Of course, it takes *some* power to force current through *any* conductor, but *wasted* power can be held to a minimum. For efficient operation, voltage drop *must* be held to a minimum. This can usually be done by using wire size suitable for the current and distance involved. All this will be explained later.

Operating Voltage. Previous paragraphs explained the fact that the greater the current in a wire, the greater the voltage drop and the greater the power loss in the form of heat. From this it should be obvious that to carry high current without undue loss, large sizes of wire are required. The greater the distance, the larger the wire must be. Therefore it is distinctly advantageous to keep the current as low as is practical.

This, at least in theory, is simple, for any given load in watts may consist of a low voltage with high current, or of a high voltage with low current. Therefore for low current, relatively high voltages must be used.

In practice, the actual voltage depends on the amount of power to be transmitted and the distance. In an automobile, while the current is very great at times, a battery of only 12 volts is used, even though the current flowing through the starting motor when it is cranking the engine is often over 250 amp (3000 watts); this is practical only because the distance is so short.

For ordinary residential lighting the voltage is usually 120 volts,[7] but for ranges and water heaters it is 240 volts. For industrial purposes, where the power demands are great, 480 volts are usually used. The distribution lines that run down city alleys and from farm to farm are usually 2400 to 13 800 volts, but the main distribution lines are at still higher voltages until, for long-distance cross-country transmission, the voltages are usually 345 000 volts or more.

Since it is advantageous to keep the current as low as possible to reduce voltage drop and power losses in the wires and to do away with the necessity of buying large-size wire when a smaller size will do, and since this can be done by making the voltage higher, it would appear entirely logical to use high voltages for all purposes. You might well ask: "Why not use, for ordinary house wiring, 240 volts or 500 volts or higher?"

First of all, the higher voltages require heavier insulation, so that wire of any one size becomes more expensive; the higher voltages are more dangerous in case of accidental contact. Another important consideration

[7] The nominal voltages generally used in residential and farm work are 120 and 240 volts. Except in Arts. 430 (Motors), 550 (Mobile Homes), and 551 (Recreational Vehicles), where 115 and 230 volts are used, the Code uses 120 and 240 volts in the text; and in the examples in Chap. 9, some of which are reproduced in the Appendix of this book, the Code requires that load calculations be based on 120 and 240 volts.

Note too that many people still speak of 110- and 220-volt current, which *long ago* were the usual voltages. In many foreign countries, the lowest voltage in use is 220 volts (not 110/220).

is that in the manufacture of devices consuming relatively low power, under 100 watts, the wire used inside the device is often of almost microscopic dimensions, even where the device is for a voltage as low as 120 volts. For example, the tungsten wire in the filament of a 60-watt 120-volt lamp as manufactured today is only 0.0 018 in. in diameter; in a 3-watt lamp it is about 0.00 033 in.[8] in diameter. If the device were for 240 volts or for an even higher voltage, the wire would have to be still smaller, making factory production and uniformity decidedly difficult. The device would also be more fragile, and it would burn out more easily. The present common level of 120 volts is a compromise for lowest over-all cost of installation, operation, and purchase of devices to be operated.

However, since the same home that has small devices consuming from 5 to 1000 watts also has appliances like electric ranges that may consume over 10 000 watts, it would be desirable to have available two different voltages, one relatively low for lighting and small appliances and one relatively high for large appliances. Fortunately this is practical.

Three-Wire Systems. The 3-wire system in common use in homes today provides both 120 and 240 volts. Only three incoming wires are used and only a single meter. The 3-wire 120/240-volt system constitutes the ordinary system as installed in practically all houses and farms and many other installations. The higher voltage is usually used for any single appliance consuming 1800 watts or more.

Figure 2-9 shows two generators,[9] each delivering 120 volts; the two combined deliver 240 volts. Any load connected to either wires *A* and *B*

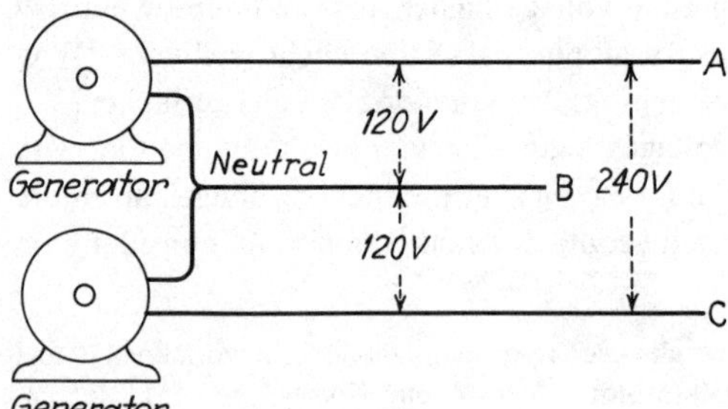

Fig. 2-9 With only three wires, two separate voltages are available.

[8] To cover a space of 1 in, 3000 such filaments would have to be laid side by side.

[9] Two generators are not actually used. This will become clear in the next chapter concerning transformers.

or to wires *B* and *C* will be connected to 120 volts. Any load connected to wires *A* and *C* will be connected to 240 volts. In actual wiring the central or neutral wire *B* is white; the outer two or "hot" wires are black or some other color, but never white or green. Connect any device operating on 120 volts to one black and one white wire, and any device operating on 240 volts to the two black wires. (These colors are correct only on grounded neutral systems; this will be explained in Chap. 9.)

Effects of Electricity. The endless assortment of things that electricity does can, in great part, be broken down into forms or combinations of three basic effects: thermal, magnetic, and chemical.

The thermal effect of electricity is heat. A current cannot flow without causing some heat. Sometimes heat is not desired, as, for example, in the case of the unavoidable power loss referred to in the examples in this chapter. In an ordinary lamp, over 90% of the current is wasted as heat, and less than 10% is converted into light; but the light is not possible without the heat. In a toaster or flatiron, only the heat is desired.

The magnetic effect can be stated very simply. When a current flows through a wire, the wire is surrounded by a magnetic field—the area immediately around the wire becomes magnetized. Bring a small compass near a wire that is carrying current and the needle will move just as it will when you bring it near an ordinary magnet. Wrap a wire a number of times around a piece of soft iron that is not magnetic; during the time that a current flows through the wire, the soft iron becomes a magnet, weak or powerful depending upon such factors as the number of turns of wire and the number of amperes flowing. *The moment the current stops flowing, the iron ceases to be magnetic.* It is this magnetic effect that causes doorbells to ring, motors to run, and telephones and radio speakers to operate.

The chemical effects are of great variety, including the electroplating of metals, the charging of storage batteries, and the electrolytic refinement of metals. In a dry-cell battery we have the reverse effect. A chemical action produces an electric current.

3

AC and DC; Power Factor; Transformers

You will frequently encounter the words "direct current," "alternating current," "cycles," "cycles per second" (or "hertz," abbreviated Hz), "single-phase," "2-phase," "3-phase," and "polyphase." These terms may seem formidable at first, but they are easily understood if you will pay close attention to their explanation.

Direct Current. If an ordinary direct-current voltmeter is connected to a battery, the pointer will swing either to the right or to the left, depending on how the two terminals on the meter are connected to the corresponding terminals of the battery. The two terminals of the meter are marked "+" and "−," "P" and "N," or "pos." and "neg.," indicating positive and negative; the battery terminals are similarly marked. Only when the positive terminal of the meter is connected to the positive terminal of the battery, and the negative terminal of the meter is connected to the negative terminal of the battery, will the meter pointer swing in the right direction. For any source of electricity, whether battery, generator, or other apparatus, if one terminal is positive and the other is negative and they never change, the current is known as "direct current," or "dc." Current from any type of battery is always direct current.

Alternating Current. Instead of an ordinary voltmeter, which has the zero at one end of the scale, a zero-center voltmeter of the type shown in Fig. 3-1 may be used. This meter is the same as the first except that the

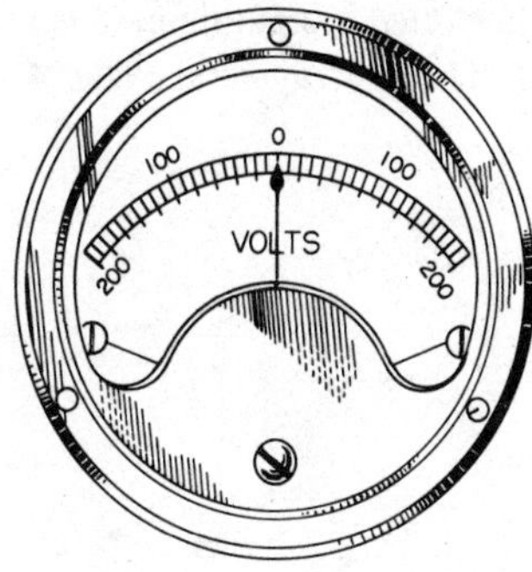

Fig. 3-1 This voltmeter is the same type as that shown in Fig. 2-1 except that the zero is in the center of the scale.

terminals are not marked "pos." and "neg." Connect the terminals of this meter to the two terminals of a battery and note which way the needle swings. Then reverse the two battery leads, and the needle will swing in the opposite direction. The meter is equally easy to read whether the pointer swings to the right or to the left, and it provides the additional convenience that it is not necessary, before connection is made, to investigate carefully which is the positive and which is the negative terminal.

We can perform an experiment with this zero-center voltmeter. Connect its two terminals to a source of electricity, the type of current being unknown. The pointer of the voltmeter performs in a peculiar fashion. It never comes to rest, but keeps on swinging from one end of the scale to the other, and back again, with great regularity. Watch that pointer carefully, starting from zero in the center.

It starts swinging toward the right, first rapidly, then more slowly, until it reaches a maximum of about 170 volts in exactly 15 seconds (s). Then it starts dropping back toward 0, first slowly, then rapidly, until in 15 s more it is back at 0. It does not stay there but keeps on swinging toward the left, and in 15 s more it reaches the extreme left at 170 volts, the same relative position as it originally had at the right. Again it swings back toward the right, and in 15 s more, 1 min from the starting point, it is back where it started from—at 0.

It repeats this same procedure indefinitely every minute. From observing the pointer it is evident that each wire is first positive, then negative, then positive, then negative, and so on, alternating between positive and negative continuously. The voltage is never constant, but is always changing from 0 to a maximum of 170 volts, first positive, then negative. Current in which any given wire regularly changes from positive to negative, not suddenly but gradually as outlined above, is known as "alternating current," or "ac." If the actual voltage is plotted against time, it will produce a chart such as is shown in Fig. 3-2. This portrays one cycle of alternating current.

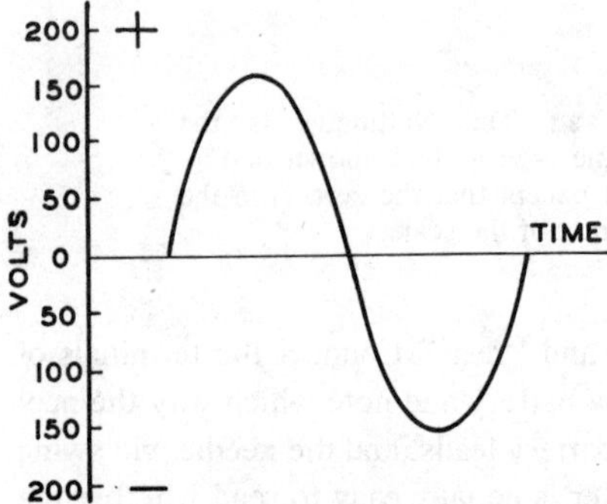

Fig. 3-2 This shows one cycle of alternating current. The voltage fluctuates regularly and continuously from 0 to maximum to 0, and each wire alternates regularly between positive and negative.

Alternating current may be defined as a current of regularly fluctuating voltage and regularly reversing polarity.

Frequency. Alternating current that takes a full minute to go through the entire cycle (from no voltage to maximum voltage on the positive side, back to 0, to maximum voltage on the negative side, back to 0) would have a frequency of 1 cycle per min. There is no such current in actual use. The ordinary ac used in the United States goes through the changes described above at the rate of 60 times per second, much too fast to be observed by an ordinary voltmeter. Such current has a frequency of 60 cycles per second, properly called 60 Hz. (Hz is the abbreviation for hertz).[1]

In the United States all standard commercial current is 60 Hz, as it is in Canada and most of Mexico. In other foreign countries most installations

[1] The hertz is named after Prof. Heinrich R. Hertz, a nineteenth-century German scientist who discovered the cyclical nature of electrical phenomena; radio waves were originally called hertzian waves.

are 50 Hz, but there are many at 60 Hz, with the trend toward 60 Hz. In the United States, 180 Hz is in use for equipment in a few industries, and fluorescent lighting installations operating at 400 Hz are in use. Military equipment often operates at 400 Hz. The advantage of the higher frequencies is that motors, transformers, and similar equipment can be smaller in physical size as the frequency increases.

Remember that "kilo-" means "thousand." When your radio receiver is tuned to a station operating at 1250 kilocycles (abbreviated kc), it means that the signal coming into the receiver is of 1 250 000 Hz. If the receiver is tuned to an FM station operating at 90 megacycles ("mega" has been adopted to designate "millions"), it means that the signal is 90 000 000 Hz.

Voltages of Alternating Current. In the curve of Fig. 3-2 the voltages range between 0 and 170 volts. If a 120-volt lamp is connected to an ac circuit of 1 cycle per min, it will burn far more brightly than normal while the voltage is above 120 volts, less brightly than normal while under 120, and not at all a part of the time because the voltage is very low, even zero twice during the cycle. Flickering would be extreme and unendurable. However, in the case of the ordinary 60 Hz, all this change of voltage takes place twice per cycle, 120 times every second. The filament of a lamp does not have time to cool off during the very short periods of time when no voltage is impressed on it, which is the reason for the lack of observable flicker. Very small lamps with very thin filaments that can cool off quickly will have an annoying flicker if operated on 25 Hz.

The rated voltage of an ac circuit is a value between 0 and the peak voltage, and in the case under discussion, it is 120 volts.[2] An ac voltmeter connected to the circuit will read 120 volts. A 120-volt ac source will light a 120-volt lamp to the same brilliancy as a 120-volt dc source.

Alternating Current and Motors. Alternating current as discussed up to this point is "single-phase" alternating current. Remember that when applied to a motor, alternating current magnetizes the steel poles of the motor every time it builds up from zero to peak voltage, or in other words 120 times per second as shown in Fig. 3-3, which shows three consecutive cycles of 60 Hz. At the top is indicated the time between cycles, or $^1/_{60}$ s. At the bottom is indicated the time between alternations, or $^1/_{120}$ s. You might say that a motor is given a push 120 times a second, just as a

[2] The rated voltage is 0.707 of the peak voltage; 0.707 is $^1/_2\sqrt{2}$. So the peak voltage is the rated voltage times $\sqrt{2}$.

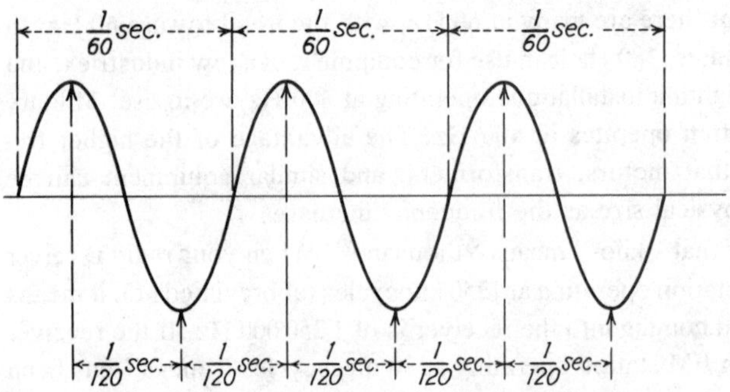

Fig. 3-3 This shows three cycles of 60-Hz alternating current.

gasoline engine is given a push every time there is an explosion in the cylinder. Offhand, 120 times per second may seem fast enough for any purpose, but remember that an ordinary motor runs at 1800 r/min, which means that the rotor (the rotating part) makes 30 revolutions every second. In turn this means that the 120 pushes per second become only 4 pushes per revolution; if the rotor or armature of the motor, is, say, 12 in. in diameter (over 36 in. in circumference), any given point on the rotor has to turn about 9 in between pushes. These pushes are not abrupt, sudden impacts. They are gradual pushes that start slowly and build up to a maximum as the voltage builds up to a maximum value.

In an ordinary 1-cylinder 4-cycle gasoline engine, running at 1800 r/min, there is an explosion in the cylinder every other revolution, or 900 times every minute, or 15 times every second. The crankshaft gets a push 15 times every second. If more pushes are needed every second to secure smoother operation, or more power, more cylinders are used: two or four or as many as needed. How is this to be done in the case of an electric motor? Fortunately it is rather simple.

Three-Phase Alternating Current. It can be done by putting into the motor three separate windings not connected to each other in any way, but each connected to one of three separate sources of single-phase alternating current. The three separate sources of current must be so designed that the peak voltage of one does not coincide with the peak voltage of another. The voltages of the three sources reach their peak in regular fashion, one after the other. Then the motor receives three times as many pushes as before. Figure 3-4 shows the voltage curves of the three separate sources. At the top is indicated the time between cycles in each separate source: $^{1}/_{60}$ s. At the bottom is indicated the time between pushes

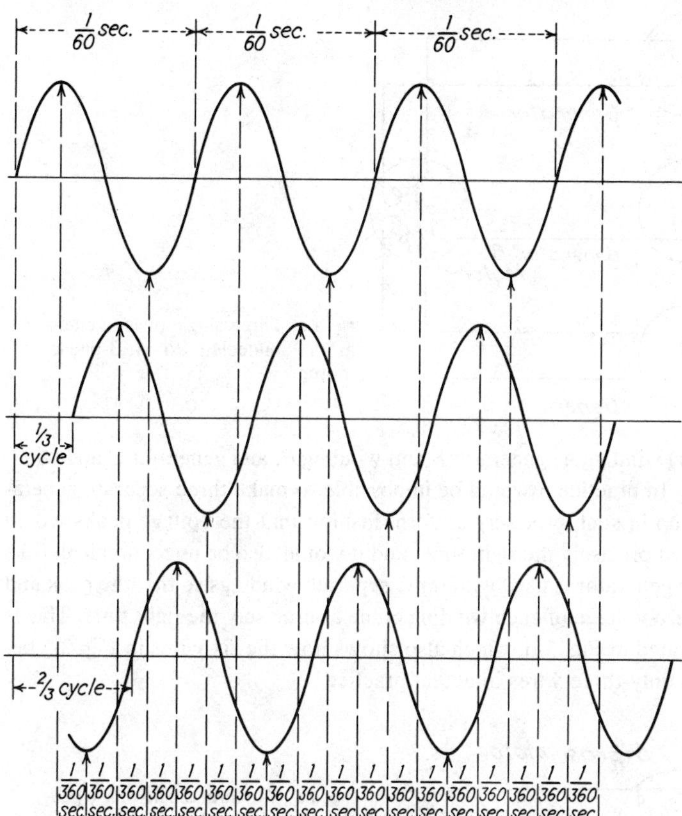

Fig. 3-4 Three single-phase currents combine to form 3-phase current.

from the three separate sources combined: $^1/_{360}$. So 3-phase alternating current consists simply of three separate sources of single-phase alternating current so arranged that the voltage peaks follow each other in a regular, repeating pattern.

Note that for each phase the duration of a cycle is $^1/_{60}$ s, but there are two pushes per cycle, so that the time between pushes is $^1/_{120}$ s. But look at the bottom of the diagram and you will see that the pushes from the three phases combined are only $^1/_{360}$ s apart. The pushes are imparted by each of the three windings in turn, as shown by the dashed lines from the peaks to the bottom of the diagram.

Figure 3-5 shows this diagrammatically: generator *A* and (inside the

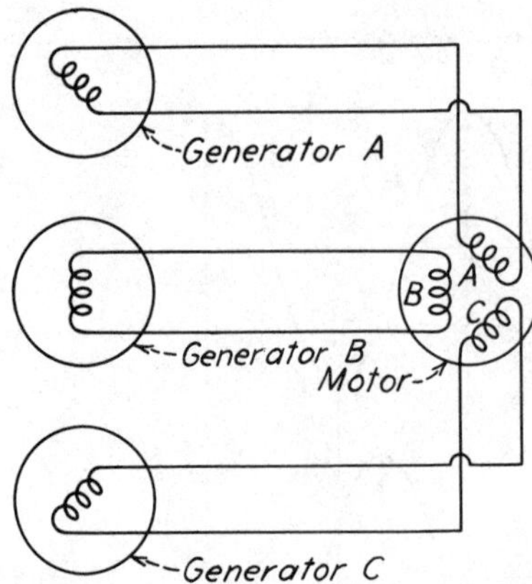

Fig. 3-5 Three single-phase generators connected to a 3-phase motor.

motor) winding *A*, generator *B* and winding *B*, and generator *C* and winding *C*. In practice it would be impossible to make three separate generators run in such precisely uniform fashion that the voltage peaks would come at precisely the right time; and it would also be uneconomical. So a single generator is used with three separate windings so that the peak and the zero voltage of each winding come at precisely the right time. This is illustrated in Fig. 3-6, which also shows how the six wires in Fig. 3-5 become only three wires in actual practice.

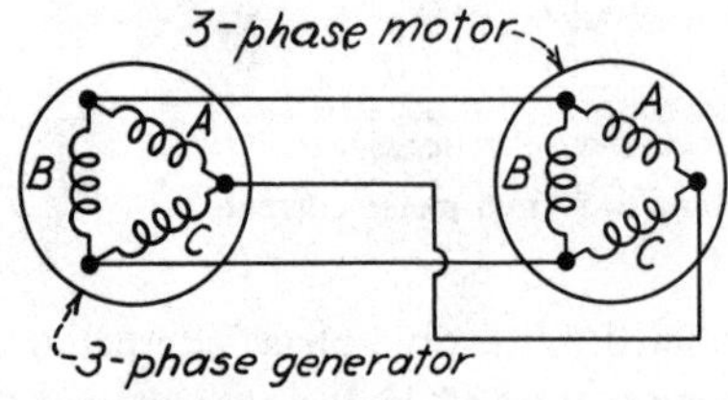

Fig. 3-6 The three single-phase generators of Fig. 3-5 combined into one 3-phase generator to supply a 3-phase motor, using only three wires instead of six.

Two-Phase Alternating Current. A 2-phase system, which is similar in principle but inferior to a 3-phase system, is used in so few localities that it does not warrant space in this book.

Polyphase Current. Both 2- and 3-phase currents are called "polyphase" current (*poly-*, a form derived from the Greek, means "many"). With a few exceptions, 3-phase current is not used in homes, but it is widely used in commercial and industrial establishments for operating motors and similar equipment. Even in those establishments, single-phase current is used for lighting and small miscellaneous loads.

Abbreviation. The word "phase" is usually abbreviated ϕ, the Greek letter phi.

Volt-Amperes. The previous chapter contained the formula volts × amperes = watts. That formula is always correct for direct current; but for alternating current, it is not always correct because of reactance, which results in a power factor (abbreviated pf) of other than 100%. This will be discussed after volt-amperes (abbreviated VA) have been further explained.

For single-phase:

$$\text{Volt-amperes} = \text{volts} \times \text{amperes}$$

For 3-phase:

$$\text{Volt-amperes} = 1.732 \times \text{volts} \times \text{amperes}$$

In 3-phase work you will frequently encounter the number 1.732, which equals $\sqrt{3}$.

Kilovolt-Amperes. One thousand volt-amperes is one kilovolt-ampere (abbreviated kVA). If the power factor of the load is 100%, then and only then is one kilovolt-ampere the same as one kilowatt (kW).

Power Factor. A detailed explanation of power factor is beyond the scope of this book. However, an explanation of how to measure it and a general idea of its importance will be covered here.

For a dc circuit consisting of the load and an ammeter, a voltmeter, and a wattmeter, the product of the volts and the amperes is without exception equal to the reading of the wattmeter. But if the same experiment is made with an ac circuit, sometimes the same is true and sometimes it is not.

For an ac circuit, if measurements show that the product of the volts and the amperes is exactly equal to the wattmeter reading, the load is said to have a power factor of 100%. Loads of this type include incandescent lamps, heating appliances of the type that have ordinary resistance-type heating elements, and in general all noninductive loads, that is, equipment that does not include windings or coils of wires such as transformers, motors, and fluorescent lighting fixture ballasts.

If the product of volts times amperes is greater than the reading of the wattmeter, then the power factor is less than 100%.

Power factor, which is also referred to as the "cos θ,"[3] is defined as the

[3] Cosine of theta.

percentage ratio of the real or measured watts (also known as "effective power") to the volt-amperes (also known as "apparent watts"). The formula is

$$\text{Power factor} = \frac{\text{watts}}{\text{volt-amperes}}$$

Measuring Power Factor. To measure power factor you need only a voltmeter, ammeter, and wattmeter. Assume a small single-phase motor on a 120-volt circuit consuming 5 amp, as indicated by an ammeter, and 360 watts, as indicated by a wattmeter. The formula then becomes

$$\text{Power factor} = \frac{360}{5 \times 120} \text{ or } \frac{360}{600} \text{ or } 0.60 \text{ or } 60\%$$

For a 3-phase 240-volt motor consuming 12 amp, the volt-amperes are 1.732 × 240 × 12, or 4988. If the watts as indicated by a wattmeter are 3950,

$$\text{Power factor} = \frac{3950}{4988} \text{ or } 0.79 \text{ or } 79\%$$

Generally speaking, the power factor of a motor improves (increases in percentage) with an increase in horsepower of the motor. It also varies considerably with the type and quality of the motor in question. It may be as low as 50% for small fractional-horsepower motors, and over 90% for a 25-hp motor.

Watts in Alternating-Current Work. The formula for ac power in watts is

$$\text{Watts} = \text{volt-amperes} \times \text{power factor}$$

For 3-phase power, multiply the volts by 1.732.

Desirability of High Power Factor. Assume that a small factory is using 100 amp at 240 volts, single-phase, or a total of 24 000 VA or 24 kVA. If the power factor is 100%, this is equivalent to 24 kW. At 6 cents per kilowatthour, the power supplier receives $1.44 per hour for the power.

Now assume a second factory also using 100 amp at 240 volts, but with a power factor of only 50%. That is still 24 kVA but only 12 kW, and at 6 cents per kilowatthour, the power supplier now receives only 72 cents per hour.

Since it is the kilovolt-ampere load that determines wire size, transformer and generator size, and similar factors and since each factory uses the same 24 kVA, the power supplier must furnish wires just as big for the factory where they are paid 72 cents per hour as for the one where they are paid $1.44 per hour. They tie up just as much transformer capacity, generator capacity, and all other equipment for the one as they do for the other.

It is natural, therefore, that power suppliers, when furnishing power to establishments where the power factor is low, not only charge for the kilowatthours consumed but also add a penalty based on the kilovolt-amperes used during the month or period in question, as compared with the kilowatthours used. Since the loss in wasted power decreases as the power factor increases, it is definitely worthwhile to improve the power factor. Few installations attain 100% power factor, and rarely does one fall as low as 50%. The over-all power factor in an industrial establishment is generally determined by the electric motors in use, although other equipment also contributes its share.

Power-Factor Correction. The theory of power-factor correction is entirely beyong the scope of this book, but the actual correction is accomplished by means of capacitors or synchronous motors; the required calculations should be made by one thoroughly familiar with the subject. Correcting the power factor not only reduces the penalty charges for power consumed but also has many other advantages, including higher efficiency of electric equipment because of reduced voltage drop, and reduced heating and power loss.

Transformers. Where it is necessary to transmit thousands of kilowatts over a considerable distance, wire large enough to transmit it at 120 or even 240 volts would have to be so big that the cost would be prohibitive. If a relatively small wire and a much higher voltage are used, the voltage will be so high as to be dangerous to use with ordinary equipment.

It would be convenient, therefore, to have a way of changing from one voltage to another as needed. In the case of direct current there is no simple, efficient means of doing this for everyday needs, but for alternating current there is a simple and efficient means of doing just that—the transformer.

A transformer changes one ac voltage to a different ac voltage. It has no moving parts. Basically it consists of a core made of many thin sheets of silicon steel, with two coils of wire or windings, as shown in Fig. 3-7. The windings are electrically insulated from each other and from the

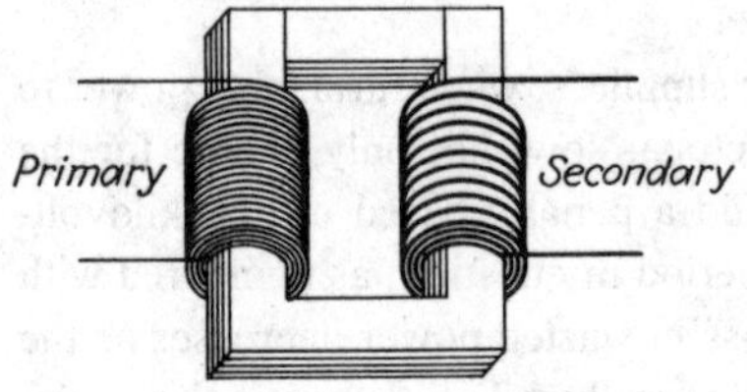

Fig. 3-7 This shows the basic construction of a transformer.

core. Electric energy is *supplied to* one winding called the *primary* winding. The other winding is called the *secondary* winding, which *delivers* the energy at a different voltage. (For some special purposes, the voltage can be the same in both windings.)

A transformer can be a *step-up* transformer that receives energy at one voltage and delivers it at a *higher* voltage, or a *step-down* transformer that receives energy at one voltage and delivers it at a *lower* voltage.

When the primary winding is energized by an ac source (which may be an ac generator or another transformer), an alternating magnetic field, called "flux," is established in the transformer core. The alternating flux surrounds both windings and *induces* a voltage in both. So once a produced electromotive force (emf) is available, an alternating flux that rapidly fluctuates (rises to maximum and falls to zero 120 times a second for 60-Hz ac) can be used to *cut through the windings* and *induce a voltage in each of them*.

Ratio of Number of Turns to Voltage. Since the same flux cuts both windings, the same voltage is induced in *each turn* of each winding. Hence, the induced voltage is *proportional* to the *total number of turns* in *each winding*. The voltage in the primary will be the same as that received from the source. But the voltage in the secondary will depend on the number of turns in the secondary winding *in proportion* to the number of turns in the primary winding. If the secondary has the same number of turns as the primary, the voltage will be the same in both windings; if the secondary has twice as many turns as the primary, the secondary voltage will be twice as high as the primary voltage; and if the secondary has half the number of turns of the primary, the secondary voltage will be half as high as the primary voltage; etc. So the formula is

$$\frac{E_p}{E_s} = \frac{N_p}{N_s}$$

where E_p = primary voltage, E_s = secondary voltage, N_p = number of primary turns, and N_s = number of secondary turns. Therefore, $(E_p)(N_s) = (N_p)(E_s)$, and $[(E_p)(N_s)]/N_p = E_s$, etc.

Transformer Current and Loads. Where there is no connected load (where the switch in the secondary conductors is open), a transformer has only a very small no-load current, which is called the ''exciting'' or ''magnetizing'' current. The magnetizing current produces the magnetomotive force that produces the transformer-core flux. But when the switch is closed and a secondary load (lamps, motors, etc.) is connected, then just as much current will flow in the primary as is required to deliver the required power to the secondary, but no more (assuming, of course, that the capacity of the transformer is adequate for the load connected to it).

Experiment shows that if the primary and secondary have the same number of turns, the voltage of the primary and secondary will be the same, but the current flowing in the primary from the power line adjusts itself to the current demanded of the secondary by the nature of the particular load connected to it.

If the secondary has twice as many turns as the primary, the *voltage* of the secondary will be *twice* that of the primary, but the *current* will be only *half* as great. If the secondary has ten times as many turns as the primary, the voltage in the secondary will be ten times that in the primary, but the current will be only one-tenth as great. By reversing the proportions and having fewer turns in the secondary than in the primary, it is equally simple to step the voltage down instead of up; but the current will then go up as the voltage goes down. The *volt-amperes* in the primary are always *equal* to the *volt-amperes* in the secondary, plus a small percentage, depending on the efficiency of the transformer.

The minimum number of turns must be kept within the limits that good engineering has shown lead to the greatest efficiency, and wire sizes in both the primary and the secondary must be adequate to carry the current involved. The smallest transformer usually found is the ordinary doorbell type, which steps 120 volts ac down to about 8 volts for operating a small doorbell and similar equipment. Chimes and some doorbells operate at a slightly higher voltage. The largest transformers are so big that it is hard to find a railway car sturdy enough to transport a single transformer.

Well-built transformers are very efficient, and, generally speaking, the larger the transformer, the greater the efficiency. In very large transformers, it is possible to obtain from the secondary over 99% of the power supplied to the primary.

In a large generating station (power plant), generators produce various voltages, 13 800 volts being typical. Transformers step up this generated voltage to a much higher voltage for transmission over great distances. Transmission voltages range from 46 to 345 kV and even higher, depending on the transmission distance. At various distribution substations, the transmission voltage is stepped down to distribution voltages, which may range from 13 800 to 34 500 volts (or 13.8 to 34.5 kV). At strategic points, it is again stepped down to utilization voltages, which range from 120 to 4160 volts and sometimes higher. See Fig. 3-8. (Single-phase transform-

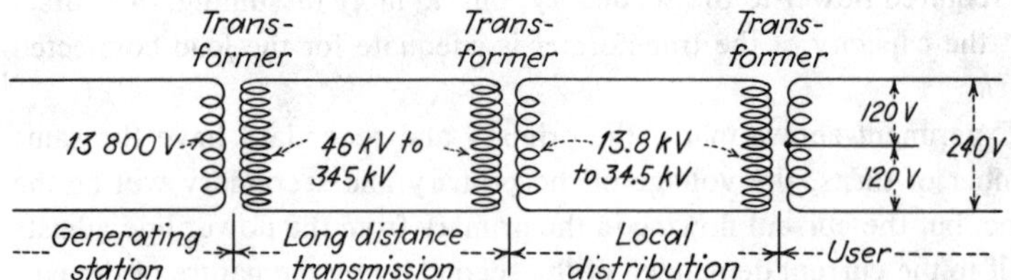

Fig. 3-8 Power is generated at a relatively low voltage, then stepped up to a much higher voltage for transmission over long distances, then stepped down to lower distribution and utilization voltages.

ers are shown here for simplicity, but 3-phase transformers are actually used for most high voltages.) Sometimes more than one distribution system is used in a single locality. For instance, 34.5 kV may be distributed throughout a general area, after which it is again stepped down at various substations for local or plant distributions at 4160 volts, and then stepped down again for various utilization voltages, such as 480Y/277 volts and 120/240 volts.

The voltage at which power is transmitted depends on many factors, but the principal factors are the distance and the amount of power involved. Several 500- and 765-kV transmission lines are in use. Still higher voltages are being developed.

Series-Parallel Connections. Power transformers for power and lighting usually have two identical coils in each primary winding, and two identical coils in each secondary winding. If the two primary coils are connected in series as shown in *A* of Fig. 3-9, the primary will be suitable for

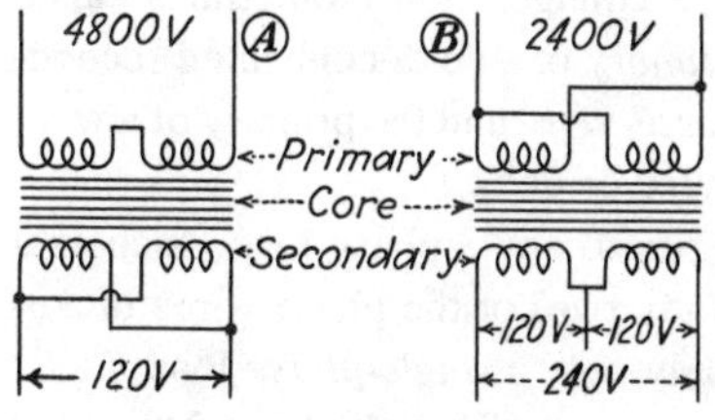

Fig. 3-9 The windings can be connected in series or in parallel, for two different voltages.

connection to a 4800-volt line. If connected in parallel as shown in *B*, the primary is suitable for connection to a 2400-volt line. Similarly, the two secondary coils can be connected in parallel to deliver 120 volts as shown in *A*, or in series as shown in *B* to deliver 240 volts. Usually, the two secondary coils are connected in series with a tap at the midpoint, forming the common 3-wire 120/240-volt system, as shown in *B*. If connected for 120 volts, the available current in amperes will be twice that available if connected for 240 volts; but the secondary volt-amperes will be the same regardless of which way it is connected. However, the available power [(VA)(pf)] at the connected loads will usually be slightly higher at 240 volts than at 120 volts because of lower voltage drop in the conductors. In addition, this type of connection provides a dual-voltage 3-wire system for supplying both 120- and 240-volt circuits.

Three-Phase Transformers. For a 3-phase system, sometimes a 3-phase transformer is used, and at other times three single-phase transformers are interconnected to form a 3-phase bank of transformers. A 3-phase transformer consists of three separate single-phase transformers that are interconnected within a single enclosure.

Figures 3-10 and 3-11 show only the secondaries of the transformers. A

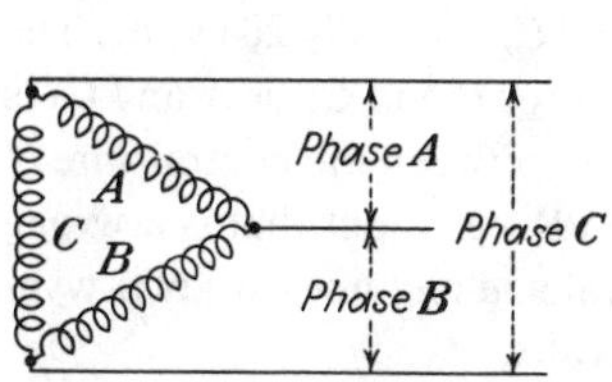

Fig. 3-10 Delta-connected 3-phase transformer.

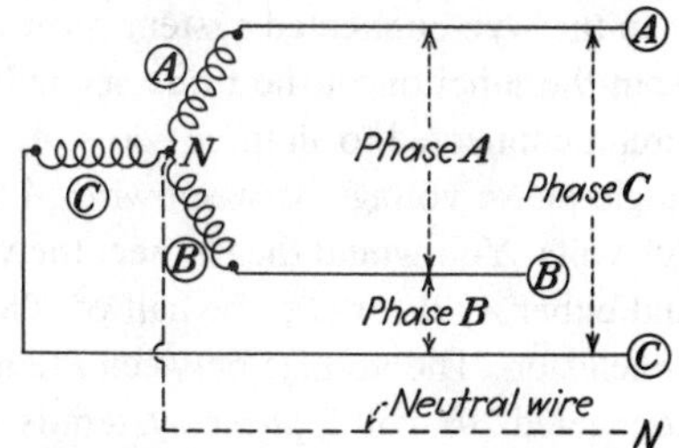

Fig. 3-11 Wye-connected 3-phase transformer.

3-phase transformer (or transformer bank) may be connected in the delta configuration (the name is derived from the Greek capital letter delta: Δ) as shown in Fig. 3-10, or in the wye configuration (sometimes called "star") as shown in Fig. 3-11. (The *primary* of a delta-connected secondary may be connected either as delta or as wye, and the primary of a wye-connected secondary may be either wye or delta.)

All three of the phase wires must run to any 3-phase load, such as a 3-phase motor, but where only *two* (any two) of the phase wires of a 3-phase system are used, they can supply only a *single-phase* load.

In the delta system of Fig. 3-10, the power will usually be at 240 or 480 volts. The single-phase power then available is one of those voltages. Single-phase power at only 240 volts isn't very practical for most purposes; we want 120/240-volt single-phase power. Figure 3-12 shows the same

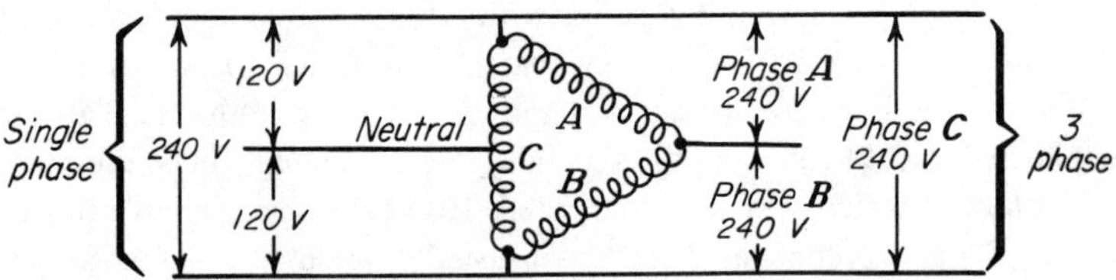

Fig. 3-12 A 3-phase delta-connected transformer can deliver both 3-phase and dual-voltage single-phase power.

transformers as in Fig. 3-10, but with one of the secondaries tapped at the midpoint. Assuming the basic voltage is 240 volts, the secondary that is tapped at the midpoint in Fig. 3-12 will deliver 120/240-volt single-phase power, which may be used at the same time as the 3-phase power. Although not shown in Fig. 3-12, the midpoint where the tap is made is also grounded to a grounding electrode. This will be explained in detail in Chap. 9.

In the wye-connected system shown in Fig. 3-11, a neutral wire is run from the junction of the three secondaries. The 3-phase voltage of any circuit connected to all three wires, *A*, *B*, and *C*, is usually 208 volts. The single-phase voltage between wires *A* and *B* (or *B* and *C*, or *A* and *C*) is 208 volts. You would then expect the voltage between the neutral wire *N* and either *A*, *B*, or *C* to be half of 208, or 104 volts, but that is a wrong conclusion. The voltage between the neutral and any hot wire of a wye-connected 208-volt 3-phase system is 120 volts.

At first glance this may seem all wrong, for if the voltage between wires *A* and *B* in Fig. 3-11 is 208 volts, the voltage between the neutral wire and either *A* or *B* might be expected to be one-half of 208, or 104 volts, instead of 120 volts as previously stated. Remember, however, that in a 3-phase circuit, the voltage comes to a peak or maximum at a different time in each phase. At the instant the voltage in secondary *A* is 120 volts, that in *B* is 88 volts, so that across wires *A* and *B* there is a voltage of 120 + 88, or 208 volts.[4] The system therefore has the advantage of making it possible to transmit over only four wires (including a grounded neutral) 3-phase power at 208 volts, single-phase power at 208 volts, and single-phase power at 120 volts. Occasionally in a home, instead of providing the usual 120/240-volt 3-wire system, three wires of the wye-connected system (the grounded conductor and any other two wires of Fig. 3-11) are provided, thus furnishing 120 volts for lighting and 208 volts (instead of the usual 240 volts) for water heaters and similar large loads. In larger installations, all four wires (the three phase wires and the neutral) are used. This is a 3-phase 4-wire 208Y/120-volt system. Although not shown in Fig. 3-11, the neutral point of the windings (where all three windings are connected together) is grounded. This will be further explained in Chap. 9.

Instead of 208Y/120 volts, some wye-connected transformer secondaries supplying commercial and industrial establishments provide power at 480Y/277 volts. More will be said about this later.

Autotransformers. An autotransformer can be defined as a transformer in which a portion of the turns of a single winding is common to both primary and secondary (see Fig. 3-13.) Let there be a tap at the midpoint of

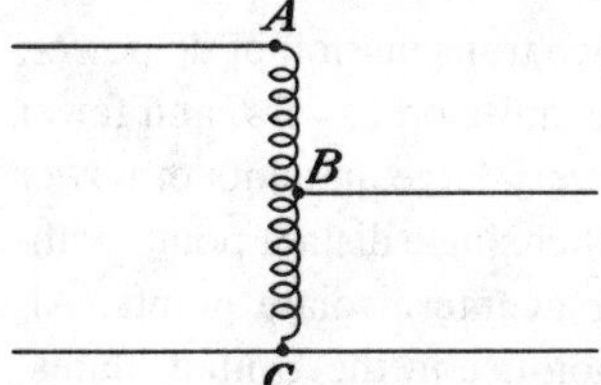

Fig. 3-13 An autotransformer has only one winding.

the coil so that although there are, for example, 1000 turns of wire between *A* and *C*, there are only 500 between *B* and *C*. If the voltage is to be stepped down, the entire coil *A* and *C* will constitute the primary, and those turns from *B* to *C* will constitute the secondary. Whatever the voltage across *A* to *C*, the voltage across *B* to *C* will be exactly half. The tap

[4] Note that $208 = 120\sqrt{3}$.

may be at any point in the coil. The voltage across B to C, as compared with the total voltage across A to C, will always be proportional to the number of turns from B to C, as compared with the total number of turns from A to C. If the voltage is to be stepped up (instead of down), a part of the winding is used for the primary and the entire winding is used for the secondary. This type of transformer, being far less expensive than the two-coil type, is often used for transmission lines and similar purposes, but it is seldom used in buildings, since NEC Sec. 210-9 has restrictions on its use with branch circuits. So, except as used in some types of lighting fixture ballasts and for some types of large motor starters, an autotransformer is usually used in buildings only to boost the voltage of an existing 208-volt branch circuit for supplying a 240-volt appliance, or to reduce the voltage of an existing 240-volt branch circuit for supplying a 208-volt appliance.

High-Voltage DC Transmission. As you now know, transformers operate only on *alternating* current. Yet, you may have read about high-voltage *direct*-current transmission lines, first used in foreign countries, at voltages up to 1 000 000 volts (1000 kV) or more. How is this possible without transformers? AC generators are used, the generated voltage is stepped up to the desired transmission voltage by transformers, and then "rectifiers"[5] are used to change the alternating current to direct current. After it is transmitted, converters are used to change the direct current back to alternating current. Then transformers are used to step the ac voltage down again. The procedures for doing this are beyond the scope of this book.

There are many advantages to long-distance transmission of dc power, such as narrower rights of way, smaller transmission towers, and fewer conductors. The method is practical only if very large amounts of power are transmitted from one single point to another single distant point, without need for tapping off any of the power at intermediate points. Although dc transmission is not yet in common use in the United States, there are several lines in use, and others are in the planning stage. Therefore dc transmission will become more common in the future.

[5] A rectifier is a device that permits current to flow in one direction, but not in the other. Four of them in combination convert single-phase alternating current to direct current. More complicated combinations of rectifiers convert 3-phase alternating current to direct current.

4
Basic Devices and Circuits

In order to properly and intelligently assemble the great number of available electrical devices, fittings, materials, and equipment to form a complete wiring system, you must understand the basic principles regarding them and electric circuits.

If electric current is to produce an effect, it must flow not only to, but also through, the equipment to be operated. In other words there must be two wires from the starting point (the source of power) to the equipment. The electric current can be compared to a series of messengers that start from some given point (such as a transformer), make a trip to their destination (the equipment to be operated), and return to the starting point before their errand is completed. The wires can be considered the streets over which they travel, but they must be one-way streets; the messengers must go out on one and return over a different street (wire), because there are millions of billions of them.

Lamps. Probably the most common electrical equipment is what is often called a "light bulb." The correct name is "lamp"; the glass part of the lamp is the bulb. The lamp consists essentially of a filament, which is a wire made of tungsten, a metal having a very high resistance and a very high melting point. That makes it possible to heat the filament to a very high temperature (over 4000°F in ordinary lamps) without its burning out. The filament is suspended on supports in the bulb, from which the air has been exhausted and into which, in most sizes, some inert gas like argon has been introduced to prolong the life. The ends of the filament are brought out to a convenient *base,* which makes replacement simple. In the base the center contact is insulated from the outer metal part of the base, thus producing two terminals for the two wires leading up to the filament.

Any lamp which operates by means of a heated filament, as described, is called an incandescent lamp.

Lamps operate in what the Code calls lampholders, but which most people just call sockets. The simplest socket is the cleat lampholder shown in Fig. 4-1; a cross section of it is shown in Fig. 4-2. One terminal

Fig. 4-1 A "cleat lampholder"—the simplest form of socket. (*General Electric Co.*)

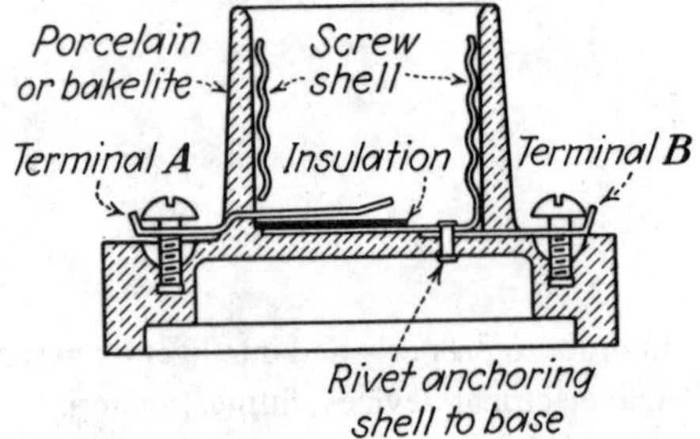

Fig. 4-2 Cross section of the cleat lampholder shown in Fig. 4-1.

A is connected to the center contact corresponding to the center contact of the lamp base; the other terminal *B* is connected to the screw-shell terminal (which is carefully insulated from the center contact and terminal *A*), corresponding to the outer screw shell of the base on the lamp. When a lamp is screwed into such a socket, the current flows in at *A,* through the filament, and out at *B*. Such sockets with exposed terminals are seldom seen today, being used with open wiring.

But note that while all sockets are lampholders, not all lampholders are of the type with a screw shell to engage a similar screw-shell base on a lamp. Other types are shown in Chaps. 14 and 29.

Circuits. An electric circuit consists of a complete path for the electric current from the source of supply, through the connected load (equipment or apparatus), and back to the source of supply. Circuits are either branch circuits or feeder circuits.

The Code defines a *branch* circuit as "the circuit conductors between the final overcurrent device protecting the circuit and the outlet(s)." It defines a *feeder* as "all circuit conductors between the service equipment, or the generator switchboard of an isolated plant, and the final branch-circuit overcurrent device."*

In the wiring of ordinary residences or other small buildings, there may be no feeders where the overcurrent devices (circuit breakers or fuses) are installed at the service equipment (all the equipment installed where the power enters the building) and all branch circuits begin at that point. In larger buildings it is not at all practical to begin all the branch circuits at that point, for the wires would be impractically long, so overcurrent devices are installed at the service equipment to protect *feeders*. Feeders are heavy wires running to other locations where panelboards containing the overcurrent devices for the smaller branch-circuit wires are installed. If the building is very large, there may be subfeeders and even sub-subfeeders. This may seem very complex, but later chapters will make these things clear.

The wiring diagrams in this chapter and in other parts of this book are mostly parts of branch circuits, rather than complete circuits. Only the basic devices and equipment necessary to make the circuit work will be discussed in this chapter. Other materials, fittings, raceways, boxes, etc., will be covered in other chapters.

Devices, Fittings, and Boxes. You will often see the term "wiring device." The Code in Art. 100 defines it as a component that *carries* current but does not *consume* it. Examples are sockets, switches, overcurrent devices, push buttons, etc. Receptacles are wiring devices in that they do not consume power, but are used only to permit power-consuming loads such as lamps and toasters to be plugged into them. Anything that *consumes* power is *utilization equipment* and constitutes the *load* on the circuit.

The Code definition of a "fitting" is "an accessory such as a locknut, bushing, or other part of a wiring system that is intended primarily to perform a mechanical rather than an electrical function."*

In installing wiring, all connections of one wire to another, or to a ter-

*Reprinted with permission from NFPA 70-1984, National Electrical Code®, Copyright© 1983, National Fire Protection Association, Quincy, Massachusetts 02269. This reprinted material is not the complete and official position of the NFPA on the referenced subject, which is represented only by the standard in its entirety.

minal, are made inside of *boxes,* usually metal, sometimes nonmetallic. There are many kinds of boxes, and they will be described in Chap. 10.

Outlets. Every point where power is taken from wires to be *consumed* is an outlet. Receptacle outlets are outlets where one or more receptacles are to be installed; the receptacle itself consumes no power, but whatever is plugged into it does consume power. Boxes where *switches* are installed are not outlets but are called "connection points" for switches. But boxes where receptacles, sockets, etc., are to be installed *are* outlets, because power-consuming loads are connected there. Sometimes the term "outlet" is loosely and improperly used to designate *any* point where a device such as a switch is to be installed, this being sometimes done in contracting work when the cost of an installation is estimated on a "per-outlet" basis.

SOURCE. In all the diagrams in this book, the word SOURCE will mean the generator, transformer, battery, or other SOURCE of supply. Actually, it may be the point where the wires enter the building, or the point where the particular circuit under discussion begins.

Basic Circuit. Figure 4-3 shows a wire running from the SOURCE to the socket with a lamp and another wire from the lamp back to the SOURCE.

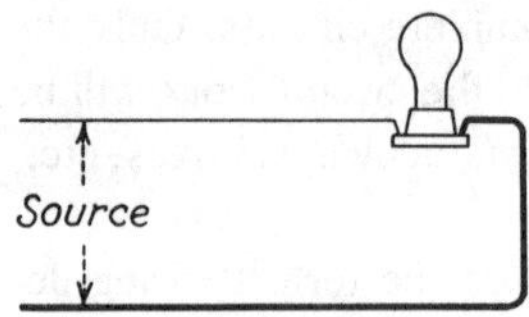

Fig. 4-3 The simplest circuit. There is no way of turning the lamp on or off.

In all such diagrams in this book, the grounded wire will be shown as a light line like this ______; the ungrounded (hot) wire will be shown as a heavy line like this ______. This does *not* mean that one wire is larger than the other; both are the same size. This is done just to make it easier for you to follow the diagram.

The current flows outward through one wire (the heavy line), through the lamp, and back through the other wire (the light line). This makes a complete circuit, and as long as the SOURCE furnishes power, the lamp will light. This is not a practical circuit, since it is necessary to disconnect one wire from the socket, or to unscrew the lamp, to turn the light off. Such a circuit would not be very sensible, so a switch must be included. This has been done in Fig. 4-4, the switch being the open porcelain-base type. Opening the switch is the same as disconnecting or cutting a wire;

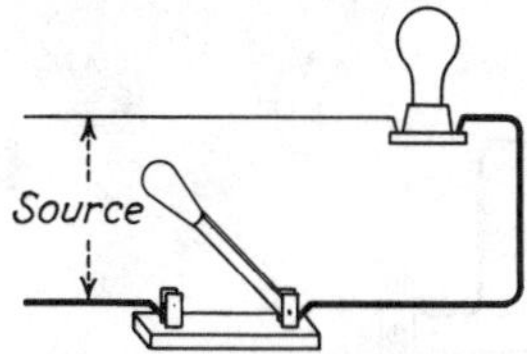

Fig. 4-4 A switch has been added to the circuit to control the lamp.

or, comparing it to the one-way street, it is the same as opening a drawbridge in the street.

Toggle Switches. In actual wiring you would not use a switch with exposed live parts as shown in Fig. 4-5. Instead, use a neat toggle switch of the type shown in Fig. 4-6, concealed in the wall, with only the handle

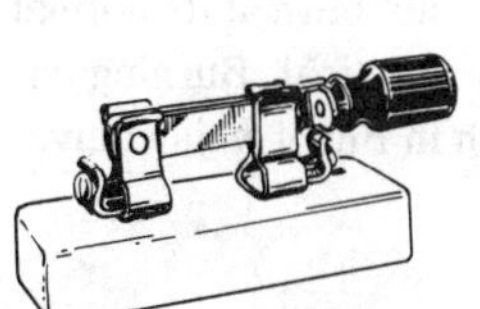

Fig. 4-5 The switch opens one wire.

Fig. 4-6 A toggle switch. The mechanism is completely enclosed. It does exactly what the switch in Fig. 4-5 does—it opens one wire. (*General Electric Co.*)

showing. It has two terminals just like the knife switch in Fig. 4-5. The mechanism is small and compact, but it does exactly what the knife switch does; in one position of the handle the switch is open, in the other it is closed. Any switch that opens only one wire is known as a "single-pole" switch. A single-pole toggle switch is identified by its two terminals, and the words ON and OFF on the handle. Obviously this style of switch is very much safer than one with exposed live parts.

Series Wiring. The circuit of Fig. 4-4 controls only one lamp; often a switch must control two or more lamps. In drawing a diagram for this, many beginners will connect several sockets as shown in Fig. 4-7. The

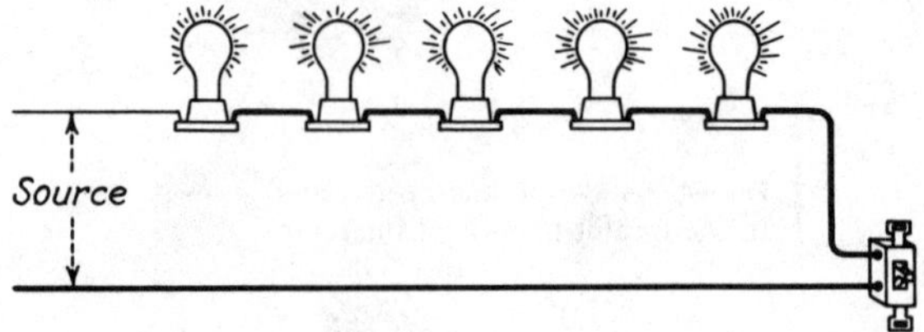

Fig. 4-7 This type of wiring is known as "series" wiring.

current can be traced from the SOURCE along the one-way street (wire) to the first lamp, to the second, to the third, to the fourth, to the fifth, and then along the other one-way street (wire) back through the switch to the SOURCE; consequently the lamps should light. They will light if not too many are used and if all are of the same size and of the proper voltage. However, assume that each lamp is a different size and that all are rated for the source voltage; since all the current that flows through one must flow through all, and the current is dependent on the total resistance, the smaller lamps (having the higher resistance) will limit the current, and the larger lamps will burn dimly, or not at all if the current is insufficient to heat up their filaments. Even the smallest lamp will not burn at its normal brilliancy. So far, the scheme does not seem very practical. Burning out one lamp or removing it from its socket, as shown in Fig. 4-8, is equiva-

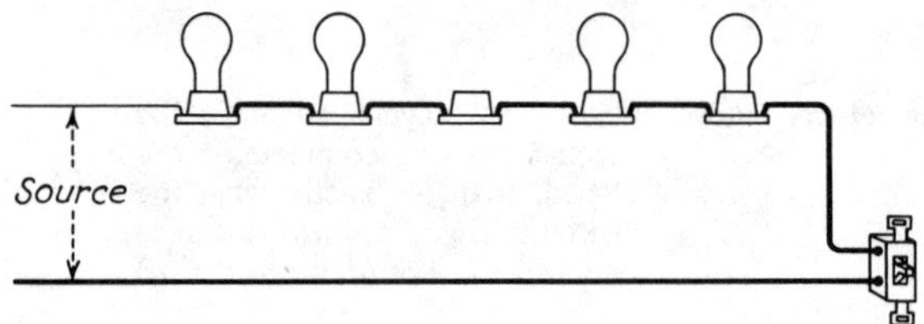

Fig. 4-8 In series wiring, when one lamp goes out, all go out.

lent to opening a switch in the circuit. All the lamps go out. This type of wiring is known as "series" wiring and is impractical for ordinary purposes.[1]

[1] The series circuit was used in very old-style Christmas-tree outfits, where eight identical lamps were used and consequently all burned at the same brilliancy. Each lamp was rated at 15 volts; they could be used on 120-volt circuits because each lamp received one-eighth of the total of 120 volts, or about 15 volts. At one time street-lighting lamps were also connected in series.

Fig. 4-9 In illustrations from this point onward, the symbol above will be used to indicate a lamp and its socket.

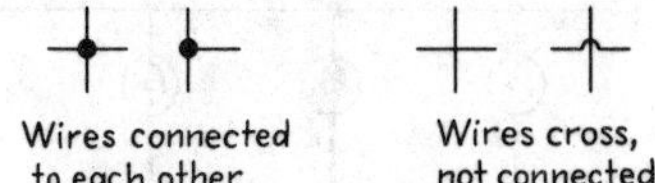

Fig. 4-10 Note carefully the designations above, which show whether crossing wires are connected to each other or not.

Instead of a picture of a lamp in a socket being used as in past diagrams, from this point onward the symbol of Fig. 4-9 will be used to denote a lamp and its socket. Note also the diagrams of Fig. 4-10, indicating whether wires that cross each other in diagrams are connected to each other or not.

Parallel Wiring. The scheme used in ordinary wiring is known as "parallel" wiring; see Fig. 4-11. When one lamp burns out or is removed, the

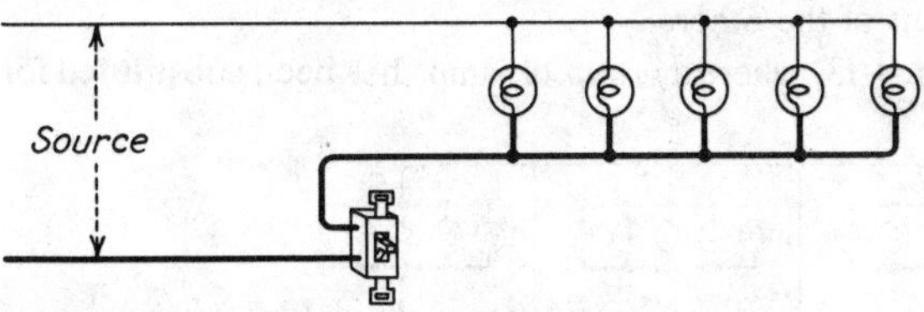

Fig. 4-11 One switch here controls five lamps.

current can still be traced from the SOURCE directly to *each* lamp whether there are five of them, or a dozen or more. From the other terminal of each lamp the current can be traced back along the wire, through the switch, to the SOURCE. Try it; cover one or more of the lamps in the drawing with a narrow strip of paper, leaving the wires exposed. The circuit will operate, regardless of the number of lamps in place, and the switch will always turn the lamps on and off. This is the way the sockets in a five-light fixture (or in five separate one-lamp fixtures) are wired, controlled by a single switch in the wall.

Using Several Switches. The circuits covered up to this point might serve in a one-room summer cottage, or an outbuilding on a farm, but all the lights in a house would never be controlled by a single switch. It is equally simple to wire a number of sockets with separate switches. Figure 4-12 will be recognized as the same as Fig. 4-11, except that in place of one switch there are now five switches; these have been numbered 1 to 5; the lamps have also been numbered 1 to 5. With a piece of paper, cover

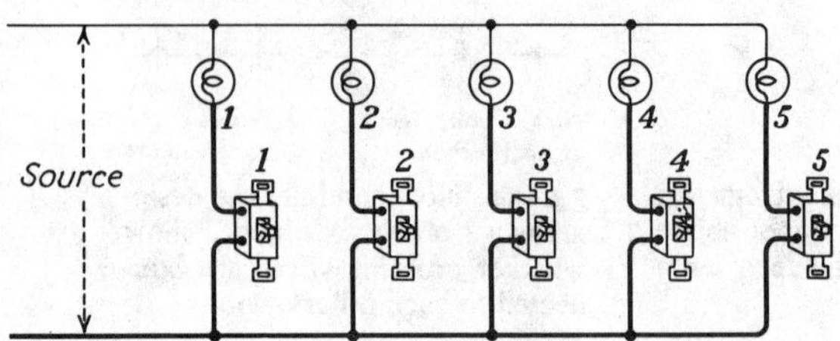

Fig. 4-12 Now each lamp is controlled by a separate switch.

both lamps and switches 2, 3, 4, and 5, leaving 1 exposed; immediately it becomes the simple circuit of Fig. 4-4. Cover lamps and switches 1, 2, 3, and 4, and again it becomes Fig. 4-4. Cover *any* four switches and lamps, and it becomes Fig. 4-4. Trace the current from the SOURCE to *any* lamp; it can be traced through the lamp to the switch for that lamp, and back to the SOURCE. This can be done whether one or two or all switches are on; each is independent of the others.

Turn now to Fig. 4-13, where a *group* of lamps has been substituted for

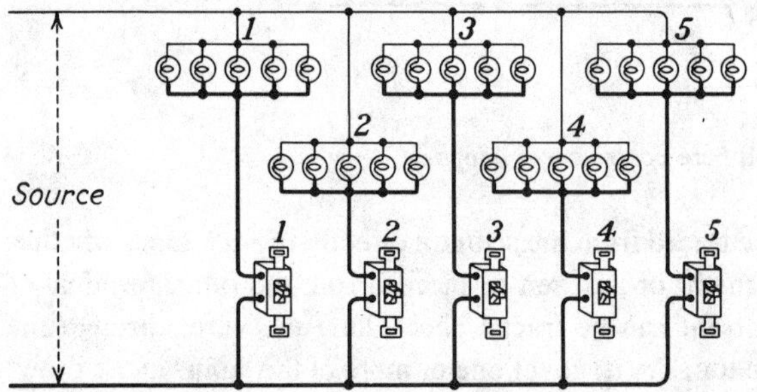

Fig. 4-13 This is the same as Fig. 4-12, except that each switch controls five lamps.

each single lamp, so that now there are five *groups* of lamps and five switches, numbered 1 to 5. With a piece of paper cover groups 2, 3, 4, and 5 with their switches, and immediately the simple circuit of Fig. 4-11 appears—five lamps controlled by a single switch. Cover any four groups, and in each case the current can be traced from the SOURCE to any of the lamps and through the switch controlling the group, back to the SOURCE.

Figure 4-13 is the basic wiring diagram for a five-room house with a five-light fixture in each room, controlled by one switch for each room. Actually the wires would run more as shown in Fig. 4-14, which is more

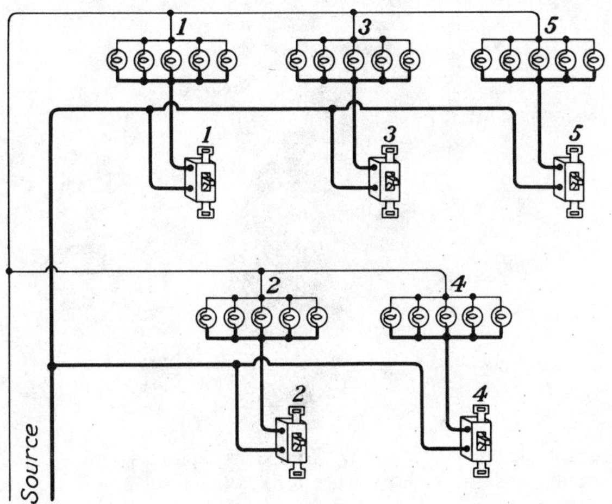

Fig. 4-14 The circuit of Fig. 4-13, but rearranged.

pictorial, with wires coming into the basement, then running to two rooms on the first floor and three rooms on the second floor.

Receptacles. Floor lamps, toasters, radios, and similar equipment must be portable; receptacles are used to plug in these items as required. The basic idea is shown in Fig. 4-15: a pair of metal contacts, one connected to each of the two wires from the SOURCE, a plug that has two corre-

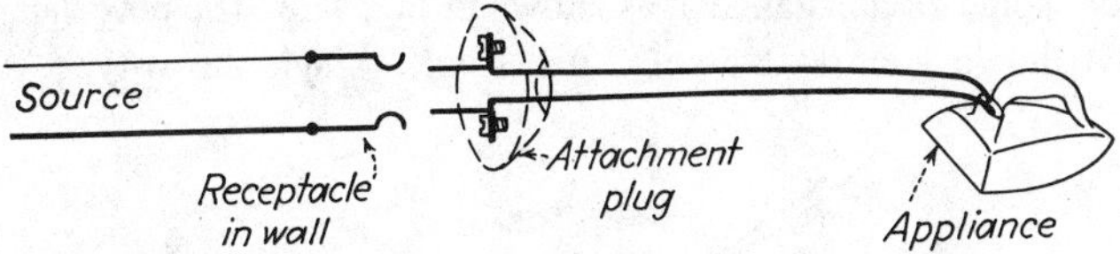

Fig. 4-15 The principle of a plug-in receptacle outlet.

sponding contacts that can be brought into contact with the first pair, and a pair of wires running to the appliance. Figure 4-16 shows a duplex receptacle (so-called because it has *two* pairs of openings that will accommodate two plugs at the same time). The single receptacle of Fig. 4-17 is seldom used in new installations.

In any wiring diagram, a receptacle can always be substituted for a socket; if, however, the socket is controlled by a wall switch, then whatever is plugged into the receptacle substituted for a socket will also be

Fig. 4-16 A duplex receptacle permits two different appliances to be plugged in at the same time. (*General Electric Co.*)

Fig. 4-17 Single receptacles are little used today. (*General Electric Co.*)

turned on and off by the switch. In any diagram or circuit, connect the receptacle in such a way that, if it were a lamp, it would always be on.[2] If in doubt go back to the one-way-street idea, and see if the messengers can go from the SOURCE to the receptacle and back again to the SOURCE even if all switches are in the open or off position.

Double-Pole Switches. While opening one of the two wires to a lamp turns it on and off, still both wires can be opened if desired (and as required under some circumstances), as shown in Fig. 4-18. The porcelain-base switch shown is a "double-pole" or "2-pole" single-throw type. A

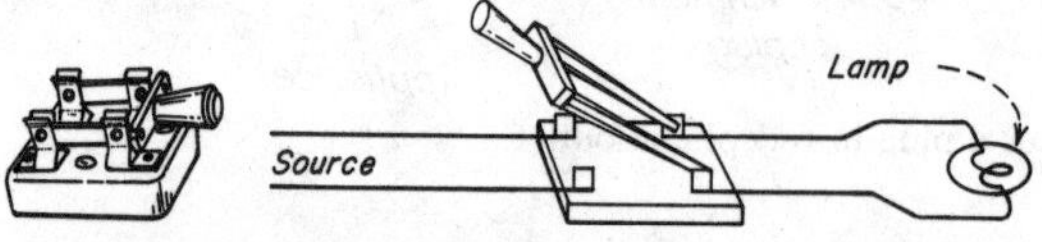

Fig. 4-18 Both wires are disconnected when a lamp is turned off with a double-pole switch.

corresponding flush toggle switch of the type shown in Fig. 4-6 but with two poles instead of one is usually referred to simply as a double-pole or

[2] In actual wiring, it is often desirable to connect a receptacle so that it is *not* permanently on, but rather controlled by a wall switch. That will be discussed in Chap. 12.

2-pole switch. It has *four* terminals for wires, *and* the words ON and OFF on the handle.

Double-pole switches are required by the Code when neither of the two wires is grounded. In practice this means you must use double-pole switches for 240-volt motors or appliances.

Three-Way Switches. Often it is convenient to be able to turn a light on or off from two different places, for example, a hall light from either upstairs or downstairs, or a garage light either from the garage door or from inside the house. Fortunately this is easily done by using switches known as "3-way" type, which are actually "single-pole double-throw" as shown in the porcelain-base type in Fig. 4-19. Figure 4-20 shows the diagram; call the two switches *A* and *B*. Tracing the circuit will show that

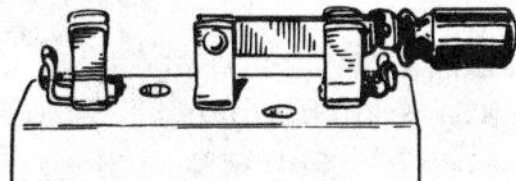

Fig. 4-19 A single-pole double-throw switch. An enclosed toggle switch that performs the same operation is called a "3-way" switch.

when the handles of both are *up,* the lamp will light; when they are both *down,* the lamp will also light. If either one is up and the other down, the lamp cannot light. Careful study of this diagram will also show that if the light is on (regardless of whether the handles of the switches are both up or both down), it can be turned off by throwing the handle of either *A* or *B* to the opposite position. The light can be controlled by either switch *A* or switch *B,* regardless of the position of the other switch in the pair.

In actual wiring, a switch that looks like the switch in Fig. 4-6 is used, except that it has *three* terminals instead of two and the words ON and OFF do *not* appear on the handle. Switches of this kind are known as 3-

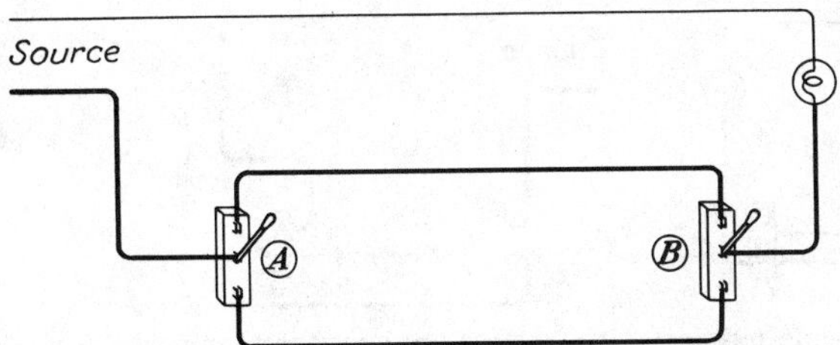

Fig. 4-20 The basic diagram for 3-way switches, which are used to control a light from two different points.

way switches, a name that is misleading because it implies that by the use of such switches a light can be controlled from three points instead of only two. The name is no doubt derived from the three terminals on the switch. The terminal that corresponds to the center terminal of the porcelain-base switches *A* and *B* of Fig. 4-20 is usually marked by being of a different color, usually a dark or oxidized finish. What happens inside the switch as it is turned from one position to the other is shown in Fig. 4-21. In one position of the handle, the current enters by terminal *A*

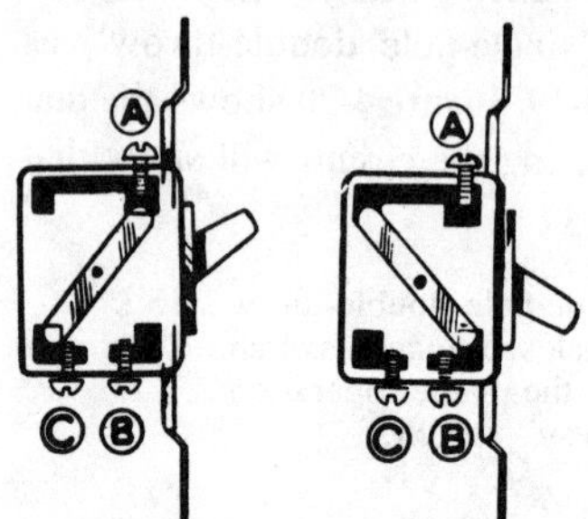

Fig. 4-21 This shows what happens inside a 3-way switch when the handle is thrown from one position to the other.

and leaves by terminal *C*. When the handle is turned to the opposite position, it enters by *A* and leaves by *B*. Analyzing Figs. 4-22 and 4-23 carefully will show that the wiring of 3-way switches is really very simple. On one of a pair of such switches, run the wire from the SOURCE to the marked or common terminal; on the other switch, run a wire from the light to the marked terminal. Then run two wires from the two remaining terminals on one switch to the two remaining terminals on the other. The wires that run between the switches are called "travelers," "runners," or "jockey legs."

The mechanical construction of 3-way switches varies among manufac-

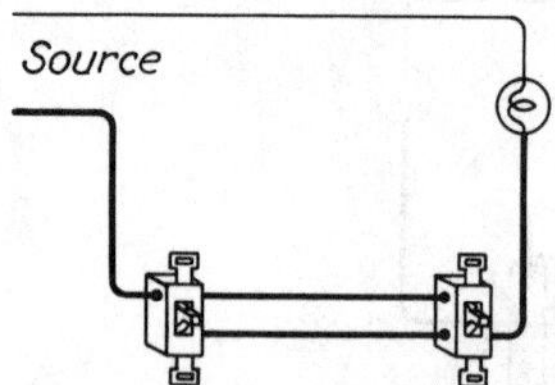

Fig. 4-22 If the common terminal on 3-way switches is alone on one *side,* use this diagram.

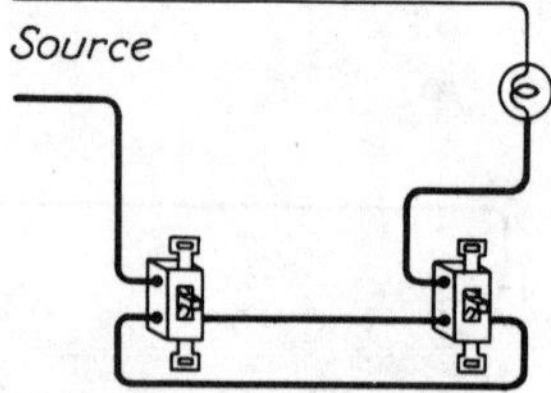

Fig. 4-23 If the common terminal on 3-way switches is alone on one *end,* use this diagram.

turers, so that the marked terminal is sometimes alone on one end of the switch, sometimes alone on one side. Therefore the pictorial diagram will be either that of Fig. 4-22 or that of Fig. 4-23, depending on the brand of switch. Fortunately no harm is done if the wrong terminals are selected, except that the circuit will not work, and if there is any doubt as to which are the correct terminals, proceed by trial and error until a combination is found that works properly. For the purposes of this book, whenever a pictorial diagram involves 3-way switches, the terminal that is alone on one *side,* as in Fig. 4-22, is the common or marked terminal.

Four-Way Switches. The preceding paragraphs show how to control a light from two points. What about three different points? It is a bit more complicated, although still relatively simple. At the point nearest the SOURCE, and also at the point nearest the light, use the 3-way switches just described. At the in-between point use a 4-way switch, the construction of which is such that it performs the operations shown in Fig. 4-24.

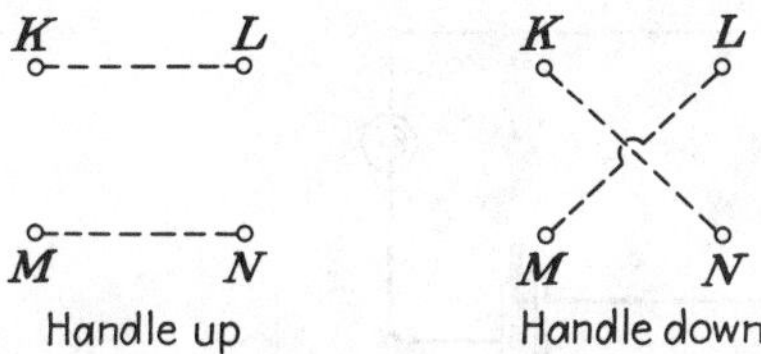

Fig. 4-24 This shows what happens inside a 4-way switch when the handle is thrown from one position to the other.

In one position of the handle, terminal *K* is connected to terminal *L;* and terminal *M* is connected to terminal *N*. When the handle is thrown, *K* is connected to *N,* and *M* is connected to *L,* as the diagram shows.

With this operation clearly in mind, note Fig. 4-25, which shows a light with three switches: a 3-way at *A,* another at *B,* and a 4-way at *C* in the center. As long as the 4-way switch *C* is in the position shown, the current flows through the switch from *K* to *L* and from *M* to *N*. The wires from *A* to *B* might just as well be continuous wires without the switch *C*. In this picture the handles of switches *A* and *B* are both in the up position, and of course, the light is then on. If, then, the wires from *A* to *B* are considered as continuous wires (forgetting for a moment that switch *C* is there), Fig. 4-25 becomes identical with Fig. 4-20, merely a light controlled from two points by two 3-way switches.

Now see Fig. 4-26, which is exactly the same as Fig. 4-25 except that

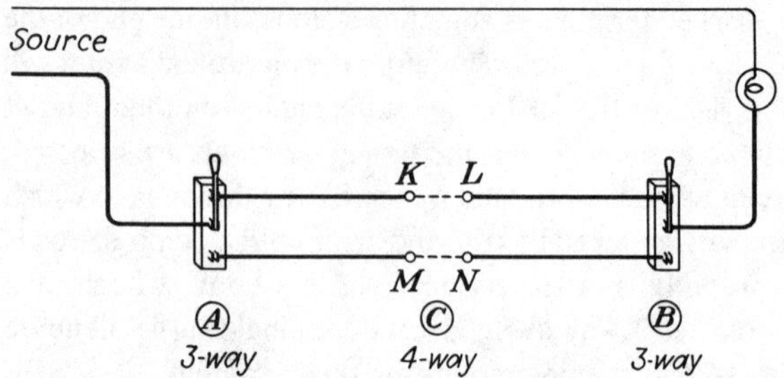

Fig. 4-25 The basic diagram for a 4-way switch, used with a pair of 3-way switches, to control a light from three different points.

the handle of the 4-way switch *C* has been thrown to the opposite position. Trace the circuit. Chase the messengers any way at all; they cannot get through and the light is off. Draw a few diagrams similar to Fig. 4-26, but with the handles of switches *A, B,* and *C* in different positions; the

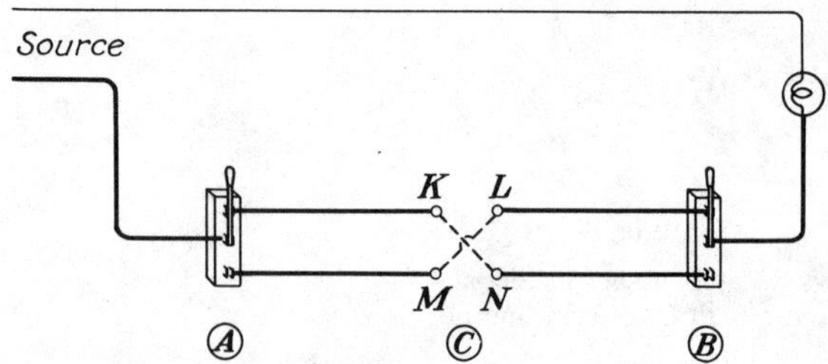

Fig. 4-26 The same as Fig. 4-25, but with the handle of the 4-way switch thrown to the opposite position.

diagrams will show that the light can be controlled from any one of the three switches. To control a light from three positions, use two 3-way switches and one 4-way switch. The flush switch of Fig. 4-6 in the 4-way type is identified by its *four* terminals and the fact that it does *not* have the words ON and OFF on the handle (double-pole switches also have four terminals but *do* have the words ON and OFF on the handle).

Some manufacturers make their 4-way switches so that the internal connections, when the handle is thrown, change as shown in Fig. 4-27. In that case the diagram of Fig. 4-26 becomes that of Fig. 4-28—simply cross two of the wires as shown. As in the case of 3-way switches, no harm can be done by wrong connections, except that the circuit will not work. If there is doubt as to the internal wiring of the switch, you can determine the internal connections by the use of a continuity tester such as the one shown in Fig. 18-23. Draw two diagrams of the switch on

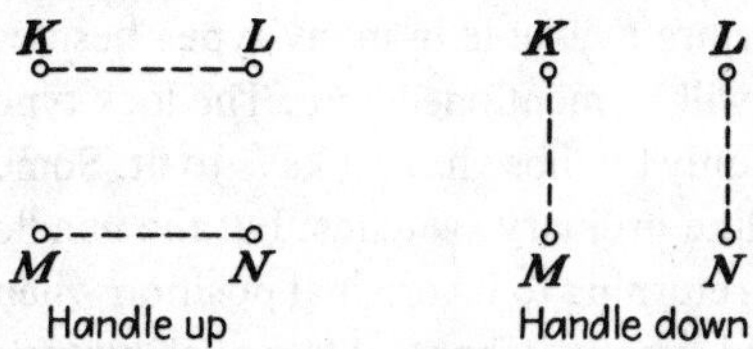

Fig. 4-27 On some brands of 4-way switches, the connections inside the switch change as shown above when the handle is thrown.

paper, one for each position of the handle. Test from each terminal to the other three in both positions of the switch handle, and record on your diagrams the internal connections by lines joining the terminals which cause the test lamp to light. You will end up with a diagram the same as Fig. 4-24 *or* Fig. 4-27.

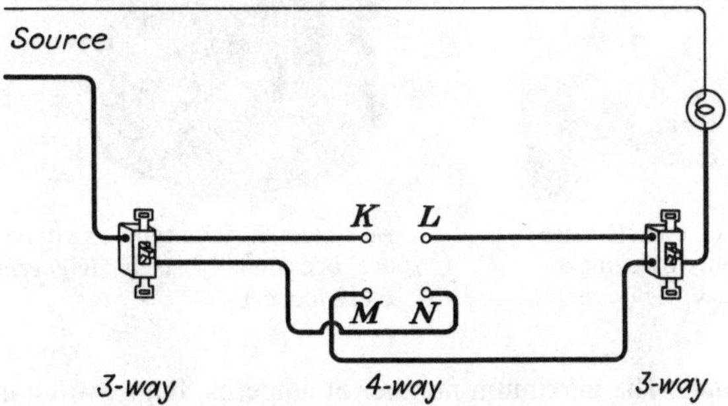

Fig. 4-28 With 4-way switches of the type shown in Fig. 4-27, use this diagram instead of the one shown in Fig. 4-25.

To control a light from four, five, or any number of points, use a 3-way switch at the point nearest the light, another at the point where the wires come from SOURCE, and 4-way switches at each of the other points; connect as shown in Fig. 4-29.

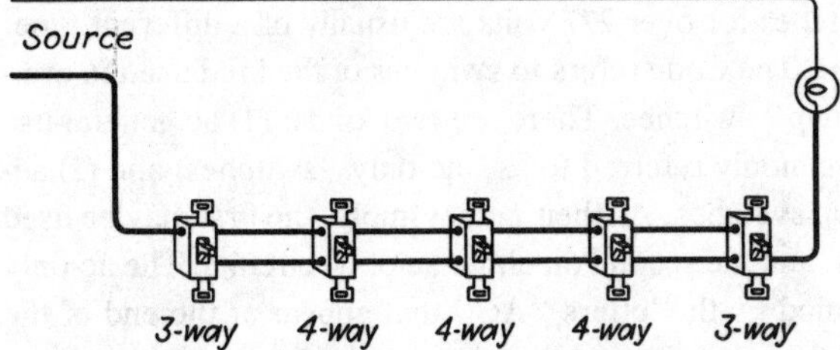

Fig. 4-29 When a light must be controlled from more than three points, use this diagram.

Miscellaneous Switches. Switches are available in many types besides the types described; some of them will be mentioned here. The lock type shown in Fig. 4-30 can be operated only by those having keys to fit. Some momentary-contact switches look like ordinary switches, but the handle is held in one position by a spring, returning to its original position when the operator releases the handle. Old-time switches had two push buttons instead of a handle. The surface type of Fig. 4-31 is used mostly in surface wiring, as in some garages and farm buildings.

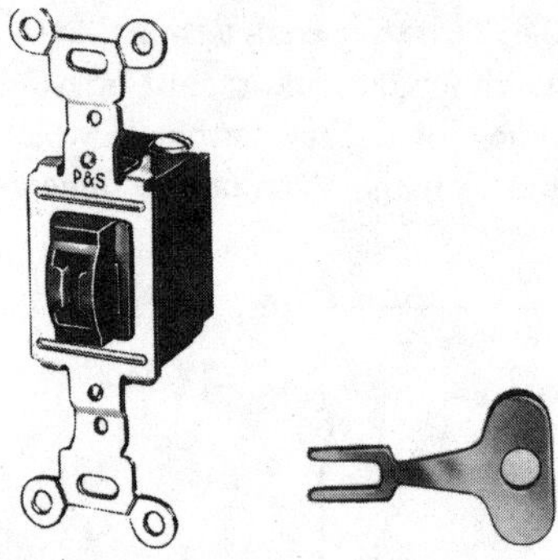

Fig. 4-30 This type of switch can be turned on or off only by using a special key. (*Pass & Seymour, Inc.*)

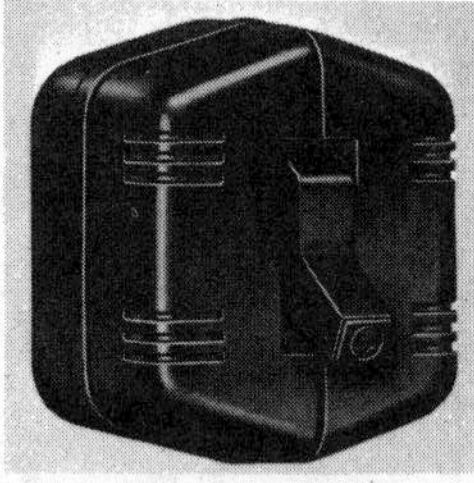

Fig. 4-31 Surface-type switches are occasionally used. (*General Electric Co.*)

Ratings of Switches. The maximum number of amperes that a switch is capable of handling, and the maximum voltage at which it may be used, are stamped into the metal mounting yoke of the switch. This might be a single rating, such as "15A 277V," indicating that the switch may be used to control up to 15 amp at not over 277 volts. Another common rating is "10A 125V—5A 250V," indicating that the switch may be used to control up to 10 amp at not over 125 volts, or up to 5 amp at not over 250 volts. Of course switches with higher ampere and/or voltage ratings are available, but switches for over 277 volts are usually of a different type.

Kinds of Switches. The Code refers to switches of the kind used in ordinary wiring as "snap" switches. There are two kinds: (1) ac general-use snap switches (commonly referred to as "ac-only" switches) and (2) ac-dc general-use snap switches. As their names imply, the first may be used only on ac circuits, and the second on either ac or dc circuits. The ac-only type can be identified by the letters "AC" that appear at the end of the rating stamped on the mounting yoke of the switch; the ac-dc type, however, does *not* have the letters "AC-DC" on the yoke. If the letters "AC" do not appear, the switch is the ac-dc type. Study the Code on the subject of switches: Art. 100 (Definitions) and Sec. 380-14 (Rating and Use of Snap Switches).

AC-Only Switches. These are the most common type being installed. They may be used anywhere to control any type of load up to their full ampere and voltage rating, except that if used to control a motor load, they must have an ampere rating of at least 125% of the ampere rating of the load. AC-only switches have a minimum rating of 15 amp—some at 120 volts, others at any voltage up to 277 volts. However, even if rated 277 volts, they may not be used to control *incandescent* lamps at a voltage above 120 volts. AC-only switches are quiet in operation and do not have the annoying click of the ac-dc type.

AC-DC Switches. At one time this was the only kind of snap switch made. However, direct current is now a genuine rarity, although it is still found in a few buildings in larger cities. However, many ac-dc switches are still in use, having been installed in the past. As they fail, there is no reason why they cannot be replaced by the newer ac type, on ac circuits.

Ordinary ac-dc switches are usually rated at 10 amp at not over 125 volts, or 5 amp at not over 250 volts. However, there are two subtypes: (1) those that are "T-rated" and (2) those that are not T-rated. If they are T-rated, the letter T appears at the end of the ampere and voltage rating stamped on the mounting yoke.

What is a T rating? When an ordinary incandescent lamp is turned on, for a tiny fraction of a second the current consumed by the lamp is from eight to twelve times higher than it consumes while burning normally. A 100-watt lamp when first turned on consumes more nearly 1000 watts for a small fraction of a second, then consumes its normal 100 watts; a 300-watt lamp momentarily consumes at least 3000 watts. The duration of this very high current is so short that it will not blow a fuse or show on an ammeter.

This high momentary current is known as the "cold inrush" of the lamp. If a switch is used on a group of lamps totaling 1000 watts, this inrush may be as much as 10 000 watts (over 80 amp). This is a severe stress for ac-dc switches. AC-DC switches specially designed to handle such loads are designated as T-rated, the "T" standing for tungsten, the material in the filaments of the lamps.

Note that ac-only switches are never T-rated, the letter T never appearing on the mounting yoke.

AC-DC switches that are *not* T-rated may be used to control resistive loads such as heaters up to their full rated capacity or inductive loads up to 50% of their ampere rating. They may *not* be used to control incandescent lamps under any circumstances.

If T-rated they may be used in any location to control lamps or other loads up to their full ampere ratings, unless the loads are inductive, in which case they must be rated at 200% of the ampere rating of the load involved.

Type of Switch to Use. In any location (except for dc circuits) it makes sense to use the ac-only switches. The quietness of their operation in itself commends them for general use.

Face Plates. Switches and receptacles cannot be mounted in walls leaving openings around them, nor can terminals be left exposed, for that would be unsafe. Therefore they are covered by face plates after installation. Figure 4-32 shows several face plates. The smaller ones are used for

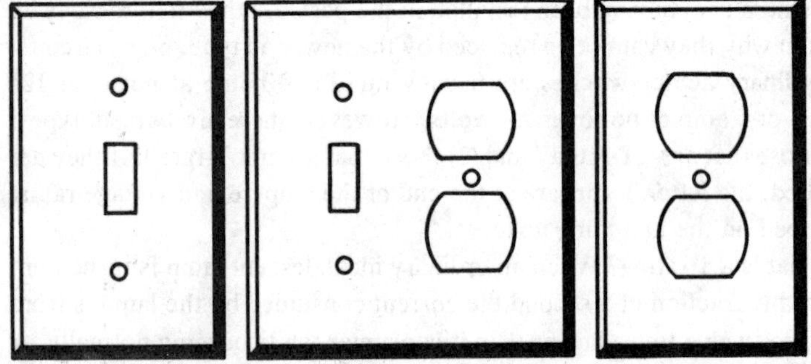

Fig. 4-32 Face plates must cover all switches, receptacles, and similar devices.

single devices. Sometimes it is necessary to install two or three or more devices side by side, requiring wider plates known as "2-gang," "3-gang," or wider, depending on how many devices the plate covers. They are also available in combinations so that switches, receptacles, and other devices can be mounted side by side, as the same figure shows.

Face plates are made of a great variety of materials, such as plastic in brown or ivory, or of brass or other metals in many different finishes to suit the user. The nonmetallic plates are generally used.

Sockets. Sockets are available in a great variety of types. The cleat lampholder that was shown in Fig. 4-1 is not actually used in house wir-

Fig. 4-33 A typical brass-shell socket. Besides the pull-chain type shown, there are several other types. (*General Electric Co.*)

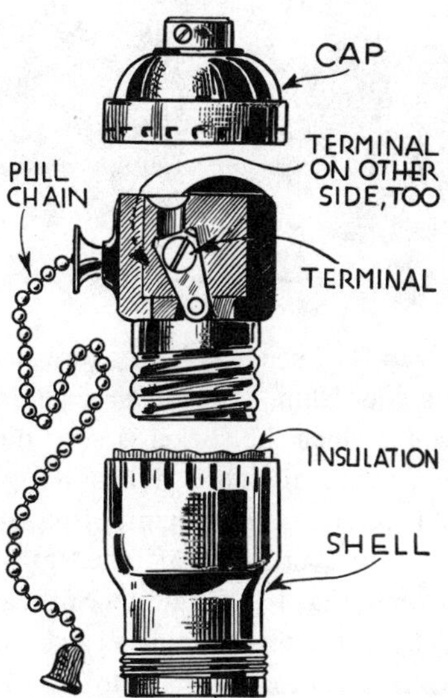

Fig. 4-34 Exploded view of the socket shown in Fig. 4-33.

ing. The commonly used brass-shell socket may be either keyless, or have a switching mechanism (key, push-through, or pull-chain) to turn the lamp on or off. One of the pull-chain type is shown in Fig. 4-33. The socket consists of the brass shell, an insulating paper liner to insulate metal parts from the shell, the mechanism proper with two terminals, and

Fig. 4-35 The socket shown fits directly on an outlet box. (*General Electric Co.*)

Fig. 4-36 The weatherproof socket shown is intended for outdoor use. (*Leviton Mfg. Co., Inc.*)

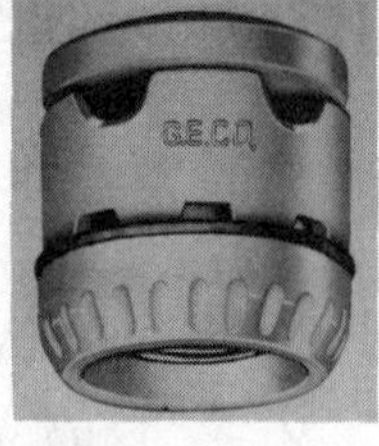

Fig. 4-37 These types of sockets are used mostly in the manufacture of lighting fixtures. (*General Electric Co.*)

the cap. The cap may have a threaded hub, used when the socket is used on a floor lamp, a fixture, or similar device, or it may have an insulating bushing when the socket is used on the end of a cord as a pendant. A cross-sectional view is shown in Fig. 4-34. Instead of brass for the outer shell, plastic or porcelain is frequently used. Other sockets are of the type shown in Fig. 4-35, used at outlet boxes; the weatherproof type shown in Fig. 4-36 for outdoor use; and the type shown in Fig. 4-37, used chiefly in the manufacture of lighting fixtures and signs.

Other Devices. There are dozens of other devices, and many will be described in later chapters of this book as their use is discussed.

5

Overcurrent Devices

It is impossible for an electric current to flow through a wire without heating the wire. As the number of amperes increases, the temperature of the wire also increases. For any particular size of wire, the heat produced is proportional to the *square* of the current. Doubling the current increases the heat four times, tripling it increases the heat nine times, and so on.

Need for Protective Devices. As the temperature of a wire increases, its insulation may become damaged by the heat, leading to ultimate breakdown. With sufficient current, the conductor itself may get hot enough to start a fire. It is therefore necessary to carefully limit the current to a maximum value, one that is safe for a given size and type of wire. The maximum number of amperes that a wire can safely carry continuously is called the "ampacity" of the wire. Ampacity is defined by the Code as "the current in amperes a conductor can carry continuously under the conditions of use without exceeding its temperature rating."* As will be discussed later, the Code specifies the ampacity of each type and size of wire under various conditions.

*Reprinted with permission from NFPA 70-1984, National Electrical Code®, Copyright© 1983, National Fire Protection Association, Quincy, Massachusetts 02269. This reprinted material is not the complete and official position of the NFPA on the referenced subject, which is represented only by the standard in its entirety.

Any device that opens the circuit when the current in a wire reaches a predetermined number of amperes is called an "overcurrent device" in the Code. There are several kinds, and they all can be considered the safety valves of electrical circuits. The two types that will be discussed here are fuses and circuit breakers. (The abbreviation for circuit breakers is CB or cb.) See Fig. 5-1, which shows how fuses and breakers are designated in wiring diagrams.

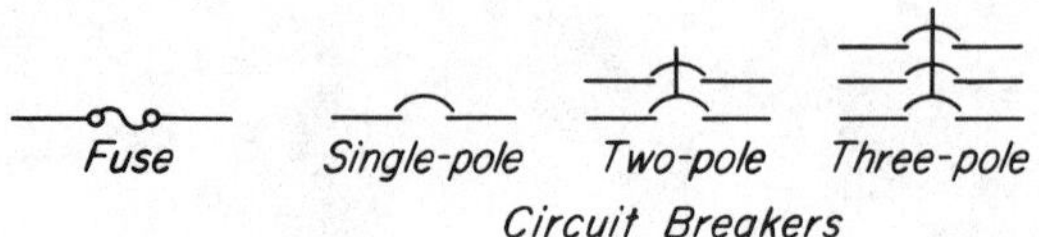

Fig. 5-1 At left, symbol for a fuse; at right, symbols for circuit breakers.

Besides being used to protect *wires* from too great a current, overcurrent devices are also used to protect electrical *equipment*. For example, a motor may require 15 amp to deliver the horsepower stamped on its name-plate. Although motors can deliver more than their rated horsepower, while doing so they consume more than the normal number of amperes. If the overload is continued long enough, the motor will burn out. To protect the motor against such overload currents, a rather special overcurrent device called an "over*load* device" is installed in the motor circuit. Motor overload protection will be explained in Chaps. 15 and 30.

Fuses. A fuse is basically a short length of metal ribbon made of an alloy with a low melting point, and of a size that will carry a specified current indefinitely, but which will melt when a larger current flows. When the ribbon inside the fuse melts, the fuse is said to "blow." When it blows, the circuit is open, just as if a wire had been cut, or a switch opened, at the fuse location.

Plug Fuses. The common plug-type fuse is shown in Fig. 5-2. The fusible link is enclosed in a sturdy housing that prevents the molten metal from spattering when the fuse blows. There is a window through which you can see whether the fuse has blown. The largest plug fuse is rated at 30 amp; smaller standard sizes are 15, 20, and 25 amp. (Sizes smaller than 15 amp are also available.) The Code requires that plug fuses rated at 15 amp or less be of hexagonal shape, or have a window or other prominent part of hexagonal form; those rated at more than 15 amp are round.

The Code limits the use of *plug* fuses to circuits of not over 150 volts *to ground*. If the premises are served by a 120/240-volt line, the voltage be-

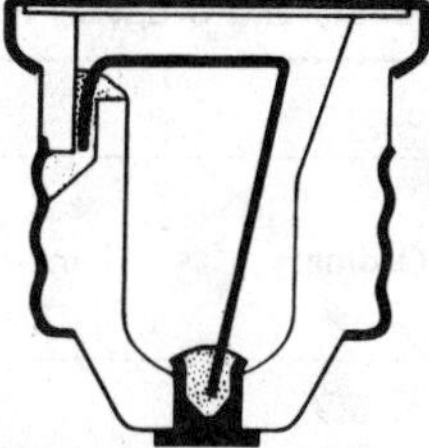

Fig. 5-2 Plug fuses are made only in ratings up to 30 amp. (*Bussmann Manufacturing Division, McGraw-Edison Co.*)

tween either hot wire and the grounded wire is 120 volts (120 volts *to ground*); therefore plug fuses may be used on 240-volt circuits for water heaters, motors, and so on, although cartridge fuses are more often used.

Time-Delay Fuses. Consider a lighting circuit in a home, wired with No. 14 copper wire that has an ampacity of 15 amp and protected by a 15-amp fuse. Most of the time the wire will be carrying less than 15 amp, and the temperature of the wire and its insulation will be well within safe limits. If the current is increased to 30 amp, the fuse will blow in a few seconds. On the other hand, 30 amp flowing for even half a minute will not heat the wire or its insulation to the danger point, especially if the current was very low before being increased to 30 amp.

In practice, there are often conditions just as described. Perhaps 5 amp is flowing in the wire, representing 600 watts of lights. Then a motor is turned on, such as a workshop motor. The motor consumes perhaps 30 amp for some seconds while it is starting; after that it drops to a normal of around 6 amp. Often the fuse blows during this starting period, although the wire and its insulation were not in any danger whatever.

Time-delay fuses (sometimes incorrectly called time-lag) have been developed that *do not blow* like ordinary fuses on large but *temporary* overloads, but do blow like ordinary fuses on small *continuous* overloads, and instantly on short circuits. The difference between the two types can be seen from Table 5-1, which indicates typical blowing times in seconds.

All fuses, ordinary and time-delay types, are tested in open air to carry 110% of their rated current indefinitely without blowing. The *continuous* load on a fuse in an enclosure should not exceed 80% of the fuse rating. But all fuses installed in a very hot location will blow faster on any given current than when installed in locations having ordinary temperature.

A plug fuse of the time-delay type has an external appearance just like

TABLE 5-1 Comparison of Time-Delay and Ordinary Fuses

	15-amp fuse		30-amp fuse	
Actual current, amp	Time-delay, s	Ordinary, s	Time-delay, s	Ordinary, s
30	31	3.9		
45	10	0.8	140	22
60	5	0.3	27	4.4
75	1.5	0.2	11	1.8
90	0.5	0.1	5.4	1.0

that of an ordinary fuse, but is made differently inside. A cross-sectional view of such a time-delay fuse is shown in Fig. 5-3. Such fuses are often referred to as Fusetrons, which actually is the trade name of one particular manufacturer. The use of time-delay fuses is especially desirable when motors are used. A large percentage of blown fuses on motor circuits could have been prevented by using time-delay fuses.

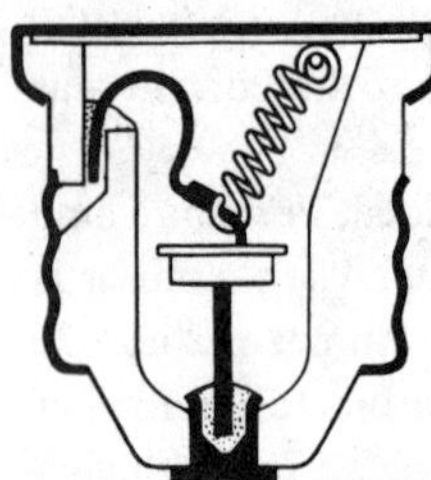

Fig. 5-3 Cross section of a typical time-delay fuse. Time-delay fuses carry *temporary* overloads safely without blowing. (*Bussmann Manufacturing Division, McGraw-Edison Co.*)

Edison-Base Fuses. Fuses of the type shown in Figs. 5-2 and 5-3 have screw-shell bases of the same kind and size as used on ordinary incandescent lamps. Such fuses are known as "Edison-base" type.

Type S Nontamperable Fuses. Since each size of wire has a very definite ampacity (the maximum safe current-carrying capacity in amperes), the Code requires that the overcurrent device selected to protect the wire must have a rated capacity in amperes no higher than the ampacity of the wire it is to protect. For example, No. 14 copper wire used in a large percentage of ordinary house wiring has an ampacity of 15 amp and should be protected by fuses rated at not over 15 amp. But all Edison-base fuses up to 30 amp are interchangeable. Nothing prevents the homeowner from substituting a 30-amp fuse for the 15-amp fuse that he should be using. That of course defeats the purpose of the fuse and is a foolish thing to do.

To prevent overfusing and tampering, nontamperable fuses that the Code calls "Type S" were developed. Type S fuses can be used only with adapters that are screwed into ordinary Edison-base fuseholders. Both fuse and adapter are shown in Fig. 5-4, while Fig. 5-5 shows a cross-sectional view of the fuse.

Fig. 5-4 A nontamperable fuse and its adapter. (*Bussmann Manufacturing Division, McGraw-Edison Co.*)

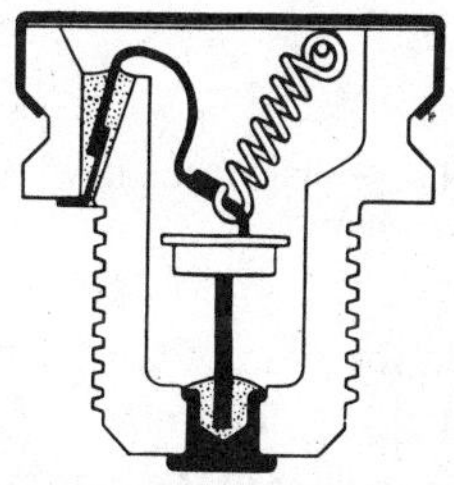

Fig. 5-5 Cross section of the fuse shown in Fig. 5-4 (*Bussmann Manufacturing Division, McGraw-Edison Co.*)

The adapter once inserted into the fuseholder cannot be removed without damaging the fuseholder. Each adapter is rated in amperes. The 15-amp adapter accepts only 15-amp or smaller fuses; the 20-amp adapter accepts only 20-amp fuses; the 30-amp adapter accepts only 25- or 30-amp fuses. Obviously, then, if a contractor installs 15-amp adapters, he makes it impossible to use fuses larger than the 15-amp size, thus making overfusing impossible by those who know no better, or are inclined to take chances. This eliminates one of the greatest causes of electrical fires.

Type S fuses are required by NEC Sec. 240-52 where plug-type fuseholders are being installed. The ordinary Edison-base plug fuse is permitted only in *existing* installations, and only if there is no evidence of overfusing or tampering. Type S fuses are usually called Fustats, although that is the trade name of a particular manufacturer. While the Code does not require it, all Type S fuses now being made are of the time-delay type.

In using Type S fuses, one caution is in order. When inserting the fuse into its adapter, turn it firmly some more after it *appears* to be tightly in place. There is a spring under the shoulder of the fuse, and this spring must be flattened or the fuse will not "bottom" (make contact) in the adapter; in that case the circuit will be open, just as if the fuse had blown.

Blown Fuses. When an ordinary Edison-base fuse blows, it is sometimes difficult to tell whether it blew because of an overload or a short circuit. Usually, however, the cause can be determined; see Fig. 5-6.

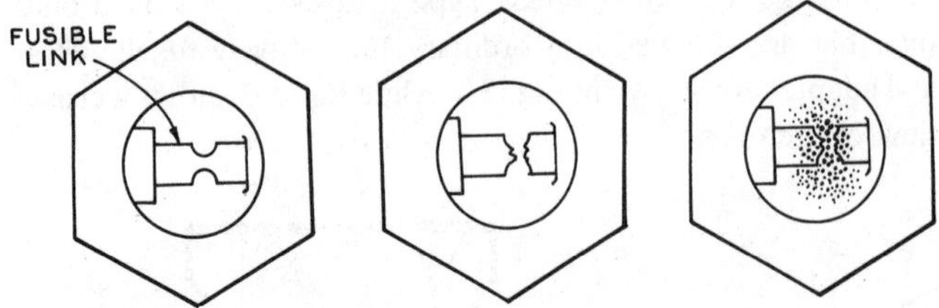

Fig. 5-6 Ordinary plug fuse: new, blown because of overload, and blown by short circuit.

However, in the case of time-delay fuses as shown in Figs. 5-3 and 5-5, the cause is easily determined. The fusible link has one end of it embedded in a bit of solder in the bottom of the fuse, with a *stretched* coil spring anchored at the top of the fuse, pulling on it. If the fuse blew because of a short circuit, the window will be blackened just as in the case of the Edison-base type. But if it blew because of an overload, the solder in the bottom of the fuse softened, permitting the stretched coil spring to pull the fusible link upward; the coils of the spring will touch each other, instead of being separated as in an unblown fuse.

Cartridge Fuses. If the fuse is rated at more than 30 amp, there is no choice but to use a cartridge fuse. Cartridge fuses are, however, available in all ratings, so that they can be used even if currents smaller than 30 amp are involved. There are two basic types of cartridge fuses: the ferrule-contact type shown in Fig. 5-7 and the knife-blade-contact type shown in Fig. 5-8. Fuses rated at 60 amp or less are of the ferrule-contact type; those rated at more than 60 amp are of the knife-blade-contact type. Cartridge fuses are made in both ordinary and time-delay types.

Fig. 5-7 The ferrule-contact type of fuse is made only in ratings up to and including 60 amp. (*Bussmann Manufacturing Division, McGraw-Edison Co.*)

Fig. 5-8 The knife-blade-contact type of fuse is made only in ratings above 60 amp. (*Bussmann Manufacturing Division, McGraw-Edison Co.*)

Cartridge fuses are made in many UL "classes," depending not only on their ampere and voltage rating, but also on the degree to which they are current-limiting, their ability to interrupt high fault currents, their time-delay characteristics, and their dimensions.[1]

The fault current available from the utility system may require fuses with a high interrupting rating (typical interrupting ratings are 10 000, 50 000, 100 000, and 200 000 amp). If the equipment in which the fuses are installed, and/or the equipment downstream cannot also withstand these high available fault currents, then current-limiting fuses may be used. A current-limiting fuse will safely open and clear a circuit, within its interrupting rating, before the current reaches the maximum available, thus limiting the energy which is let through to downstream wire and equipment. [Fault currents above 10 000 amp (the interrupting rating of ordinary fuses and breakers) are not likely to occur on 120/240-volt single-phase systems except at or near the service location, and then only under certain conditions. See Chap. 27 for a further discussion.]

Table 5-2 shows the more common UL classes of fuses and some of their characteristics. A complete coverage of the proper application of fuses is beyond the scope of this book, but you should be aware that each class of fuse is designed for a specific purpose; that each is significantly different than any other class; and that substitution of one class or size of fuse for another should never be done without full knowledge of the possibly unsafe results. In many cases, but by no means all, such unsafe interchangeability is inherently prevented by the design of the fuses and fuseholders.

Table 5-3 shows the dimensions of Class H fuses. These are also the dimensions of Class K and Class RK fuses; however, Class RK fuses also have a groove in one ferrule, or a slot in one knife blade, to make them noninterchangeable with Class H or K when Class RK fuseholders are used. In addition to the power, lighting, and control-circuit fuses shown in Table 5-2 there are also supplementary, miscellaneous, and special-purpose fuses, each with its own combination of characteristics.

In this chapter we will talk mainly of Class H fuses, the kind most often used in homes, on farms, and in most small installations of other types. Cartridge fuses are available in two voltage ratings, 250 and 600 volts,

[1] Fault current: the current, determined by the system voltage and the impedance between the source and the fault, which will flow through a short circuit.

TABLE 5-2 Characteristics of Control-Circuit, Power, and Lighting Cartridge Fuses

UL Class	CC	G	H	J	K	L	RK	T
Renewable			X					
Nonrenewable	X	X	X	X	X	X	X	X
250-volt			0–600 A		0–600 A		0–600 A	
300-volt		0–60A						0–600 A
600-volt	0–20A		0–600 A	0–600 A	0–600 A	601–6000 A	0–600 A	0–600 A
Noninterchangeable	X	X*		X		X	X†	X
Current-limiting	X	X		X	‡	X	X	X
Maximum interrupting rating, amp	200 000	100 000	10 000	200 000	200 000	200 000	200 000	200 000
Time delay	Optional	Optional	Optional§	Optional	Optional	Optional	Optional	

* 16–20- and 21–30-amp Class G fuses are interchangeable with some miscellaneous or supplementary fuses but not with any of the other fuses in this table.

† Class RK fuses will fit in Class H fuseholders, but only Class RK fuses will fit in Class RK fuseholders.

‡ Class K fuses have current-limiting characteristics but are not permitted to be so labeled, as they are interchangeable with Class H. See NEC Sec. 240-60(b).

§ Time delay not available in renewable type.

TABLE 5-3 Class H Fuse Dimensions, Inches

Fuse rating, amp	250-volt			600-volt		
	Diameter of ferrule	Width of blade	Overall length	Diameter of ferrule	Width of blade	Overall length
0–30	9/16		2	13/16		5
31–60	13/16		3	1 1/16		5 1/2
61–100		3/4	5 7/8		3/4	7 7/8
101–200		1 1/8	7 1/8		1 1/8	9 5/8
201–400		1 5/8	8 5/8		1 5/8	11 5/8
401–600		2	10 3/8		2	13 3/8

which define the maximum voltage of the circuit on which they are to be used. They are available in many ampere ratings up to 600 amp.

Renewable Fuses. Class H cartridge fuses come in two types: nonrenewable or one-time type, and the renewable kind. Nonrenewable fuses once blown have no further value. Since in most cases only the fusible link is destroyed when the fuse blows, renewable fuses permit the fusible link to be replaced after blowing. Extreme care must be used to thoroughly tighten the screws or bolts in the fuse when replacing the fuse link. Figure 5-9 shows a cross-sectional view of a renewable fuse. In external appearance there is no difference between the two types except that the renewable type is made so it can be taken apart.

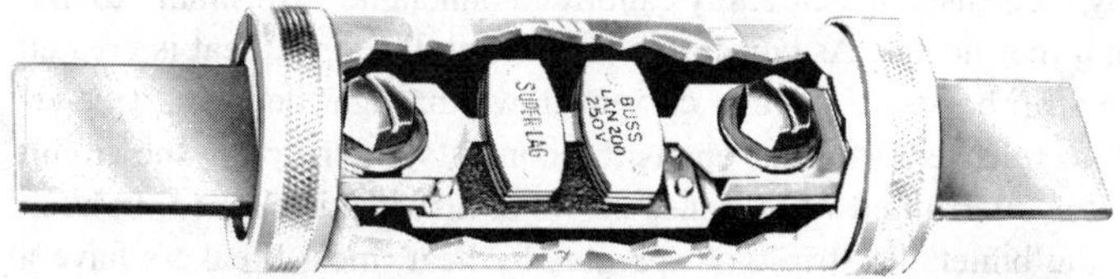

Fig. 5-9 A renewable type of fuse can be disassembled for replacement of the fusible link. (*Bussmann Manufacturing Division, McGraw-Edison Co.*)

As a practical matter, one-time fuses usually run cooler and therefore do not overheat the panelboard clips as much as the renewable kind. They are generally considered more satisfactory and therefore are recommended. If a circuit is not overloaded, fuses don't blow very often. When one does blow, the cause for its blowing should be corrected before replacing the fuse. Many feel there is little justification for renewable fuses in properly maintained installations.

In either time-delay or non-time-delay cartridge fuses, the fusible link melts on a short circuit. On the non-time-delay, it also melts on an overload. In the time-delay type shown in Fig. 5-10, an overload causes a bit of solder to melt, and a spring then opens the circuit, just as in time-delay plug fuses.

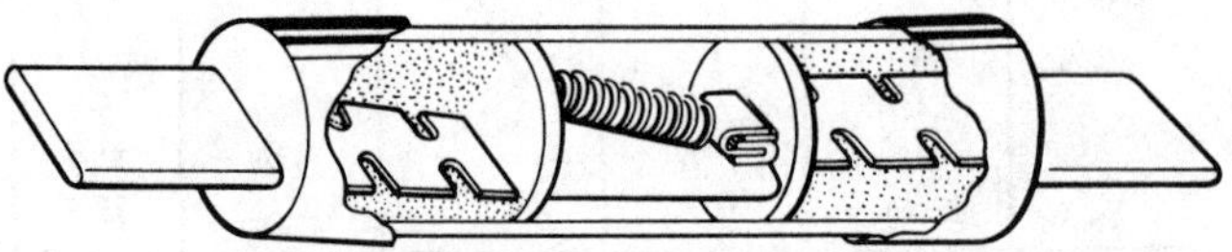

Fig. 5-10 Cartridge fuses are also made in the time-delay type. They are especially useful in protecting motors, which usually require several times as many amperes while starting as while running. (*Bussmann Manufacturing Division, McGraw-Edison Co.*)

Circuit Breakers. The Code defines a circuit breaker as "a device designed to open and close a circuit by nonautomatic means, and to open the circuit automatically on a predetermined overcurrent without injury to itself when properly applied within its rating."* So a circuit breaker is a combination device composed of a manual switch and an overcurrent device. Its basic function is similar to that of a combination switch and fuse.

A circuit breaker of the type used in homes looks something like a toggle switch used to turn a light on and off. One is shown in Fig. 5-11. Essentially it consists of a carefully calibrated bimetallic strip similar to that used in a thermostat. As current flows through the strip, heat is created and the strip bends. If enough current flows through the strip, it bends enough to release a trip that opens the contacts, interrupting the circuit just as it is interrupted when a fuse blows or a switch is opened. In addition to the bimetallic strip that operates by heat, most breakers have a magnetic arrangement that opens the breaker instantly in case of a short circuit. A circuit breaker can be considered a switch that opens itself in case of overload.

Circuit breakers are rated in amperes,[2] just as fuses are rated. Like

*Reprinted with permission from NFPA 70-1984, *National Electrical Code*®, Copyright© 1983, National Fire Protection Association, Quincy, Massachusetts 02269. This reprinted material is not the complete and official position of the NFPA on the referenced subject, which is represented only by the standard in its entirety.

[2] Large adjustable-trip breakers are available but rarely if ever used in the kinds of installations discussed in Parts 1 and 2 of this book.

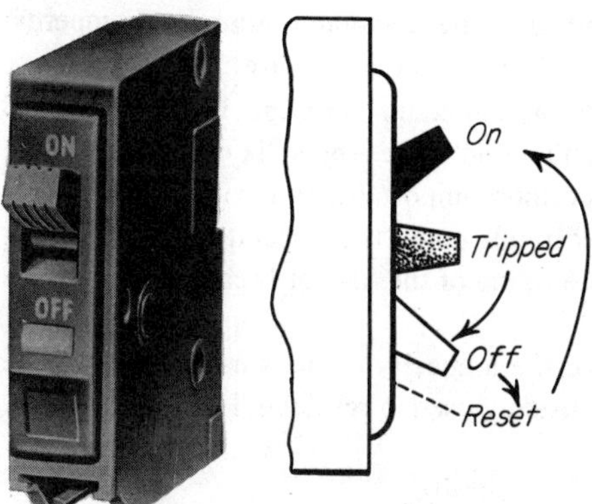

Fig. 5-11 A typical single-pole circuit breaker. If it trips, correct the reason for the overload, then reset by pressing the handle *beyond* the off position, then move to on. (*Square D Co.*)

fuses, breakers are tested in open air to carry 110% of their rated loads indefinitely without tripping. Manufacturers of breakers furnish curves showing how long breakers of a particular variety will carry specific overloads. Most breakers will carry 150% of their rated load for perhaps a minute; 200% for about 20 s, and 300% for about 5 s, long enough to carry the heavy current required to start a motor.

The trend is rapidly away from fuses and toward circuit breakers, for the latter have many advantages. When a fuse blows, spare fuses may or may not be on hand. When a circuit breaker trips, correct the cause of the overload, then reset it as shown in Fig. 5-11. The circuit breaker provides good protection and does not trip on large but temporary overloads. Modern homes are usually equipped with circuit breakers.

Standard Ratings. Both fuses and breakers are available in standard ratings of 15, 20, 25, 30, 35, 40, 45, 50, 60, 70, 80, 90, 100, 110, 125, 150, 175, and 200 amp, and of course larger sizes (up to 6000 amp) for use where required. Additional standard ratings for fuses only are 1, 3, 6, and 10 amp, mainly for the protection of small motor circuits. See NEC Sec. 240-6.

Determining Proper Rating of Overcurrent Device. The fuse must blow, or the breaker open, when the current flowing through it exceeds the number of amperes that is safe for the wire in the circuit. The larger the wire, the greater the number of amperes it can safely carry.

The Code specifies the ampacity (the maximum number of amperes) that can be safely carried by each size and type of wire. The ampacity of any size and kind of copper or aluminum wire can be found in NEC Tables 310-16 through 310-19, all found in the Appendix of this book. The Notes following the tables are most important. But with a few exceptions that will be covered later in this book, the ampacity of the wire determines the maximum ampere rating of the fuse or breaker that may be used to protect the circuit.

You may wish to memorize the ampacity of the smaller sizes of the Types T and TW copper wire usually used in residential and farm wiring. These ampacities are:

No. 14	15 amp
No. 12	20 amp
No. 10	30 amp
No. 8	40 amp
No. 6	55 amp

The ampacities shown are for wires in conduit, in cable, or buried directly in the earth. If installed in free air, the ampacities are higher. These and many more details will be discussed in Chap. 7. Strictly speaking, the *ampacity* of No. 14, No. 12, and No. 10 is greater than shown above because the figures given actually represent the maximum permitted overcurrent protection. The difference is significant only when applying derating factors for continuous loads, for more than three conductors in a raceway, or for ambient temperatures over 30°C (86°F). To keep it simple, we will use the figures shown above.

Joining of Different Sizes of Wire. If two different sizes of wire are joined together, as in Fig. 5-12, then the ampere rating of the overcurrent device must be no greater than that permitted for the *smaller* wire. Since in this case the smaller No. 14 has an ampacity of 15, that is the maximum ampere rating of the overcurrent device used. In practice, a situation of this kind is sometimes found, especially in farm wiring, where a No. 8 wire is used in an overhead wire for mechanical strength and to avoid excessive voltage drop. Although the No. 8 wire has an ampacity of 40, the circum-

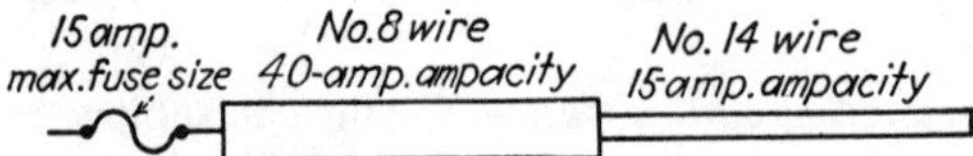

Fig. 5-12 When two different sizes of wire are connected in series, the largest fuse that may be used is one that protects the *smaller* wire.

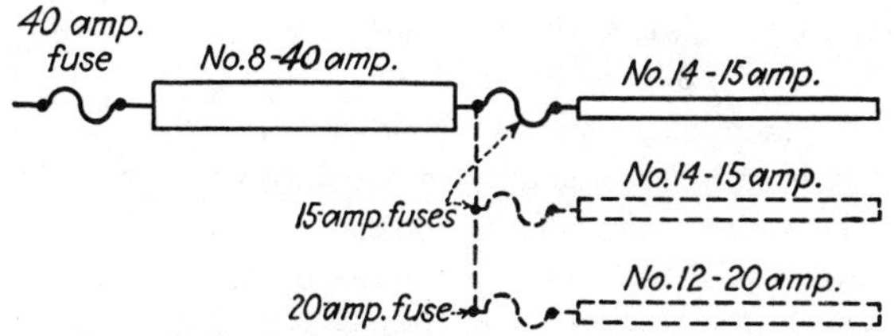

Fig. 5-13 Fuses or breakers are used to protect the smaller wires where the wire is reduced in size.

stances may be such that more than 15 amp is never required, and the maximum rating of 15 amp for the overcurrent device will not be inconvenient.

On the other hand, there may be a situation, as shown in Fig. 5-13, where No. 8 is used but where more than one 15-amp load is to be served. In that case a 40-amp overcurrent device is used at the starting point, but one or more smaller overcurrent devices are installed at the point where the wire size is reduced. Each of these overcurrent devices must have an ampere rating not more than the ampacity of each of the smaller wires installed.

There are a number of exceptions to these general requirements, one of them being the "25-ft tap" rule permitted by NEC Sec. 240-21 Exc. 3. No overcurrent protection is required at the point where the wire size is reduced if *all four* of the following conditions are met:

1. The smaller wire is not over 25 ft long.
2. The smaller wire has an ampacity of at least one-third that of the larger wire.
3. The smaller wire ends in a *single* overcurrent device having an ampere rating not greater than the ampacity of the *smaller* wire.
4. The smaller wire is protected against physical damage and is enclosed in a raceway.

All this is illustrated in Fig. 5-14, which shows a No. 1 wire to which a

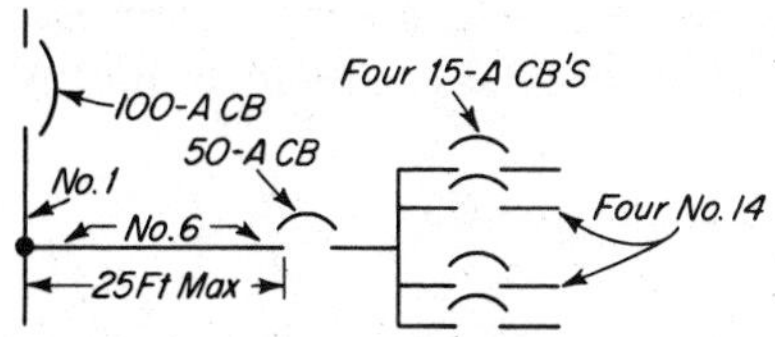

Fig. 5-14 This illustrates the "25-ft rule," which permits a wire size to be reduced without fuse or breaker at the point where reduced. See text.

No. 6 wire is connected. (In this case the No. 1 wire is a "feeder" and the No. 6 wire is a "subfeeder.") Beyond the single overcurrent device at the end of the No. 6 wire, there may be as many circuits of smaller size as desired, but each one must be protected by an overcurrent device rated not higher than the ampacity of the wire it protects.

Another exception is the "10-ft rule" permitted by NEC Sec. 240-21 Exc. 2. A smaller wire may be tapped to a larger one without an overcurrent device where the size is reduced, if the tap wires meet *all* the following conditions:

1. They must be not over 10 ft long.
2. They must have an ampacity not less than (a) the combined computed loads on the circuits supplied by the tap conductors, and (b) not less than the ampere rating of the panelboard or other load supplied by the tap wires.
3. They may not extend beyond the switchboard, panelboard, or control device supplied by the tap wires.
4. They must be installed in a raceway (such as conduit) from the tap to the panelboard or other overcurrent-device enclosure.

A third and common-sense exception is covered by NEC Sec. 210-19(c), which permits taps not over 18 in long to be made from circuit wires, to serve an individual outlet, provided only that the short wire (a) has sufficient ampacity to serve its specific load and (b) has a minimum ampacity of 15 amp on circuits smaller than 40 amp and of 20 amp on 40- or 50-amp circuits. However, No. 18 fixture wires up to 50 ft long, No. 16 fixture wires up to 100 ft long, and No. 14 and larger fixture wires of any length may be used on circuits protected by overcurrent devices not exceeding 20 amp. A flexible cord attached to or for use with a specific Listed appliance is considered to be protected by a 20-amp overcurrent device. An extension cord (a cord with a male plug cap on one end and a female body or receptacle on the other) of No. 16, 25 ft or less in length, and larger than No. 16 of any length, is considered to be protected by a 20-amp overcurrent device. Otherwise, No. 18 fixture wires and cords must be protected at 7 amp, and No. 16 fixture wires and cords at 10 amp.

6

Types and Sizes of Wires

Conductors are used to conduct electric power from the point where it is generated to the point where it is used. Wires are the most commonly used conductors, and copper is the material most commonly used, although aluminum is also widely used. The Code seldom uses the word "wire" but frequently uses the word "conductor," which may be a wire, busbar, or any other form of metal suitable for carrying current. All electric wires therefore are electric conductors, but not all conductors are wires. Copper busbars, for example, are conductors but not wires.

Previous chapters showed that all wire has resistance that prevents an unlimited flow of current and causes voltage drop. For any given load, you must select a size of wire or other conductor that limits voltage drop to a reasonable value.

Current flowing through a wire causes heat; the heat varies as the square of the current (amperes). There is a limit to the degree of heat that various types of insulation can safely withstand, and even bare wire must not be allowed to reach a temperature that might cause a fire. The Code specifies the ampacity (the maximum current-carrying capacity in amperes) that is safe for wires of different sizes with different insulations and under different circumstances. These ampacities will be given later.

Circular Mils. In order to discuss intelligently the different sizes of wires, you must understand the scheme used in numbering these sizes. The units used are mils and circular mils. A mil is 1/1000 (0.001) in. A circular mil (abbreviated cm, CM, or cmil) is the area of a circle that is 0.001 in or 1 mil in diameter. Thus a wire that is 0.001 in or 1 mil in diameter has a cross-sectional area of 1 cmil. Since the area of a circle is always proportional to the square of its diameter, it follows that the cross-sectional area of a wire 3 mils (0.003 in) in diameter is 9 cmil; that of a wire 10 mils in diameter is 100 cmil; that of a wire 100 mils in diameter is 10 000 cmil, etc. The cross-sectional area of any round wire in circular mils is the area of the metal only, and is found by squaring the diameter in mils or thousandths of an inch (multiplying the diameter by itself).

Wire Sizes. Instead of referring to common sizes of wire by their areas, sizes or numbers have been assigned to them. The gauge commonly used is the American Wire Gauge (AWG); it is the same as the Brown and Sharpe (B&S) gauge. This gauge is not the same as that used for steel wires used for nonelectrical purposes, for example, fence wires.

Number 14 wire, commonly used for ordinary house wiring, has a copper conductor 0.064 in, or 64 mils, in diameter. Wires smaller than No. 14 are Nos. 16, 18, 20, and so on. Number 40 has a diameter of approximately 0.003 in, as small as a hair; many still smaller sizes are made. Sizes larger than No. 14 are Nos. 12, 10, 8, etc. Note that the bigger the number, the smaller the diameter of the wire.

In this way, sizes proceed until No. 1/0 is reached; the next sizes are No. 2/0, then 3/0, and finally No. 4/0, sometimes shown as Nos. 0, 00, 000, and 0000, in either case called one-naught, two-naught, etc. Wires larger than No. 4/0 are not designated by a numerical size but simply by their cross-sectional area in circular mils, beginning with 250 000 cmil (250 kcmil, sometimes abbreviated MCM, as is done in the Code) up to the largest recognized size of 2 000 000 cmil (2000 kcmil or thousands of circular mils).

Figure 6-1 shows the approximate actual sizes of typical sizes of wire, without the insulation. If a wire is *stranded,* it is a trifle larger in over-all

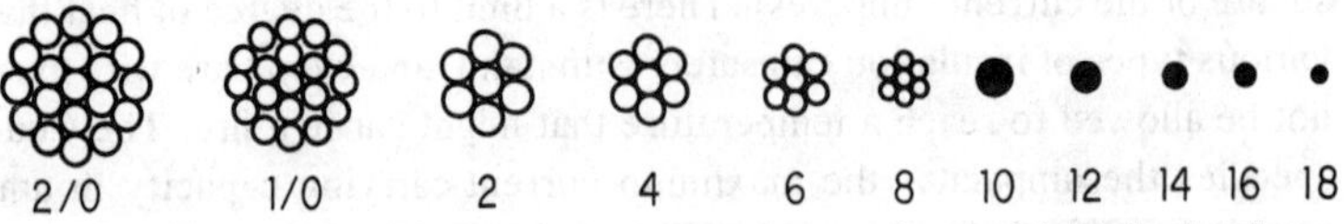

Fig. 6-1 Actual diameters of typical sizes of electric wires, without the insulation.

diameter than a *solid* wire of the same size would be, but the number of circular mils in any one size is the same whether the wire is stranded or solid. The sizes from No. 50 (less than $^1/_{1000}$ in. in diameter) to No. 20 are used mostly in manufacturing electrical equipment of all kinds. Numbers 18 and 16 are used chiefly for flexible cords, signal systems, and similar purposes where small currents are involved. Numbers 14 to 4/0 are used in ordinary residential and farm wiring, and of course in industrial and commercial work, where still larger sizes are also used. Number 14 is the smallest size permitted for ordinary wiring. The even sizes, such as Nos. 18, 16, 14, 12, 8, are commonly used; the odd sizes, as Nos. 15, 13, 11, 9, (with the exception of Nos. 3 and 1 in service-entrance cables), are seldom used in wiring. The odd sizes, however, are very commonly used in the form of magnet wire for manufacturing motors, transformers, and similar equipment, for which fractional sizes such as No. 15½ are not at all uncommon.

In Fig. 6-2 is shown the usual gauge used in measuring wire size. The

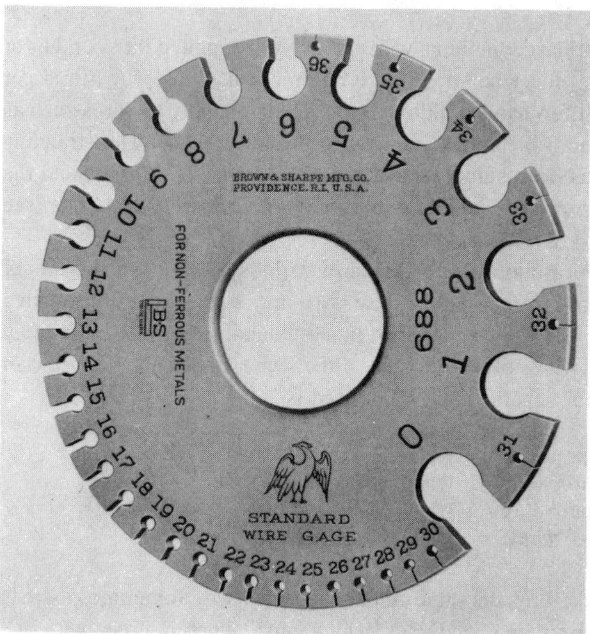

Fig. 6-2 A wire gauge. Measure the wire by the slot into which it fits. This illustration is actual size. (*Brown & Sharpe Mfg. Co.*)

wire is measured by the slot into which it fits, not by the hole behind the slot. You will not actually need this gauge, since the Code requires all building wire to be continuously marked with its size. If it is not properly marked, don't use it.

Table 8 in Chap. 9 of the Code (see Appendix) shows all the NEC-recognized sizes of wire, their areas in circular mils, their resistance in ohms per thousand feet, and their dimensions in fractions of an inch.

You may find it useful to remember that any wire that is three sizes larger than another will have a cross-sectional area twice that of the other. For example, No. 11 has an area exactly twice that of No. 14; No. 3 has an area twice that of No. 6. Any wire that is six sizes larger than another has exactly twice the diameter, and four times the area, of the smaller wire. Number 6 wire has exactly four times the area of No. 12.

Stranded Wires. Where considerable flexibility is needed, as in flexible cords, the conductors consist of many strands of fine wire twisted together. The number assigned to such conductors is determined by the total cross-sectional areas of all the fine wires, the individual strands, added together.

Number 6 and larger building wires (also No. 8 if pulled into conduit or other raceways) must be stranded to be practical. Solid wires in larger sizes are too stiff to handle, although this does not apply to wires such as the weatherproof type installed only overhead in free air. The stranding of each size has been standardized, and the number of strands, and the size of each strand, can be found in Table 8 of Chap. 9 of the Code (see Appendix).

Colors of Wire. Building wires come in various colors, and there is of course a purpose in this. Only white wire may be used for a grounded wire in wiring; this will be explained in more detail later. White wire may not be used for any other purpose. Other wires must not be white or green. The color scheme most often used is

Circuit of two wires	White, black
Circuit of three wires	White, black, red
Circuit of four wires	White, black, red, blue
Circuit of five wires	White, black, red, blue, yellow

In cables and cords, the same color scheme is used. Sometimes there is an additional grounding conductor, which must be green, green with one or more yellow stripes, or in some cases, bare, uninsulated. This will be explained in more detail later.

In outdoor, overhead runs this color scheme is not required. Weatherproof wire is often used for this purpose and is available only in black.

Colors of Terminals. The color of terminals on electrical equipment identifies the kind of wire that may be connected to them. Natural copper or brass terminals are for hot wires. Terminals of a whitish color such as nickel, tin, or zinc-plated are for ground*ed* wires only. Green terminals are for ground*ing* wires only. (The distinction between ground*ed* and ground*ing* will be explained in Chap. 9.) On switches all terminals are natural copper or brass, since only hot wires are switched. However, on *3-way* toggle switches, one of the terminals is a dark, oxidized finish or color, to identify it as the common terminal, as was explained in Chap. 4.

Types of Wires. The NEC recognizes many different types of wire that may be used in wiring buildings. The more ordinary ones will be described in this chapter. Still other types less frequently used in the kinds of buildings discussed in this book are described in the Code in Table 310-13 (see your copy of the Code).

All wires normally used in wiring are suitable for use at any voltage up to and including 600 volts, the only kind of wiring discussed in this book. However, some flexible cords, and some fixture wires used in the internal wiring of fixtures, are suitable for use only up to 300 volts.

Kinds of Locations. The Code limits some wires to use in dry locations; others may be used in dry or wet locations. In damp locations, use wires approved for wet locations. It is important to understand the definitions of these different locations. The Code defines them in Art. 100, as follows:

> DAMP LOCATION: Partially protected locations under canopies, marquees, roofed open porches, and like locations, and interior locations subject to moderate degrees of moisture, such as some basements, some barns, and some cold-storage warehouses.
>
> DRY LOCATION: A location not normally subject to dampness or wetness. A location classified as dry may be temporarily subject to dampness or wetness, as in the case of a building under construction.
>
> WET LOCATION: Installations underground or in concrete slabs or masonry in direct contact with the earth, and locations subject to saturation with water or other liquids, such as vehicle washing areas, and locations exposed to weather and unprotected.*

Plastic-Insulated Wires. Most kinds of wire used in the wiring of buildings have thermoplastic insulation, the thickness of which depends on the size of the conductor. The wire is clean and easy to handle, and the insulation is easy to strip off. There are several types.

Types T, TH, THW. The most commonly used types of plastic-insulated wires are Types T and THW. Type T is shown in Fig. 6-3; it may be used

Fig. 6-3 Types T and TW are used for general wiring. Numbers 6 (and sometimes No. 8) and larger are stranded. (*Crescent Insulated Wire & Cable Co.*)

only in dry locations and only where the temperature of the wire when loaded to its full ampacity will not exceed 60°C. But wire must often be installed in wet locations and often in locations where its temperature will exceed 60°C.

If the Type designation includes a "W" the wire may be used in a dry, damp, or wet location. So in damp or wet locations, use wire with a "W" in its Type designation, such as Types TW, THW, etc. Similarly, if the Type designation has an "H," its insulation can safely operate at a temperature higher than allowed for Type T; if it includes an "HH" it can operate at a still higher temperature. Therefore if the ambient temperature is above 86°F or 30°C, use a wire that has an "H" or "HH" in its Type designation. (Of course, wire without the "H" or "HH" can be used if its ampacity is derated to a lower figure, as will be explained.)

The most commonly used types of wire are listed in Table 6-1, which shows where they may be used and their temperature ratings. In damp locations use only wire that is suitable for a wet location.

Practical Advantages of -H and -HH Wires. Sometimes it is to your advantage to install an -H or -HH wire even if the ambient temperature where it is installed is *not* above the normal 86°F. Such -H or -HH wires in the larger size have a higher ampacity than the corresponding size of Type T or TW. The more expensive Type THW, because of its greater ampacity,

TABLE 6-1 Temperature Ratings of Ordinary Types of Wire

Type	Locations	Sizes	Temperature ratings
RHH	Dry only	Nos. 14, 12, 10	75°C or 167°F
		No. 8 and larger	90°C or 194°F
RHW	Dry or wet	All sizes	75°C or 167°F
T	Dry only	All sizes	60°C or 140°F
TW	Dry or wet	All sizes	60°C or 140°F
THW	Dry or wet	All sizes	75°C or 167°F
THWN	Dry or wet	All sizes	75°C or 167°F
THHN	Dry only	Nos. 14, 12, 10	75°C or 167°F
		No. 8 and larger	90°C or 194°F
XHHW	Dry only	Nos. 14, 12, 10	75°C or 167°F
	Dry only	No. 8 and larger	90°C or 194°F
	Wet only	All sizes	75°C or 167°F

often becomes less expensive for a specific amperage than Type T or TW, especially where installing smaller-diameter Type THW also permits a smaller size of conduit to be used.

Now all that has been said pertains to wires installed in locations with a normal ambient of not over 86°F. If the ambient is higher, the use of -H or -HH wire is just about essential to avoid excessive derating of the ampacity of ordinary wire when used in a higher ambient, as has been explained.

Types THWN and THHN. Type THWN may be used in dry or wet locations and has a temperature rating of 75°C. Type THHN may be used only in dry locations and has a temperature rating of 90°C, except that in Nos. 14, 12, and 10 it is considered a 75°C wire. These two types are oil-resistant or gasoline- and oil-resistant *when so marked on the jacket*. They are essentially the basic Types THW and THH plus a final layer of extruded nylon or thermoplastic polyester, both of which are exceedingly tough mechanically and have excellent insulating qualities. The final layer replaces the much thicker outer layer of Types THW and THH. The construction leads to an over-all *outside* diameter and cross-sectional area (including the insulation) much smaller than that of ordinary wires with thicker insulation, especially in the smaller sizes. These are expensive wires, but their use is often justified because for a given ampacity smaller conduit may be used than in the installation of wires with thicker insulation.

Type XHHW. This type of wire insulation is a cross-linked synthetic polymer. It has no outer braid and in appearance is substantially like Type T. The insulation is tough, moisture- and heat-resistant, and of high quality. The over-all outside diameter is smaller than Types T, TW, etc., but not as small as Types THHN, THWN, etc. In dry locations it has a 90°C temperature rating and corresponding ampacity; in wet locations its temperature rating is 75°C, with a correspondingly lower ampacity.

Types FEP and FEPB. These are relatively new types of wire with insulation of fluorinated ethylene propylene. If there is no outer braid it is Type FEP, and has a temperature rating of 90°C. If it has an outer braid (glass in Nos. 14 to 8, asbestos in larger sizes), it becomes Type FEPB. The normal temperature rating of Type FEPB is the same as for Type FEP, 90°C, but on special applications (such as switchboard wiring), it may be approved by the inspector for use up to 200°C. Both types may be used only in dry locations.

Rubber-Covered Wire. At one time, all wire for general use had rubber insulation. Some of it is still being installed. The insulation is not necessarily natural rubber but may be made from neoprene, synthetic rubber, or other vulcanizable materials. Its construction is shown in Fig. 6-4.

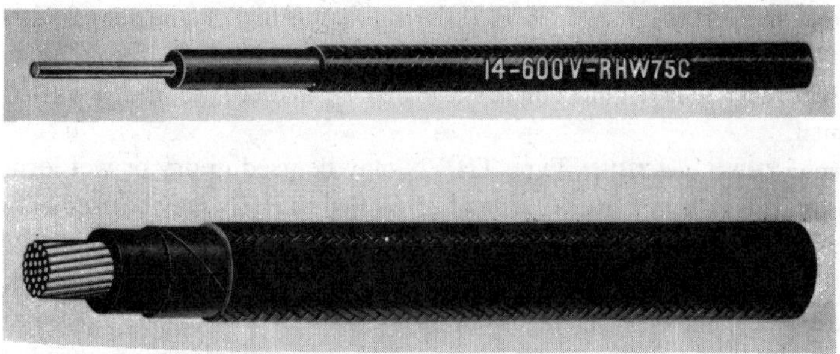

Fig. 6-4 Rubber-covered wire has a protective braid over the insulation. This kind of wire is used less and less each year. (*Crescent Insulated Wire & Cable Co.*)

It consists of the copper conductor, tinned to make it easy to remove the insulation, and an outer moisture- and flame-resistant covering. Type RHW may be used in dry or wet locations. Types RH and RHH have insulation that withstands more heat and therefore have a higher ampacity in the larger sizes than Type RHW, but they are permitted only in dry locations.

Type -L. If the last letter in a Type designation is an "L," it shows that the wire or cable is covered with a final seamless layer of lead; it may then be used in a wet location. Lead-covered wires or cables are now very rarely used, for many insulations are now available that may be used in wet locations without lead covering.

Style RR. Although you might see "Type RR" mentioned in magazine articles or advertising, or on shipping cartons, it is not a Code type. It is a wire with rubber insulation of varying grades, plus a final layer of very tough and moisture-resistant rubber or neoprene. That in a general way also describes Type USE service-entrance cable (described in Chap. 20). But if a wire is tagged only "Type RR" it is not a UL Listed wire. However, if it is labeled "Type USE," the manufacturers may if they wish also add their private designation "*Style* RR," but there is no such thing as a *Type* RR.

Cables. For many purposes it is desirable to have two or more wires grouped together in the form of a cable. This is easy to install in any case, but especially in rewiring a building, for the cable lends itself well to being fished through hollow wall spaces. A cable that contains two No. 14 wires is known as "14-2" (fourteen-two); if it contains three No. 12 it is called "12-3"; etc. If cable has, for example, two insulated No. 14 wires and a bare uninsulated ground*ing* wire, it is called "14-2 with ground." If a cable contains two insulated wires, one is white, the other black. If it contains three, the third is red. A few cables are single-conductor type, such as Type USE, for direct burial in the earth, and Type UF, which may also be buried but subject to restrictions. Both types are discussed in Chap. 17.

Sometimes ordinary wires in sizes larger than No. 4/0 are called "cables," but the term is properly applied only to an assembly of two or more wires.

Nonmetallic-Sheathed Cable. This cable is available in two types—Type NM and Type NMC—and contains two or three insulated conductors bundled together. It costs less than other kinds of cable, and is lightweight and easy to install. It is very popular and widely used.

Type NM is the ordinary kind that has been available for many years; it may be used only in permanently dry locations. It is often called Romex, which is the trademark of one manufacturer. Figure 6-5 shows the construction. The outer jacket of moisture- and flame-resistant material may be either thermoplastic as shown, or of a woven fiber. The wires may be individually wrapped with a spiral layer of paper, and there may be fiber fillers in the space between the wires in some constructions.

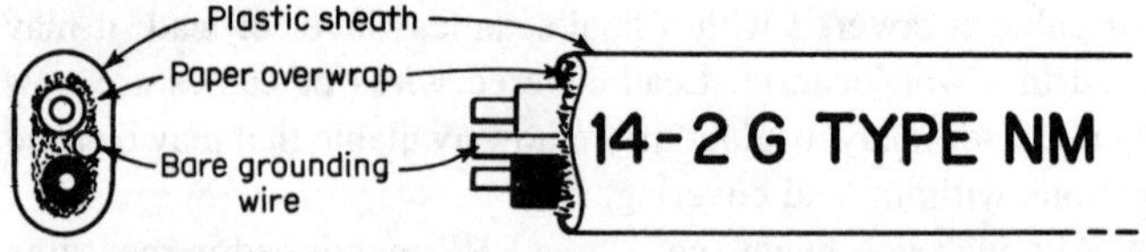

Fig. 6-5 Nonmetallic-sheathed cable is popular for ordinary wiring. This is NEC Type NM and may be used only in dry locations.

At one time, Type NM was the only kind of nonmetallic-sheathed cable made. Its use at that time was not restricted to any particular type of location. It proved quite satisfactory in permanently dry locations, but millions of feet were installed in barns, which normally have considerable moisture and highly corrosive conditions. It proved totally unsuitable in such locations; the fibrous materials then used in the overwrap of the individual wires as well as in the jacket acted as wicks, pulling moisture into the inside of the cable. The result was rotting of both insulation and all other parts of the cable, which quickly led to dangerous shock and fire hazards. For that reason, Type NM has for many years been restricted to use in permanently dry and noncorrosive locations, and a different kind of cable has been developed for damp and corrosive locations.

Type NMC is shown in Fig. 6-6. The individually insulated wires are embedded in solid plastic; no fibrous, wicklike materials are used. There-

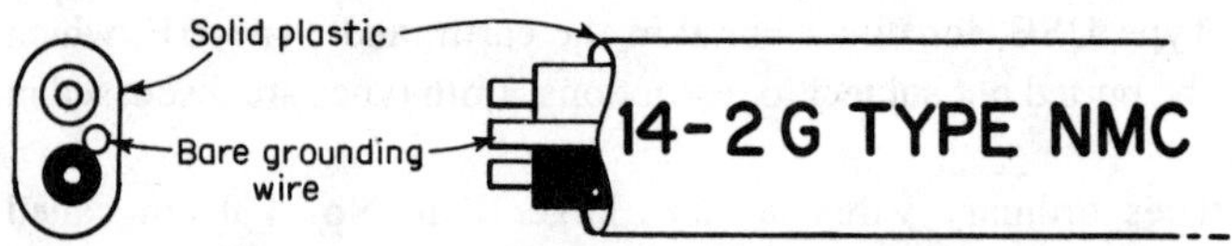

Fig. 6-6 Type NMC nonmetallic-sheathed cable may be used in dry or damp locations.

fore it may be used in dry, damp, or corrosive locations (such as barns with corrosive materials from the excreta of animals), but it must not be buried directly in the ground. This will be discussed in more detail in Chap. 11.

Most nonmetallic-sheathed cable comes with an additional bare wire in addition to the insulated wires. This is used for equipment grounding purposes, and will be discussed later.

Type UF Cable. In some localities Type NMC cable is hard to find, while Type UF (*U*nderground *F*eeder) is available. Type UF is very similar in appearance to Type NMC, and costs very little more. It may be used wherever Type NMC may be used. In addition, it may be buried directly in the ground, provided it has overcurrent protection at its starting point. So, it may not be used in a service entrance. However, on farms, if it starts at the yard pole and runs to various buildings, it may be used, but again, only if it has overcurrent protection at the pole. Where it enters a building, it may continue into the building and be used for interior wiring.

Armored Cable. Another common type of cable is armored cable, shown in Fig. 6-7. It is usually called BX, although that is the trade name of one

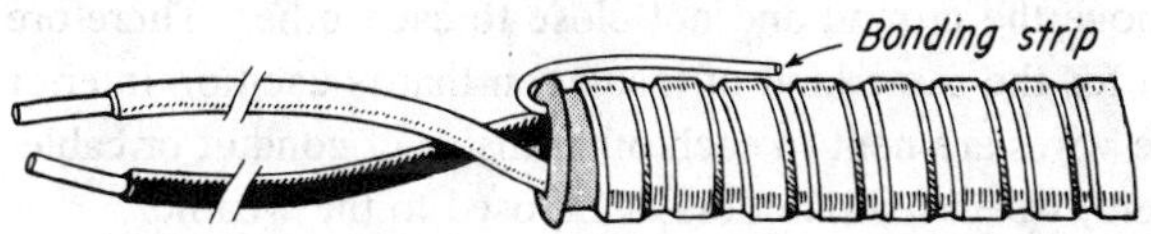

Fig. 6-7 Armored cable has metal armor for its final protection.

particular manufacturer. The Code calls it Type AC or ACT, depending on the type of conductor insulation, etc. Each wire is wrapped in a spiral layer of tough paper, and over all there is a spiral armor of galvanized steel (occasionally aluminum). The paper protects the wires against abrasion by the metal armor. Inside the armor, outside the paper and parallel with the wires, there is a bare aluminum strip to improve the equipment grounding capability of the armor. Its use will be discussed in Chap. 11.

Armored cable is covered in Art. 333 of the Code, which includes the Types AC and ACT just discussed. (Also in Art. 333 is Type ACL cable, with lead-covered conductors, which is rarely used.)

Service-Entrance and Underground Cables. There are several kinds of cable specially designed for bringing wires into a building as service-entrance conductors (or between buildings), either overhead or underground. These will be discussed in Chap. 17.

Other Types of Wire and Cable. No attempt has been made to describe all the types of wires, cables, and wiring methods that are recognized by the Code. Rather, the intent has been to provide at least some guidance, a sort of preview, to better enable you to understand and use the Code.

Identification of Different Kinds of Wire. In most cases it is impossible to determine what type a wire is just by looking at it. For example, there is no difference in appearance between Types T, TW, THW, and some others. So most kinds of wires are required by the Code to be plainly identified by printing the type designation on their surface, at regular intervals. Other required markings are the size, the voltage limitation, and the manufacturer's name or trademark. Other markings, not required but usually shown, include the month and year of manufacture. The UL Mark may be shown if the wire is UL Listed. If the conductor is aluminum, the word "aluminum" or its abbreviation "AL" must be shown on the wire. Where it is not practical to surface-mark a wire, the tag, coil, reel, or shipping carton must contain the information.

Overhead Spans. When wires are run overhead, there is no likelihood that anyone will ever touch them. They are usually installed a considerable distance above the ground and not close to each other. Therefore there is no need for the same kind of insulation that is used on interior wires, where the wires are next to each other inside of conduit or cable. On the other hand, wire used outdoors is exposed to the weather.

A special kind of wire called "weatherproof" may be used for outdoor overhead spans but must *never* be used for any other purpose, since its covering is not suitable for insulation. The Code does not recognize it as a wiring method or as an *insulated* conductor. It is considered merely a *covered* wire, which is defined in NEC Art. 100 as "a conductor encased within material of composition or thickness that is not recognized by this Code as electrical insulation."* The covering usually is a layer of neoprene, so that in appearance the wire greatly resembles Type T, shown in Fig. 6-3.

But there is no reason why weatherproof wire *must* be used in overhead spans. The Code does not at all prohibit the use of other wires (so long as they have a "W" in their Type designation) from being used outdoors in overhead spans, and Type TW, for example, costs less than weatherproof wire.

When wires are installed in overhead spans, any heat that develops is radiated into the air. For that reason, any kind and size of wire is assigned a higher ampacity in free air than when installed in conduit or in the form of cable. This will be discussed in Chap. 7.

Aluminum Wire. In this book, where a reference is made to a wire size, it always refers to a *copper* wire unless otherwise noted.

But the use of aluminum wire is increasing, and its use has introduced new problems. Two types of aluminum wire are in general use: ordinary

*Reprinted with permission from NFPA 70-1984, National Electrical Code®, Copyright© 1983, National Fire Protection Association, Quincy, Massachusetts 02269. This reprinted material is not the complete and official position of the NFPA on the referenced subject, which is represented only by the standard in its entirety.

or all-aluminum, and copper-clad[1] aluminum. The copper-clad variety has a thin layer of copper bonded to the outer surface of the aluminum.

Aluminum has a higher resistance than copper, so an aluminum wire must be larger than a copper wire, to carry the same number of amperes. Many follow a rule of thumb: Use aluminum larger than copper by two size numbers (No. 8 aluminum instead of No. 10 copper); however, that is not always a dependable method of meeting Code requirements. The Code contains four tables that define the ampacity of different sizes and kinds of wire under various conditions: Tables 310-16 through 310-19; all four tables appear in the Appendix of this book. When using aluminum wire you should use the tables instead of following the rule of thumb mentioned above. The ampacity of copper-clad aluminum is the same as that of all-aluminum wire.

The real problems in using aluminum wire lie in making connections of wire to terminals or connectors. When heated, aluminum expands much more, and faster, than copper. Hence, if an aluminum wire is clamped in a brass or copper terminal screw or connector, the heat from the current flowing in the wire causes the aluminum to expand more than the copper or brass of the connector, which causes the aluminum to cold-flow out of the connector. When the connection cools, it is a little loose. Repeated heating and cooling as the current in the wire is turned on and off leads to still more aluminum being extruded. In due course of time, the connection badly overheats and fails. That can be not only annoying but dangerous from the standpoint of fire. (In a damp location, there is also electrolytic action between the two metals, hastening the failure.)

When aluminum wire was first introduced, it was connected just as copper wires were connected, but that led to the conditions just described. UL then imposed restrictions on the kinds of terminals required for aluminum, but in due course of time it was found that these requirements were not sufficient. These problems have been basically solved by still newer UL requirements, if proper installation methods are followed, as will be discussed in detail in Chap. 8. Do carefully note and understand the statements that follow.

Before September 1971, switches, receptacles, and other devices *rated at 15 or 20 amp,* and intended for copper only, were not marked. Those

[1] Aluminum wire may also be nickel-clad. All statements in this book pertaining to copper-clad aluminum are also correct for nickel-clad aluminum, except that the nickel-clad variety may be used only in dry locations.

intended for copper or aluminum were marked "AL-CU."[2] It gradually became evident, however, that the "AL-CU" devices were not as suitable for aluminum in the 15- and 20-amp ratings as had been thought when they were first used. Since September 1971, UL has required still different terminals (in the 15- and 20-amp sizes), and these are marked "CO/ALR."[3] The UL also requires special chemical and physical properties for the aluminum now used in the wire.

To recap: unmarked 15- and 20-amp devices should be installed only when using copper wire; if marked "AL-CU," they should now be installed only when using copper wire; if marked "CO/ALR," they may be installed when using copper, copper-clad aluminum, or all-aluminum wire. But do note that devices with push-in or screwless terminals, as will be described in Chap. 8, may *not* be used with all-aluminum wire but are acceptable for copper or copper-clad aluminum.

Do note also that devices rated at 30 amp or higher (whether made before or after September 1971), if *not* marked, may be used with copper only. If marked "AL-CU," they may be used with copper, copper-clad aluminum, or all-aluminum. Devices rated at 30 amp or more are *not* marked "CO/ALR."

Devices, connectors, and connection methods will be discussed in considerable detail in Chap. 8.

Flexible Cords. Where wires are installed permanently, they need be only sufficiently flexible to permit reasonably easy installation, and for some flexible connections, to allow for slight movement or normal vibrations. If the wires must be moved about, as on a floor lamp, a vacuum cleaner, or a portable tool, they must be very flexible. This is necessary for convenience and to prevent the conductors from breaking, which would be more likely if they were solid copper, especially if of considerable diameter. Flexible wires of this kind are called "flexible cords" in the Code. There are a great many different kinds, the more common of which will be described here. Others are covered by NEC Table 400-4 (see your copy of the Code).

Use of Cords. Flexible cords should be used only as intended and as permitted by Art. 400 of the Code, especially Secs. 400-7 and 400-8. Flexible

[2] "Cu" is the chemical abbreviation for copper, "Al" for aluminum, and "Ni" for nickel.

[3] "CO" became an arbitrary mark in place of the usual symbol "CU" for copper, and "ALR" (meaning "aluminum revised") in place of the usual "AL."

cords are intended primarily for use as an integral part of a portable tool, appliance, or light. Cords are *not* intended to be used as extension cords except for *temporary* duty, as indicated briefly in Chap. 1 under "Identifying Listed Products." So, do not use a flexible cord as a permanent extension of the fixed wiring.

When receptacles are installed as required by recent Codes, it should be possible to connect floor or table lamps, clocks, and similar equipment without using an extension cord. But there are millions of older homes with too few receptacles; really old homes often had only one receptacle in a room. If you do use an extension cord under those conditions, *never* run it under a rug or carpet. Keep it close to walls, or hidden behind furniture, so that nobody can step on it and damage it, or trip on it. Treat an extension cord with respect, as a dangerous device. Don't abuse it.

Where an extension cord is used temporarily, be sure to unplug it when you have finished using the portable equipment. Do not go off and leave it plugged in. See Sec. 240-4 for overcurrent protection of flexible cords. Use only extension cords with a suitable female member of a separable connector, *not* with a lamp socket, not even a socket made of insulating material, let alone a brass socket. Be sure that all portable lights (as distinct from extension cords) are equipped with a suitable handle and lamp guard; some portable lights may also need a glass globe for additional protection. Figure 20-7 shows a portable light suitable for use in a residential garage.

Types SP and SPT. This is the cord commonly used on lamps, clocks, radios, and similar appliances. As shown in Fig. 6-8, the wires are embedded in a solid mass of insulation, so no outer jacket is required. Type SP

Fig. 6-8 Types SP and SPT cords are used mostly on floor and table lamps, clocks, and similar equipment.

cords have rubber insulation; Type SPT have plastic insulation. Often the cord is made with a depression between the two conductors, for ease in separating the two conductors to make connections.

Some cords have suffixes such as -1, -2, and -3, for example, SP-1 or SPT-2. The "-1" is the most ordinary and is made only in No. 18; the "-2" has heavier insulation and is made in Nos. 18 and 16; the "-3" has still heavier insulation and is made in Nos. 18 to 12.

Types S, SJ, SV. The cords described in the preceding paragraphs are designed for ordinary household devices, which are generally speaking moved about very little once they are plugged into a receptacle. They will not withstand a great deal of wear and tear. A cord that has a sturdier construction is needed for portable appliances and tools, such as vacuum cleaners, washing machines, electric tools.

A Type S cord is shown in Fig. 6-9. It consists of two or more stranded conductors with a serving of cotton between the copper and the insulation to prevent the fine strands from sticking to the insulation. Jute or similar "fillers" are twisted together with the conductors to make a round assembly that is held together by a fabric overbraid. This assembly has an outer jacket of high-quality rubber. Type SJ is similar in construction but has a lighter jacket. Type SJ is for ordinary hard usage, while Type S is for extra-hard service such as in commercial garages. Type SV has a still lighter outer jacket and is used *only* on vacuum cleaners.

Fig. 6-9 Type S cord is very tough and durable.

Types ST, SJT, SVT. If the outer jacket is made of plastic materials instead of rubber, Types S, SJ, and SV become Types ST, SJT, and SVT.

Oil-Resistant Cords. When cords of the types just described are exposed to oil, the oil attacks the outer rubber or plastic jacket, which swells, deteriorates, and falls apart. For that reason ordinary cords cannot be used where exposed to oil or gasoline, such as in commercial garages. But cords are also made with a special oil-resistant jacket, made of neoprene or similar material, in which case the letter O is added to the type designation; Type S, for example, becomes Type SO.

Heater Cords. Cords used on flatirons, toasters, portable heaters, and similar appliances that develop a lot of heat are called "heater cords." The construction shown in Fig. 6-10 was at one time the only kind used,

Fig. 6-10 One type of heater cord. It has a layer of asbestos over the insulation. It is used on toasters, irons, and similar heat-producing appliances.

and it is still being used. The basic insulation is rubber, with a serving of cotton between the copper and the rubber to keep the copper clean. Over the rubber there is a serving of asbestos or the equivalent to withstand the heat in case of accidental contact with a hot surface. Over all, there is a braid of cotton or rayon. This describes the Type HPD illustrated. If the outer layer is rubber, it becomes Type HSJ.

Today Type HPN is far more popular. It has neither rubber nor asbestos in its construction, but the conductors are embedded in solid neoprene; there is no outer braid. In appearance it is quite similar to Type SP shown in Fig. 6-8.

Ampacity of Cords. The ampacity of the more ordinary cords is shown in the table below; for others see NEC Table 400-5. Cords having two

	A	*B*	*C*
No. 18	10 amp	7 amp	10 amp
No. 16	13 amp	10 amp	15 amp
No. 14	18 amp	15 amp	20 amp

current-carrying conductors (with or without a third grounding conductor not normally carrying current) have the ampacities shown in Col. *A*. If three conductors carry current (as in the case of a 3-conductor cord to a 3-phase load) the ampacities are slightly reduced to those in Col. *B*. Heater cords (any with an "H" in the Type designation) have higher ampacities as shown in Col. *C*.

Fixture Wire. For the internal wiring of lighting fixtures, special wire known as "fixture wire" is used. There are many types, and the particular type used depends to a great extent on the temperature that exists in the fixture itself (while it is in use) and therefore in the wire itself. Wires with rubber or plastic insulation, as shown in Fig. 6-11, may be used only

Fig. 6-11 Fixture wire is used only in the internal wiring of fixtures. This shows one of several different constructions.

if the temperature of the wire *while carrying current* in the fixture does not exceed 140°F (60°C). Type CF, which has only cotton in its insulation, may be used up to 194°F (90°C). Type AF with asbestos insulation may be used up to 302°F (150°C). Type SF with silicone insulation is good for 392°F (200°C). Other types are covered in NEC Table 402-3 (see your copy of the Code).

Fixture wires may be used only in the internal wiring of fixtures, and from the fixture up to the circuit wires in the outlet box on which the fixture is mounted. They may never be used as branch-circuit wires leading to an outlet.

Other Types. There are many other types of wires, cables, and cords. Some are rarely used in the kinds of wiring discussed in this book and will not be mentioned. Other types are used for specific purposes, such as underground wiring, and will be discussed in the chapters pertaining to that kind of wiring.

Low-Voltage Wiring. Some kinds of equipment, such as doorbells, chimes, and thermostats, operate at low voltages, usually under 30 volts. The power for operating such equipment comes from transformers of very limited power capacity, so that even under short-circuit conditions the total available power is unlikely to cause a fire; the voltage is too low even to produce a shock. Usually No. 18 or No. 16 wire with a thin plastic insulation is used for the purpose. Several such wires bundled together become "thermostat cable."

However, in other cases (seldom if ever in homes) power for low-voltage circuits is supplied by transformers that are *not* of limited capacity, and such cases will be discussed in Chap. 28 under "Remote-Control and Signaling Circuits."

7
Selection of Proper Wire Sizes

You must always use wire (1) that has insulation suitable for the voltage, temperature, and location (wet, dry, corrosive, direct burial, etc.); (2) that has an ampacity rating adequate for the load; and (3) that is of sufficient size to avoid excessive voltage drop (which is wasted power). The Code lists insulation types and ampacities in NEC Tables 310-16 through 310-19. But it is up to you to determine voltage drop. This chapter will show you how.

It is impossible to prevent *all* voltage drop. Sometimes it is difficult to hold the voltage drop to a desired level. But you must hold it to a practical minimum. A drop of over 3% on the branch-circuit conductors at the farthest outlet, or a total of over 5% on both feeder and branch-circuit conductors, is definitely excessive and inefficient. It is usually possible and practical to hold the voltage drop to *less* than that.

Advantages of Low Voltage Drop. Voltage drop is simply wasted electricity. Moreover, all electrical equipment operates most efficiently at its rated voltage. If an electric motor is operated on a voltage 5% below its rated voltage, its power output drops almost 10%; if operated at a voltage 10% below normal, its power output drops 19%.

If an incandescent lamp is operated on a voltage 5% below its rated voltage, the amount of light it delivers drops about 16%; if the voltage is 10% below normal, its light output drops over 30%. (Fluorescent lamps will not operate at all if the voltage is considerably below their rated voltage.) So in all cases, the output drops much faster than the reduction in voltage. It is apparent then that the voltage *drop* must be limited to as small a value as is practical.

Practical Voltage Drops. In ordinary residential wiring, there are no feeders, or very short ones, since the branch circuits are usually connected to the service equipment panelboard which also contains the overcurrent devices (circuit breakers or fuses). In farm wiring, however, there usually are feeders, sometimes quite long, such as between the meter pole and the various buildings. The drop in the feeders then should not exceed 2%, and in the branch circuits not over 3%. A *lower* drop is desirable. A commonly accepted figure is 2% from the beginning of the branch-circuit wires to the farthest outlet, with an additional 1 to 2% in the feeders, depending on their length.

This means that on a 120-volt circuit the voltage drop from the branch-circuit panelboard (in the service entrance) to the most distant outlet should not exceed 2.4 volts; on a 240-volt circuit it should not exceed 4.8 volts. In residential wiring, if No. 14 wire is used for ordinary branch circuits, the voltage drop will not usually exceed 2%. But lamps are getting bigger; 1000- to 1500-watt appliances are becoming more common; people are using more electric power every day, so that circuits are becoming loaded closer to their limits. Therefore, there is good reason for the trend toward using No. 12 wire for residential wiring, especially if the circuits are long. The Code already requires No. 12 for the small appliance circuits and special laundry circuit (which will be discussed in Chap. 13) in dwellings. Some local codes already require No. 12 as the minimum, and engineers and architects often specify a minimum of No. 12 copper. In commercial and industrial establishments, where branch circuits are longer than in dwellings, it is often essential to use No. 12 for ordinary branch circuits to avoid excessive voltage drop.

Calculating Voltage Drop by Ohm's Law. The actual voltage drop can be determined by Ohm's law, which was discussed in Chap. 2: $E = IR$, or voltage drop = amperes × ohms.

For example, assume that a 500-watt floodlight is to be operated at a distance of 500 ft from the branch-circuit circuit breaker or fuse. This requires 1000 ft of wire. At 120 volts a 500-watt lamp draws about 4.2 amp.

If No. 14 wire is used, NEC Table 8 (see Appendix) shows that it has a resistance of 3.07 ohms per 1000 ft at 75°C. The voltage drop then is (4.2)(3.07), or 12.89, far over 2%. It is nearly 11%.

When we try other sizes, No. 6 with 0.491 ohms per 1000 ft has a drop of 2.06 volts, less than 2%, while No. 8 with 0.764 ohms per 1000 ft has a drop of 3.2 volts, or nearly $2^3/_4$%. Therefore, if the floodlight is to be used a great deal, use No. 6 wire; if it is to be used relatively little, use No. 8 wire; if it is to be used only in emergencies, use No. 10 wire, with 5.08 volts drop, about $4^1/_4$%, or even No. 12 wire with 8.1 volts drop, about $6^3/_4$%. The amount of power wasted when used rarely would be insignificant. Normally, however, a drop of $6^3/_4$% should not be tolerated.

If in this example the distance had been 400 ft instead of 500 ft, the length of the wire would have been 800 ft instead of 1000 ft. The voltage drop then would have been $^{800}/_{1000}$ or 80% of what it is for 1000 ft.

Now assume that the same floodlight is to be operated at the same distance of 500 ft but at 240 volts instead of 120 volts. The amperes then would be about 2.1 instead of 4.2. By making the same calculations, No. 14 wire now involves a drop of only about 6.5 volts, or about $2^3/_4$%, while No. 12 with a resistance of 1.93 ohms per 1000 ft has a drop of 4.05 volts, less than $1^3/_4$%. This emphasizes the desirability of using higher voltages where a considerable distance is involved, as well as where considerable power is involved.

Desirability of Higher Voltages. For any given *number of watts* at any given *distance,* the voltage drop *measured in volts* on any given size of wire is always twice as great on 120 volts as it is on 240 volts. Doubling the voltage reduces the voltage drop *in volts* exactly 50% if the watts, the distance, and the wire size remain the same. It is the power in *watts,* and not the current in *amperes,* that must remain unchanged for this statement to be correct.

When the voltage drop *in percentage* is considered, remember that in the case of the 240-volt circuit the initial voltage is twice as high, but the voltage drop measured in *volts* is only half as much as in the case of the 120-volt circuit. From this you can see that the voltage drop measured *in percentage* will be only one-fourth as much for 240 volts as it is for 120 volts—if the watts, the wire size, and the distance remain the same.

All the foregoing can be simply restated: Any size of wire will carry any given amount of power in watts at 240 volts *four* times as far as it will at 120 volts, with the same *percentage* of voltage drop. This statement should not be confused with the previous statement that any size wire will carry any given amount of power in watts twice as far at double the voltage with the same *number of volts* drop.

Another Method of Calculating Voltage Drop. The preceding method of computing voltage drop was explained to emphasize the difference in drop at different voltages. A more practical and more frequently used formula is

$$\text{Circular mils} = \frac{\text{distance in feet} \times \text{amperes} \times 22}{\text{volts drop}}$$

or

$$\text{cmil} = \frac{(DI)(22)}{\text{Vd}}$$

where D = distance in feet one way, I = amperes, and Vd = volts drop.

This formula is based on the resistance per circular mil–foot[1] of copper wire, which is 10.8 ohms (at 25°C). Since the wire has to go to and from a load, either the distance or the resistance has to be doubled. To make it simple, the resistance was doubled to 21.6 ohms, and to compensate for the added resistance of joints, the number was rounded off to 22. Therefore the distance *one way* is used with 22 ohms per cmil·ft in this formula. (For aluminum wire, substitute 36 for 22 in the formula.)

For 3-phase circuits, the formula is[2]

$$\text{cmil} = \frac{(DI)(22)(0.866)}{\text{Vd}}$$

For aluminum wire, again substitute 36 for 22 in the formula.

Applying this more practical formula to the floodlight example, which involves a distance of 500 ft, a current of 4.2 amp, and a drop that is to be limited to 2.4 volts (2%), the formula becomes

$$\text{cmil} = \frac{(500)(4.2)(22)}{2.4} = \frac{46\ 200}{2.4} = 19\ 250 \text{ cmil}$$

In other words, to limit the drop to exactly 2.4 volts, the wire must have a cross-sectional area of 19 250 cmil. NEC Table 8 (see Appendix) shows that there is no wire having exactly this cross-sectional area, which falls

[1] A "circular mil–foot" means a wire 1 ft long and 1 mil or 0.001 in. in diameter.

[2] Note that 0.866 is ½ $\sqrt{3}$.

about halfway between No. 6 and No. 8. Therefore use No. 6 with a little under 2.4 volts drop if the floodlight is to be frequently used, or No. 8 with a little more than 2.4 volts drop if infrequently used.

If, instead of determining the size of wire that will produce a specific voltage drop, the actual voltage drop with a given size of wire is to be determined, merely transpose the formula to read

$$\text{Volts drop} = \frac{(DI)(22)}{\text{cmil}}$$

To determine the number of feet any given size of wire will carry any given current in amperes at a specific voltage drop, transpose the formula once more to read

$$\text{Distance} = \frac{(\text{Vd})(\text{cmil})}{(\text{amperes})(22)}$$

This is the formula that was used to compile Tables 7-1 and 7-2 in this chapter.

Voltage-Drop Tables. For most purposes, it is not necessary to use formulas to determine proper wire sizes. Use Tables 7-1 and 7-2 (for *single-phase* circuits only). Table 7-1 is for 120 volts, and Table 7-2 is for 240 volts. Both tables are for *copper* wires only. Each is based on a 2% drop. Under each size is shown the *one-way* distance that the wire will carry the current in amperes, shown in the left-hand column, while producing a 2% drop. The tables are only for Types T or TW copper wire in conduit or cable, except that if a distance is marked with a dagger (†) it means that Types T and TW do not have enough ampacity to carry the number of amperes involved. In that case, refer to NEC Tables 310-16 through 310-19 and select a type of wire that has the required ampacity. If the wires are installed overhead in free air, Types T and TW will have enough ampacity so that they may be used for *some* of the distances marked with a dagger. NEC Tables 310-16 through 310-19 are found in the Appendix of this book.

Note that Tables 7-1 and 7-2 are based on a voltage drop of 2%. If, under certain circumstances, other voltage drops are to be permitted, the tables are easily converted, as follows:

For a voltage drop of 1%, decrease all distances by 50% (to 50%).
For a voltage drop of 2½%, increase all distances by 25% (to 125%).
For a voltage drop of 3%, increase all distances by 50% (to 150%).
For a voltage drop of 4%, increase all distances by 100% (to 200%).
For a voltage drop of 5%, increase all distances by 150% (to 250%).

Suppose you use any wire size for a circuit longer (or shorter) than the distance shown for a 2% drop for that size of wire; what will the drop be? It is easy to determine. For example, the table for 120 volts shows that by using No. 8 wire with a load of 20 amp the drop will be 2% on a circuit 90 ft long. But you are going to use it on a circuit 120 ft long. What will the

TABLE 7-1 Wire Table (Copper)—120 Volts Single-Phase—2% Voltage Drop

Amperes	Volt-amperes* at 120 volts	No. 14	No. 12	No. 10	No. 8	No. 6	No. 4	No. 2	No. 1/0	No. 2/0
5	600	90	142	226	360	573	911			
10	1 200	45	71	113	180	286	455	704		
15	1 800	30	47	75	120	191	304	483	768	965
20	2 400	22†	36	57	90	143	228	362	576	726
25	3 000	18†	28†	45	72	115	182	290	461	581
30	3 600	15‡	23†	38	60	95	152	241	384	494
40	4 800		18‡	32†	45	72	114	174	288	363
50	6 000			27†	36†	57	91	145	230	290
60	7 200			22‡	30†	48†	76	121	192	242
70	8 400				26†	40†	65	104	165	207
80	9 600					36‡	57†	90	144	181
90	10 800					32‡	51†	80	128	161
100	12 000					29‡	46‡	72†	115	145

In this table, the figures below each wire size represent the maximum distance in feet that that size wire will carry the current in the left-hand column with 2% drop. All distances are one-way.

* The figure in this column is the volt-amperes in the circuit. If the power factor is 100%, as is the case with incandescent lamps and heating appliances, it is also the watts.

† For these distances, Types T and TW must not be used because they do not have enough ampacity. For all distances with a dagger, select a wire with enough ampacity (depending on whether it is in conduit or cable or in free air) from NEC Tables 310-16 through 310-19 (see Appendix).

‡ Allowed in free air only for ordinary wiring.

TABLE 7-2 Wire Table (Copper)—240 Volts Single-Phase—2% Voltage Drop§

Amperes	Volt-amperes* at 240 volts	No. 14	No. 12	No. 10	No. 8	No. 6	No. 4	No. 2	No. 1/0	No. 2/0
5	1 200	179	285	453	720					
10	2 400	90	142	226	360	527	911			
15	3 600	60	95	151	240	351	607	965		
20	4 800	45†	71	113	180	264	455	724		
25	6 000	36†	57†	91	144	211	364	579	922	
30	7 200	30‡	50†	75	120	176	304	483	768	968
40	9 600		41‡	56†	90	132	228	362	576	726
50	12 000			45†	72†	105	182	290	461	581
60	14 400				60†	88†	152	241	384	484
70	16 800				50†	76†	130	207	329	415
80	19 200					66†	114†	181	288	363
90	21 600					59†	101†	161	256	323
100	24 000					53†	91†	145	230	290
125	30 000						73†	116†	184†	232
150	36 000							97†	154†	194†
175	42 000							83†	132†	166†
200	48 000								115‡	145‡

§The footnotes under Table 7-1 apply to this table also.

drop be? Since 120 is 133% of 90 ($^{120}/_{90}$ = 1.33 = 133%), the drop will be 133% of 2% or 2.66%.

If you use any size of wire for a load *smaller* than shown in the left-hand column, the drop will be correspondingly smaller. Again, No. 8 wire will carry 20 amp 90 ft with a 2% drop at 120 volts. But if instead of 20 amp the wire is to carry only 17 amp, the drop will be 85% of 2% ($^{17}/_{20}$ = 85%), or 1.7%. Or you can use one of three formulas given to find what you want.

Cost of Voltage Drop. Assume you want to operate a motor consuming 20 amp at 120 volts (2400 VA) at the end of a 100-ft circuit. Per Table 7-1, at 120 volts a 20-amp load on No. 8 wire will result in a 2% drop if the circuit is 90 ft long; at 100 ft the drop will be about 2.2%. Suppose the circuit is in use 3 h per day, about 1000 h per year. In 1000 h, that circuit will consume about 2400 kWh, which at 6 cents per kilowatthour will cost you $144.00; of that, 2.2%, or about $3.17 is wasted. That does not seem an unreasonable total. But suppose again that, in order to reduce your initial investment, you had elected to use No. 12 wire. Carrying 20 amp, the same table tells you that the drop will be 2% if the circuit is 36 ft long, or about 6.1% for a 100-ft circuit. Then 6.1% of your $144.00, or $8.78, is wasted. That is a difference of $5.61, which in five years becomes $28.05, a sum which would have paid for the difference between the small and the larger wire. More important: your motor would perform more efficiently and deliver more horsepower with the larger wire.

Now, instead of operating the motor at 120 volts, you decide to operate it at 240 volts. As pointed out in earlier paragraphs, when the watts remain unchanged and the voltage is changed from 120 to 240 volts, *without changing the wire size,* the drop at the higher voltage is one-quarter of the drop at the lower voltage, in percentage. In other words, by using No. 8 wire, the drop changes from 2.2 to 0.55%, and by using No. 12 wire it changes from 6.1 to 1.53%. You can use the smaller wire, and still have less voltage drop.

In debating whether to use the smaller or the larger of two sizes of wire, remember that the labor for installing the larger size is little if any more than for the smaller size. The difference in cost is basically the difference in cost of the wire only. Consider also that with a larger wire and less voltage drop, your motor (or other load) will operate more efficiently than when using the smaller wire.

Overhead Spans. In running wires overhead, there is one additional factor that must be considered—mechanical strength. The wires must be large enough to support not only their own weight but also the strain imposed by winds, ice loads, and so on. In northern areas, it is not unusual to see a layer of ice an inch thick around outdoor wires, after a severe sleet storm. The Code in Sec. 225-6(a) requires a minimum of No. 10 for spans up to 50 ft, and No. 8 for longer spans. If the distance is over 100 ft, use No. 6; if it is over 150 ft, it is best to use an extra pole.

In installing outdoor overhead wires in northern areas, take into consideration the expansion and contraction that take place with changes in the temperature. A 100-ft span of copper wire will be almost 2 in shorter when it is 30° below zero than on a hot summer day when the temperature is 100°. Therefore, if the wires are installed on a cold winter day, they may be pulled as tight as practical. If installed on a hot day, allow considerable sag in the span so that when the wires contract in the winter, no damage will be done.

By consulting Tables 7-1 and 7-2, you can determine whether a given size of wire used in a circuit of a specific length will or will not result in an acceptable voltage drop. Remember, though, that the ampacity of the wire must not be less than that permitted by the Code for the current involved.

Cautions about Ampacity. You must always use a size and type of wire that has an ampacity rating at least as great as the number of amperes to be carried (see NEC Tables 310-16 through 310-19 in the Appendix of this book). For *short* runs, voltage-drop tables *may* show a wire size smaller than is permitted for the current involved: Use wire with sufficient ampacity. Voltage drop for any given current, at any given distance, depends *solely* on the circular-mil area of the wire: ampacity of any size and kind of wire *varies* with the kind of insulation, and other factors. Using wire that has a higher ampacity in any given size, than another type of wire of the same size, will *not* reduce the voltage drop.

What Determines Ampacity of Wire. Before entering into a discussion of Code provisions concerning wire, it will be well to analyze the reasons why the ampacity of any given size of wire varies with the kind of insulation and the method of installation.

Conductor insulation can be damaged by excessive heat in various ways, depending on the kind of insulation and the degree of overheating. Some insulations melt, some harden, some burn. In any event, insulation loses its usefulness if overheated, leading to breakdowns and fires.

The specified ampacity for any particular kind and size of wire is the current that it can carry continuously *without increasing the temperature of its insulation beyond the danger point*. The insulation of Types T and TW wire withstand the least heat; consequently these types have lower ampacities than other kinds. Asbestos will withstand far more heat; consequently asbestos-insulated wire has a much higher ampacity. The temperature of asbestos-insulated wire carrying its rated current will be much higher than the temperature of plastic-insulated wire carrying its rated current, but its insulation will not be damaged by the higher temperature.

The rated ampacity of each kind and size of wire is based on an ambient temperature[3] of 30°C or 86°F. NEC Table 310-13 in your copy of the Code shows the maximum temperature the insulation of each kind of wire

[3] Ambient temperature is the normal air temperature in an area while there is no current flowing in the wire. When current does flow, heat is created, and the surrounding air temperature (as well as the conductor temperature) will increase above the ambient.

is permitted to reach, and other information about each kind of wire. That temperature will be reached when a wire is carrying its full ampacity where the ambient temperature is 30°C or 86°F. The maximum permitted temperature is called the *temperature rating* of the wire. Note that while the temperature rating of any wire can be found in Table 310-13 as already mentioned, you can find the temperature ratings of wires ordinarily used in wiring buildings (but not all other types used for other purposes) at the top of the various columns in NEC Tables 310-16 through 310-19, all of which are found in the Appendix of this book.

To repeat: If a wire has a temperature rating of, say, 60°C or 140°F (which is the lowest temperature rating assigned to any kind of wire), it does *not* mean that the wire may be used where the ambient temperature is 140°F. It means that the temperature of the wire itself may not exceed 140°F. It will reach that temperature when carrying its rated current in a room where the ambient temperature is 30°C or 86°F. If the room ambient temperature is higher than 86°F and the wire is carrying its full rated current, its actual temperature will exceed its temperature rating of 60°C or 140°F. Therefore if any kind of wire is installed in hot locations, its ampacity is reduced from that shown in NEC Tables 310-16 to 310-19.

Correction Factors, High Ambients. Do not use wires in such a way that their temperature ratings will be exceeded while they are carrying their rated ampacity. Remember these tables assume an ambient temperature of 30°C or 86°F. If the ambient temperature is higher than that, the ampacity of any size and kind of wire must be calculated in accordance with the correction factors at the bottom of the tables.

For example, if the ambient temperature is 50°C or 122°F (which is 20°C or 36°F more than the basic ambient assumed in these Code tables), then if you are using 60° wire (any kind listed at the top of the 60° column in the tables), you must limit the load current to 58% of the value shown in the tables. For example, No. 12 Type T or TW wire in conduit has an ampacity of 25; if you install it in a 50°C or 122°F ambient, its maximum ampacity becomes 14.5 amp (58% of 25 is 14.5). If you use No. 12 Type RH or THW (which has a 75°C temperature rating), you must derate to 75% of its normal ampacity, or 18.75 amp (75% of 25 is 18.75).

Wires installed in conduit, or in the form of cable, or buried directly in the ground cannot radiate the heat developed by the current flowing through them as easily or quickly as when installed as single conductors in free air. Therefore, when installed in free air, any type and size of wire has a higher ampacity than when installed in other ways.

Ampacity vs. Size of Wire. From Chap. 6 you will remember that of two wires six sizes different from each other (for example, No. 6 as compared with No. 12) the larger will have *twice* the diameter,[4] the circumference, and the surface area but *four times* the cross-sectional area of the smaller wire. It might seem logical that a wire that has four times the cross-sectional area of another wire would have four times the ampacity of the smaller wire, but that is not the case, as the data in Table 8 and the ampacities for 75°C wire in Table 310-16 (see Appendix) will demonstrate.

The No. 12 wire has an ampacity of 20, and No. 6, with four times the cross-sectional area, might be expected to have four times the ampacity of No. 12, or 80, but its actual ampacity is only 65; No. 1/0, with four times the cross-sectional area of the No. 6, might be expected to have four times its ampacity, or 260, but its actual ampacity is only 150. The ampacity *per thousand circular mils* of No. 12 is 3, of No. 6 is 2.5, of No. 1/0 is 1.4, and of 2000 kcmil is 0.30. The larger the size, the lower the ampacity per thousand circular mils.

Why? A wire that has four times the *cross-sectional area* of another wire has only *twice the surface* area of the smaller wire, and the heat developed in the wire can be dissipated only from its surface. Compare No. 1/0 with No. 12. The No. 1/0 has sixteen times the cross-sectional area, but only four times the surface area, from which heat can be dissipated. The ampacity of a wire, while not directly in proportion to its surface area, is more nearly related to its surface area than to its cross-sectional area. Another factor, especially in the larger sizes, is the ''skin effect'' discussed in the next paragraph.

Resistance of Wires, AC vs. DC. In a dc circuit, the current flows at a uniform density throughout the entire diameter of a wire. In an ac circuit involving the larger sizes of wire the current does not flow uniformly throughout the diameter of the wire, as in direct current, but flows mostly in the outer layer of the wire, with a much lower density in the inner portions of the conductor. This phenomenon is known as the skin effect, and in practice it means that the resistance of a very large wire is higher when used in an ac circuit than when used in a dc circuit.

Because of this, in ac circuits you must not use the resistance values shown in Table 8 of NEC Chap. 9 (for example, in calculating voltage drop), but must take the square root of the sum of the squares of the ac resistance and inductive-reactance values given in Table 9 of NEC Chap. 9.

[4] The diameters of all sizes of wire, as well as their area and their resistance, can be found in Table 8 of Chap. 9 of the Code (see Appendix).

The figures in Table 9 are for three conductors, 3-phase, at an assumed operating temperature of 75°C. The columns headed "magnetic" are for conductors inside a steel armor or raceway. Other combinations of conductors or other temperatures will give different values. In any case, the ac skin effect is not significant in conductors smaller than 4/0.

Continuous Loads. If a load on a feeder or branch circuit is likely to continue for 3 h or longer (as is usual in stores, offices, factories, and similar locations), NEC Art. 100 defines it as a continuous load. The maximum load on any feeder or branch circuit so classified *must not exceed 80%* of the nominal rating of the feeder or branch circuit. On a 15-amp circuit (one with a 15-amp breaker or fuse) the continuous load may not exceed 12 amp (80% of 15 is 12); on a 20-amp circuit (one with a 20-amp breaker or fuse) it may not exceed 16 amp, and so on. The reason for this derating is that the quite considerable heat generated in panelboards by the breakers or fuses cannot be dissipated rapidly enough if the circuits are *fully* loaded at their nominal rating, and if the load is *continuous*. Unless a load has been derated to 80%, it is not unusual for breakers to trip or fuses to blow, even if they are carrying somewhat *less* than their rated capacity. The greater the number of breakers or fuses in a single panelboard, the greater the likelihood of this happening.

Using NEC Tables 310-16 through 310-19. These tables show the ampacity of different sizes and kinds of wire under different installation conditions. They also show the temperature rating of each kind of wire. All these tables are shown in the Appendix of this book.

If the wires are installed as overhead spans, or on insulators (so that air circulates freely around the wires), use Tables 310-17 and 310-19. In all other cases (including cables and underground wires) use Tables 310-16 and 310-18. The Notes following the tables in the Appendix are self-explanatory, but comments concerning some of them will follow, to more fully explain them.

The dagger (†) footnotes to Tables 310-16 and 310-17 limit overcurrent protection for certain No. 14, No. 12, and No. 10 wires. The table *ampacity* is employed as the basis for calculation when the wire is used in a high ambient temperature or when applying Note 8 for four or more current-carrying wires grouped together in one raceway or cable, but for everyday application the dagger footnotes establish the ampacity, and the values in those notes are often referred to as the ampacity for these wires.

Note 5: Bare Wires. If a cable or raceway contains both insulated wires and one or more bare wires (such as a bare neutral), you must consider the ampacity of the bare wire to be the same as that of whatever kind of insulated wires are used. In other words, if the insulated wires are Type THW, consider the bare wire to be Type THW also, so far as ampacity is concerned.

Note 8: Over Three Wires in Raceway or Cable. Code Tables 310-16 and 310-18 show the ampacity of wires, provided there are no more than three in a cable or raceway. If there are more than three, more heat develops in the cable or conduit, so Note 8 to the tables requires that you reduce the ampacity as follows:

4 to 6 wires, derate to 80% of their normal ampacity.
7 to 24 wires, derate to 70% of their normal ampacity.
25 to 42 wires, derate to 60% of their normal ampacity.
Over 42 wires, derate to 50% of their normal ampacity.

Exception No. 1 indicates that only current-carrying power and lighting conductors need be counted in applying Note 8. Equipment-grounding, power-limited, control, and signal conductors need not be counted for derating purposes (but *must* be counted for conduit fill, as they occupy space whether carrying current or not). For neutrals, see comment on Note 10, below.

Double Derating. Do note that when loads must be limited because there are more than three wires in a cable or a raceway, the wires do *not* need to be derated again if the load is continuous (3 h or more). However, the ampacity must be derated both for high ambient temperature and for more than three wires in a cable or raceway if both conditions exist.

Note 10: Neutrals. The requirements of Note 8 seem simple enough, but if neutral wires are involved, they need further explanation, as is given in Note 10. A neutral wire [which carries only the unbalanced current of a circuit (such as the neutral wire of a 3-wire single-phase circuit, or of a 4-wire 3-phase circuit)] is not counted for derating purposes. But remember: Grounded wires are not always neutrals; the grounded wire of a 2-wire circuit carries the same current as the hot wire and therefore is *not* a neutral. So if you install two such 2-wire circuits in a conduit, they must be counted as four wires. The same is true in a 3-wire circuit of two phase wires (hot wires) and a grounded wire, all supplied by a 4-wire 3-phase wye-connected system; the neutral carries approximately full load current, and it is counted for derating purposes. So if you have two such 3-wire circuits in one conduit, count six wires and derate to 80%.

But there is still another exception. If the 4-wire 3-phase wye circuit supplies mostly electric-discharge lighting (fluorescent, mercury, and similar types discussed in Chap. 29), or data processing equipment, you must count the neutral.[5]

Many types of control wires, such as motor control wires and similar wires that carry only intermittent or insignificant amounts of current, need not be counted.

Caution: Conduit Size. In determining the size of conduit to use for any number of wires, you *must count all* wires whether they carry current or not, whether they are bare or insulated. In other words include in the total any wires that you may ignore when determining whether derating is necessary because of more than three wires in a raceway.

Number of Wires in Conduit. In the larger sizes, wires such as Types THHN or THWN have a considerably higher ampacity than the same sizes of ordinary Types T or TW. Assume that you need a 3-wire 120-amp circuit. If you use Types T or TW, you will need No. 1/0, requiring 1½-in conduit per NEC Table 3A. If you use Type THWN, which has a higher ampacity than Types T or TW, you can use No. 1 wire, and since it is a small-diameter wire, you can use 1¼-in conduit. Another example: For four No. 2/0 Type T with an ampacity of 145, you will need 2-in conduit; for four No. 1/0 Type THWN with an ampacity of 150, you will need only 1½-in conduit. Similar information for other wires can be easily determined from Tables 3A, 3B, or 3C of NEC Chap. 9, all found in the Appendix.

Two points of caution are in order: First, if the runs are long, using small-diameter wires with their higher ampacities will lead to more voltage drop than will using somewhat larger sizes of ordinary Types T or TW, which have lower ampacities. Consider this carefully if the runs are long and voltage drop is a factor. On the other hand, if the ambient temperature is higher than normal, you will have less derating to do if you use a wire type that has a high temperature rating. Second, in general, switches, circuit breakers, panelboards, appliances, and similar equipment are tested with 60°C-rated wire (T, TW, UF, etc.) in circuits rated 100 amp or less and with 75°C-rated wire (THW, RH, THWN, USE, etc.) in circuits rated *over* 100 amp. Thus, such equipment and appliances may not be suitable for connection to a wire selected for higher ampacity because of its insulation type. If the equipment or appliance has been tested with higher-rated wire, it will be marked with the temperature

[5] In such circuits, the neutral carries a "third harmonic" or 180-Hz current. The explanation of why such circuits produce such a 180-Hz current is far beyond the scope of this book. But be aware that the neutrals of such circuits or feeders must be counted for derating purposes.

rating or with both the temperature rating and the size of wire to be used. Higher-temperature-rated wires may, of course, be used in any case, such as when the wires will pass through an area of high ambient temperature, *provided* their *size* is based on the ampacity of 60°C wire at 100 amp or less or 75°C wire at over 100 amp.

If you are going to install two or more different sizes of wire in the same conduit, you will have to calculate the size of conduit required by the hard, or long, way. From NEC Chap. 9, Table 5 (see Appendix), determine the total cross-sectional area of all the wires you intend to install. Compare this with the area of fill permitted for each size of conduit, as shown in NEC Chap. 9, Table 4 (see Appendix). For example, assume that the total cross-sectional area of all the wires you intend to install is 0.66 sq in. Reference to Table 4 shows that you will have to use 1$^{1}/_{2}$-in conduit, in which wires with a total cross-sectional area of not over 0.82 sq in may be installed.

Wires in Parallel. Instead of using a very large wire, NEC Sec. 310-4 permits two *or more* smaller wires to be connected in parallel for use as a single wire. There are half a dozen conditions, *all* of which must be met. All the wires that are to be paralleled to form the equivalent of one larger wire must (1) be No. 1/0 or larger, (2) be of the same material (all copper, all aluminum, all copper-clad aluminum, etc.), (3) have precisely the same length, (4) have the same cross-sectional area in circular mils, (5) have the same type of insulation, and (6) be terminated in the same manner (same type of lug or terminal on all).

As an example, assume you have to provide a 3-wire circuit carrying 490 amp. Per NEC Table 310-16 (75°C column) you will need 800-kcmil (mcm, MCM) wire. In place of each 800-kcmil wire, you can use two smaller wires each having half the required ampacity, or 245 amp. Per Table 310-16, two 250-kcmil wires each with an ampacity of 255 would appear to be suitable.

But if you plan to run all six wires through one conduit, 250-kcmil wire is not suitable because you will have six wires in one conduit, and therefore must limit the maximum allowable load current to 80% of the normal ampacity of 225 (Note 8 of Table 310-16). The ampacity of the 250-kcmil wire then becomes 80% of 255, or 204. Use 350-kcmil wire with an ampacity of 80% of 310, or 248.

But instead of running all wires through one conduit, use two[6] separate

[6] Instead of two, you may use three or even more conduits, provided all the requirements outlined are met.

conduits with three wires in each; then load reduction is not required, and 250-kcmil wire is suitable. But do note that the conduits or other raceways through which the wires run must have the same physical characteristics; all might be rigid steel conduit, or all might be EMT. It would not be permissible to use one steel and one aluminum conduit.

When running parallel wires through a raceway, special care must be taken to meet the requirements of NEC Sec. 300-20. One wire of each phase must be run through *each* conduit. See Fig. 7-1, which shows six wires in two conduits. Wires *A* and *a* are paralleled with each other, but both must *not* run in the same conduit. Likewise wires *B* and *b* are paralleled and again must not run in the same conduit. The same applies to *C* and *c*. If you ran, for example, *a*, *b*, and *B* in one conduit and *c*, *A*, and *C* in the other, each conduit would contain wires of only two of the three phases. In that case considerable induced and eddy currents would be set up in the conduits themselves, resulting in excessive heating and power losses.

The neutral wire of a feeder or circuit may often be smaller than the hot wires. If that is the case in your installation, the neutral also must be paralleled even if its ampacity is less than that of the hot wires, just so that their combined ampacity is equal to that which would be required if the neutral were not paralleled. This is shown in Fig. 7-2, which is the same as Fig. 7-1 except that the neutral has been added, labeled *N* and *n*. If the wires are installed in a nonmetallic raceway, the raceway of course cannot serve as the equipment grounding conductor, and in that case a full-size equipment grounding conductor must usually be installed in each raceway.

Conductors in parallel are quite common if using one wire would require a very large wire. Copper wires larger than 500 kcmil are very difficult to handle, and the smaller wires have much higher ampacity *per*

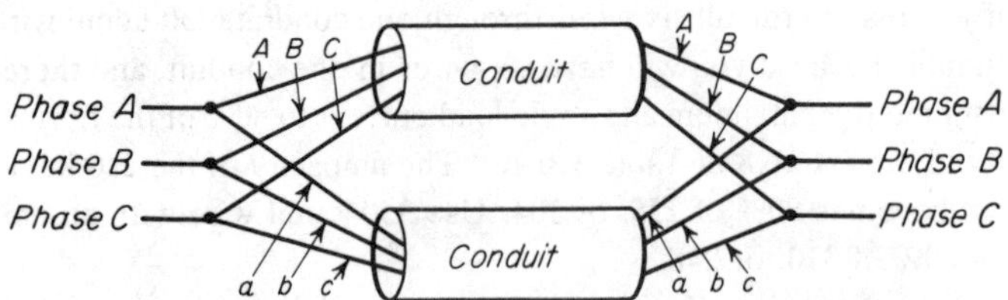

Fig. 7-1 Where paralleled wires are run through conduit, each conduit must contain one wire from *each* phase. The installation shown does not include a neutral.

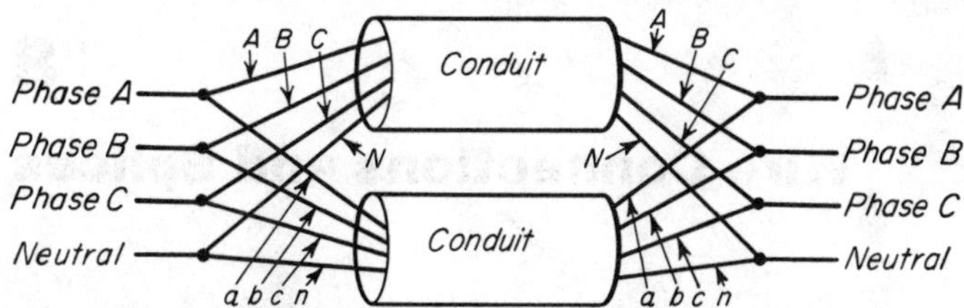

Fig. 7-2 Same as Fig. 7-1 except that a neutral has been added. Each conduit must contain one wire of the neutral. In each illustration, while three ungrounded wires are shown, there might be any number, but each must appear in every conduit.

thousand circular mils than the larger one. Thus much less copper by weight is used when two smaller wires are paralleled to replace one larger one. The total installation cost is reduced. However, when you use two smaller wires in place of one larger wire, the combined circular-mil area of the two smaller wires will be less than that of one larger wire.

That will result in a somewhat higher voltage drop than you would have using one wire of the larger size. If the circuit is long, you may have to increase the size of the paralleled wires to offset this.

8
Wire Connections and Splices

Wires must be properly connected to switches, receptacles, and other devices. Wires must be spliced to each other. All this is easy to do but is often done improperly. This chapter explains how to do these things correctly.

Safety Tip. This being the first place in the book in which specific work procedures are discussed, it is appropriate to bring up the subject of safety. In a discussion of splicing and terminating wires it might seem strange to treat the use of safety glasses, but many an eye has been injured by a small bit of copper cut from the end of a wire. Whatever kind of electrical work you are doing, *use eye protection.* Soldering, although not covered in this book, must sometimes be done, and flux or hot solder can unexpectedly fly toward your eye. You should work only on de-energized systems, but some testing *must* be done when circuits are energized, and should you make a mistake which starts an electric arc, molten metal can be thrown out so quickly that you cannot escape it. Even when you are cutting openings in walls and ceilings, plaster dust will almost certainly find its way into your eyes unless you protect them. *Wear safety glasses at all times.*

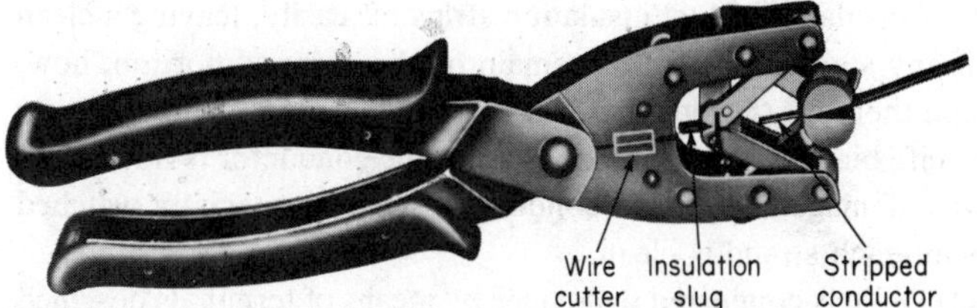

Fig. 8-1 A wire stripper saves much time in removing insulation. (*The Eraser Co., Inc.*)

Removing Insulation. Figure 8-1 shows a wire stripper, a very handy tool for removing insulation from wires. Ordinary strippers are for smaller wires up to No. 10 in size; larger strippers are needed for larger wires. Always follow the manufacturer's instructions when using a stripper. Another easy way to strip off the insulation is to use a pair of side-cutter pliers. Place the wire to be stripped between the handle grips close *behind* the cutter hinge, and squeeze the insulation enough to soften it and break it down, but not hard enough to damage the wire. Do this from the point where the insulation is to be stripped, to the end of the wire. Then place the cutter jaws over the wire at the point where the insulation is to be stripped, squeeze just hard enough for the jaws to grasp the insulation (but not hard enough to touch the conductor), then slide the insulation off.

If you have no stripper and have the wrong kind of pliers, use a knife to cut through the insulation down to the conductor, holding the knife not at a right angle but at about 60 deg, so that it will cut the insulation as shown in Fig. 8-2. This precaution reduces the danger of nicking the conductor, which weakens it and sometimes leads to breaks. After the insulation has been cut all around, pull it off, leaving the conductor exposed far enough to suit the purpose.

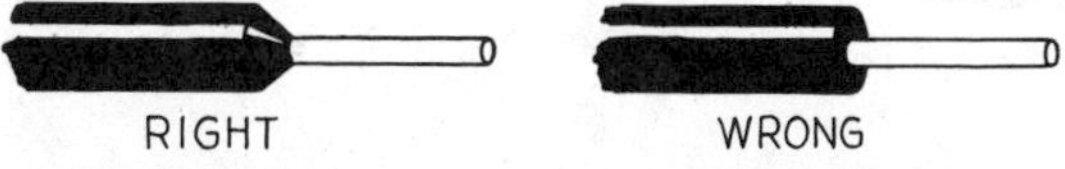

Fig. 8-2 When using a knife to remove insulation from a wire, hold the knife at an angle of about 60 deg.

About 6 in of each wire is needed to allow a switch or similar device to be connected while it is still outside the box. About 4 in more are needed for making connections rapidly and efficiently, as will be explained later in this chapter. So leave about 10 in of free wire at each box, and strip off about 4 in of insulation. Most insulation strips off easily, leaving a clean surface. If any spots of insulation remain on the stripped portion, however, scrape them off thoroughly for a clean connection, using the *back* edge of a knife blade for the purpose, so that the conductor is not nicked or otherwise damaged. Stranded conductors must be specially watched to make sure each strand is clean.

Terminals. Wire is connected to devices by means of terminals designed for the purpose. For No. 10 and smaller wires, screw terminals of the type shown in Fig. 8-3 are usually used. The brass base of the terminal or

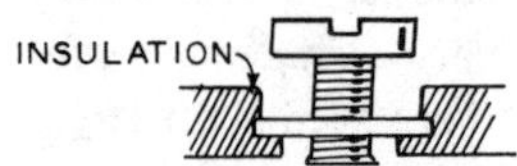

Fig. 8-3 A terminal for connecting No. 10 or smaller wires.

the surrounding insulation is shaped to prevent the wire from slipping out from under the terminal screw, which is usually "upset" so that it cannot be entirely removed. While the Code permits No. 10 wire with screw-type terminals, it is difficult to make a good connection if the wire is *stranded.* In all sizes of stranded wires, it is best to twist the strands tightly together before connecting them under a terminal screw.

The Code in Sec. 110–14 prohibits more than one wire under a single screw, yet many times there is an apparent need to do so. Don't be tempted to do this. As shown in Fig. 8-4, make a pigtail by connecting the two ends of wire plus another short piece to each other, using a solderless connector that will be described later in this chapter. Connect the short wire to the terminal screw.

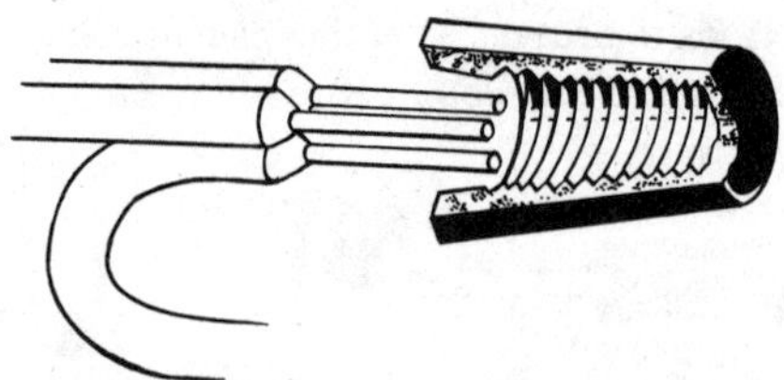

Fig. 8-4 Two wires may not be connected under a single screw terminal. Use the method shown above; see text.

After about 4 in of the wire has been cleanly stripped of its insulation, and the terminal screw has been backed out until the upset prevents further loosening, wrap the wire clockwise around the screw, as in Fig. 8-5, in such a way that (1) the insulated portion of the wire is near enough to the screw to allow the head of the screw, when tightened, to clear the

Fig. 8-5 Insert wire under a terminal screw so that tightening the screw tends to *close* the loop.

Fig. 8-6 Don't leave a long bare wire next to a terminal screw.

insulation by not over ¼ in, as shown in Fig. 8-6; and (2) the stripped portion of the wire is brought around the screw so that it nearly closes the gap but does *not* overlap—not less than about two-thirds nor more than three-fourths of a turn. Then tighten the screw so that its head fits flat and snugly against *all* the wraparound portion of the wire. Next, tighten the screw about another half-turn so that it is firm and tight. If you use a torque screwdriver, tighten to 12 lb·in. Finally, take the excess "tail" of bare wire and twirl it round and round until it breaks off near the screw head. It should take only three or four twirls to break it.

Figure 8-7 shows the steps UL stipulates for connecting *aluminum* wire to 15- and 20-amp screw terminals, and Fig. 8-8 shows methods to be avoided. The same procedures should also be used with copper wire. There is one exception: UL lists some screw terminals (for copper or copper-clad aluminum only) that permit a wire to be inserted straight, without wrapping it around the screw, but only when the terminal is specifically designed with a groove to lay the wire in. Many inspectors do not look with favor on this type of connection.

If the wire in the box is too short to allow 4 in or so for twisting off the "tail" after the screw has been tightened, then form the end of the wire into a loop, using long-nose pliers, and insert the loop under the head of the screw, closing it completely with the same pliers. Be sure the loop fits flat under the screw head.

Some switches and receptacles are made without terminal screws. The straight end of a wire is pushed into small holes in the device, as shown in Fig. 8-9. Strip the insulation off the wire as far as indicated by a strip gauge on the device. If it becomes necessary to remove a wire, push a

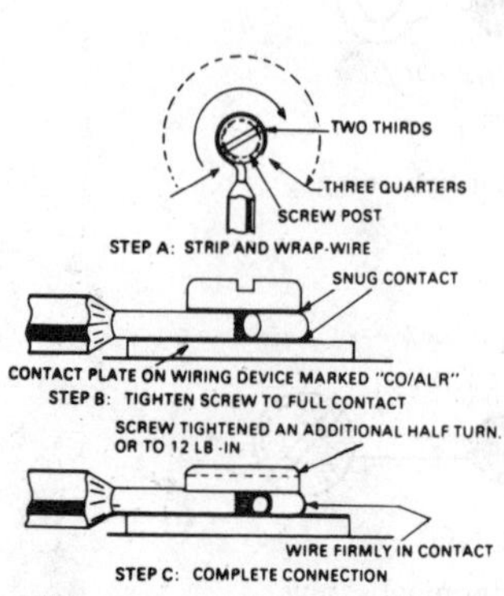

Fig. 8-7 Be sure to observe the suggestions shown in this illustration. (*Underwriters Laboratories Inc.*)

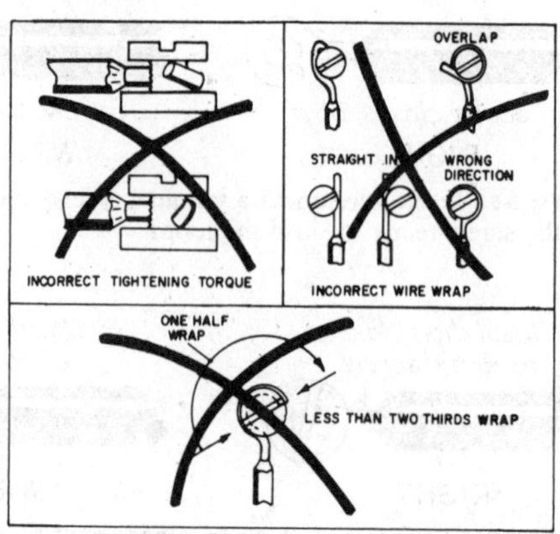

Fig. 8-8 Avoid these common errors when connecting wires to terminal screws. (*Underwriters Laboratories Inc.*)

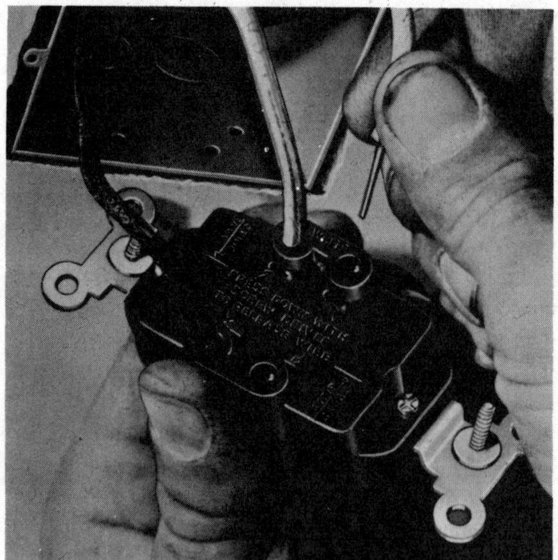

Fig. 8-9 Some devices do not have terminal screws. For a permanent connection, push the bare wires into openings in the device. The device has a "strip gauge" showing how much insulation to remove from the wire. (*General Electric Co.*)

small screwdriver blade into slots on the device to release the wire. These devices may *not* be used with all-aluminum wire but are acceptable for copper or copper-clad aluminum. Some devices have both terminal screws *and* push-in terminals. Many inspectors do not favor push-in terminals.

Remember that if the receptacle is installed on one side of a 3-wire circuit, the white circuit wire must be continuous or spliced, so that its continuity will not be interrupted by the removal of the receptacle. In such cases, pigtail the white wire as shown in Fig. 8-4, and connect *one* white wire to the receptacle.

Terminal screws as discussed up to this point may not be used for wires larger than No. 10. For No. 8 and larger wires, solderless connectors or terminals of the general type shown in Fig. 8-10 are used. Simply insert the stripped and cleaned end of the conductor into the connector, and tighten the nut or screw.

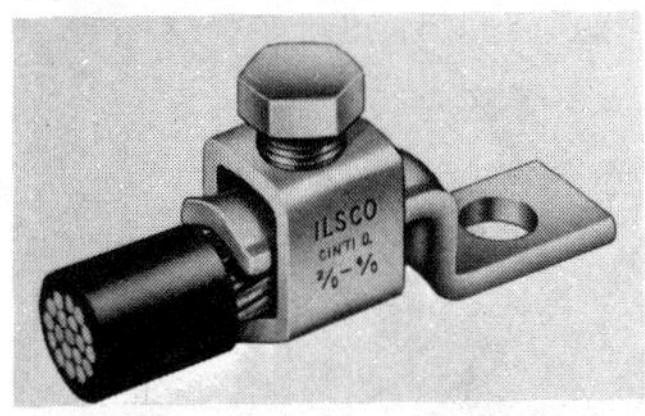

Fig. 8-10 A typical solderless connector for larger wires. (*Ilsco Corp.*)

Splices. In many cases the Code does not permit spliced wires; these cases will be mentioned as the related work is discussed in this book. In other cases splices are necessary; it is most important that you make them properly. A splice must be as strong mechanically and as good a conductor as a continuous piece of wire. After the splice is completed its insulation must be equivalent to the original insulation. To accomplish these three things, it is necessary to remove the insulation where the wires are to be spliced, make the mechanical joint using solderless connectors, and then take care of the insulation. Most solderless connectors have insulation shells that completely cover the bare ends of the wire, if properly installed. If your connectors do not have insulating shells, you must insulate the splice using electrical insulating tape.

For wires No. 8 and smaller, one of the most popular splicing devices is the screw-on connector, often called a Wire Nut, although that is the trade name of one particular manufacturer. Two types are shown in Fig. 8-11. One type has a removable metal insert; the ends of the wires to be spliced are pushed into this insert, the setscrew of which is then tight-

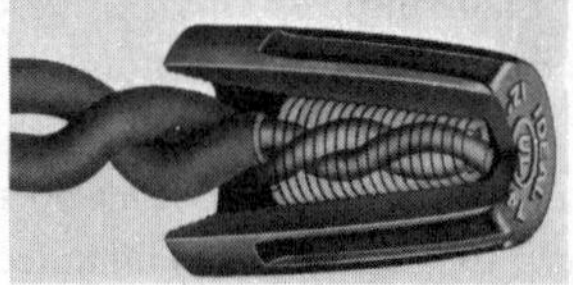

Fig. 8-11 Two types of solderless connectors for smaller wires, often called Wire Nuts. (*Ideal Industries, Inc.*)

ened, and the plastic cover screwed on over the insert. The other type has no removable insert but has an internal tapered thread; lay the wires to be spliced parallel to each other (if one is much smaller than the others, let it be a bit longer than the larger wires), then screw the connector over the bare ends of the wires. If the ends of the wires have been stripped to the proper length, no bare wire will be exposed and no taping is needed. Figure 8-12 shows the use of such connectors. Of course there are other

Fig. 8-12 This shows the use of solderless connectors. (*Ideal Industries, Inc.*)

connectors, for example, the Scotchlok shown in Fig. 8-13. Within the tough outer insulating shell there is a steel sleeve, and inside that a coiled tapered spring that expands as the connector is screwed onto the wires; after you release the connector, the spring holds the wires in com-

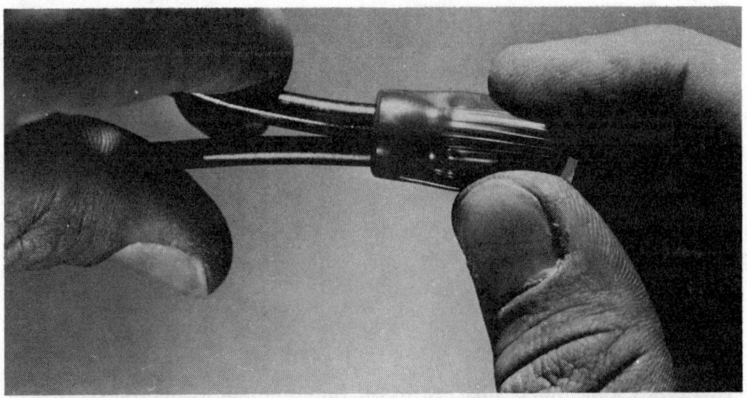

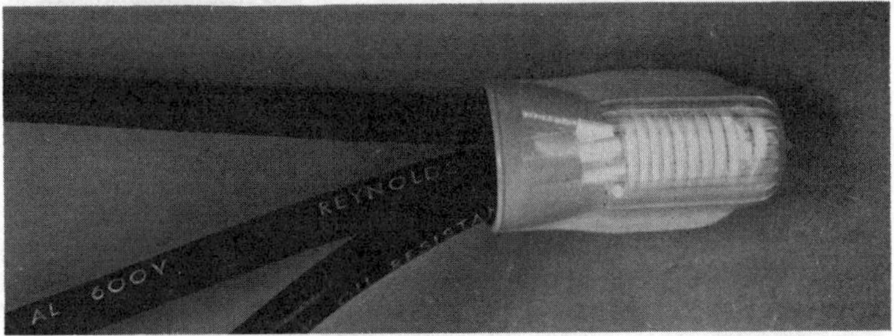

Fig. 8-13 Spring-type solderless connector, showing ease of installation and, below, an x-ray photograph of a completed splice. (*Minnesota Mining & Mfg. Co.*)

pression, forming a good permanent connection. Be sure that no bare wire is visible after the connector has been installed.

Another commonly used splicing device is the tool-applied tube of copper (or plated steel) shown in Fig. 8-14. These are available noninsulated, as shown, and also insulated, and with separate insulating covers. Unless the manufacturer specifies otherwise, wires should be first twisted together and then inserted in the tube, which is then crimped using a tool similar to that shown in Fig. 8-15. Wire ends should then be cut off close to the end of the tube.

Always make sure that the connector you are using is the right size for the number and size of the wires you are connecting, and that it is UL Listed as a pressure cable connector and *not* for fixture taps only, as the

Fig. 8-14 Crimp-type connector, for copper wire only. (*Thomas & Betts Corp.*)

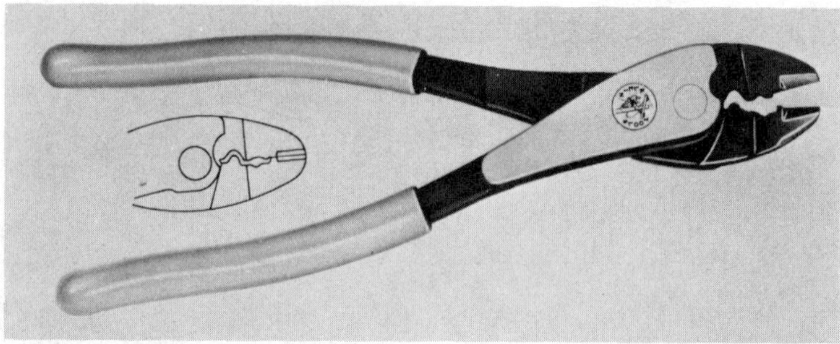

Fig. 8-15 Crimping tool for both insulated and noninsulated connectors. (*Klein Tools, Inc.*)

two types look much alike. Strip off only enough insulation so that no bare wire will be exposed after installation. If you use a connector without an insulating shell, you must of course insulate the splice with insulating tape (unless it is a splice in a bare grounding conductor).

Another splicing device, requiring only parallel-jawed pliers to install, is shown in Fig. 8-16. The plastic "clam shell' forms its own insulation.

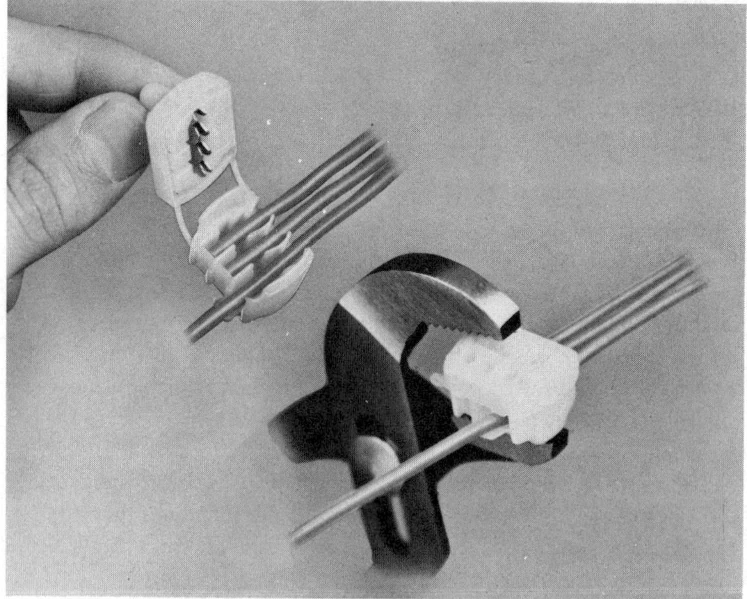

Fig. 8-16 This "clam-shell" connector is suitable for either taps or pigtail splices up to No. 12 wire. Check UL Marking for use with aluminum wire. (*Thomas & Betts Corp.*)

The wires should *not* be stripped, as the wire insulation positions the wires under the U-shaped elements in the metal blade, which cuts through the insulation when the device is squeezed closed.

For heavier sizes of wire, use connectors of the general type shown in Fig. 8-17. Being made of bare copper, the splice must be insulated with tape. However, such connectors are also available with insulating clamp-on shells, which do away with the need for tape.

Taps. Often it is necessary to connect a wire to another *continuous* wire, this being called a "tap connection" or just a "tap." The simplest way is to use a split-bolt solderless connector of the general type shown in Fig. 8-18. Unless the connector has a clamp-on insulating cover, the connection must be taped.

Fig. 8-17 Use a connector of this type to splice two large wires. The splice must be taped.

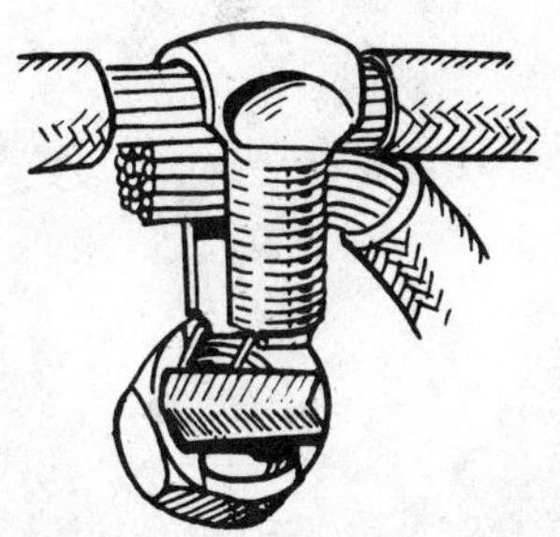

Fig. 8-18 A split-bolt connector of this type permits one wire to be tapped to another *continuous* wire.

For larger-size wires and particularly for aluminum wires, a tool-applied pressure connector is often used. The connector marking (which may be on the smallest shipping carton) will specify the tool to be used with the connector, for some manufacturers' connectors cannot be applied properly with another manufacturer's crimping tool. The investment in such a tool cannot be justified unless you will be making a great many terminations. A typical connector and tool are shown in Fig. 8-19.

Terminals and Splicing Devices for Aluminum Wire. As already discussed in Chap. 6, plain all-aluminum No. 12 and No. 10 wires may be connected to the terminals of 15- and 20-amp devices only if they are marked "CO/ALR." Push-in terminals as shown in Fig. 8-9 may *not* be used with all-aluminum wire but may be used with copper-clad aluminum.

Solderless connectors of any kind may not be used for all-aluminum wire unless they are marked "AL-CU" or "AL." Especially on smaller sizes, such markings may appear only on the carton holding the connec-

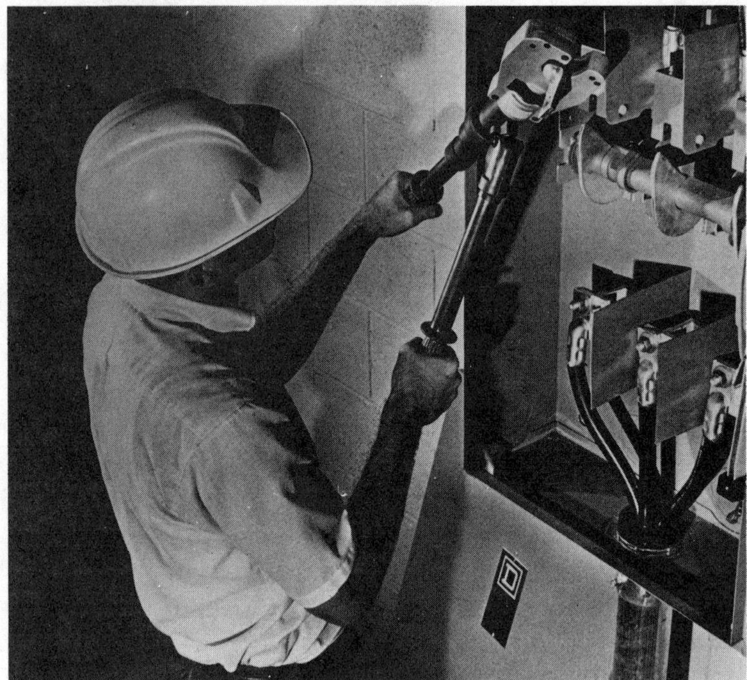

Fig. 8-19 Tin-plated aluminum lug being applied to an aluminum conductor. This compression hand tool crimps all sizes from No. 10 to 500 MCM without the need for separate dies or adjustments. (*VERSA-CRIMP, Square D Co.*)

tors. Any kinds of solderless connectors or terminals may be used for copper or copper-clad aluminum wire unless they are marked "AL." These markings and applications are summarized in Table 8-1.

One of the characteristics of aluminum when exposed to air is the rapid formation, on the surface, of aluminum oxide, a poor conductor. Most aluminum wire connectors are designed to crush and bite through this oxide layer, but some require that the exposed aluminum be wire-brushed to remove the oxide and immediately coated with a pastelike oxide inhibitor to keep the air away before insertion into the connector. For stranded aluminum conductors this is a good practice whether the connector being used requires it or not.

If it is necessary to splice a copper or copper-clad aluminum wire to an all-aluminum wire, a combination, or two-compartment, connector must be used. One end is suitable for aluminum; the other, for copper. Such connectors must be marked that they are combination splicing devices.

TABLE 8-1

Device, connector, or terminal	Marking	Recognized for use with
15 and 20 amp	CO/ALR	Copper, copper-clad aluminum, or aluminum
30 amp and larger	AL-CU	Copper, copper-clad aluminum, or aluminum
Any	AL	Aluminum *only*
Any	None	Copper *only*
Screwless, push-in	None	Copper or copper-clad aluminum *only*

Some connectors are listed for a combination of copper and aluminum wires in contact with each other, but they may be so used only in dry locations. Also, note that for equipment such as enclosed switches and panelboards the suitability for use with aluminum must be marked *on the equipment itself,* for not only must the terminals be suitable for aluminum but the enclosure space must be adequate to accommodate the larger aluminum wires.

Taping. At one time it was necessary to use two kinds of tape to insulate a splice: rubber tape, properly called splicing compound, to replace the original insulation, and friction tape to protect the rubber tape. Today only one kind of tape is necessary, a plastic electricians' tape made for the purpose. It is very strong mechanically, and has a very high insulating value per mil ($^1/_{1000}$ in) of thickness, so that a comparatively thin layer is sufficient, thus doing away with clumsy, bulky splices. This results in a less crowded condition inside a box, especially where there are several splices within the same box.

In applying this tape on a splice, start at one end, laying the tape over the original insulation, then wind it spirally toward the other end, letting the successive turns slightly overlap each other. Keep the tape stretched so that wherever the turns overlap, they will fuse to each other. Work back and forth in this fashion until the several layers of tape are as thick as the original insulation.

Soldering. The Code prohibits soldered connections in the service equipment, in the ground wire, and in all ground*ing* wires. At one time, soldering was commonly used for other splices and taps, but it is rarely if ever used today. Accordingly, the art of soldering will not be discussed in this book.

9

Grounding: Theory and Importance

In all discussions of electrical wiring, you will regularly meet the terms "ground," "grounded," and "grounding." They all refer to deliberately connecting parts of a wiring installation to a grounding electrode or electrode system. Grounding falls into two categories: (1) system grounding, or grounding one of the current-carrying wires of the installation, and (2) equipment grounding, or grounding non-current-carrying parts of the installation, such as the service equipment[1] cabinet, the frames of ranges or motors, or the metal conduit or metal armor of armored cable.

The purpose of grounding is *safety*. If an installation is not properly grounded, it can be exceedingly dangerous in that it may cause shocks, fires, and damage to appliances and motors. Proper grounding reduces such dangers, and also minimizes danger from lightning, especially on

[1] In any installation, the service equipment will include one or more main fused switches, or one or more main circuit breakers. A variety of switches and circuit breakers are available and will be discussed later in this book. But throughout this book, unless otherwise stated, "service equipment" will refer to either fused switches or breakers, whichever are installed.

farms. Grounding is a most important subject; it is so important that many points will be repeated throughout this book, for emphasis. Study the subject thoroughly; *understand it*.

The Code rules for grounding are quite complicated, and at times appear ambiguous. However, for installations in homes and farm buildings they are relatively simple. In this chapter only the basic principles of grounding will be discussed; other chapters will discuss ground clamps, ground rods, sizes of ground wire, and similar details.

Terminology. In this book the term "the ground" will mean what the Code calls the "grounding electrode system"—the metal pipe of an underground water system bonded to a driven pipe or rod—or the "grounding electrode," a driven pipe or rod that is used where there is neither any metal underground piping nor any other grounded component available. The term "to ground" will mean connecting something to the ground.

The term "ground wire" will mean what the Code calls the "grounding electrode conductor": the wire that runs from the service equipment to the ground. In the service equipment enclosure (cabinet) the ground wire is connected to the neutral wire, thus grounding it. The metal cabinet is grounded (bonded) to the grounded neutral wire; thus the cabinet is also grounded, and so is the conduit, or the armor of armored cable, either of which are connected (bonded) to the cabinet.

The term "equipment ground*ing* wire" refers to a wire which does not carry current at all during normal operation; it is connected to non-current-carrying parts of the installation, such as the frames of motors or clothes washers, and the outlet boxes in which switches or receptacles are installed. The ground*ing* wire is in the same cable or conduit and runs with the current-carrying wires. In other words, it is connected to parts that normally do not carry current, but do carry current in case of damage to or defect in the wiring system, or the connected appliances. In the case of wiring with metal conduit, or cable with metal armor, it is not necessary in most cases to install a separate grounding wire, for the metal conduit or armor serves as the ground*ing* conductor.

Grounded Neutral Wire. In ordinary residential or farm wiring, the power comes into the premises from the power supplier's line over three wires. See Fig. 9-1, in which the wires are marked *N, A,* and *B*. Wire *N* is grounded (both at the service equipment and also at the transformer supplying the power) and is called the grounded neutral wire, and wires *A* and *B* are ungrounded and are called "ungrounded wires," "phase wires," or more usually just "hot" wires. Note that the voltage between

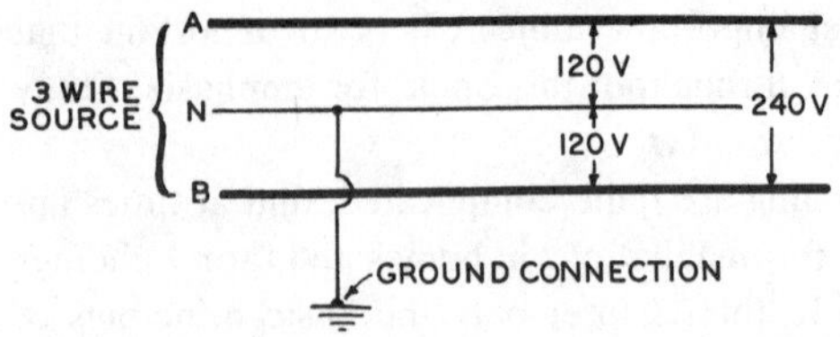

Fig. 9-1 By using only three wires, two different voltages are available. The neutral wire *N* is grounded.

the neutral *N* and either *A* or *B* is 120 volts; between the two hot wires *A* and *B* it is 240 volts.

If you touch both *A* and *B*, you will receive a 240-volt shock. If you touch either *A* or *B* and also *N*, you will receive a 120-volt shock. But note that *N* is connected to the ground; so to receive a 120-volt shock, you don't have to actually touch *N*. Touching *A* or *B* while standing on the ground is the same as touching *A* or *B* and also *N*, and you will receive the 120-volt shock.

The two hot wires may be any color except white, natural gray, green, or green with one or more yellow stripes. One is usually black, and the other red, and sometimes both are black.

Is It a Neutral Wire? Where there are three service wires, as in Fig. 9-1, the grounded wire is definitely a neutral wire. What about the wires that begin at the service equipment and run to various circuits in the building? If the circuit is a 3-wire 120/240-volt circuit, the grounded wire is definitely a neutral wire. But if the circuit is a 2-wire 120-volt circuit, the grounded wire, even if it does connect to the neutral wire in the service equipment, is *not* a neutral wire; it is called just the grounded wire. Many people do call the grounded wire in a 2-wire circuit a "neutral wire," but there can*not* be a neutral in a *2-wire* circuit. All this will be explained in more detail in Chap. 20.

White Grounded Wire. In this book, the grounded wire will be called just the "white wire" or "grounded wire" unless it is in fact a neutral wire, in which case it will be called that. The Code calls it the "identified wire" in either case. In Sec. 200-6 the Code requires a grounded wire to be identified by a white or natural gray finish throughout its length if it is No. 6 or smaller. If larger than No. 6, it must (1) have the same white or gray color, or (2) be painted white at each terminal, or (3) be taped with white or gray tape at each terminal. The Code, with the exception of switch legs in cable wiring (which will be explained in Chap. 18), requires that a white wire must never be used for any purpose other than that of the grounded wire.

The grounded wire is never interrupted by a circuit breaker, fuse, switch, or other device unless the device used is so designed that in opening the grounded wire it at the same time opens all the ungrounded wires. But such devices are not used in residential or farm wiring. So, the white wire with the exception noted always runs from its source to the equipment where the current is finally consumed, without being switched, and without overcurrent protection. However, it may be spliced when necessary, or connected to terminals. This simple fundamental requirement of wiring must at all times be kept firmly in mind.

The grounded wire and one hot wire must run to equipment operating at 120 volts. But two hot wires without a grounded white wire must run to 240-volt loads.[2] Some appliances such as ranges have both 120- and 240-volt loads, so the white wire and both hot wires must run to the appliance. A separate ground*ing* wire runs to the non-current-carrying parts of most loads (unless metal conduit or the armor of cable serves as the ground*ing* conductor). In the case of appliances such as ranges and dryers having both 120- and 240-volt loads, the grounded neutral wire under some circumstances is permitted to serve also as a ground*ing* wire. These cases will be explained in Chap. 21. The white wire, whether it is a neutral or not, is grounded at the service equipment cabinet. The power supplier also grounds the neutral service wire at the transformer serving the premises.

If the neutral wire is *properly* grounded both at the transformer and at the service equipment at the building, it follows that if you touch an exposed white wire at a terminal or splice, no harm follows, no shock, any more than if you touch a water pipe or a faucet, for the white wire and the piping are connected to each other. Any time you touch a pipe, you are in effect touching the white grounded wire.

Since the white wire is actually grounded and there is no possible danger in touching it, why put insulation on that wire? The white wire is a current-carrying wire; if it is uninsulated it will touch piping and the like, and part of the current that the wire normally carries will flow through the piping and the like, to travel all over the building and possibly start fires.

[2] Anything that is connected to a circuit and consumes power constitutes a "load" on the circuit. The load might be a lamp, a toaster, a motor—anything *consuming* power. Switches and receptacles do not consume power and therefore are not loads (but anything plugged into a receptacle is a load).

Unless otherwise stated, it is necessary to use the same kind of insulation, and the same careful splices, for the white wire as for the hot wires.

How Grounding Promotes Safety. See Fig. 9-2, which shows a 115-volt[3] motor with the white wire grounded, and a fuse in the hot wire (actually the fuse and the ground connection might be a considerable distance from the motor, although there might be an additional fuse near the motor). Assume the fuse blows (which is the same as cutting a wire at the fuse location); the motor stops. You may inspect the motor, or accidentally touch one of the wires at a terminal. What happens? Nothing! The circuit is hot only up to the fuse location. Between the fuse and the motor the wire is dead, just as if the wire had been cut at the fuse location. The other wire to the motor is the grounded wire, so it is harmless. But do note that if the fuse is *not* blown and you touch the hot wire, you will receive a 120-volt shock, for the current will flow through your body to the earth, and through the earth back to the grounded wire, and thence to its source.

Now see Fig. 9-3, which shows the same motor except with the fuse wrongly placed in the grounded wire instead of the hot wire. The motor

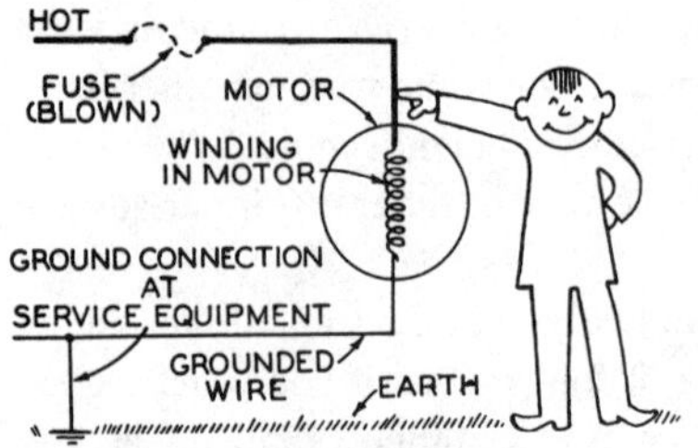

Fig. 9-2 A 115-volt motor properly installed except for a grounding wire. It is a safe installation so long as the motor remains in perfect condition.

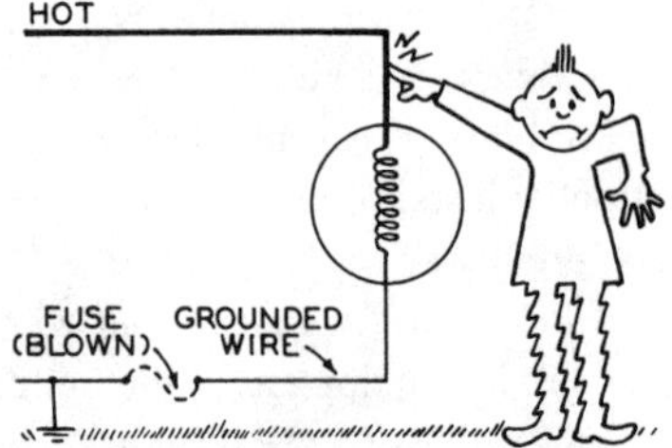

Fig. 9-3 A 115-volt motor installed with a fuse in the grounded wire. This is a dangerous installation.

will operate properly, and if the fuse blows it will stop. But the circuit is still hot, through the motor, up to the blown fuse. You touch one of the wires at the motor. What happens? You complete the circuit through your body, through the earth, to the grounded wire ahead of the fuse. You are directly connected across 120 volts; as a minimum you will receive a shock, and at worst you will be killed. The degree of shock and

[3] Instead of the usual 120 and 240 volts, the Code still classifies motors as 115- and 230-volt equipment.

danger will depend on the surface on which you are standing, on your general physical condition, and on the condition of your skin at the point of contact. If you are on an absolutely dry nonconducting surface, you will notice little shock; if you are on a damp surface as in a basement, you will experience a severe shock; if you are standing in water, you will undergo maximum shock, or electrocution.[4]

Now see Fig. 9-4, which again shows the same 115-volt motor as in Fig. 9-2. But suppose the motor is defective, so that at the point marked *G* the

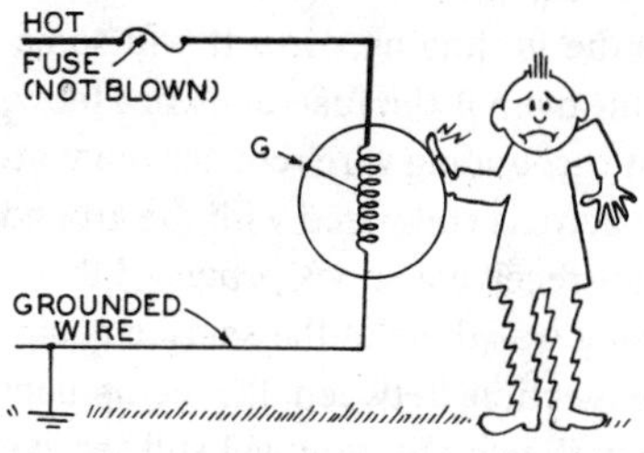

Fig. 9-4 The motor of Fig. 9-2, but the motor is defective. Its windings are accidentally grounded to the frame. It is a dangerous installation.

winding inside the motor accidentally comes into electrical contact with the frame of the motor; the winding becomes "grounded"[5] to the frame. That does not prevent the motor from operating. But suppose you touch just the frame of the motor. What happens? Depending on whether the internal ground between winding and frame is at a point nearest the grounded wire, or nearest the hot wire, you will receive a shock up to 120 volts, for you will be completing the circuit through your body back to the grounded wire. It is a potentially dangerous situation; shocks of a whole lot less than 120 volts can be fatal.

Now see Fig. 9-5, which shows the same motor, with the same accidental ground between winding (or cord) and frame, but protected by a ground*ing* wire connected to the frame of the motor and running back to

[4] Per NEC Sec. 110-16(a), concrete, brick, or tile walls are considered as grounded. Most inspectors will consider a concrete floor, even if tiled as in a basement amusement room, the same as the ground.

[5] Breakdowns in the internal insulation of a motor, so that an electrical connection develops between the winding and the frame of the motor, are not uncommon. The entire frame of the motor becomes "hot" electrically. The same situation develops if the motor is fed by a cord that becomes defective where it enters the junction box on the motor, so that one of the bare wires in the cord touches the frame. If there is no cord but the motor is fed directly by the circuit wires, a sloppy splice between the circuit wires and the wires in the junction box of the motor can lead to the same result. The frame of the motor becomes hot.

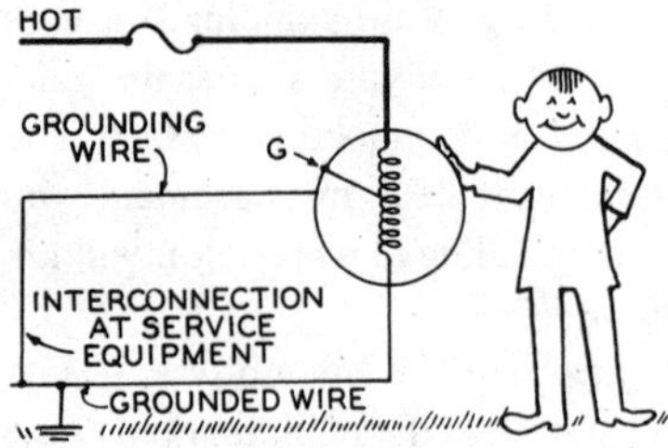

Fig. 9-5 The motor of Fig. 9-4, but a grounding wire has been installed from the frame of the motor to ground. It is a safe installation.

the ground connection at the service equipment. When the internal ground occurs, current will flow over the grounding wire. It will sometimes but not always blow the fuse. But even if the fuse does not blow, the grounding wire will protect you. The grounding wire reduces the voltage of the motor frame to substantially zero as compared with the ground on which you are standing; you will not receive a shock *provided* that a really good job of grounding and bonding was done at the service equipment. If there is a poor bonding connection between the equipment ground*ing* wire and the grounded wire in the circuit, you will still receive a shock. Actually you would *feel* the shock in any case, but it would be less dangerous where the frame is well grounded and bonded to the grounded circuit conductor, *at the service*.

Now refer to Fig. 9-6, which shows a 230-volt motor installed with a fuse in each hot wire (all hot wires must be protected by fuses or circuit breakers). Remember that in 240-volt installations, the white wire does not run to the motor, but the white wire is nevertheless grounded at the service equipment. If you touch both hot wires, you will be completing the circuit through your body, and you will receive a 240-volt shock. But if you touch only one of the hot wires, you will be completing the circuit through your body, through the earth, back to the grounded neutral at the service equipment, and you will receive a shock of only 120 volts: the same as touching the white wire and one of the hot wires of Fig. 9-1. The difference between 120- and 240-volt shocks may be the difference between life and death.

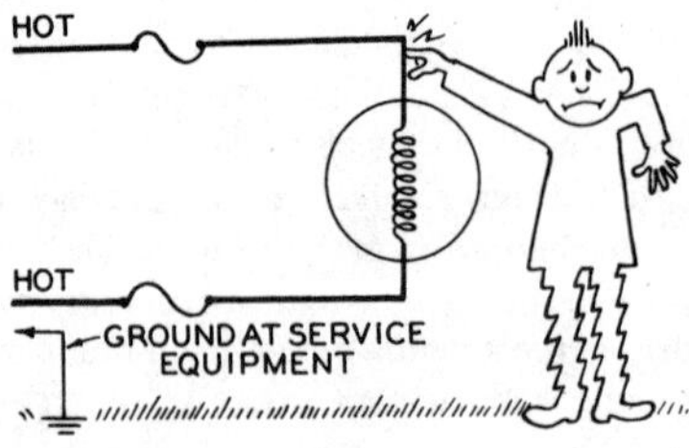

Fig. 9-6 A 230-volt motor properly installed except for a grounding wire. It is a safe installation so long as the motor remains in perfect condition.

But assume that the motor becomes defective, that the winding in the motor becomes accidentally grounded to the frame, as shown in Fig. 9-7. You will recognize this as the same as Fig. 9-4, except that the motor is supplied by 240 volts instead of 120 volts. Touching the frame will produce a 120-volt shock. But if the frame has been properly grounded, as shown in Fig. 9-8, one of the fuses will probably blow; even if it does not blow, touching the frame will not produce as dangerous a shock because the frame is grounded, again assuming that a really good grounding and bonding job was done at the service equipment.

Fig. 9-7 The motor of Fig. 9-6, but the motor is defective; its winding is accidentally grounded to the frame. It is a dangerous installation.

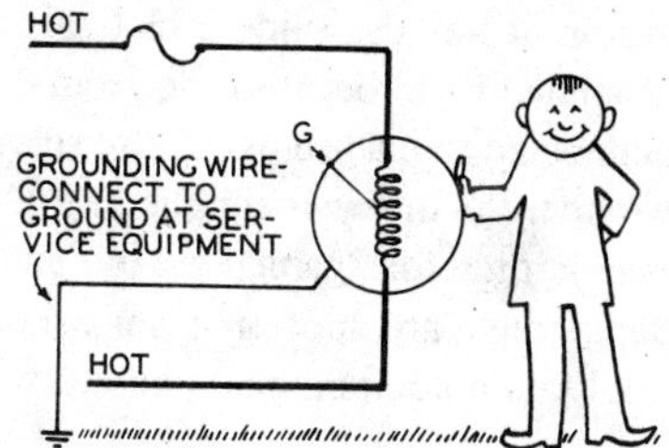

Fig. 9-8 The motor of Fig. 9-7, but a grounding wire has been installed from the frame of the motor to the service equipment, where it is bonded to the ground wire.

Green Wire. The ground*ing* wire from the frame of the motor (or from any other normally non-current-carrying component) may be green, or green with one or more yellow stripes,[6] or in many cases bare, uninsulated. If metal conduit or cable with metal armor is used, the conduit or armor serves as the grounding wire. Green wire may not be used for any purpose other than the ground*ing* wire.

Advantages of Conduit or Armored Cable. Now consider a different aspect of the situation. In any of the illustrations of Figs. 9-2 to 9-8, if there is an accidental contact between any two wires of the circuit, a fuse will blow, regardless of whether the contact is between two hot wires, or between a hot wire and the grounded wire. For this to happen, there must be bare places on two different wires touching each other, and that does not happen very often in a properly installed job. But consider a system

[6] Green with one or more yellow stripes is basically the European color scheme. Using green with yellow stripe(s) permits American appliances to be exported without modification for the European market.

in which all the wires are installed in a metal raceway such as the steel pipe called conduit, or in a cable with metal armor such as Type AC cable (often called BX, which is the trade name of one particular manufacturer). The raceway or armor is grounded at the service equipment. It is also connected to the motor itself. (If the motor is connected by a 3-wire cord and 3-prong plug, the metal conduit or cable armor is connected to the motor frame through the grounding wire in the cord.) No separate grounding wire is required inside the metal conduit or metal armor. If there is now an accidental ground from winding to frame, or if a hot wire at a defective bare spot comes in contact with the metal raceway or armor, it has the same effect as a short between the hot wire and the grounded wire, because the grounded wire and the raceway or armor are connected to each other at the service equipment. A fuse then will blow whether the motor is supplied by 120 or 240 volts. A considerable advantage is therefore gained by the use of metal raceway or armor, since a dangerous bare spot on a hot conductor might otherwise go undetected until you made personal contact with it. This is shown in Fig. 9-9.

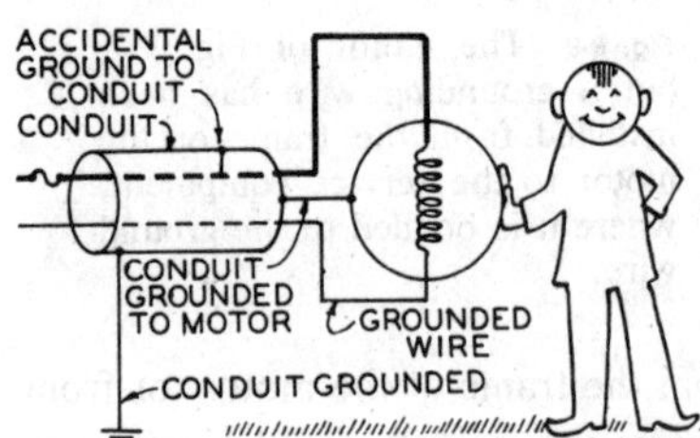

Fig. 9-9 A 115-volt motor properly installed, now using wires in metal conduit or armor. The conduit is bonded to the steel cabinet of the service equipment, which is bonded to the grounded conductor, and also to the frame of the motor. It is a safe installation.

In spite of the advantages of a metal raceway system, in some locations (especially on farms) there are conditions, which will be explained later, that make the use of a cable *without* metal armor more desirable. Then nonmetallic-sheathed cable is used, which, in addition to the usual insulated wires, has a bare equipment ground*ing* wire that runs with the insulated circuit wires.

Continuous Grounds. As will be explained in Chap. 10, at every location where there is an electrical connection, a metal[7] outlet box or switch box is used. Where a grounded neutral wiring system is used (which is just about 100% of the time for residential and farm wiring as well as for most other occupancies) and you use metal conduit or cable with metal armor, you must ground not only the white wire but also the conduit or cable

[7] Nonmetallic boxes are also used and will be discussed in Chap. 10.

armor. The white wire is grounded only at the service equipment, but the conduit or metal armor must be securely grounded to *every* metal box or cabinet. When fixtures are mounted on metal boxes they become grounded. When conduit or metal armor is connected to motors or appliances, they become grounded. This is also the case where a motor or appliance is connected by means of a flexible cord that includes a ground*ing* wire, a 3-prong plug, and a grounding receptacle, all of which will be explained later in this chapter.

It is of the utmost importance that you solidly drive down the locknuts on the conduit (or on the connectors of armored cable) so that the locknut bites down into the metal of the box. In that way you provide a continuous metallic circuit (entirely independent of any of the wires installed), which is such an important part of grounding.

If you are using nonmetallic-sheathed cable, it will contain an extra wire, a bare ground*ing* wire, which must be carried from outlet to outlet, (as will be explained in Chap. 11), providing a continuous ground. Regardless of the wiring method used, a *continuous* ground is of the utmost importance.

Other Advantages of Grounding. Suppose a 2400-volt line accidentally falls across your 120/240-volt service during a storm, as shown in Fig. 9-10. If the service is not grounded, you can easily be subject to 2400-volt

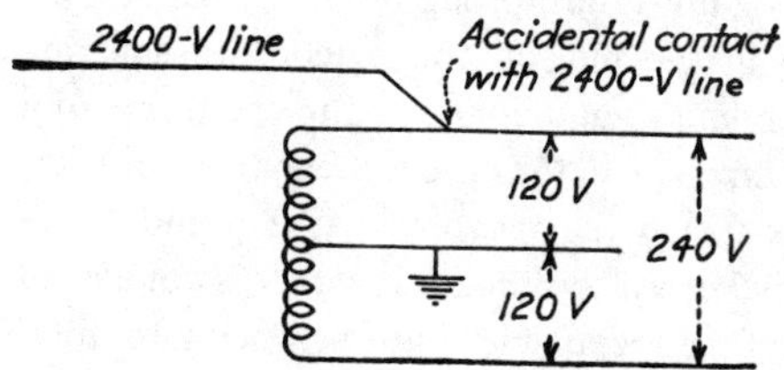

Fig. 9-10 In an ungrounded system, if a 2400-volt line falls across the 120/240-volt wires, it raises their voltage to 2400 volts; in a grounded system, the voltage will be considerably above 120/240 volts, but much less than 2400 volts.

shocks, and wiring and appliances will be ruined. If the system is *properly* grounded, the highest voltage of a shock will be much more than 240 volts, but very much less than 2400 volts.

Lightning striking on or even near a high-voltage line can cause great damage to your wiring and your appliances, and can cause fire and injuries. Proper grounding throughout the system will largely eliminate the danger.

Receptacles. Many people are used to the original receptacle having only two parallel openings for the blades of the plug, as shown in Fig. 9-11. Anything plugged into it duplicates the condition of Fig. 9-2. If the appliance is defective and the owner handles it, he or she could receive a shock, as shown in Fig. 9-4. This led to the development of what are called "grounding receptacles," shown in Fig. 9-12. Note that such a receptacle has the usual two parallel slots for two blades of a plug, plus a third round or U-shaped opening for a third prong on the corresponding plug. In use, the third prong of the plug is connected to a separate grounding wire in the cord, running to and connected to the frame of a motor or appliance. The grounding wire in the cord is green (with or without one or more yellow stripes). On the receptacle, the round or U-shaped opening leads to a special *green* terminal screw that is connected to the yoke or mounting strap of the receptacle. In turn, when a receptacle is being installed, this green terminal must be connected to ground. If conduit or cable with armor is used, the conduit or the metal armor becomes the grounding conductor. Cable without armor will contain an extra wire, bare, in addition to the insulated conductors, and this must be grounded to the boxes. In this way, the frame of the motor or appliance is effectively grounded, leading to extra safety as shown and discussed in connection with Fig. 9-5. The details of how to connect the ground*ing* wire for the green terminal will be discussed in Chap. 11.

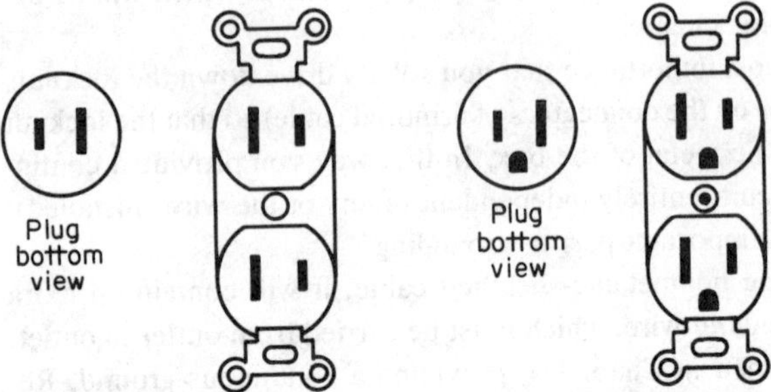

Fig. 9-11 An ordinary receptacle and plug. Only plugs with two blades will fit it.

Fig. 9-12 A grounding receptacle and plug. Plugs with either two blades or two blades plus a grounding prong will fit it.

It is not required, but it is becoming common practice to install grounding receptacles so that the opening for the grounding prong of the plug is at *the top*. If a plug is carelessly inserted into a receptacle only part way, so that part of the prongs is exposed, and if a metal object were to fall into the space between plug and receptacle, what would happen? If the grounding prong is at the top, nothing would happen. But if the "hot" prongs of the plug are at the top, the metal object also would be "hot," and if the face plate is metal a short circuit would occur.

For residential occupancies, the Code in Sec. 250-45(c) requires every *cord-connected* refrigerator, freezer, air conditioner, clothes washer, clothes dryer, dishwasher, and sump pump, and every hand-held motor-driven tool such as a drill,[8] saw, sander, hedge trimmer, lawn mower and similar items, to be equipped with a 3-wire cord and 3-prong plug. It is not required on ordinary household appliances such as toasters, irons, radios, TV sets, razors, lamps, and similar items. This might lead you to think you must install two different kinds of receptacles, one for use with 2-prong plugs, and others for use with 3-prong plugs. Not so: *The grounding receptacles are so designed that either plug will fit.*

Section 210-7(a) of the Code requires all receptacles on 15- and 20-amp circuits, at 120 volts, to be of the grounding type, for all *new* installations. If you replace a 2-wire receptacle, the new one must be of the grounding type, except that if a grounding means does not exist at the box, either a nongrounding or a ground-fault circuit-interrupter (GFCI) receptacle must be used.

Three-to-Two Adapters. The question will naturally arise: What if you have something with a 3-prong plug but have only ordinary receptacles that accept only 2-prong plugs? The best procedure would be to replace the old receptacles with new grounding receptacles, if you can ground them properly.

Although replacing receptacles is the preferred procedure, you may also use a "three-to-two" adapter as shown in Fig. 9-13. Note the adapter has a green ground*ing* lug on it, with a hole in it. Remove the screw that holds the face plate to your receptacle, but do not remove the plate. Plug the adapter into the receptacle with the grounding lug to-

[8] In the case of ordinary portable tools, the Code does not require grounding if the tool is constructed with what is called *double insulation*. Ordinary tools have only what is known as functional insulation, only as much as is required for proper functioning of the tool; this includes the insulation on the wires used to wind the motor, an insulated switch, and so on. Double insulation goes much farther: Usually the case of the tool is made of insulating material rather than metal; the shaft is insulated from the steel lamination of the rotor of the motor; switches have handles of an especially tough insulating material; there are many other points of specially durable insulation.

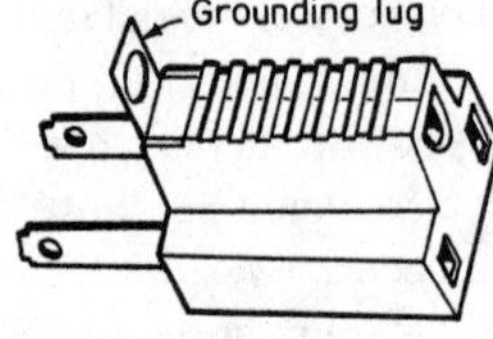

Fig. 9-13 This adapter permits a 3-prong grounding plug to be used in an ordinary 2-slot receptacle. The use of the adapter, however, does not provide the same protection as would be had if a grounding receptacle had been properly installed.

ward the center of the plate. Reinstall the screw that holds the face plate, through the hole in the grounding lug. Then you can plug either 2-prong or 3-prong plugs into the adapter. But be warned: Unless the existing circuit provides a grounding connection to the screw in the mounting yoke of the receptacle, you will *not* have grounding protection when the adapter is used.

How Dangerous Are Shocks? Most people think it is a high voltage that causes fatal shocks. This is not necessarily so. The amount of current flowing through the body determines the effect of a shock. A milliampere (mA) is one-thousandth of an ampere. A current of 1 mA through the body is just barely perceptible. One to eight milliamperes causes mild to strong surprise. Currents from 8 to 15 mA are unpleasant, but usually victims are able to free themselves, to let go of the object that is causing the shock. Currents over 15 mA are likely to lead to "muscular freeze," which prevents the victim from letting go and often is fatal. Currents over 75 mA are likely to be fatal. Much depends on the individual involved.

Of course the higher the voltage, the higher the number of milliamperes that would flow through the body, under any given set of circumstances. We must distinguish between shocks resulting from touching *two* hot wires, and those resulting from touching *one* hot wire. In the latter case, a shock from a relatively high voltage while the victim is standing on a completely dry nonconducting surface will result in fewer milliamperes than a shock from a much lower voltage while standing in water. Many deaths have been caused by shock on circuits considerably below 120 volts; many people have survived shock from circuits of 600 volts and more.

Another determining factor in danger is the number of milliamperes a source of power can produce. When you walk on the dry carpeting of a house in the winter when the humidity is low, you pick up a static charge, so that you feel a mild shock when you touch a grounded object. The voltage is high, but the current is infinitesimal. If you touch the spark plug of a car while the motor is running, you receive a shock of at least 20 000 volts, but the current is exceedingly small, so no harm is done. The same principle applies to "electric fences" used on farms: a voltage high enough to be uncomfortable, but a very small current, for a very short period of time, that does no harm.

It should be noted that farm animals are much less able to withstand shocks than are human beings. Many cattle have been killed by shocks that would be only uncomfortable to people.

Voltage to Ground. This term is often used in the Code. If one of the current-carrying wires in the circuit is grounded, the voltage *to ground* is the maximum voltage that exists between that grounded wire and any hot wire in the circuit. If no wire is grounded, then the voltage to ground is the maximum voltage that exists between *any* two wires in the circuit (Code, Art. 100).

Faults and Fault Currents. A fault is the accidental breakdown of insulation between two or more hot wires, or between a hot wire and the grounded or the ground*ing* wire, or between a hot wire and metal conduit or armor of cable, or the frame of equipment. If the breakdown is between two hot wires it is called a short circuit, and the current that flows as a result is called "short-circuit current." If the breakdown is between a hot wire and the grounded wire it is often called a "ground fault," but in effect it is still a short circuit, and the current that flows as a result is usually called "fault current" but many call it "short-circuit current."

If the breakdown is between a hot wire and grounded conduit or cable armor, or a ground*ing* wire, or between the hot wire and the frame of a tool, motor, appliance, or other equipment, it is known as a *ground* fault, and the current that flows as a result is called "ground-fault current"; many call it just "fault current." When there is a short circuit between two hot wires, or between a hot wire and the grounded wire, the fault current is usually high enough to blow a fuse or trip a breaker. This is also true where a *low-resistance* accidental contact is made between a hot wire and properly installed conduit, cable armor, or a ground*ing* wire.

But a *high-resistance* or arcing contact between a hot wire and such non-current-carrying equipment, including the frame of a motor, tool, or other plug-in equipment, could, and often does, result in a fault current too small to blow a fuse or trip a breaker. Yet the current could be high enough to cause a fire. It would very definitely constitute a shock hazard, for anybody holding or touching such a tool and also standing on the ground or a grounded surface would have the full voltage of the circuit across his or her body, as shown in Fig. 9-4. Such shocks could be, and often are, fatal.

While the use of 3-prong plugs on 3-wire cords with properly installed grounding receptacles does substantially eliminate or greatly reduce the danger of shocks, there are millions of poorly or improperly grounded tools and equipment that are unsafe to use. This includes equipment with 2-wire cords and 2-prong plugs; there can be defects in 3-wire cords and 3-prong plugs, especially where a worn cord has been replaced. Some people even stupidly cut off the third grounding prong on a 3-prong plug because they have only ordinary receptacles that accept only 2-prong plugs.

Ground-Fault Circuit-Interrupters. Under normal conditions the current in the hot wire and that in the grounded wire are absolutely identical. But if the wiring, tool, or appliance is defective and allows some current to leak to ground, then a ground-fault circuit-interrupter (abbreviated GFCI) will sense the difference in current in the two wires. If the fault current exceeds the trip level of the GFCI, which is between 4 and 6 mA, the GFCI will disconnect the circuit in as little as $^1/_{40}$ s.

The fault current, much too low to trip a normal breaker or blow a fuse, could possibly flow through a person in contact with the faulty equipment and a grounded surface. The use of a GFCI is a most desirable safety precaution, especially when using electrical equipment outdoors, for standing on the ground (especially if it is wet) greatly increases the likelihood and especially the severity of a shock. The GFCI you install must be rated in amperes and volts to match the rating of the outlet or circuit it is to protect.

The GFCI is not an inexpensive device, but it should be considered insurance against dangerous shocks. It is not to be considered as a substitute for grounding, but as supplementary protection which senses leakage currents too small to operate ordinary branch-circuit fuses or circuit breakers. The GFCI will not prevent a person who is part of a ground-fault circuit from receiving a shock, but it will open the circuit so quickly that the shock will be below levels which would inhibit breathing or heart action, or the ability to "let go" of the circuit.

Three types of GFCI are available. The *first* or separately enclosed type is available for 120-volt 2-wire, and 120/240-volt 3-wire circuits up to 30 amp. This type is most often used in swimming pool wiring, installed at any convenient point in the circuit. The *second* type combines a 15-, 20-, 25-, or 30-amp circuit breaker and a GFCI in the same plastic case, is installed in place of an ordinary breaker in your panelboard, and is available in 120-volt 2-wire, or 120/240-volt 3-wire types (which may also be used to protect a 2-wire 240-volt circuit). It provides protection against ground faults *and overloads* for all outlets on the circuit. Each GFCI circuit breaker has a white pigtail which you must connect to the grounded (neutral) busbar of your panelboard. You must also connect the white (grounded) wire of the circuit to a terminal provided for it on the breaker. A plug-in GFCI circuit breaker is shown in Fig. 9-14. The *third* type, a receptacle and a GFCI in the same housing, provides only ground-fault protection to the equipment plugged into that receptacle or, if this GFCI is of the "feed-through" type, to equipment plugged into other ordinary

Fig. 9-14 GFCI circuit breaker, 120-V, 1-pole, 20-A. The white pigtail connects to the neutral bar in the panelboard. Note the test button for periodic testing by simulation of a ground fault. (*Square D Co.*)

receptacles installed ''downstream'' on the same circuit. When you desire GFCI protection on an existing circuit or outlet or when such protection is required, it is a simple matter to replace existing circuit breakers or receptacles with GFCIs.

Regardless of the type or brand of GFCI you install, it is most important that you very carefully follow the installation and periodic testing instructions that come with the unit. Every GFCI has a test button for easy verification of its functional operation.

Even where wiring, tools, and appliances are in perfect condition and there is no ground fault, be on the lookout for these installation problems which will cause tripping of a GFCI: a 2-wire type (other than an end-of-run GFCI receptacle) is connected in a 3-wire circuit; the white circuit conductor is grounded on the load side of the GFCI; or, the protected portion of the circuit is excessively long (250 ft, as a rule of thumb; longer circuits may develop a capacitive leakage to ground). Of course, the GFCI is *designed* to trip if the cords or tools supplied are in poor repair and provide a path for current to leak to ground.

Remember: the GFCI will not prevent shock, but it will render shocks relatively harmless; it will *not,* however, protect a person against contact with both conductors of the circuit at the same time, unless there is also a current path to ground. The GFCI may be pictured as an electronic adding machine which constantly monitors the current out to a load and back again. The GFCI acts to quickly disconnect the circuit only when the current out to the load and the current returning differ by 0.005 amp or more.

The Code requires GFCI protection as follows:

1. In dwellings, on all 15- and 20-amp outdoor receptacles where there is direct grade-level access, and on all bathroom receptacles. In garages the *required* receptacles [NEC Sec. 210-52(f)] must be GFCI-protected, but receptacles dedicated to serve specific stationary appliances such as clothes washers, freezers, or garage-door openers need not be so protected. The GFCI may be installed to protect individual receptacles or an entire branch circuit (Sec. 210-8).

2. On construction sites, on all 15- and 20-amp 120-volt single-phase receptacles that are not a part of the permanent wiring of the building or structure. Here again the GFCI may be installed to protect either individual receptacles or the entire circuit supplying the receptacles. GFCI protection may be omitted where the grounding integrity of construction site receptacles, cords, and portable tools is tested at specified intervals in ac-

cordance with a written procedure that is acceptable to the inspector (Sec. 305-4).

3. Swimming pools: There are very stringent requirements that will be discussed separately in Chap. 20 (Art. 680).

Grounding of Fixed Equipment. In this chapter we have discussed the equipment gounding wire running to a motor. But many other items must have an equipment grounding wire (unless the wiring is in rigid metal conduit or if EMT or armored cable is used—any of which will serve the purpose of a separate equipment grounding wire). In general, follow the requirements of NEC Sec. 250-42, which specifies that exposed non-current-carrying metal parts of fixed equipment likely to become energized shall be grounded where within 8 ft vertically or 5 ft horizontally from a grounded surface, where in a wet or damp location, where in contact with metal, where in a hazardous location, where supplied by a metallic wiring method, or where the equipment operates with any terminal at over 150 volts to ground.

Do note that this equipment includes metal face plates. If installed on metal switch boxes, they are properly grounded if the boxes are properly grounded. But if a *metal* face plate were to be installed on a *nonmetallic* box, it would have to be grounded. Also note that toggle switches with a green grounding screw terminal are available, as mentioned in Chap. 11.

System Grounding. The details of how to ground the entire wiring system will be discussed in Chap. 17.

10

Outlet and Switch Boxes

In the interest of safety from both fire and shock hazards, switch and outlet boxes or equivalent enclosures must be installed at every point where wires are spliced or connected to terminals of electrical equipment.

Purpose of Boxes. Boxes house the ends or splices in wires at all points where the raceway (or cable armor, shield, or jacket) has been terminated or interrupted, and where the insulation of the wire has been removed. A continuous piece of wire properly protected against physical damage is substantially safe, but a poorly made splice or terminal connection may lead to short circuits, grounds, or overheating. Dust, cobwebs, many kinds of thermal insulation or its wrapping, many types of construction or finishing materials in the hollow spaces of walls and ceilings, all can be readily ignited by electric arcs, which are hotter than gas flames. The same can be said about furnishings and stored materials. Therefore there is some danger of fire at splices and terminals, but when these are enclosed in a box, this danger is practically eliminated. Moreover, metal boxes provide a continuity of ground for a metal wiring system, as explained in the preceding chapter. The Code requires that boxes be supported according to definite standards that provide mechanical strength for supporting lighting fixtures, switches, and other equipment, eliminating the danger of mechanical breakdown.

Knockouts. Around the sides and in the backs of boxes there are "knockouts"—sections of metal that can easily be knocked out to form openings for conduit or cable to enter. The metal is completely severed around these sections except at one small point that serves to anchor the metal until it is to be removed. It is a simple matter to remove these knockouts—usually a stiff blow with a pair of pliers on the end of a heavy screwdriver held against the knockout will start it, and with a pair of pliers the metal disk is then easily removed. If the knockout is near the edge of a box, a pair of pliers is the only tool needed.

On some brands of boxes, the pry-out type of knockout is used. The pry-out type has a small slot near or in the knockout; use a screwdriver to pry out the disk. Concentric knockouts will be discussed in Chap. 13.

Remove only as many knockouts as are actually needed for proper installation. If you accidentally remove a knockout that won't be used, you must close the opening with a knockout seal. The most common type, shown at *A* of Fig. 10-1, is a metal disk with tension clips around the edge

Fig. 10-1 Any unused opening must be sealed with a knockout closure. (*RACO, Inc.*)

on one side. Push the clips into the unused opening with your thumb, then tap the disk in with a pair of pliers or any other suitable tool. Another type is shown at *B* of the same Fig. 10-1; this consists of two disks, one for the inside and the other for the outside of the box. A screw through the disks holds the seal in place.

Types of Boxes. This chapter will discuss both metal and nonmetallic boxes. The Code in Sec. 370-14 recognizes boxes $^{1}/_{2}$ in deep, but these may be used only for mounting fixtures where the fixture canopy provides additional space. A box enclosing a flush device must have an internal depth of not less than $^{15}/_{16}$ in.

The kind and size of the box you are going to use will depend on the purpose of the box and, more importantly, on the number of wires that will enter the box; all of these will be discussed in this chapter.

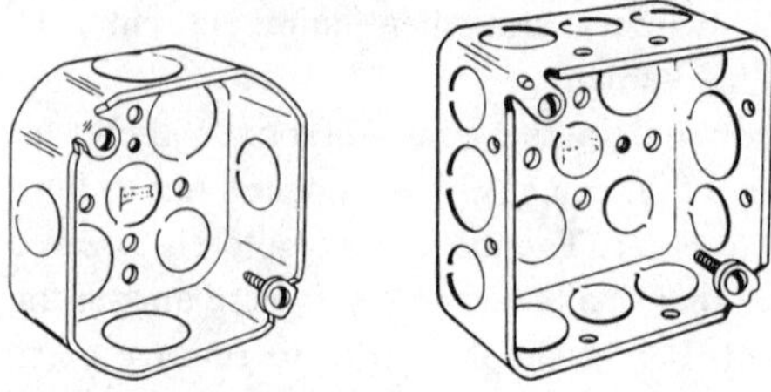

Fig. 10-2 Typical octagonal and square outlet boxes. (*RACO, Inc.*)

Steel Outlet Boxes. Figure 10-2 shows the common $1^1/_2$-in-deep octagonal outlet box, made in $3^1/_4$-, $3^1/_2$-, and 4-in sizes, and used mostly at lighting outlets. Figure 10-2 also shows the roomier square box, made in both 4- and $4^{11}/_{16}$-in sizes, and in $1^1/_2$- and $2^1/_8$-in depths. The square box can be used for either one or two switches or similar devices.

Sometimes it is necessary to install three or more switches or similar devices side by side. Boxes similar to square boxes are made in wider sizes to permit this. Figure 10-3 shows a 4-in box with two switches, and also a wider 4-gang box and cover to permit the mounting of four devices.

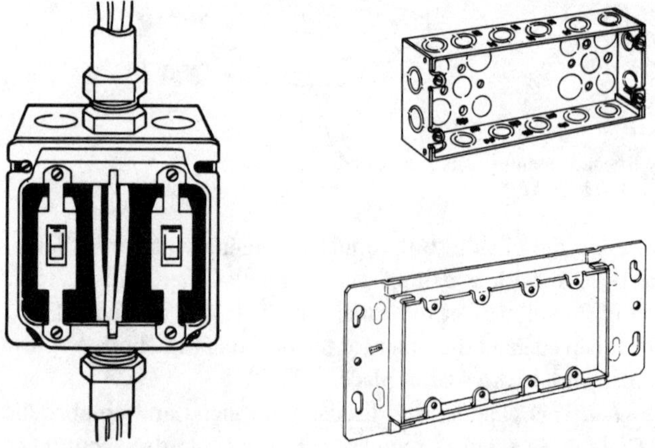

Fig. 10-3 A square box with two switches, and a 4-gang box and cover for it. (*RACO, Inc.*)

Nonmetallic Outlet Boxes. Figure 10-4 shows a representative assortment of nonmetallic boxes. These boxes may be made of thermosetting or thermoplastic materials, and they are used mainly with nonmetallic-sheathed cable. Accessories available for nonmetallic boxes include grounding terminals, cable clamps, and box supports for use in old work

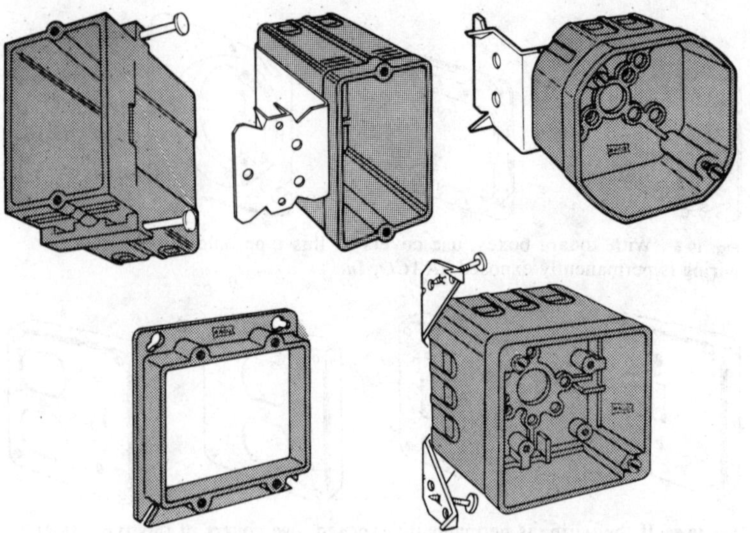

Fig. 10-4 Nonmetallic boxes and 2-gang plaster ring. (*RACO, Inc.*)

(Chap. 22). The grounding terminals serve to terminate (or to tap, if the cable feeds through the box) the grounding conductor in nonmetallic-sheathed cable to a terminal that also receives one of the mounting screws for a fixture or fixture strap, thus automatically grounding the fixture. In new work, nonmetallic-sheathed cable need not be secured to a wall-mounted single-gang nonmetallic box, provided the cable is supported within 8 in of the box, measured along the cable sheath, and the sheath projects into the box at least $^1/_4$ in. Thus, single-gang boxes normally do not have cable clamps. For use in old work, however, such boxes are available with cable clamps, and with box supports similar to the one shown in Fig. 22-9.

Covers. All boxes must be covered; a fixture mounted on a box and completely concealing it is considered a cover. For square boxes, use covers of the type shown in Fig. 10-5 if the boxes are installed in walls or ceilings; the fronts of the covers will be flush with the wall surface. Use covers of the type shown in Fig. 10-6 if the boxes are installed on the surface of a wall, in permanently exposed wiring.

With octagonal boxes, use the covers shown in Fig. 10-7. At *A* is shown a blank cover used only on pull or junction boxes. At *B* is shown a drop-cord cover used with pendant lights or fixtures; the hole is bushed to do away with sharp edges that might otherwise injure the insulation on the cord. At *C* is shown a "spider cover" sometimes used to mount sur-

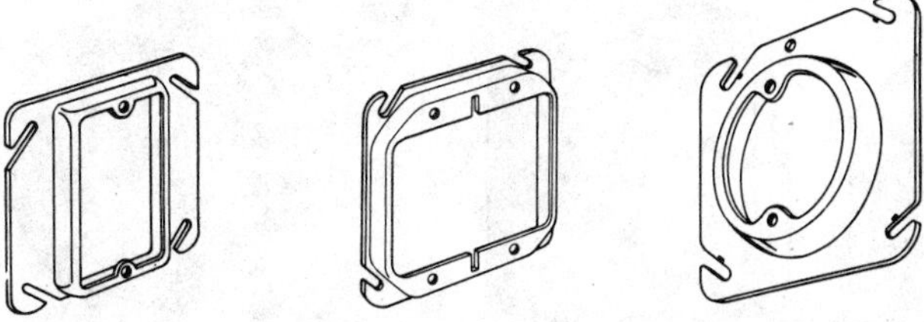

Fig. 10-5 With square boxes, use covers of this type unless the wiring is permanently exposed. (*RACO, Inc.*)

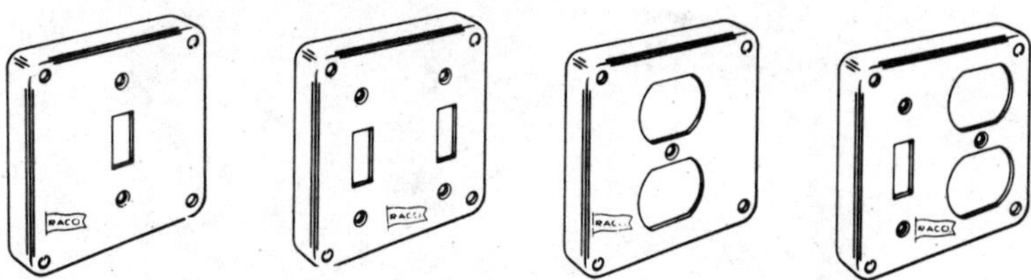

Fig. 10-6 If the wiring is permanently exposed, use covers of this type. (*RACO, Inc.*)

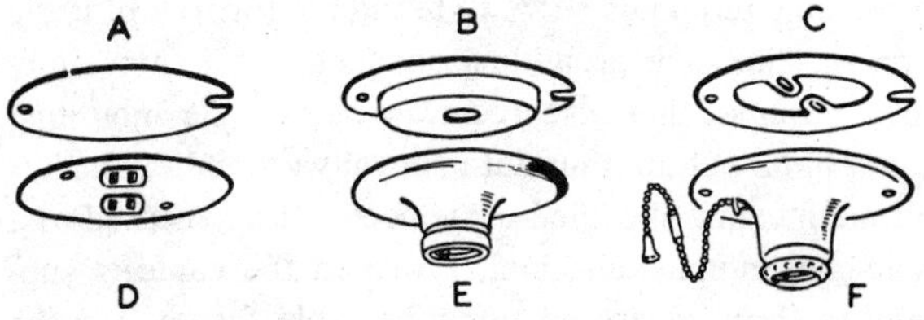

Fig. 10-7 Covers for octagonal boxes are made in dozens of types.

face-type switches. At *D* is shown a cover with a duplex receptacle used mostly in basements, workshops, and similar locations. At *E* and *F* are shown lampholders, one unswitched and the other provided with a pull-chain switch, that really are inexpensive lighting fixtures used in closets, attics, basements, farm buildings, and similar locations. There are many other types of covers.

Figure 10-8 shows how box, cover, device, and final face plate all fit together. The distances between mounting holes have been well standardized.

Switch Boxes. Boxes known as switch boxes are commonly used for switches, receptacles, and similar devices. See Fig. 10-9, which shows the ordinary switch box; note that the sides are removable. This makes it easy to make a double-size, or "2-gang," box out of two single boxes as

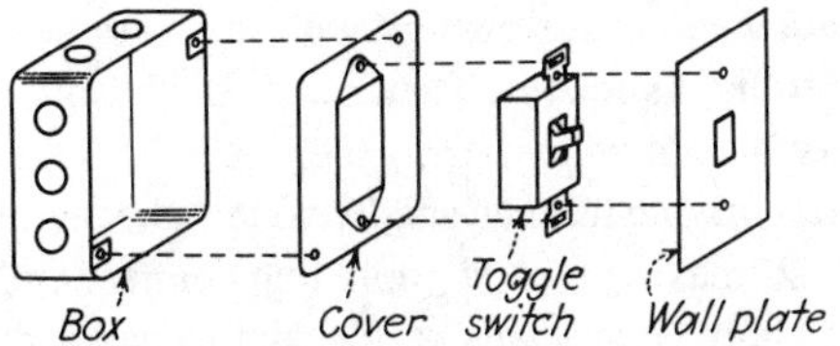

Fig. 10-8 All parts of an electrical outlet are standardized in size so as to fit properly and easily.

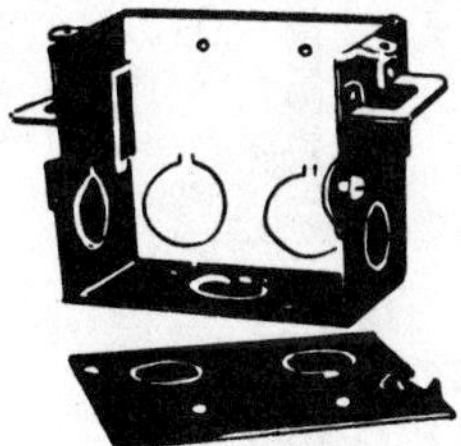

Fig. 10-9 A typical single-gang switch box. The sides are removable.

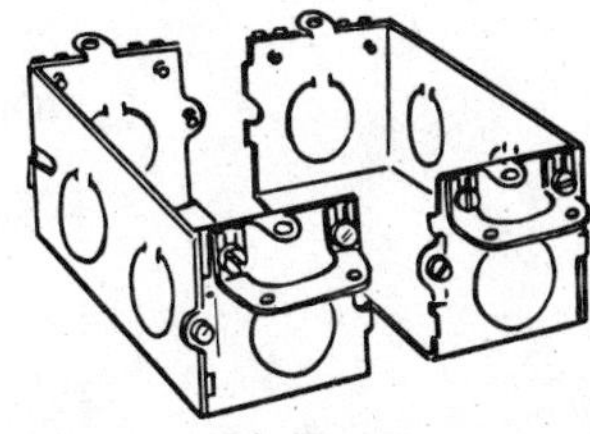

Fig. 10-10 Two single-gang boxes joined to form one 2-gang box. Any number of boxes may be ganged to form a box of any necessary size.

shown in Fig. 10-10. Simply throw away one side of each box and then join the boxes together as shown in the picture. No additional parts are needed. In similar fashion it is a simple matter to join several boxes together to mount three or more devices. In most cases, however, it is better to use a single box made for the purpose, since this box is sturdier and easier to support.

Switch boxes range in depth from $1^1/_2$ to $3^1/_2$ in. Use the shallower ones only when not more than two or three wires are to enter the box.

Securing Conduit and Cables to Boxes. To provide a good, safe continuous ground throughout an installation, it is absolutely necessary that every length of metal conduit or armored cable entering a box be firmly and solidly anchored to the box. In the case of conduit, this is done by means of a locknut and bushing, both shown in Fig. 10-11. Note that the locknut is not flat, but is dished so that the lugs around the circumference are actually teeth that will dig into the metal surface of the box. In the case of bushings, the smallest internal diameter is slightly less than the internal diameter of the conduit. This causes the wires, where they emerge from the conduit, to rest on the rounded surface of the bushing; this prevents damage to the insulation of the wires that might occur if the wires were to rest against the sharp edge of the cut end of the conduit.

Screw a locknut on the threaded end of the conduit with the teeth facing the box. Slip the conduit into the knockout. Then screw the bushing on the conduit inside the box, as far as it will go. Only then tighten up the locknut on the outside of the box, driving it home solidly so that the teeth will bite into the metal of the box, making a good grounding connection. All this is shown in Fig. 10-12. Many electricians prefer, and some local

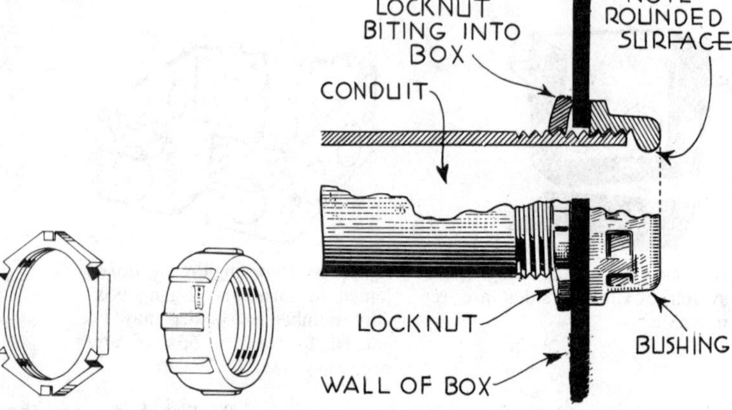

Fig. 10-11 Locknuts and bushings are used at all ends of conduit. They anchor the conduit to the box and also provide a continuous grounded raceway. (*RACO, Inc.*)

Fig. 10-12 Cross section showing how a locknut and bushing are used to anchor conduit to any box or cabinet.

codes require, that two locknuts be used, one on each side of the box wall, plus a bushing. The Code requires double locknuts and no concentric or eccentric knockouts where circuits are over 250 volts to ground. Detailed instructions for cutting and using conduit will be given in Chap. 11.

Where the wires are No. 4 or larger, the bushing must be of the insulating type, shown in Fig. 10-13. Such bushings may be made of metal with an insulated throat. Bushings made entirely of insulating material are also available, but if these are used a locknut must also be installed between the box and the bushing.

In the case of cable, use connectors of the type shown in Fig. 10-14. After the connector is solidly anchored to the cable, slip the connector into a knockout, and then install the locknut inside the box, running it home tightly as in the case of conduit, so that the teeth bite into the metal

Fig. 10-13 Insulated bushings are used with larger sizes of conduit. (*RACO, Inc.*)

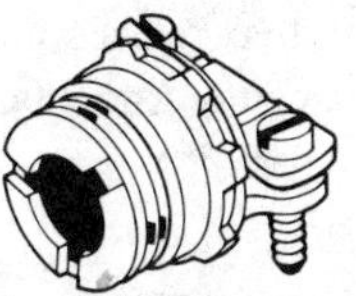

Fig. 10-14 Cables are anchored to boxes with connectors of this general type.

of the box. The shoulder of the connector acts as a locknut on the outside of the box.

Clamps. Boxes having built-in clamps for cable, which eliminate the need for separate cable connectors, are commonly used. Typical boxes of this kind are shown in Fig. 10-15. Some clamps are suitable for armored cable, others for nonmetallic-sheathed cable. The illustration should be self-explanatory; the ends of the wires would of course be much longer than shown.

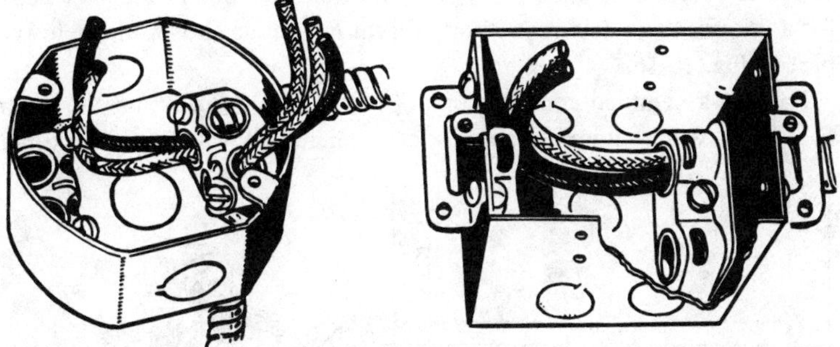

Fig. 10-15 Boxes are available with clamps for cable, making the use of separate connectors unnecessary.

Boxes in Ceilings or Walls. If the wall or ceiling is combustible (if it can burn), the front edge of the box (or the cover installed on the box) must be flush with the surface. If the wall material is noncombustible, the front edge may be as much as $^1/_4$ in below the surface, but it is better to install boxes flush in any case. You will have to find out the thickness of the ceiling or wall materials from the builder. If the box is installed in an unfinished area that may later be finished, mount the box so that its front edge will be flush with the future wall or ceiling.

Supporting Outlet Boxes. The usual method of supporting a ceiling box for a lighting fixture is by means of a bar hanger shown in Fig. 10-16. Such hangers consist of two telescoping parts, and thus are adjustable in length. Remove the center knockout in the back of a box, slip the fixture

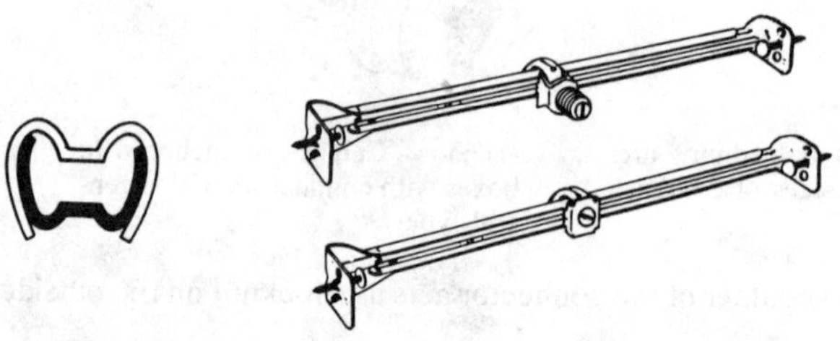

Fig. 10-16 Adjustable hangers for supporting outlet boxes. (*RACO, Inc.*)

stud that is part of the hanger into the opening, as shown in the upper part of the illustration, and then install the locknut on the inside of the box, after having adjusted the length of the hanger to fit between two joists. Then nail the hanger to the joists as shown in Fig. 10-17. The fixture stud can later be used to support a lighting fixture. If the box is not to be used for a fixture, use a hanger without a fixture stud, as shown in the lower part of Fig. 10-16.

Factory-assembled combinations of boxes with hangers, as shown in Fig. 10-18, will be found to be most convenient and will save much time.

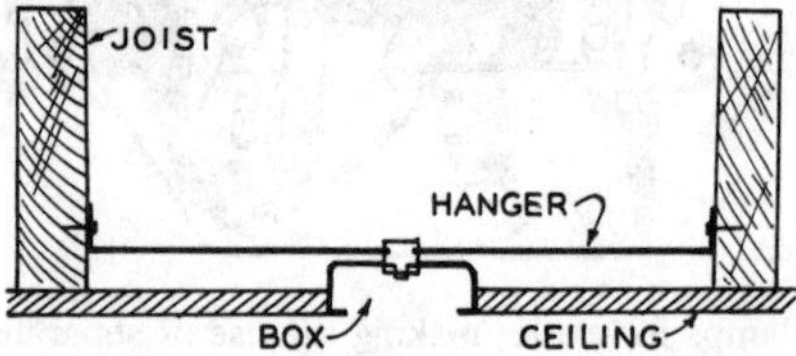

Fig. 10-17 This shows how a hanger is used to support a box in a ceiling.

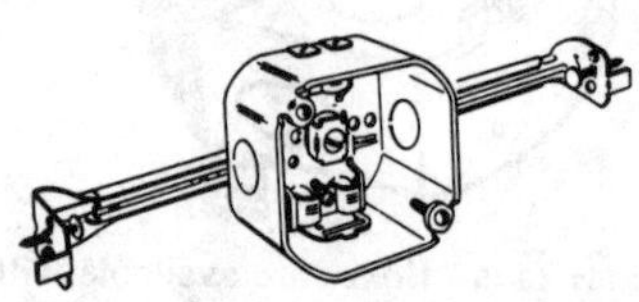

Fig. 10-18 Preassembled boxes with hangers save time in installation. (*RACO, Inc.*)

While boxes may be installed in walls in the same way as in ceilings, it will usually save much time to use boxes with mounting brackets, several types of which are shown in Fig. 10-19. Simply nail the brackets to the studs and the installation is finished. Naturally the same bracket boxes may be used in ceilings, if a location next to a joist is suitable.

Supporting Switch Boxes. Ordinary switch boxes as shown in Fig. 10-9 can be nailed directly to studs, at the proper depth to be flush with the final surface. Some boxes of this type have nails already installed in the boxes, staked so that they can't fall out. (Where mounting nails pass

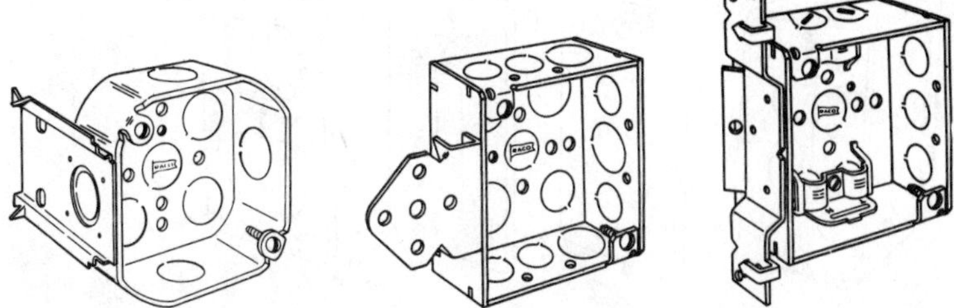

Fig. 10-19 Outlet boxes with mounting brackets. Nail the brackets to studs. (*RACO, Inc.*)

through the box interior, they must be within 1/4 in of the back or ends of the box.) But it saves a great deal of time to use boxes with mounting brackets as shown in Fig. 10-20. The brackets can be nailed directly to the studs. Be sure to select boxes with the brackets at the proper location for the thickness of the particular wall materials that will be used.

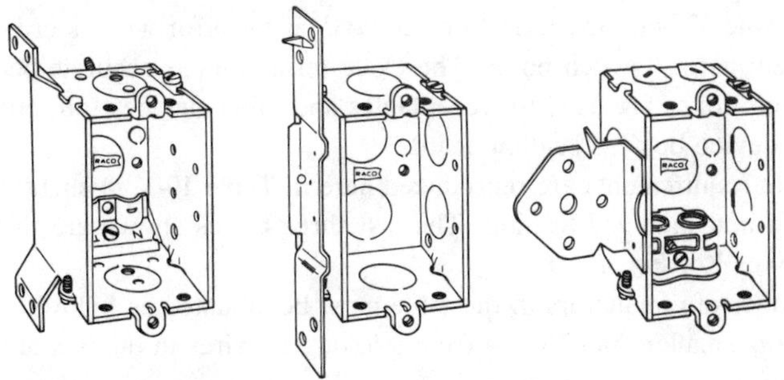

Fig. 10-20 Switch boxes, too, are available with mounting brackets. (*RACO, Inc.*)

Surface Boxes. When the wiring is on the surface of a wall so that it will be permanently exposed, ordinary boxes are impractical because of the sharp corners, both on the boxes and on covers for switches or similar devices. Use a special box known as a utility box or "handy box"; both box and covers are shown in Fig. 10-21. Square boxes and appropriate covers, as shown in Fig. 10-6, are also often used for surface wiring.

Other Boxes. Instead of wooden studs in buildings, metal studs and door bucks are becoming popular. Special brackets are available for use in connection with standard boxes to mount them to such metal structural members without requiring drilling or tapping of metal members.

Number of Wires in Box. The Code in Sec. 370-6 and Table 370-6(a) limits

Fig. 10-21 The "handy box" and covers shown are used for surface wiring.

the number of wires that may enter a box, depending on the cubic-inch capacity of the box, and the sizes of the wires. Besides being unsafe because of possible damage to the insulation, crowding a box with wires makes it difficult to install devices in the box. Shorts and grounds may develop as a result of forcing a switch or receptacle into a box already crowded with wires. Use a box that is of adequate size.

In NEC Table 370-6(a) the term "device box" is used for what is ordinarily called just a "switch box." The Code term is more accurate because switch boxes are used for receptacles and other devices too, but the term "device box" is seldom heard.

The Code requirements are reproduced here in Table 10-1, in slightly abbreviated and modified fashion. The last three boxes shown are the "handy boxes" of Fig. 10-21.

The number of conductors in the table must be adjusted as follows:

1. Wires smaller than No. 14 from a fixture to wires in the box are not counted.

2. The number of conductors in the table must be reduced by one for each fixture stud, one for each hickey, and one for one or more cable clamps (as in Fig. 10-15). Connectors external to the box, of the type shown in Fig. 10-14, are *not* cable clamps for this purpose.

3. Deduct one from the numbers in the tables for each switch, receptacle, or similar device, or a combination of them if mounted on a single strap.

4. A wire originating in a box and ending in the same box (for example, the wire from the green terminal of a receptacle, grounded to the box) is not counted.

5. If one *or more* equipment grounding wires enter the box, deduct one from the numbers in the tables. Do not deduct one for *each* such wire; deduct one regardless of the number of such wires.

TABLE 10-1 Number of Wires Entering Boxes*

Box dimensions, in. trade size, or type	Min. cubic inch cap.	Maximum number of conductors			
		No. 14	No. 12	No. 10	No. 8
4 × 1¼ round or octagonal	12.5	6	5	5	4
4 × 1½ round or octagonal	15.5	7	6	6	5
4 × 2⅛ round or octagonal	21.5	10	9	8	7
4 × 1¼ square	18.0	9	8	7	6
4 × 1½ square	21.0	10	9	8	7
4 × 2⅛ square	30.3	15	13	12	10
4 11/16 × 1¼ square	25.5	12	11	10	8
4 11/16 × 1½ square	29.5	14	13	11	9
4 11/16 × 2⅛ square	42.0	21	18	16	14
3 × 2 × 1½ device	7.5	3	3	3	2
3 × 2 × 2 device	10.0	5	4	4	3
3 × 2 × 2¼ device	10.5	5	4	4	3
3 × 2 × 2½ device	12.5	6	5	5	4
3 × 2 × 2¾ device	14.0	7	6	5	4
3 × 2 × 3½ device	18.0	9	8	7	6
4 × 2⅛ × 1½ device	10.3	5	4	4	3
4 × 2⅛ × 1⅞ device	13.0	6	5	5	4
4 × 2⅛ × 2⅛ device	14.5	7	6	5	4

*Reprinted with permission from NFPA 70-1984, National Electrical Code®, Copyright© 1983, National Fire Protection Association, Quincy, Massachusetts 02269. This reprinted material is not the complete and official position of the NFPA on the referenced subject, which is represented only by the standard in its entirety.

6. A wire running through a box without interruption (without splice or tap), as is often the case in conduit wiring, is counted as only one wire.

Boxes not listed in the table, and all nonmetallic boxes, are marked with their cubic-inch capacity. For these boxes, and where different wire sizes enter the same box, follow NEC Sec. 370-6(b). The essence of this section is that each wire must have a specific number of cubic inches within the box, as follows:

No. 14	2 cu in
No. 12	2.25 cu in
No. 10	2.5 cu in
No. 8	3 cu in
No. 6	5 cu in

Factors 1 through 6 apply, as they do for boxes in the table.

11 Different Wiring Methods

The Code recognizes many different wiring methods. Some of these are used only in large buildings and will not be discussed here. Except for outdoor overhead spans, the most commonly used wiring methods use conductors installed in some form of raceway, or assembled into some type of cable.

Article 100 of the Code defines a raceway as "an enclosed channel designed expressly for holding wires, cables or busbars, with additional functions as permitted in this Code."* Although this Code definition does not so state, raceways are installed so that wires can be placed in them or be removed after installation; the metal armor of armored cable, while containing wires, does not permit removal of the wires and therefore is not a raceway. Raceways can be made of metal, such as rigid metal conduit, electrical metallic tubing, surface raceways, and many others, or of nonmetallic materials such as nonmetallic conduit. Although a raceway is

*Reprinted with permission from NFPA 70-1984, National Electrical Code®, Copyright © 1983, National Fire Protection Association, Quincy, Massachusetts 02269. This reprinted material is not the complete and official position of the NFPA on the referenced subject, which is represented only by the standard in its entirety.

used exclusively for containing electric conductors, a metal raceway is generally also used as an *equipment grounding* conductor.

Although the word "cable" is not defined in the Code, a cable is usually an assembly of two or more wires within an outer metal or nonmetallic enclosure, such as metal armor or a nonmetallic sheath. However, some types of cable for direct burial in the earth, which will be discussed in later chapters, contain a single conductor in a tough, outer, nonmetallic jacket.

Each type of raceway or cable and its use are further defined in one Code article covering that type of wiring method. Some of these methods will be further discussed in other chapters of this book. In this chapter, however, only the methods most commonly used in the wiring of residential and farm buildings will be covered.

Conduit. There are many types of conduit, the most usual being (1) rigid metal conduit; (2) intermediate metal conduit; (3) electrical metallic tubing, also called thin-wall or EMT; (4) flexible metal conduit; (5) liquid-tight flexible metal conduit; (6) electrical nonmetallic tubing, or ENT; and (7) rigid nonmetallic conduit.

Rigid Metal Conduit. In this method, all wires are enclosed in steel (sometimes aluminum) pipe called conduit. Conduit differs from water pipe in that the interior surface is carefully prepared so that wires can be pulled into it with a minimum of effort and without damage to the insulation on the wires. The chemical composition of the steel in conduit is also carefully controlled so that it will bend easily. Steel conduit usually has a galvanized or similar finish inside and outside. Some galvanized-steel conduit and some aluminum conduit have an *additional* enamel or plastic coating for *additional* protection against corrosion. This type may be used in wet, corrosive locations. Steel conduit having only a galvanized finish may be used indoors or outdoors, above ground or underground, in ordinary soils. If the soil has high acidity or is otherwise highly corrosive, steel conduit should have the additional coating already mentioned. Aluminum conduit should always have an additional protective coating when installed in contact with the earth or when embedded in concrete. See NEC Sec. 346-1. Never use steel or aluminum conduit in cinder fill or cinder concrete unless the conduit has an additional protective coating. See NEC Sec. 346-3.

Conduit comes in 10-ft lengths with one coupling; each length bears a UL label. See Fig. 11-1. The $^1/_2$-in size is the smallest used in ordinary wiring. All sizes are identical in dimensions with the corresponding size

Fig. 11-1 Rigid metal conduit looks like water pipe, but differs in several important ways.

of water pipe, but as in the case of water pipe, the nominal size indicates the *internal* diameter; however, the *actual* diameter is larger than the *indicated* diameter. For example, "1/2-in" conduit has an internal diameter of nearly 5/8 in and an external diameter of about 7/8 in.

Cutting Rigid Metal Conduit. A hack-saw blade with 18 teeth to the inch will do a good job. *Never* use a metal-parting tool such as a plumber's pipe cutter because this will leave an expanded sharp edge that will tear the insulation of wires as they are pulled into the conduit; see Fig. 11-2. Of course a hack saw will also leave a sharp edge (but much less than a pipe cutter). All cut ends must be reamed as required by NEC Sec. 346-7.

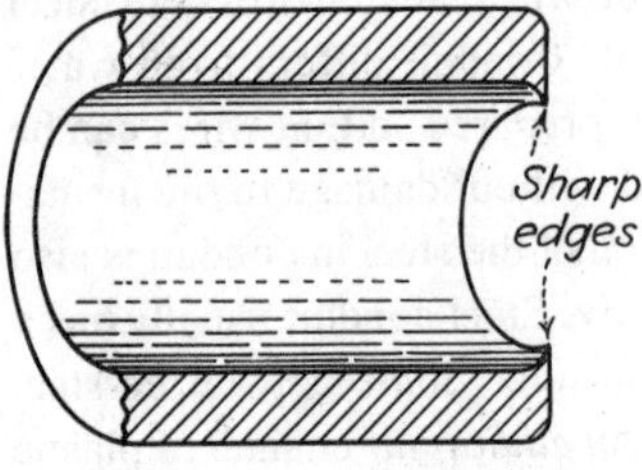

Fig. 11-2 When pipe is cut, a sharp edge is likely to be left; this must be removed before conduit is installed.

A reamer of the type shown in Fig. 11-3, which fits the brace of a brace-and-bit set, is ideal for this purpose. For 1-in or smaller conduit, the handle of a pair of side-cutting pliers, or the shaft of a large square-shaft screwdriver, can be used to do the reaming; wring the handle of the pliers or the shaft of the screwdriver round and round a few times inside the cut end. But always inspect the cut end after you have reamed it to make sure it is absolutely smooth.

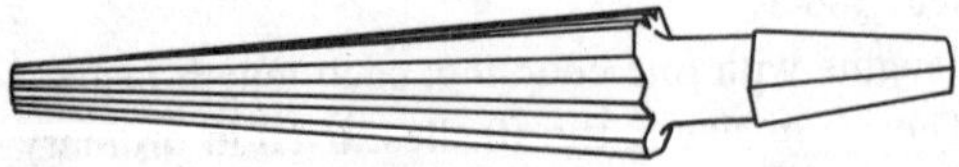

Fig. 11-3 A pipe reamer is handy in removing burrs.

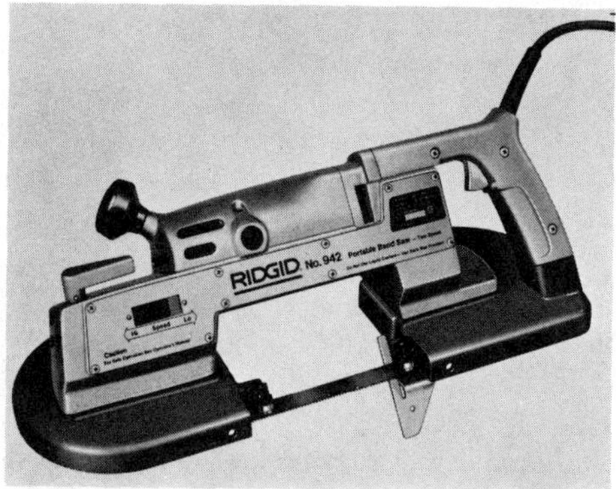

Fig. 11-4 If many cuts must be made, a portable power band saw will prove to be a good investment. (*Ridge Tool Co.*)

If you do a considerable amount of conduit work, a power saw of the general type shown in Fig. 11-4, with a metal-removing blade, will be found to be a good investment.

Threading Rigid Metal Conduit. Thread conduit using the same tools used for water pipe. Be careful not to make the thread too long, for then the conduit ends will butt before the coupling tightens up. This results in burred conduit ends that can damage wire as it is pulled in, and also causes poor grounding continuity of the run of raceway due to the loose couplings. Figure 11-5 shows a 3-way conduit threader with $^1/_2$-, $^3/_4$-, and 1-in dies, while Fig. 11-6 shows a ratchet-type threader that can be used with the same three sizes of dies in removable die heads. Use plenty of thread-cutting oil for cleaner threads and longer life for the dies. If you do a lot of conduit work, consider an electrically powered pipe-threading machine for saving labor and producing clean, uniform threads.

Fig. 11-5 This conduit threader contains dies for three separate sizes of conduit. (*Ridge Tool Co.*)

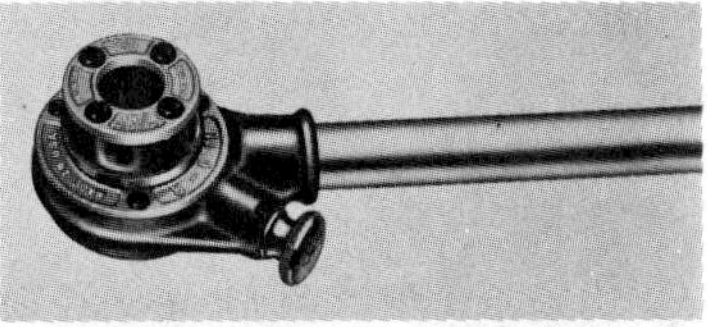

Fig. 11-6 This threader holds a die for only one size of conduit, but one size can easily be substituted for another. (*Ridge Tool Co.*)

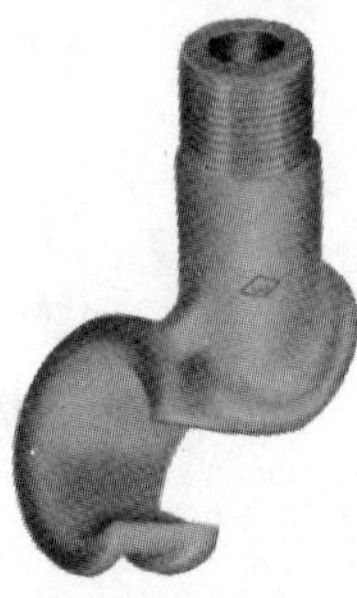

Fig. 11-7 Conduit hickey. Use a piece of 1-in conduit or pipe for a handle. (*Efcor products manufactured by Gould, Inc.*)

Bending Rigid Metal Conduit. Because the wires are pulled into conduit after it is installed, it is most important that all bending be carefully done so that the internal diameter is not substantially reduced in the bending process. Make the bends uniform and gradual. Use a conduit hickey or a bender for sizes 1 in and smaller. The hickey, shown in Fig. 11-7, is inched along the conduit so that the bend is not all in one place; considerable practice is required to make these bends without flattening the conduit. Less skill is required in using a bender of the type shown in Fig. 11-8. Select a bender suitable for the size of conduit to be bent. Assume that a right-angle (90-deg) bend is to be made and that the end of the conduit is to have a "rise" of 13 in above the floor. How far from the end must the bend begin? From the rise, subtract 5 in for the $^1/_2$-in, 6 in for $^3/_4$-in, and 8 in for 1-in conduit. In the example of Fig. 11-9, assuming

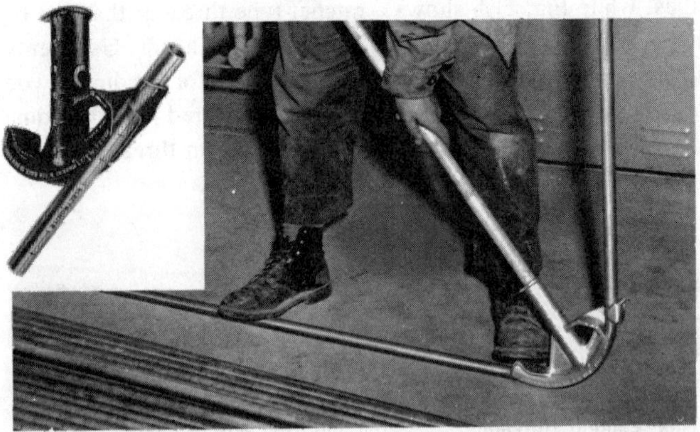

Fig. 11-8 A typical conduit bender and method of use. (*Republic Steel Corp.*)

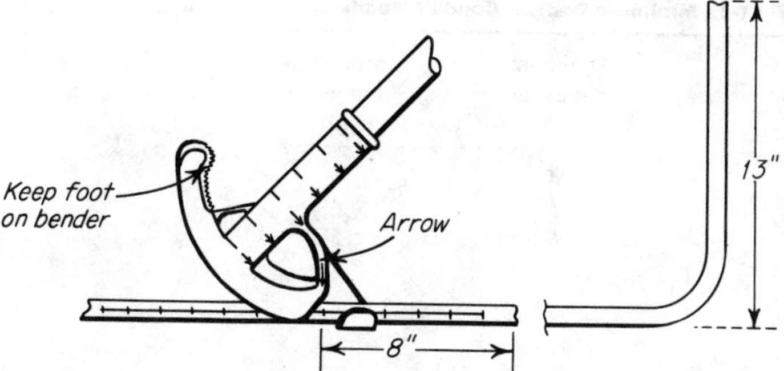

Fig. 11-9 This shows how to figure a 90-deg bend.

$^1/_2$-in conduit, subtracting 5 in from the rise of 13 in leaves 8 in. Hook the bender over the conduit so that the arrow on the bender points to a spot 8 in from the end of the conduit. Full instructions for many types of bends usually come with a bender as purchased.

Factory-made elbows are usually used for $1^1/_2$-in and larger conduit, and in many cases for all sizes. For bends other than 90 deg (and in many cases, all bends), hand benders are used for the smaller sizes and power benders for the larger sizes. Power benders are available in a variety of types, hydraulically or electrically operated. Most power benders are of the single-operation, or one-shot, type.

The Code in Sec. 346-10 specifies the minimum radius of bends, and this information is shown in Table 11-1. Use Col. 2, unless the bends are made with a one-shot power bender, in which case use Col. 3. Note that the radius in Col. 2 is from six to eight times the inside diameter of the conduit.

Number of Bends. The Code in Sec. 346-11 prohibits more than four quarter-bends or their equivalent in one "run" of conduit. A run of conduit is that portion between any two openings, such as a cabinet, a box, or conduit fitting with a removable cover. The fewer the bends, the easier it is to pull wire into the conduit.

Wire Splices in Conduit. Wires in *any* kind of conduit must be continuous and without splice in the conduit itself. See NEC Sec. 300-13. Splices are permitted only in switch or outlet boxes, junction boxes, or places where the splice is otherwise permanently accessible.

Number of Wires in Conduit. The tables in Chap. 9 of the Code (see Appendix) show how to determine the maximum number of various sizes

TABLE 11-1 Minimum Radius, Conduit Bends*

Size of conduit, in (1)	Minimum radius, in (2)	Minimum radius, in (3)
$\frac{1}{2}$	4	4
$\frac{3}{4}$	5	$4\frac{1}{2}$
1	6	$5\frac{3}{4}$
$1\frac{1}{4}$	8	$7\frac{1}{4}$
$1\frac{1}{2}$	10	$8\frac{1}{4}$
2	12	$9\frac{1}{2}$
$2\frac{1}{2}$	15	$10\frac{1}{2}$
3	18	13
$3\frac{1}{2}$	21	15
4	24	16
5	30	24
6	36	30

*Reprinted with permission from NFPA 70-1984, National Electrical Code®, Copyright© 1983, National Fire Protection Association, Quincy, Massachusetts 02269. This reprinted material is not the complete and official position of the NFPA on the referenced subject, which is represented only by the standard in its entirety.

and types of wire you may install in any kind of conduit of a given size. Where all the wires are of the same size, use NEC Tables 3A, 3B, or 3C, but note the first column of each of these tables. It shows you which *type* of wire each table covers. Use the table that lists the type of wire you intend to use. But if you intend to use two *different* types of wire in the same conduit, you may not use these tables unless the types you intend to use are all listed together in the first column of any one table. If they are not included in any one table, you must use the long procedure explained in the next paragraph.

Where some of the wires are of one size and others of a different size (or even if of the same size but not all listed in the first column of one of the tables), the procedure is as follows: NEC Table 1 shows the permissible "percentage of fill" (the total cross-sectional area in square inches of all the wires you intend to install, as compared with the internal area in square inches of the conduit). First determine the total area of all the wires you intend to install, from Table 5. Then in Table 4, find the size of conduit whose cross-sectional area in square inches, using the correct percentage-of-fill column (such as 40% for "over two wires, not lead-covered"), is at least as great as the total square-inch area of all the wires you intend to install. If the conduit is to contain a cable of two or more conductors, each cable is considered as one wire.

An equipment grounding wire, which does not carry current during normal operation, is only occasionally required in nonflexible metal conduit but is required most of the time in flexible and nonmetallic conduit. But if it is installed it must nevertheless be counted, because it does help fill the conduit. If it or a ground*ed* service wire is bare, you can find its cross-sectional area in Table 8.

A condensed tabulation showing how many wires, *all the same size*, may be installed in the smaller sizes of conduit is shown in Table 11-2. The table is basically correct only for the commonly used Types T and TW wire, but is also correct for Type RHW unless the conduit size is marked*, Δ, or †, as explained in the footnotes under the table. This table is also correct for all other kinds of metal conduit.

Some readers may remember that, at one time, if you were pulling old wires out of a conduit and installing new ones, you were permitted to in-

TABLE 11-2 Number of Wires in Conduit

Size of wire	Number of wires to be installed				
	2	3	4	5	6
14	½	½	½*	½*	½*
12	½	½	½*	½*	½Δ†
10	½	½*	½*	½Δ†	¾*
8	½*†	¾*	¾*†	1*	1*†
6	¾*	1*	1*	1¼	1¼*
4	1*	1*	1¼*	1¼*	1½*
2	1*	1¼	1¼*	1½*	2
1/0	1¼*	1½*	2	2*	2½
2/0	1½*	1½*	2*	2*	2½*
3/0	1½*	2	2*	2½*	2½*

Conduit sizes in table above are for Type T or TW wire (also Type RHW unless the size is marked *, Δ, or †). If you are using Type RHW wire, note that this is available two ways: with and without an outer cover. The kind you are using will affect the size of conduit required:

*If RHW *with* an outer cover, use conduit one size larger than shown.
ΔIf RHW *with* an outer cover, use conduit two sizes larger than shown.
†If RHW *without* an outer cover, use conduit one size larger than shown.

stall more wires than in a new installation. This provision last appeared in the 1968 Code, but the same goal can usually be accomplished by using the thinner-insulation wires now available.

Pulling Wires into Conduit. If the run is very short, the wires can be pushed through the conduit from one outlet to the next. But in most cases, especially if the conduit contains bends, "fish tape" is used. For occasional work, a length of ordinary galvanized-steel wire may be used, but a fish tape made of thin steel about 1/8 in wide is usually used. Figure 11-10 shows a fish tape on a plastic reel with a pull-eye on the end of the tape. Some steel fish tapes have a loop at the end of the tape instead of a replaceable eye. If the loop breaks off, a new one can be formed by heating the end of the tape with a blowtorch, and then forming a new loop with your pliers. If you attempt to make a new loop without heating, the tape will break. Some steel fish tapes are about 1/4 in wide, and plastic types are also available. Both types come wound on a reel that makes it easier to handle and that provides a "handle" for pulling in the wires after they have been attached to the eye or loop on the end of the tape. Push the tape into the conduit through which the wires are to be pulled, and attach the wires that are to be pulled into the conduit to the tape. This is done by removing the insulation from about 4 in of each wire, threading the wires through the eye or loop of the tape, and folding the wires back upon themselves. Then wrap friction or plastic tape over the wires to hold them fast; just a few turns of the tape are necessary. See Fig. 11-11.

Then pull the wires into the conduit. This usually requires one person pulling at one end and another feeding the wires into the opening at the other end, to make sure there will be no snarls, and in general to ease the

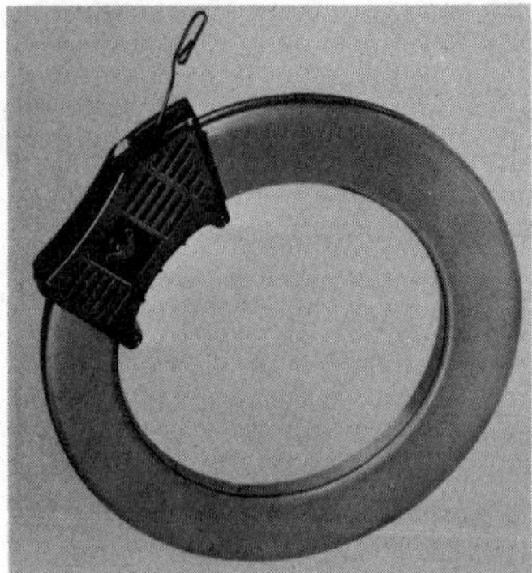

Fig. 11-10 Fish tape in its reel, with a convenient pull handle. *(Ideal Industries, Inc.)*

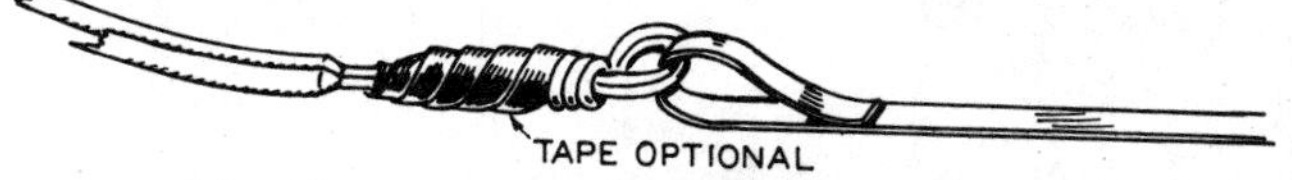

Fig. 11-11 How to attach wires to be pulled to tape.

wire in and prevent damaging it. Powdered soapstone or a UL Listed paste or liquid lubricant may be used to make it easier to pull the wires. Do *not* use a harsh soap or other lubricant that over a period of time may damage the insulation.

The Code in Sec. 300-14 requires a minimum of 6 in of wire projecting at each outlet. But as you learned in Chap. 8, about 10 in is required to make connection to a device such as a switch or receptacle. So leave plenty of wire to make a good but quick connection, and to allow room for replacing the device at a future date. Of course at a panelboard, such as service equipment, enough wire must be pulled in to form it around the enclosed equipment and still reach to the most distant terminal in the equipment.

If a wire running through a box is to be electrically continuous, but with its insulation stripped off for a short distance for connection to a receptacle or other device (as will be shown in Chap. 18), you must provide enough slack to permit easy connection of the center portion of the loop in the wire, to the device. If a wire is to run through a box without a connection or a splice in that box, then no slack is needed in that box, but a loop should be folded back in the box, if space permits, in case of future changes.

For larger jobs and longer runs, electric blower/vacuum equipment is made which will quickly either blow or pull a fish line through the conduit. The line is then used to pull through a pulling rope. For pulling large wires a power winch can be used. A small power puller is shown in Fig. 11-12. A time-saver when pulling stranded conductors is a reusable wire basket similar to that shown in Fig. 11-13. The harder the pull, the more tightly the basket grips the conductors, but without damaging the insulation.

Supporting Rigid Metal Conduit. Conduit is required to be supported within 3 ft of each box, fitting, or cabinet and, in addition, at intervals of not over 10 ft. However, if the conduit is in *straight runs* with threaded-type couplings, the distance between supports is as shown in NEC Table 346-12. Supporting straps are shown in Fig. 11-14.

Intermediate Metal Conduit This material in sizes not over 4 in is similar

Fig. 11-12 Compact portable power puller, foot-switch-operated. (*Greenlee Tool Division, Ex-Cell-O Corporation.*)

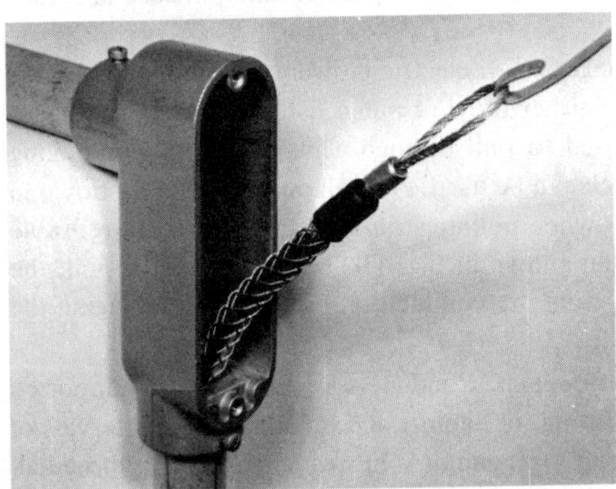

Fig. 11-13 Wire-pulling basket of galvanized-steel mesh slips on and off cable without tools. (*Kellems Division, Harvey Hubbell Incorporated.*)

Fig. 11-14 Either 1- or 2-hole straps may be used for supporting conduit.

to ordinary rigid metal conduit and, size for size, has about the same outside diameter but thinner walls. That makes the internal area in square inches a little more than in ordinary conduit, but it is not permissible to install more wires than in other conduit. IMC is installed in the same way as ordinary conduit, including the spacing of supports. Hand bending must be done with a bender similar to that in Figs. 11-8 and 11-9, as the walls of IMC must be supported during bending to avoid flattening. IMC is acceptable under the Code for all applications for which the heavier-wall rigid metal conduit is permitted, including locations classified as hazardous.

Electrical Metallic Tubing. A raceway that is similar to rigid metal conduit but of the "thin-wall" construction is called "electrical metallic tubing" (abbreviated "EMT") in the Code.

A length of it is shown in Fig. 11-15. It is made of galvanized steel, occasionally of aluminum, either of which may have an additional plastic or other protective coating. The rules that apply to rigid metal conduit, such as the number of wires permitted and bending, apply also to EMT. It is made only in sizes up to and including 4-in. It must be supported within 3 ft of every box, cabinet, or fitting and additionally at intervals not exceeding 10 ft, regardless of size.

Fig. 11-15 Electrical metallic tubing ("EMT" or thin-wall conduit) is never threaded. It is lighter and easier to use than rigid metal conduit.

The *internal* diameter in the smaller sizes is the same as in rigid conduit, but in the larger sizes it is a little larger. However, since the walls are so thin, EMT must never be threaded. Instead, all joints and connections are made with threadless fittings that hold the material through pressure. Figure 11-16 shows both a coupling and a connector; note that

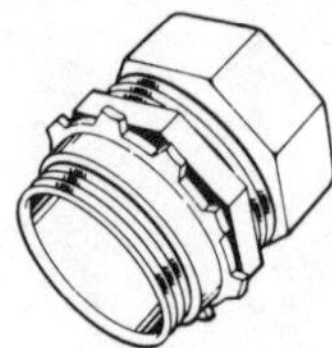
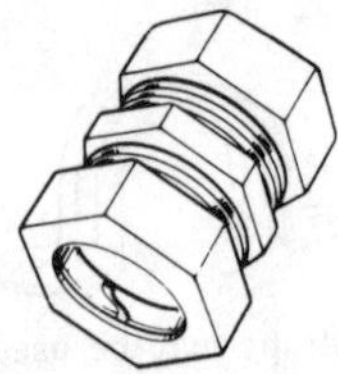

Fig. 11-16 A connector and coupling used with thin-wall conduit. (*RACO, Inc.*)

these fittings do have *external* threads. Each coupling and connector consist of a body and a split ring through which tremendous pressure is exerted on the EMT when the nut is forced home tightly. In installing these fittings, the compression nut is first loosened until the fitting will slip over the end of the EMT. Be sure that the end of the EMT goes *past* the inner ring to the shoulder inside the fitting. Then tighten the nut.

Another type of connector and coupling is shown in Fig. 11-17. This type of fitting has no threads, but the special indentor tool shown in Fig. 11-18 anchors it to the EMT. Slip the fitting over the end of the EMT, and be sure to push it all the way until the EMT butts up against the shoulder in the fitting. Use the indenting tool to indent the fitting in one position, then turn the tool about a quarter of a turn and indent the fitting in another position, so that indents will appear at four points around the fitting. The tool will deeply indent both the fitting and the EMT. If you use an EMT fitting in a wet location or in concrete, be sure it is UL Listed for the purpose. Be sure that any fitting is made up properly to provide a good continuous ground, which is so necessary for safety.

Always cut EMT with a hack saw having a blade with 32 teeth to the

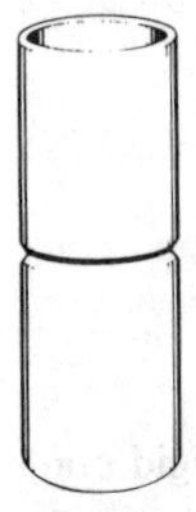
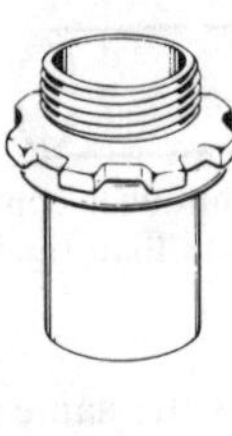

Fig. 11-17 Telescopic type of fittings shown here may also be used with thin-wall conduit. (*RACO, Inc.*)

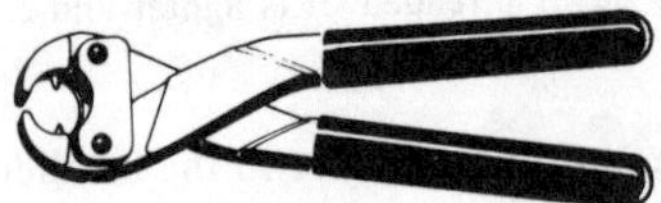

Fig. 11-18 Use this indenting tool to indent conduit and the fittings shown in Fig. 11-17. (*RACO, Inc.*)

inch, or with a metal-removing power saw. After the cut, ream the ends to remove burrs or sharp edges.

Occasionally it will be necessary to join a length of EMT to a length of rigid conduit, or to a fitting with threads for rigid conduit. Use a connector of the type shown in Fig. 11-16 or 11-17. The threaded portion of the connector will fit the internal thread of any rigid metal conduit coupling or fitting designed for the corresponding size of rigid conduit.

Flexible Metal Conduit. This material is covered by NEC Art. 350 and is shown in Fig. 11-19. It is often called "greenfield" or "flex." It is similar in appearance to armored cable but is an empty raceway, and wires must be pulled into it after it is installed.

Fig. 11-19 Flexible metal conduit. The wires are pulled into place after the conduit is installed.

The smallest size used in ordinary wiring is the $^1/_2$-in trade size, which is actually over $^3/_4$ in. in outside diameter. Except in a few scattered areas this material is little used in general wiring. Where it is used, the flexible conduit is first installed, and the wires are pulled into it later just as in the case of rigid conduit or EMT.

As the flex is accessible for removing the burr from the cut end, the connectors are similar to armored cable connectors except that peepholes are not required, since fiber bushings are not used. If you use the type of connector that threads into the flex, cut the flex off square. It need not be reamed, as the cut end butts against a shoulder on the fitting and is isolated from the wires inside.

If the flex is made of aluminum, be careful not to collapse the flex when installing connectors that clamp around the outside.

Flex must be supported within 12 in of every box, cabinet, or fitting and also at intervals not over $4^1/_2$ ft, with two exceptions: (1) a free length not over 36 in is allowed at terminals where flexibility is required, such as at a motor where vibration is a factor or where the motor location must be slightly changed for proper belt tension, and (2) a 6-ft length is allowed between a recessed lighting fixture and a junction box, as will be explained in Chap. 18.

Occasionally a short length of flexible metal conduit is used in a run of rigid metal conduit where it would otherwise be extremely difficult to bend the rigid conduit into the necessary configuration. This is not permitted in a service conduit, and should not be done at all if it can be avoided (as it usually can), since this introduces additional resistance into the equipment grounding path.

Metal pipe conduit may be used as the equipment ground*ing* conductor. Flexible metal conduit has a very much higher resistance per foot than the pipe types. A separate equipment grounding wire must be installed in the conduit, with some exceptions. Per Sec. 350-5 of the Code, the equipment grounding conductor is not required when both the conduit *and* the fittings are approved for the purpose. In effect that means "if the conduit *and* the fittings are UL Listed *for the purpose,*" but the catch is that the UL have Listed *some* fittings, but no flexible conduit, specifically for the purpose.

So, when using flexible metal conduit (even if it is UL Listed for general use) a separate equipment grounding wire must be installed. The wire may be bare or insulated (if insulated it must be green) and must be installed in the conduit along with the circuit wires. At each end, connect it as you would the bare equipment grounding wire of nonmetallic-sheathed cable as discussed later in this chapter. The size depends on the rating of the overcurrent protection in the circuit involved. With 15-amp protection use No. 14 wire; with 20-amp protection, No. 12 wire; with 30- to 60-amp protection, No. 10 wire; with 100-amp protection, No. 8; with protection over 100 amp but not over 200 amp, No. 6. All these values are from NEC Table 250-95 and are for copper wire only. In runs not over 6 ft long, you may install the equipment grounding wire *outside* the flex, but make it closely follow the flex itself, so as to protect it from physical damage.

There is an exception. In 6-ft lengths ordinary flexible conduit may be used without the equipment grounding wire, but only if the fittings are UL Listed for the purpose, *and* the circuit is protected by a breaker or fuse rated not over 20 amp. All listed fittings that clamp around the outside of the flex; all listed fittings 3/8 in through 3/4 in that are not of the clamping type; and larger fittings if marked "GRND" or the equivalent, are UL Listed as grounding fittings under this exception.

Liquidtight Flexible Metal Conduit. This material is flexible metal conduit, plus an outer liquidtight nonmetallic sunlight-resistant jacket. It is covered by NEC Art. 351. Figure 11-20 shows both the material and the special connector that must be used with it. Part of the connector goes

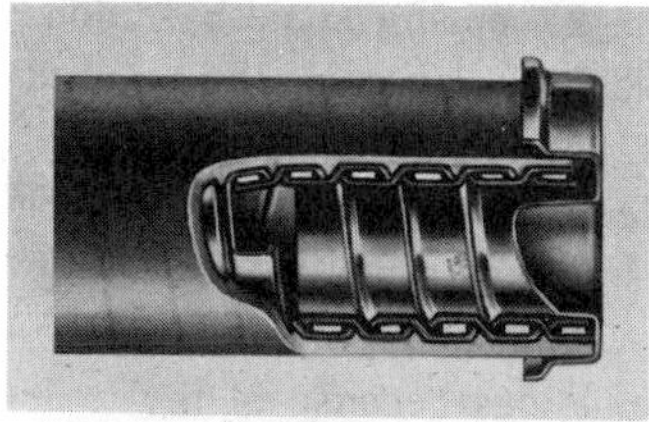

Fig. 11-20 Liquidtight flexible conduit, and the special fittings used with it. (*Ideal Industries, Inc.*)

inside the conduit, thus making a good connection for grounding continuity; and part of the connector goes over the outside, forming a watertight seal. For proper assembly of the connector, the end of the flex must be cut square. For lengths not over 6 ft, the $^3/_8$-in and $^1/_2$-in sizes are recognized as an equipment grounding means when the contained conductors are protected at not over 20 amp, as are the $^3/_4$-in through $1^1/_4$-in sizes when the contained conductors are protected at not over 60 amp. For lengths over 6 ft, or for any size of $1^1/_2$ in and larger, or when used to connect equipment where flexibility is required, a bare or green equipment grounding wire must be installed with the circuit wires. An example of a common use is in the final connection to air-conditioning equipment that is installed outdoors, exposed to the weather. The plastic jacket will not withstand continuous abrasion, and it will soften at high temperatures, so do not use this type of conduit where it will come in contact with moving machinery or with wires rated higher than 60°C such as THW or THHN. When so Listed and marked, liquidtight flex is permitted to be buried directly in the earth.

Nonmetallic Conduit. Only polyvinyl chloride (PVC) rigid nonmetallic conduit and electrical nonmetallic tubing (ENT) will be discussed here. Other types, for underground use only, include impregnated fiber, fiberglass epoxy, high-density polyethylene, and Type A PVC. Excellent moisture and corrosion resistance, light weight, and ease of installation have led to wider use of PVC. Except for some hazardous locations and for support of fixtures, rigid PVC can generally be used wherever rigid metal conduit is permitted, and for some corrosive locations it is preferred to metal. ENT is limited to use in buildings not over three floors above grade. For specific restrictions see NEC Secs. 331-3, 331-9, 347-2, and 347-3. Although PVC will not support combustion, some inspectors restrict its use within buildings because of the toxic fumes it emits when exposed to fire.

Bending PVC. ENT is a pliable corrugated raceway easily bendable by hand without the use of tools. Rigid PVC can be bent by heating, forming the bend, and cooling. An open flame should *never* be used. Hot-air and hot-liquid methods are used, but the simplest is a "hot box," which has electric heating elements and rollers to help heat the conduit all around the place where the bend is to be made. In the smaller sizes, bends can be made by hand (use gloves or rags to protect your hands from the heat), but if many bends are to be made, a simple jig can be used to simplify holding the bend in position while cooling it with a wet rag (see Fig. 11-21). Care must be taken to maintain the circular cross section at bends. Two-inch and larger sizes require internal support during bending, which can be provided by plugging the ends before heating. The expanded, heated air inside prevents the walls from collapsing while soft. Number-of-bends and bend-radius requirements are the same as for metal.

Other Considerations. Any fine-toothed saw may be used to cut PVC, and the ends should be reamed both inside and out. Fittings (couplings, connectors, etc.) for rigid PVC conduit are solvent-cement-"welded" to the outside, as shown in Fig. 11-22. Plastic connectors and couplings for ENT snap together, locking onto the corrugations in the raceway. Because of the slight flexibility of PVC it must be supported at closer intervals than metal conduit or EMT. See NEC Table 347-8 and Sec. 331-11. The wire fill in Table 11-2 applies for PVC, but remember to allow room for the grounding (green or bare) conductor that is probably required. Extra-heavy-wall (Schedule 80) PVC has a reduced internal cross-sectional area, and wire fill must be calculated by using Table 1, Chap. 9, in the Code. When long runs are exposed to wide variations in temperature, expansion joints must be used. If rigid PVC is left exposed to wide variations in temperature (as from day to night) in a trench before backfilling, allowance must be made for expansion and contraction. Expansion joints are not usually required on underground runs, for once the

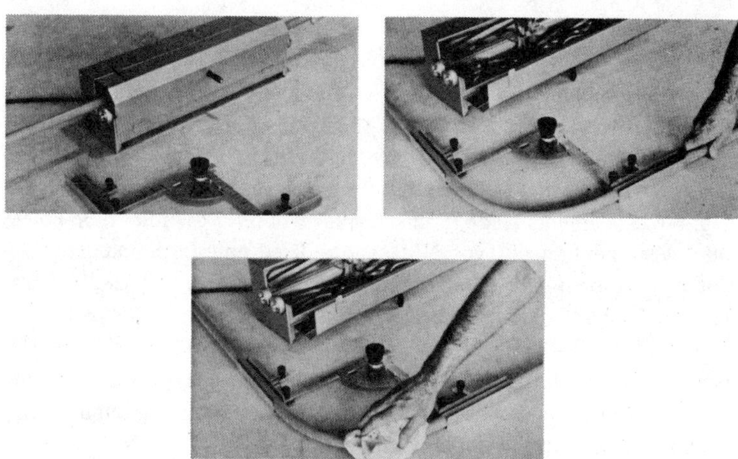

Fig. 11-21 Bending rigid PVC conduit by rotating it in an electric "hot box" to soften, forming the bend, and then cooling it with a wet rag. (*Carlon, an Indian Head company.*)

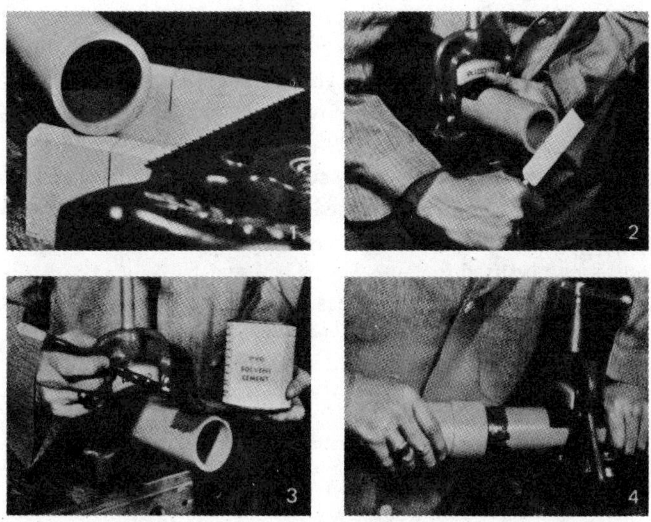

Fig. 11-22 Cutting, deburring, and applying solvent cement to rigid PVC conduit and sliding on the fitting while rotating about half a turn. (*Carlon, an Indian Head company.*)

trench has been backfilled, expansion and contraction cease to be significant. ENT is not permitted underground.

Nonmetallic-Sheathed Cable. As you have already learned in Chap. 6, there are two types of nonmetallic-sheathed cable. Type NM has wires enclosed in an outer sheath as shown in Fig. 6-5. Type NMC has individually insulated wires embedded in a solid plastic mass as shown in Fig. 6-6. Statements made in this chapter will refer to either type unless otherwise mentioned. Type NM may be used only in permanently dry indoor locations. It must not be embedded in plaster, nor used in barns and similar buildings that have ammonia fumes or other corrosive conditions. Type NMC, on the other hand, may be used wherever Type NM is acceptable, but may also be used in damp or moist locations, but not in permanently wet locations; it may not be buried in the ground. It may be embedded in plaster but must then be protected by a strip of steel not less than $^{1}/_{16}$ in thick and $^{3}/_{4}$ in wide, to guard against nails. Type NMC is the successor to the so-called "barn cable" that was used many years ago, and is well suited to such corrosive locations.

Nonmetallic-sheathed cable is easy to install and comparatively inexpensive. Hence it is widely used. Sec. 336-3 of the Code limits its use to "one- and two-family dwellings, multifamily dwellings and other structures provided that such dwellings or structures do not exceed three floors above grade."* Some local codes have restrictions in addition to those imposed by the Code, so check your local code.

Removing the Outer Cover. The outer cover must be removed at the ends for a distance of about 10 in. For Type NM this can be done by slitting the jacket with a knife. Cut off the dangling cover with a knife or pair of sidecutter pliers. Be most careful not to damage the individual wires. Using a cable ripper as shown in Fig. 11-23 will save time and avoid damage to the conductors; it can be used for both Types NM and NMC.

Anchoring Cable to Boxes. The cable is anchored to switch and outlet boxes with connectors, several types of which are shown in Fig. 11-24.

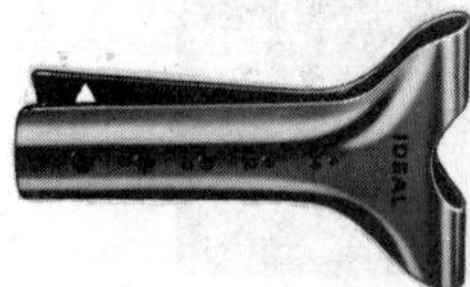

Fig. 11-23 This cable ripper saves much time in installing nonmetallic-sheathed cable. (*Ideal Industries, Inc.*)

*Reprinted with permission from NFPA 70-1984, National Electrical Code®, Copyright© 1983, National Fire Protection Association, Quincy, Massachusetts 02269. This reprinted material is not the complete and official position of the NFPA on the referenced subject, which is represented only by the standard in its entirety.

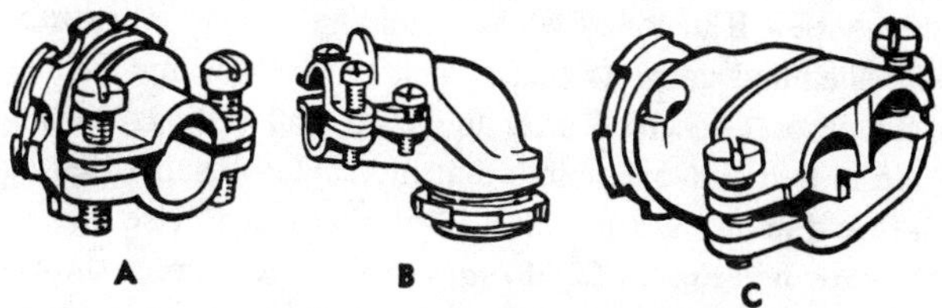

Fig. 11-24 An assortment of connectors for anchoring cable to outlet and switch boxes.

First solidly anchor the connector to the cable; remove the locknut of the connector; slip the connector into the knockout of the box; then place the locknut on the connector and drive it home solidly on the inside of the box. Other connectors, without locknuts, are shown in Fig. 11-25. These

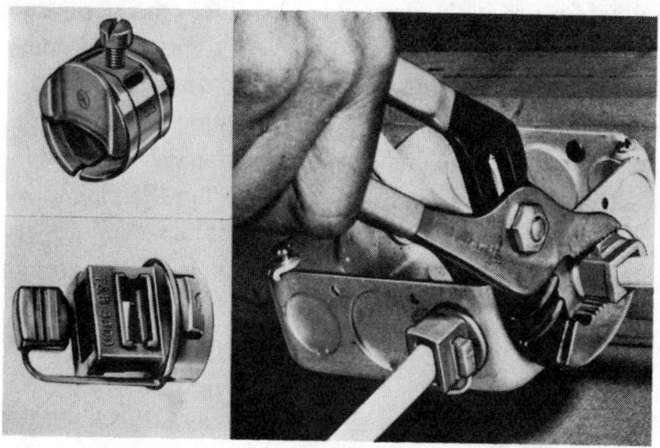

Fig. 11-25 A one-screw metal connector and a plastic connector, both used for terminating Type NM cable at boxes. Having no locknuts, these connectors can also be installed *inside* an existing box when fishing cable in old work. (*T&B/ Thomas & Betts Corp.*)

types are handy for old work, as they can be installed either inside or outside the box. As mentioned in Chap. 10, many boxes have built-in cable clamps that eliminate the need for separate connectors. Metal connectors or cable clamps should be fastened snugly to Type NM cable but not overtightened, for too much pressure can eventually force metal parts to cut through the plastic jacket and insulation and cause a short circuit or ground fault.

Splices. Splices are not permitted in nonmetallic-sheathed cable (and other cables) except in outlet boxes, where they are made as in any other wiring method.

Mechanical Installation. If installed while a building is under construction, nonmetallic-sheathed cable is placed inside the wall and ceiling spaces, where it will be concealed after the walls and ceilings are installed. The Code in Sec. 336-5 requires that the cable be supported at least every $4^1/_2$ ft and in any event within 12 in of any box. Use plastic straps of the kind shown in Fig. 11-26; do *not* use the steel staples shown

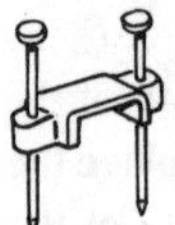

Fig. 11-26 It is best to use nonmetallic straps to support nonmetallic-sheathed cable.

in Fig. 11-40, designed for use with armored cable, unless the staples are insulated or padded and unless you use *extreme* care to prevent damage to the cable by driving the staples in too hard. If you do a great deal of work using nonmetallic-sheathed cable, handy stapling machines designed for the purpose are available and save time and may be a good investment for you. Ordinary 2-hole straps can be used, but the plastic variety shown in Fig. 11-26 and the kind dispensed by stapling machines are better and more practical. In old work where the cable is fished through wall spaces, the cable does not need to be supported except where exposed; of course connectors or built-in cable clamps must be used at all boxes.

All bends in cable must be gradual so as not to damage the cable; the requirement of NEC Sec. 336-10 is that if a bend were continued so as to form a complete circle, the diameter of the circle would be at least ten times the diameter of the cable.

Where exposed, as in basements, attics, barns, etc., cable must be protected against physical damage. This protection can be provided in a variety of ways; see Fig. 11-27. If the cable is run along the side of a joist, rafter, or stud as at *A*, or along the bottom edge of a timber as at *B*, no further protection is required. If it is run at an angle to the timbers, the cable may be run through bored holes as at *C* and no further protection is required. The holes should be bored in the approximate center of the timbers. If the cable is run across the bottoms of joists, then it must be run on substantial running boards, as shown at *D*. In unfinished basements running boards are not required if the cable contains at least three No. 8 or two No. 6 or larger conductors. Cables smaller than that, if not run

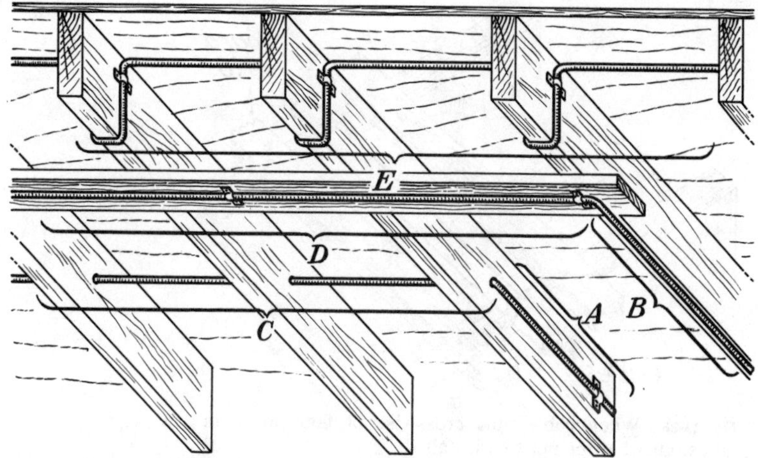

Fig. 11-27 This shows five different ways in which cable may be run on an open ceiling. Various methods are used to protect the cable, depending on the method of installation.

through bored holes or on running boards, are required by NEC Sec. 336-6(a) to follow the surface of the building structure, as shown at *E*.

If the edge of a hole in a stud is closer than $1^1/_4$ in from the nearest edge of the stud, install a $^1/_{16}$-in-thick steel plate or bushing to protect the cable against future penetration by nails or screws, as required by NEC Sec. 300-4. Be careful when feeding and pulling the cable through bored holes in wood members so as not to damage the outer jacket.

In accessible attics if the cable is run across the top of floor joists instead of through bored holes, it must be protected by guard strips at least as high as the cable, as shown in Fig. 11-28. If run at right angles to studs or rafters, it must be protected in the same way at all points where it is within 7 ft of floor joists. No protection is required under other conditions. If the attic is not accessible by means of stairs or a permanent ladder, this protection is required only for a distance of 6 ft from the opening to the attic.

In all cases the cable must follow a surface of the building or of a running board, unless run through bored holes in timbers not over $4^1/_2$ ft apart. It must *never* be run across open space, either unsupported or suspended by wires or strings.

When run through a metal stud, the cable must be protected from the sharp edge of the punched opening. Snap-in nonmetallic bushings are

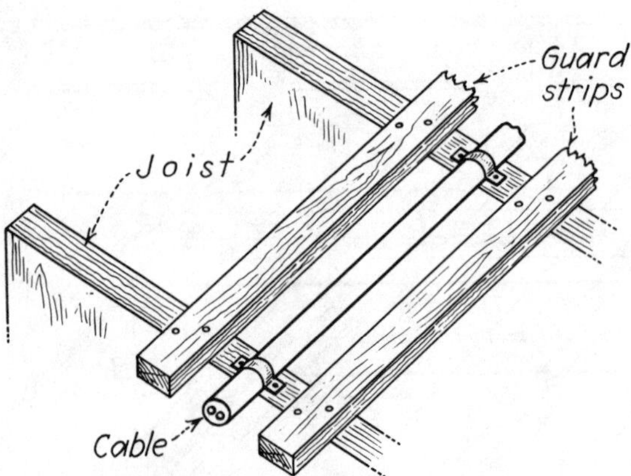

Fig. 11-28 When cable runs crosswise of floor joists in attics, guard strips must be installed.

available for this purpose for all the commonly shaped openings in metal studs.

Cable with Equipment Grounding Wire. Most nonmetallic-sheathed cable made today has a bare equipment grounding wire in addition to the insulated conductors; see Fig. 6-5. Cable with grounding wire was first required by the 1962 Code, in runs to boxes containing receptacles. It is now required to practically every box. Since all 15- and 20-amp receptacles must be of the 3-wire grounding type and since all exposed metal parts of lighting fixtures must be grounded, there is very little application for cable without an equipment grounding wire. A nonmetallic fixture such as a keyless porcelain lampholder mounted on a nonmetallic box or a switch with a nonmetallic plate held on by nylon screws and mounted in a nonmetallic box could be supplied by Type NM cable without an equipment grounding conductor, but the practical thing is to use only grounded cable and to take the equipment grounding conductor to every switch and outlet. The small difference in cost more than offsets the bother of keeping track of two types of cable on the job and the risk of using the wrong type.

The grounding wire contributes to safety. It does complicate the wiring a bit, but it is not difficult to learn how to do it. How to connect this bare grounding wire will be explained later in this chapter.

Armored Cable. The ordinary kind of armored cable, called "BX" by most people, is shown in Fig. 11-29. The Code in Art. 333 calls it "Type

Fig. 11-29 Armored cable. Steel armor protects the wires. A bare bonding strip runs inside the armor, for better equipment grounding purposes.

AC cable.''[1] It is also called "Type AC" by UL if it contains rubber-covered wires, but "Type ACT" if it contains plastic-insulated wires.

In armored cable, the insulated wires are wrapped in a layer of tough kraft paper, with a spiral outer steel armor. Between the paper and the armor there is a flat or round aluminum bonding strip, which reduces the turn-to-turn resistance of the armor itself, thus providing better continuity of ground. This grounding strip thus improves the performance of the armor in case of accidental grounds.

Cutting Cable. A hack saw can be used to cut armored cable. Do not hold the saw at a right angle to the cable, but rather at a right angle to the *strip of armor,* as shown in Fig. 11-30. You must use extreme care to

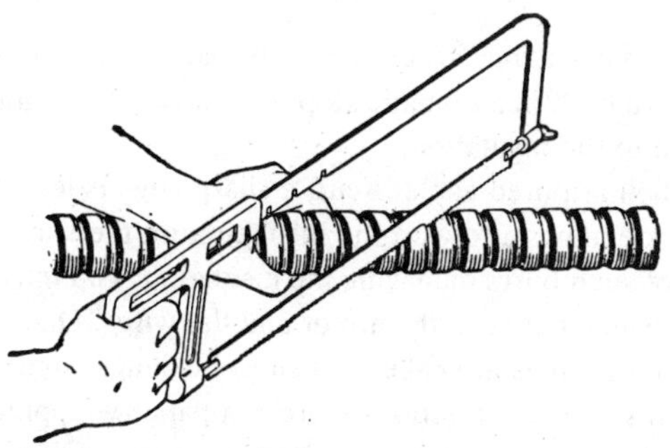

Fig. 11-30 In sawing armored cable, hold the hack-saw blade at an angle as shown.

avoid cutting past the armor into the conductor insulation or the grounding strip. After the cut is made through the armor, grasp the armor on each side of the cut and give a twist, as shown in Fig. 11-31. The end of the armor can then be pulled off. The cut should be made about a foot from the end of the cable.

[1] The Code in Art. 334 also covers another style of metal-clad cable called "Type MC," which somewhat resembles ordinary armored cable, but is rarely if ever used in residential and farm wiring.

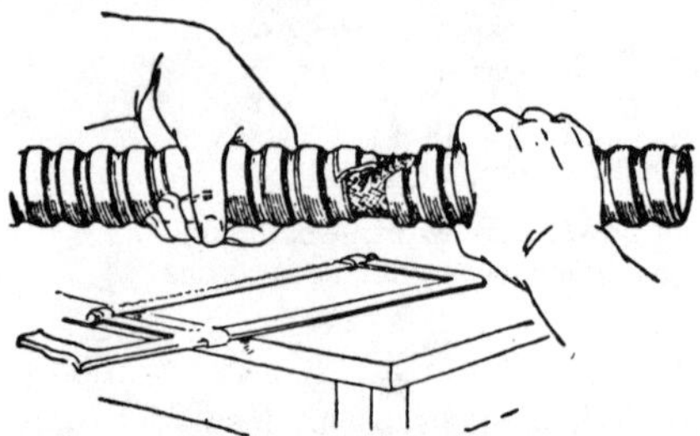

Fig. 11-31 After sawing, twist the cut end of armor to remove it.

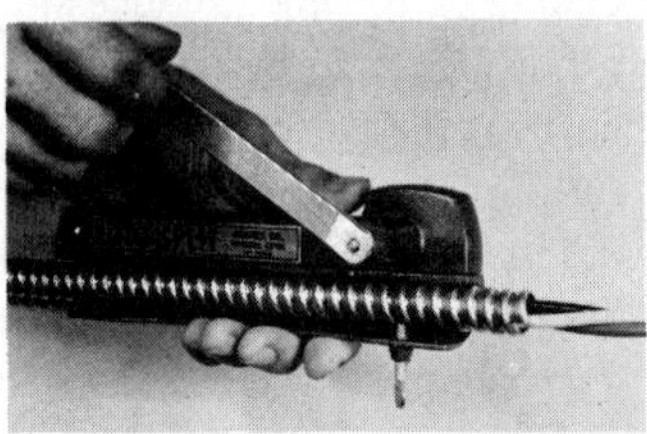

Fig. 11-32 The Roto-Split cutter provides clean-cut ends of armor and saves much time as compared with using a hack saw. (*Seatek Co., Inc.*)

Instead of a hack saw, a time-saving BX cutter can be used; one type is shown in Fig. 11-32. It has a built-in automatic stop that allows the armor to be cut without damage to the insulation.

Insulating Bushings. When armored cable is cut, a sharp edge is left at the cut end, with burrs extending in toward the insulation on the wires. To eliminate the danger of such burrs damaging the insulation, you must insert a fiber or plastic bushing between the armor and the wires. One is shown in Fig. 11-33. Such bushings are called antishort bushings, sometimes just "redheads." In some constructions there is a paper wrapping over all the conductors, which may make it difficult to insert the bushing. Space can be provided by following the steps shown in Figs. 11-34 and 11-35. Unwind the paper beneath the armor, and then with a sharp yank tear it off some distance inside the armor. Insert the bushing as shown in Fig. 11-36. Figure 11-37 shows the final assembly.

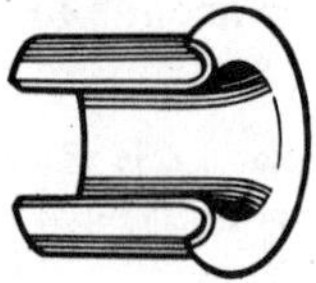

Fig. 11-33 Insert a bushing of tough fiber between the armor and the wires to protect them from the danger of their insulation being punctured by cut ends of armor.

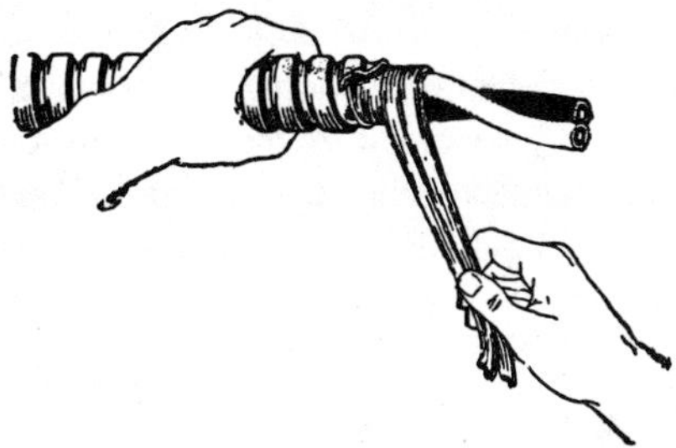

Fig. 11-34 Unwrap the paper over the wires as far as you can; do not tear off.

Fig. 11-35 Remove paper from inside the armor as far as you can; then with a sharp yank tear it off *inside* the armor.

Fig. 11-36 Insert the bushing.

Fig. 11-37 Cross section of cable, showing paper removed inside the armor and bushing in place.

Connectors. Connectors for armored cable are similar to those shown in Fig. 11-24 for nonmetallic-sheathed cable, except that those for armored cable have openings or peepholes through which the antishort bushing can be seen from the inside of the box by the inspector. Because of these peepholes such connectors are known as the "visible type."

To install the connector properly, first insert the antishort bushing inside the armor. Then bend the bare bonding strip *over* the bushing and back along the outside of the armor to help hold the bushing in place. Now slip the connector over the armor and the bent-back grounding strip. Push the cable into the connector as far as it will go so that the bushing cannot slip out of place and so that it can be seen through the peepholes in the connector. Then tighten the screw(s) on the connector to anchor it solidly on the cable. Be sure that the bonding strip is placed in the connector so that it will be solidly squeezed by the connector instead of lying loosely in a part of the connector that does not make solid contact with the armor. Remove the locknut, slip the connector into the

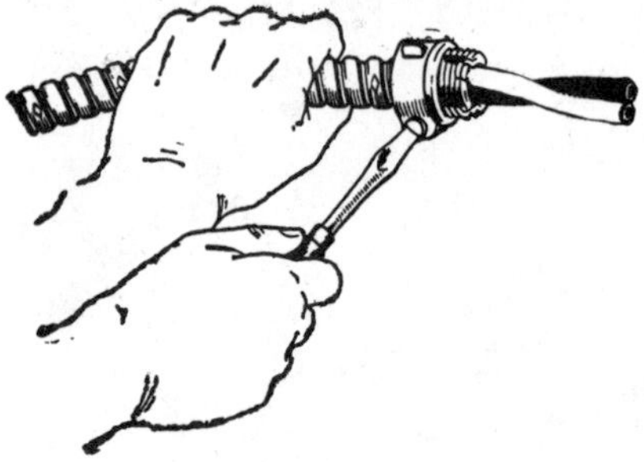

Fig. 11-38 Installing connector on cable.

knockout of the box, and drive the locknut solidly home inside the box as with other cable; see Figs. 11-38 and 11-39. If this is carefully done, with the teeth of the locknut biting into the metal of the box, all the outlets on the circuit are then tied together through the armor of the cable and the grounding strip, providing the continuity of ground discussed in Chap. 9. When installing Type AC cable in a box with built-in cable clamps, as shown in Fig. 10-15, be sure to tighten the clamp screw as tightly as you can, for the equipment grounding path depends on this connection between the armor and the box.

Fig. 11-39 Drive the locknut of the connector down tightly, so that the teeth of the locknut bite down into the metal of the box.

Bends. Avoid sharp bends. The Code requirement is that if a bend were continued into a complete circle, it would not have a diameter less than ten times the diameter of the cable.

Where Used. Like Type NM nonmetallic-sheathed cable, armored cable may be used only in permanently dry locations. For residential wiring it is usually acceptable, but it must *not* be used in barns or other locations where corrosive conditions are the rule. Check your local code for any restrictions in addition to those imposed by the NEC.

Supporting Armored Cable. Except where fished through existing wall spaces, armored cable must be supported within 12 in of each box and cabinet, and additionally at not less than $4^1/_2$-ft intervals. However, a free length not over 24 in long is permitted at terminals where flexibility is needed, such as for a motor that must be adjusted for belt tension. Staples of the type shown in Fig. 11-40 are more convenient than 2-hole straps, but such staples must have a rustproof finish as required by NEC Sec. 300-6.

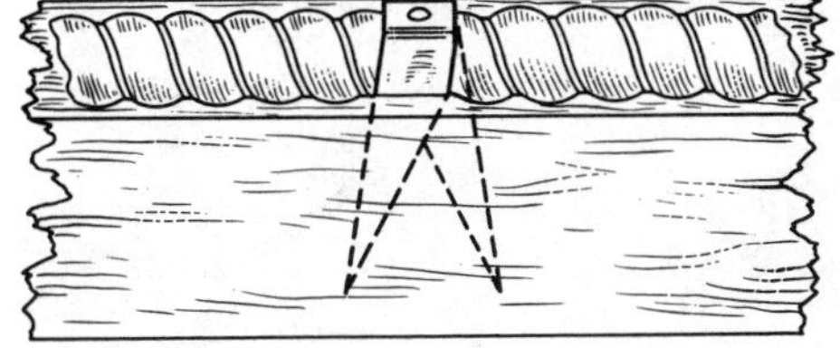

Fig. 11-40 Staples of this kind are used to support armored cable on wooden surfaces. Do not use them with nonmetallic-sheathed cable.

Grounding Receptacles. In Chap. 9 we discussed the importance and advantages for safety in using receptacles of the grounding type, with two parallel slots plus one U-shaped opening for the U-shaped prong of a 3-wire plug. The grounding wire of the cord is connected to this third prong on the plug. The U-shaped opening (contact) of the receptacle is bonded to the mounting yoke of the receptacle, and also to the receptacle grounding screw, which is identified by its *green* color and *hexagonal* shape. This green terminal screw of the receptacle must be *effectively* bonded to the box (if metal); the box is connected to and grounded by the equipment grounding conductor (which may be a wire, metal conduit, or armored cable).

Since the conduit or armor of armored cable is already bonded (through locknuts and bushings, or connectors) to outlet and switch boxes all the way back to the service equipment, it is only necessary to bond the U-shaped receptacle contact to the box. Since the yoke of the receptacle is connected to the box with steel screws, it might seem that these screws would effectively bond the grounding contact of the receptacle to the box. However, in many cases the mounting yoke is not in solid contact with the box, and the bond is dependent on $1^1/_2$ or 2 threads of the small (6-32) mounting screws; that is not a very effective bond.

For that reason, NEC Sec. 250-74 requires that an equipment bonding jumper (bare or green) be run from the green terminal of a receptacle to a flush box. Connect the bonding jumper (copper wire) to the box using the special metal clip as shown in Fig. 11-41, or install a small screw through

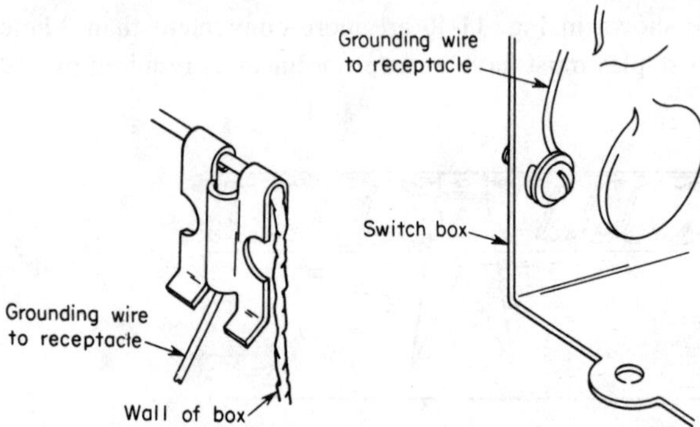

Fig. 11-41 The grounding wire to the green terminal of a receptacle may be grounded to the box using either method shown here.

an unused hole in the box, as shown in the same illustration. Most outlet and switch boxes have an extra hole tapped for a 10-32 screw for the purpose. *This screw hole may not be used for any other purpose.*

However, many UL Listed receptacles are now available with *special* screws for connecting the receptacle to flush-mounted boxes, providing an effective bond between the receptacle and the box. See Fig. 11-42.

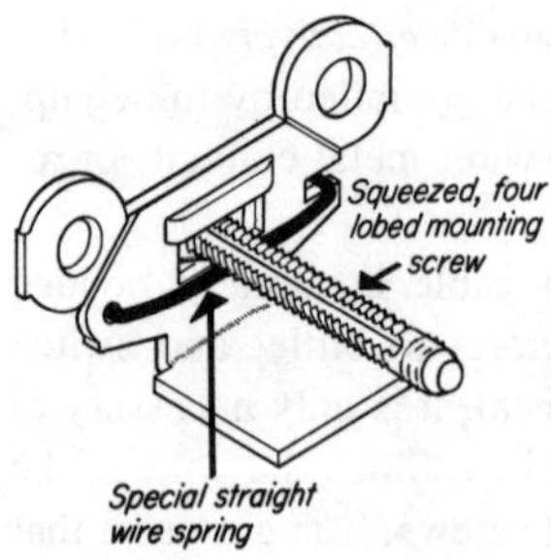

Fig. 11-42 Several brands of receptacles are now available with special construction, doing away with the need for running a grounding wire from the green terminal of the receptacle. (*Slater Electric Inc.*)

The use of such receptacles does away with the need for the bonding jumper from the green terminal to the box.

When using nonmetallic-sheathed cable Type NM or NMC, use the kind with the bare equipment grounding wire. Connect this grounding wire to the neutral busbar in your service equipment cabinet, or to the separate equipment grounding bus if the circuit originates in a panelboard not part of the service equipment. At each box, connect the ends of *all* the grounding wires together, using a solderless connector such as a wire nut. From the junction you have made, run a short wire to the box itself as described in previous paragraphs. If the box contains a receptacle, run another short wire from the junction to the green terminal of the receptacle. All this is shown in Fig. 11-43. This allows you to remove and replace a receptacle without interrupting the grounding continuity to other outlets, as required by the Code.

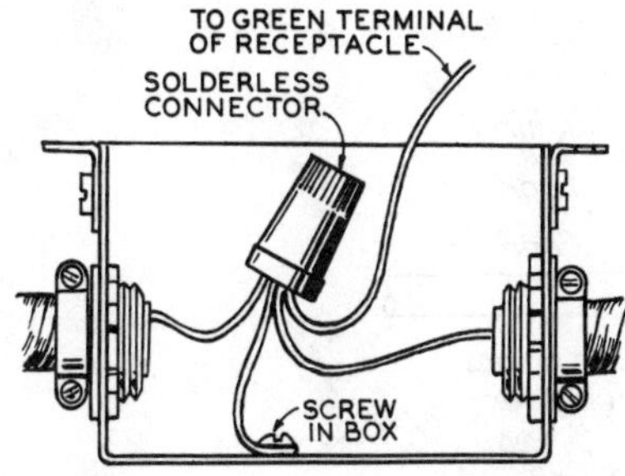

Fig. 11-43 How to connect bare equipment grounding wire of nonmetallic-sheathed cable.

If you are using nonmetallic boxes, the only difference is that the grounding wire does not need to be connected to the box.

Surface-Mounted Boxes. If you are using surface-mounted metal boxes (such as "handy boxes"), where the mounting yoke of the receptacle is in direct solid contact with the surface of the box, you need not install a bonding jumper from the green terminal of the receptacle to the box. If there are fiber washers temporarily retaining the receptacle mounting screws, be sure to remove them to allow for good metal-to-metal contact. If you are using flexible metal conduit, the equipment grounding wire must be installed as previously discussed, but it need not be connected to the receptacle. If you are using nonmetallic-sheathed cable, the grounding wire need not be connected to the receptacle but must otherwise be used as shown in Fig. 11-43.

Switches with Grounding Terminal. Switches are available with a green grounding terminal as on receptacles. In case of breakdown in the switch this provides additional safety. While not required by the Code, this type of switch should be installed in bathrooms, and especially in farm buildings if switches are installed with *metal* covers on *nonmetallic* boxes. Metal covers installed on nonmetallic boxes should always be grounded when within reach of a grounded surface.

Knob-and-Tube Wiring. In the early days of electric wiring, open wires were installed supported on porcelain insulators and run through porcelain tubes in holes through timbers. This wiring method is so rarely used today that it does not warrant the space for a discussion in this book, but you may run into it in existing work (see Chap. 22).

Changing from Conduit to Cable. At times you will find it necessary to change from conduit wiring to cable wiring, but this presents no problem. Make the change in an outlet box as shown in Fig. 11-44. Connect all the

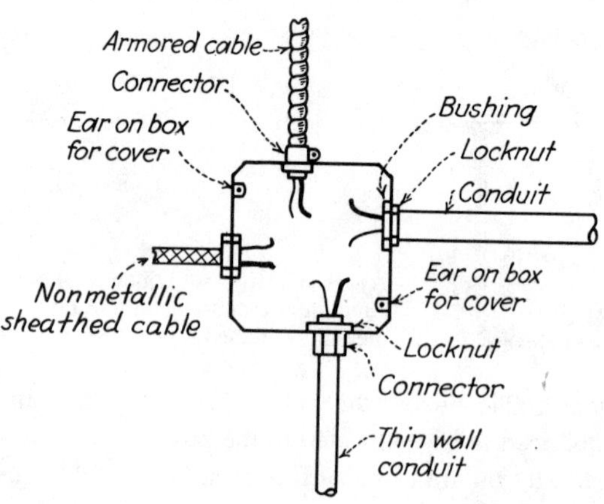

Fig. 11-44 How to change from one wiring system to another. A cover must be installed on the box.

black wires of the same circuit together, connect all white wires of the same circuit together, and connect the bare equipment grounding wire of the cable to the box, using a screw used for no other purpose. Use a metal box, since metal wiring methods are involved. Use a box large enough to avoid crowding of the wires. Leave the wires long enough to easily make the splices using solderless connectors; they are shown short in the illustration to avoid cluttering the sketch. The box must be permanently accessible and covered with a blank cover.

12
Adequate Wiring

An adequately wired home is one that has been wired so that it is completely safe, but also so that the occupants will get maximum convenience and utility from the use of electric power.

They must have light available where needed, in the amount needed, from permanently installed fixtures or portable lamps. They must be able to plug in lamps, clocks, radio, TV sets, and so on, where they please, without resorting to extension cords, even after the furniture is moved around. They must be able to turn lights on and off in any room without stumbling through darkness to find a switch, to move from basement to attic with plenty of light but without leaving an unneeded light turned on behind them.

They must be able to plug in needed appliances without unplugging others; motors must get full power and heating appliances must heat quickly, and lights must not dim as an appliance is turned on. Fuses must blow or circuit breakers trip only in case of a fault, or in case of defective appliances.

Many people feel that if a house is wired "according to the Code" it will necessarily be adequately wired. As explained in Chap. 1, however, the Code is concerned *only* with safety. Hence, as stated in NEC Sec. 90-1(b) compliance with the Code and proper maintenance will result in a reasonably safe installation "but not necessarily efficient, convenient, or adequate for good service or future expansion of electrical use."*

A house will be adequately wired only if you plan it that way. Many houses are being wired with entirely too little thought about adequacy, even though today's norm requires much more light, and a greater variety of appliances than ever before—cooking and laundry appliances, color TV sets, air conditioners—appliances that were unheard of when many existing houses were built. Who is to say what new uses for electric power will come along in the next ten or twenty years? These things must be planned for.

Factors in Adequate Wiring. In order that a house may be adequately wired, careful attention must be paid to the following details:

1. Service of sufficient capacity.
2. Wires of sufficient ampacity throughout the house.
3. Sufficient number of circuits.
4. Receptacle (plug-in) outlets in sufficient number.
5. Lighting fixtures of proper design, in sufficient number.
6. Wall switches in sufficient number for complete flexibility.
7. Miscellaneous outlets and devices for signaling, security, and so on, in proportion to the pretentiousness of the house.

Service. The Code in Art. 100 defines the service as "The conductors and equipment for delivering energy from the electricity supply system to the wiring system of the premises served."* As used in this book, "service *equipment*" will mean the wires from the point where the power supplier's wires end, to the meter, to and including the circuit breakers or fuses and similar equipment in the building, up to the point where branch circuits or feeders begin. It will also include the ground wire. The service equipment must be of sufficient size so that the maximum load in use at one time will neither overload the entrance wires, causing excessive voltage drop and wasted electricity, nor trip breakers or blow fuses. Provision must be made for future equipment that the owner will no doubt install later.

Wire Size. Remember that the Code specifies only minimum sizes. For long runs or heavy loads, be sure to use wires large enough to prevent excessive voltage drop.

*Reprinted with permission from NFPA 70-1984, National Electrical Code®, Copyright© 1983, National Fire Protection Association, Quincy, Massachusetts 02269. This reprinted material is not the complete and official position of the NFPA on the referenced subject, which is represented only by the standard in its entirety.

Circuits. If all the lights in a house were protected by a single circuit breaker or fuse, the entire house would be in darkness when a breaker tripped or a fuse blew. To avoid this the outlets are divided into groups called branch circuits, each protected by a separate circuit breaker or fuse. The greater the number of circuits, the greater the flexibility, the less the likelihood of breakers tripping or fuses blowing (because the likelihood of overloading any one circuit is reduced), and the lower the voltage drop, thus making for brighter lights and greater efficiency of appliances.

Receptacles. Sufficient receptacles do away with the need for extension cords, which are unsightly, inconvenient, and dangerous, both from the danger of possible injury caused by tripping on them, and also from the standpoint of electrical and fire hazards caused by defective or damaged cords. The Code in Sec. 210-52(a) requires that:

> In every kitchen, family room, dining room, living room, parlor, library, den, sunroom, bedroom, recreation room, or similar rooms of dwelling units, receptacle outlets shall be installed so that no point along the floor line in any wall space is more than 6 feet, measured horizontally, from an outlet in that space, including any wall space 2 feet or more in width and the wall space occupied by sliding panels in exterior walls. The wall space afforded by fixed room dividers, such as freestanding bar-type counters, shall be included in the 6-foot measurement.*

The receptacles required are in addition to any receptacles that are over $5^1/_2$ ft above the floor, or are inside a cabinet, or are part of an appliance or lighting fixture. In kitchens and dining rooms, the Code also requires a receptacle at each counter space wider than 12 in. In each one-family dwelling you must install at least one receptacle within 6 ft of the intended location of the laundry equipment; in each basement and attached garage (*in addition* to any provided for laundry equipment); outdoors (of one- *and two-family* dwellings); and (in *all* dwelling units) at every bathroom basin location. The bathroom and outdoor receptacles and at least one receptacle in the garage must be GFCI-protected (see Chap. 9).

These are Code minimums, but you may want to install additional receptacles for added convenience, for portable appliances, for workshop motors, or for a vacuum cleaner where no furniture will ever block it.

*Reprinted with permission from NFPA 70-1984, National Electrical Code®, Copyright © 1983, National Fire Protection Association, Quincy, Massachusetts 02269. This reprinted material is not the complete and official position of the NFPA on the referenced subject, which is represented only by the standard in its entirety.

Lighting Outlets. Many rooms require one or more permanently installed lighting fixtures for general lighting. Additional lighting in most rooms can be, and usually is, provided by floor or table lamps. The lighting fixtures themselves can be whatever you choose; this will be discussed in more detail in Chap. 14.

Wall Switches. Permanently installed lighting fixtures that are controlled by a pull chain, or by a switch on the fixture itself, are inexcusable today, except perhaps in closets so small that it is impossible to miss the pull chain. With this exception, every fixture should be controlled by a wall switch.

The Code requires that a wall switch must control lighting in every habitable room, hallway, stairway, and attached garage and at all outdoor entrances. In kitchens and bathrooms the switch must control permanently installed lighting fixtures; in other rooms the switch may control receptacle outlets into which floor or table lamps can be plugged.

If a room has only one door, the logical location for the switch is near that door. If, however, there are two entrances, it is equally logical that there should be a switch at each door so that the light can be controlled from either location; in other words use a pair of 3-way switches. Should there be three entrances, a switch at each of the three entrances is a touch of luxury the occupant will appreciate. In a house that has been really adequately wired, you can enter by any outdoor entrance and move from basement to attic without ever being in darkness, yet never having to retrace your steps to turn off lights.

Where it is difficult to add 3-way switching in an existing building, consider the convenience of a switch with a time delay for the "off" position such that your way will be lighted for a short preset period of time after you move the switch handle to "off," after which the switch automatically turns off the lights.

In any case, provide plenty of switches, because much energy is wasted by leaving on lights when they are not needed. Energy conservation should be a way of life, so make it easy to turn off lighting whenever it is not needed.

Miscellaneous Outlets. Every house needs a doorbell or chimes with a push button at each door. Pilot lights are desirable at switches that control lights that cannot be seen from the switch location, to indicate whether the lights are on or not; common locations for pilot lights are switches for basement, attic, or garage lights. Careful consideration of these details will make a home much more livable. Other suggestions will be found in Chap. 20.

Adequacy by Rooms. Some rooms need much more light than others; some need more receptacles than others. Provide each room with what it needs.

Living Rooms. At one time lamps bigger than 100 watts were not often used in floor or table lamps; today 300-watt lamps are common. However, floor and table lamps provide illumination only in local areas, leaving dark places in much of the room area. Therefore general illumination is needed. Ceiling fixtures are rarely seen today; cove lighting, installed on the wall some distance below the ceiling, will if properly designed provide excellent lighting. See Chap. 20.

Be generous with receptacle outlets in the living room. The Code minimum of receptacles placed so that no point along the wall will be more than 6 ft from a receptacle isn't really adequate in a living room. Cords on floor and table lamps are not always 6 ft long. Install enough receptacles so that floor and table lamps can be placed where you want them without using an extension cord. Locate one so that it will always be readily available for a vacuum cleaner, no matter how the furniture is arranged in the room.

The ordinary duplex receptacle is really two receptacles in a single housing, so that two different things can be plugged in at the same time. In this book each set of openings in these two-receptacles-in-one will be referred to as one of the "halves" of the duplex receptacle. Most duplex receptacles are constructed so that both halves are either on or off. There is also available a "2-circuit" receptacle constructed so that one of the two halves is permanently live for clocks, radios, and so on, but the other half can be controlled by a wall switch.[1] Instead of each lamp having to be turned off separately, the entire group can be turned off at one time by a single wall switch.

In planning the switches for the living room, be sure to use 3- or 4-way switches so that lights can be controlled from any entrance to the room (and from upstairs too, if there is an upstairs). Low-voltage remote-control switches, which will be discussed in Chap. 20, should be considered.

Sunroom, Den. Provide a generous number of receptacle outlets for lamps and appliances, and provide general illumination (probably a ceiling fixture) depending on how the room is to be used.

Dining Room. Be sure to provide a ceiling outlet for a lighting fixture,

[1] Many ordinary duplex receptacles are so made that they can quickly be changed to the 2-circuit type as they are installed; they will be discussed in Chap. 18, as will the wiring of such receptacles.

controlled by 3-way switches located at both entrances to the room. Visualize the arrangement of the furniture, and locate the ceiling light so that it will be over the center of the dining table rather than in the center of the room. Wall brackets are little used; they are more decorative than useful. Lots of receptacle outlets should be provided, taking into consideration the probable location of the furniture. Too many dining-room outlets are located where it is impossible to get at them easily for a vacuum cleaner and table appliances.

Kitchen. There are kitchens—and kitchens. One will be the modest kitchen in a small five-room house, with relatively few appliances; another will be the deluxe kitchen in a larger house, with several thousand dollars' worth of equipment. What is adequate in the one will be too little in the other. But whatever the nature of a particular kitchen, more of the occupant's waking hours are probably spent there than in any other room of the house. Therefore it is but logical that special attention should be paid to adequacy of wiring in that room.

For general lighting, there should be a ceiling outlet, controlled by switches at each entrance to the kitchen. A light over the sink and another over the stove are essential, for without these the occupant will be standing in his or her own shadow when working at these points. These lights should be controlled by wall switches.

The kitchen needs lots of receptacles. Many appliances consume 1000 watts or more, and several are often in use at the same time. The Code recognizes this and requires two special 20-amp "small appliance" circuits for portable appliances; no lighting outlets may be connected to these circuits. The details of these circuits will be discussed in the next chapter.

For a kitchen clock, install a clock-hanger type of receptacle of the kind shown in Fig. 12-1, at the location where a clock is to be placed. The receptacle itself is located at the bottom of a well in the device; the outlet supports the clock. Cut the cord of the clock to a few inches in length; the cord and plug will be completely concealed behind the clock. This clock-hanger outlet may be installed either on a lighting circuit, or on the 20-amp small appliance circuit.

Even if you do not intend to install an electric range immediately, it would be wise to provide a range receptacle in the initial installation, for it is quite possible that sooner or later an electric range will be installed. It will cost very much more to add the receptacle later.

If the kitchen contains the type of utilities that are more often placed in

Fig. 12-1 This outlet supports a clock. The receptacle is in a "well" so that the cord and plug of the clock are completely concealed. (*Pass & Seymour, Inc.*)

the basement (clothes washers, clothes dryers, water heater, and similar appliances), be sure to provide the individual outlets required for that purpose. To add such outlets later is usually much more expensive than to install them in the beginning.

Breakfast Room. A ceiling fixture directly above the breakfast table or bar, and controlled by a wall switch, is necessary. At least one of the receptacles should be at table height for convenience in operating a toaster or coffee maker.

Halls. Every hall should have a ceiling light and a receptacle. Many houses have rather long halls with rooms opening off either side. These should have two ceiling lights, controlled by 3-way switches at either end. A receptacle for a vacuum cleaner is an absolute necessity.

Bedrooms. Every bedroom should be provided with general illumination controlled by a wall switch; some prefer ceiling fixtures, others cove lighting, which will be discussed in Chap. 20. One of the required receptacles, or an additional one, should be located where always accessible for the vacuum cleaner. Be sure there are enough for table lamps, heating pad, radio, and so on.

It will be well to remember that sooner or later you will probably add a room-size air conditioner. While the smaller models can be plugged into an ordinary circuit *if it is not already loaded,* a larger one might require a circuit of its own. One receptacle on a separate circuit would be a wise investment for the future.

Clothes Closets. It is most exasperating to grope around trying to find something in a dark closet. This has caused people to use a match or lighter for illumination, with tragic results. A hot lamp (bulb)[2] can also

[2] The glass bulb of a lamp often reaches a temperature over 400°F.

have tragic results if it comes in contact with, or is left turned on near, combustible material in a clothes closet. Hence, NEC Sec. 410-8 has specific rules about the placement and type of lights in a clothes closet.

A pendant light (a light socket on the end of a drop cord) is not permitted because it can be moved about and laid on or against clothing. If an incandescent ceiling fixture is used (as for a walk-in closet), it must be mounted on the ceiling in a location that is unobstructed vertically all the way to the floor and that has a horizontal clearance of not less than 18 in from any storage area, such as the area on and above shelving. A recessed incandescent fixture with a solid lens or a surface-mounted fluorescent fixture may be located in or on the ceiling with a minimum horizontal clearance of 6 in from any storage area. If a bracket fixture is used (as for a shallow closet), it must be installed on the wall *above* the closet door and at least 18 in from any storage area.

Shallow closets and some walk-in closets are usually of such size that a pull-chain switch can be found without having to grope for it, though a wall switch (perferably with a pilot light) is more satisfactory. For a deluxe installation, use an automatic door switch, which automatically turns the closet light on when the door is opened and off when the door is closed. A door switch requires a special narrow box (which in some brands is provided with the switch), as shown in Fig. 12-2.

Fig. 12-2 A door switch turns a closet light on when the door is opened, off when it is closed. (*Pass & Seymour, Inc.*)

Bathrooms. To provide enough light for putting on makeup or shaving, a lighting fixture (such as a fluorescent bracket) on each side of the mirror is essential; it must be controlled by a wall switch. A ceiling fixture is needed in a large bathroom, although for the average bathroom, the lights at the mirror may be sufficient.

A touch of luxury is an infrared heating lamp fixture installed in the ceiling, or a permanently wired electric wall heater installed near the floor, well away from the tub, shower, and basin. Either one should be controlled by a wall switch. An exhaust fan is desirable in all cases. Faucets and other grounded objects should never be touched while holding a plug-in shaver or other appliance. Never, never touch an appliance such as a radio while in the tub or shower.

The Code in Sec. 210-52(c) requires a receptacle near the bathroom basin for electric shavers and similar appliances; it must be protected by a GFCI.

In general the use of portable appliances in bathrooms is to be emphatically discouraged. The occupant of a tub or shower is in direct contact with grounded plumbing, the ideal condition for a fatal shock. There are records of dozens of fatal accidents each year caused by people in bathrooms, especially while in tubs, touching a defective appliance or letting a heater or radio fall into the tub; even if the appliance is not defective, it can be just as fatal as a defective one. The same fatal result can be brought about by touching a defective cord while at the same time touching a faucet or other grounded object.

If it is necessary to use a therapeutic appliance (such as a "whirlpool" pump) while in a bathtub, be sure it is UL Listed for the purpose. Preferably, another person should be in attendance while the appliance is in use, so he or she can unplug it in case of an emergency. The appliance should be plugged into a receptacle protected by a GFCI.

Porches and Patios. If the porch is a simple stoop, a ceiling or wall light lighting the floor and the steps is sufficient; of course it must be controlled by a wall switch inside the house. An illuminated house number is a touch the owners and their friends will appreciate.

If the porch is larger, or if there is a patio that is used more or less as an outdoor living room in summer, provide a number of receptacles for radios, floor lamps, appliances, or yard tools. The Code in Sec. 210-8(a) requires that such receptacles be protected by a GFCI.

Basements. First of all there should be a light that illuminates *every step* of the stairs, controlled by 3-way switches at the top and bottom of the stairs. If one switch is in the kitchen or some other location from which the basement light cannot be seen, use a switch with a pilot light to indicate when the basement light is on. Beyond this, the requirements vary greatly, depending on how elaborate the basement is.

If there is an all-purpose "amusement room," which might be anything from a children's playroom to a second living room, wire it as you would the living room. Since the ceilings may be relatively low, recessed fixtures that fit flush or semiflush with the ceiling may be considered. One such fixture is shown in Fig. 12-3. Naturally they will not provide the light over as wide an area as conventional fixtures, but they do give good light directly below for cards, Ping-Pong, and other games. Provide receptacles generously, and provide wall switches for the lights; pull chains can be maddening.

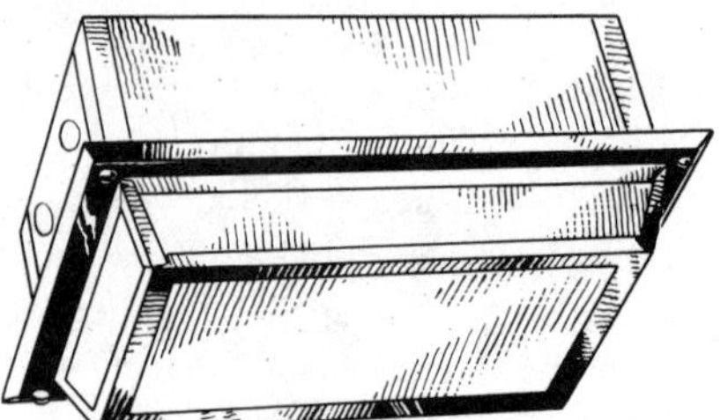

Fig. 12-3 Recessed ceiling fixtures are convenient where ceilings are low. Additional lighting from floor lamps is usually needed.

As already stated, NEC Sec. 210-52(e) requires at least one receptacle for the laundry area. Section 220-3(c) requires a 20-amp circuit for one or more receptacles, which must be within 6 ft of the intended location of the appliance. Since the clothes washer requires one receptacle, at least one more is required for an iron. Many people like to iron in the basement; a good ceiling light, preferably with a reflector, is essential in that area. Remember, a clothes dryer requires its own *separate* circuit.

At other points in the basement, install ceiling lights as required. If there is a storage area, it requires a ceiling light; surely one is needed near a furnace. Nearly every basement has at least a corner that sooner or later becomes a workshop, and a light plus a receptacle (usually several receptacles) is essential for that area.

In unfinished basements without a plastered or other reflective ceiling finish, it is wise to install reflectors of the general type shown in Fig. 12-4.

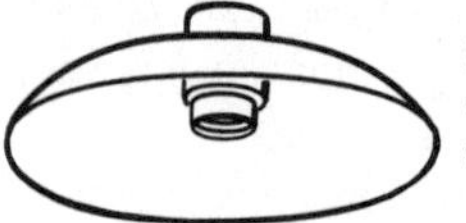

Fig. 12-4 A reflector greatly increases the amount of *useful* light obtained from a lamp.

Use of such reflectors will greatly increase the amount of useful light. Dark ceilings absorb light; reflectors, if kept clean, will throw the light downward where it is needed.

Attics. If the attic is used mostly for storage, a single light placed so that it will light up *the stairs* may be sufficient. It should be controlled by a wall switch, preferably with a pilot light, at the bottom of the stairs. If it is a large attic, install additional lights as required; they can all be controlled by the same switch. If the attic is really an additional floor unfinished but *capable of being finished later into completed rooms,* do not make the mistake of providing only a single *outlet,* from which cable will later radiate in octopus fashion to a large number of additional outlets. Bear in mind possible future construction, and provide an extra *circuit,* terminating it in a box in some convenient location from which additional wiring can be added at some future date.

Stairs. This paragraph will be partly repetition; read it in the interest of safety. Thousands of people every year are involved in serious accidents due to falls on stairs. They stumble because they cannot see some object lying on one of the steps, or fall because they think they have reached the bottom of the stairs when in fact there is one more step. A fixture that lights up the general area of the stairs, but does not light up every *single step,* is an invitation to such an accident. If necessary, provide two lights, one at the top and another at the bottom of the stairs. Stairway lights (except possibly attic lights) should be controlled by 3-way switches so that they can be turned on and off from either level.

13

Service Entrance and Branch Circuits

This chapter will discuss only residential installations. For service entrances in farm buildings, see Chap. 23. A single set of wires brings electric power from the power supplier's line into the house. Inside the house, wires are run to each outlet where power is to be used. The wiring to a group of outlets, and in many cases to a single outlet, is called a "branch circuit," which the Code in Art. 100 defines as "the circuit conductors between the final overcurrent device protecting the circuit, and the outlet(s)."*

The wires enter the house at the service equipment, which contains *main* overcurrent protection (breakers or fuses) that protect the entire building. The wires between the main overcurrent protection and the point where the branch-circuit overcurrent protection is installed are "feeders." In a residence, the *branch-circuit* overcurrent protection is often enclosed in the same cabinet that contains the *main* overcurrent protection, so that there are no feeders; or, the branch circuits may origi-

*Reprinted with permission from NFPA 70-1984, National Electrical Code®, Copyright© 1983, National Fire Protection Association, Quincy, Massachusetts 02269. This reprinted material is not the complete and official position of the NFPA on the referenced subject, which is represented only by the standard in its entirety.

nate in a separate panelboard. But do note that an over*load* device[1] installed just ahead of a motor to protect the motor and not the circuit is *not* the *branch-circuit* over*current* device, and you must disregard it when determining where a branch circuit begins.

This chapter explains how to determine the number of branch circuits to install, the size of wire to use for each circuit, the size of the service-entrance wires, and similar details. Chapter 17 will discuss the actual selection and installation of the particular components involved.

In this chapter there will be many references to overcurrent devices, which may be either circuit breakers or fuses. It would be tedious to use repeatedly long phrases such as "circuit breakers or fuses" or "the circuit breaker trips or the fuse blows." Neither is it practical to duplicate the illustrations showing circuit breakers in one set and fuses in the counterparts. Therefore the references will be mostly to fuses, but it must be understood that the references apply equally to circuit breakers and fuses, unless otherwise stated.

Advantages of Numerous Branch Circuits. Having separate circuits for groups of outlets leads to the practical result that an entire building is never in complete darkness on account of a blown fuse except on the rare occasions when a main fuse blows. There is added safety in having a considerable number of separate branch circuits. Most of the time each fuse carries but a portion of its maximum carrying capacity. However, there are times during each day when a considerable number of lights are turned on. A washing machine may be running; perhaps an iron is being used; other devices may also be put into service at the same time. With a sufficient number of circuits in a properly designed installation, the load will be fairly well divided among the various fuses with the result that none is overloaded and none blows. Where there are only a few circuits, each fuse will carry a heavier load and fuses will blow far more frequently. Fuses that blow frequently tempt the owner to use a fuse larger than permitted for the size of wire used, or perhaps to use substitutes that defeat the purpose of fuses; both are dangerous practices that can lead to fires. Another practical consideration is that the greater the number of circuits, the lower the voltage drop will be in each circuit, with less wasted power, and greater efficiency: brighter lights, toasters that heat quicker, percolators that make coffee faster, and so on.

[1] As will be explained in later chapters, sometimes devices other than fuses or ordinary breakers are installed to protect a motor; sometimes such devices are installed inside the motor.

Area Determines Number of Circuits. The Code in Secs. 220-2 and 220-3 specifies the minimum number of circuits that may be installed. Do read the explanatory note to Sec. 220-2(b): "The unit values herein are based on minimum load conditions . . . and may not provide sufficient capacity for the installation contemplated."* The starting point in determining the minimum is the floor area. NEC Sec. 220-2(b) states that

> The floor area for each floor shall be computed from the outside dimensions of the building, apartment, or other area involved. For dwelling unit(s), the computed floor area shall not include open porches, garages, or unused or unfinished spaces not adaptable for future use.*

Those last five words, "not adaptable for future use," are important. Many houses are being built that have unfinished spaces not at first used for living purposes but intended to be completed later by the owners, when their need for added living space or their financial ability suggests that this be done. This future space must be considered when planning the original electrical installation. One outlet may possibly be installed in such a space, and when the space is later finished, many more outlets are added, branching off from the one and only outlet originally installed. That overloads the existing circuit. Run a separate circuit to the unfinished space.

The question of basements is not too clear in the Code. If the basement space is to be used for ordinary basement purposes as in older and less pretentious houses, it can be safely disregarded in your watts-per-square-foot calculations, so far as *lighting* circuits are concerned. But if any part of a basement can be finished off into an amusement room, or an upper unfinished floor can be finished off into living areas, add those areas to the total square-foot area otherwise determined. Local building code requirements for ceiling height and/or window area may be helpful in determining possible future use of a space.

Houses and Apartments. The Code in Sec. 220-2 requires you to allow a minimum of 3 volt-amperes (VA) per sq ft of floor area to determine the minimum number of *lighting* branch circuits. A lighting branch circuit includes not only permanently installed lighting fixtures but also receptacle outlets for floor and table lamps as well as clocks, radios, TV, vacuum cleaner, and similar portable appliances, all of which consume comparatively small amounts of power. But this does *not* include larger appliances such as kitchen appliances and larger room air conditioners.

A house that is 25 × 36 ft has an area of 900 sq ft per floor, or 1800 sq ft for two floors. Assume that it has a space for a future finished amusement room in the basement, $12^1/_2$ × 16 ft, or 200 sq ft. This makes a total area of 2000 sq ft and will require a minimum of 3 × 2000 or 6000 VA for the lighting circuits.

The usual lighting circuit is wired with No. 14 wire that has an ampacity of 15 amp, which at 120 volts is 1800 VA (15 × 120 = 1800). For 6000 VA, $^{6000}/_{1800}$ or 3.3 lighting circuits would be required. That is more than three circuits, and since circuits must not be overloaded, four circuits must be provided. Divide the loads as evenly as practical among the four circuits.

You can reach the same answer another way. Since each circuit can carry 1800 VA and since 3 VA must be provided for each square foot, each circuit can serve $^{1800}/_{3}$ or 600 sq ft. For 2000 sq ft there will then be required $^{2000}/_{600}$ or 3.3 circuits as before.

Remember, the Code is concerned primarily with *safety,* not with convenience or adequacy. Many people feel that one 15-amp lighting circuit should be provided for every 500 sq ft of area, rather than every 600 sq ft. In the example of the preceding paragraphs, $^{2000}/_{500}$ would lead to an answer of exactly four circuits, the answer we already reached. But if the area had been 2200 sq ft, on the basis of 600 sq ft per circuit, the answer would be four lighting circuits, while on the basis of 500 sq ft it would be five circuits. The extra circuit would provide for more flexibility and allow some provision for future needs.

Small Appliance Circuits. Many years ago kitchen appliances consisted of an iron and a toaster, each consuming about 600 VA. They were plugged into ordinary lighting circuits and were rarely both used at the same time. A modern kitchen is equipped with many additional appliances: coffee maker, mixer, blender, roaster, frying pan, deep-fat fryer, to mention some of the common ones. Many such appliances consume 1000 VA, and an electric roaster even more. Often several such appliances are used at the same time. Ordinary lighting circuits no longer can handle such loads.

The Code now requires special circuits which are called "small appliance branch circuits'' for such appliances. NEC Sec. 220-3(b) requires two or more 20-amp small appliance branch circuits for all small appliance loads, including refrigeration equipment, in the kitchen, pantry, breakfast room, and dining room of dwelling occupancies. These small appliance circuits must not have other outlets.[2]

Both these special circuits must run to the kitchen so that some of the receptacles in the kitchen will be connected to each of the circuits, thus increasing the number of kitchen appliances that can be operated at one time. Either (or both) of the circuits may extend to the other rooms mentioned.

NEC Sec. 220-3(c) also requires one 20-amp circuit for the laundry area, where it may serve one or more receptacle outlets.

Note that on a multioutlet 20-amp circuit, receptacles rated at either 15 or 20 amp may be installed. Also note that a plug designed specifically for a 20-amp receptacle will not fit a 15-amp receptacle, but the ordinary plug designed for 15-amp receptacles will also fit the 20-amp receptacle.

These small appliance circuits and the special laundry circuit are strictly for cord- and plug-connected appliances. No permanently connected appliances or lighting outlets may be installed on them. These circuits must be wired with No. 12 wire and protected by 20-amp overcurrent protection. Each such circuit therefore has a capacity of 20×120, or 2400 VA, in most cases permitting two appliances to be used on it; in the kitchen supplied by both circuits, four appliances could generally be used at one time.

If there is to be a built-in counter, locate the receptacles about 6 or 8 in above the counter top, preferably not more than 36 in apart, so that appliances may be placed where wanted or where convenient and still be within easy reach of a receptacle. The Code requires one receptacle in any counter space wider than 12 in. Spaces separated by items such as range tops, sinks, and refrigerators are considered separate spaces. One or more of what the Code calls ''multioutlet assemblies,'' often called just ''plug-in strips,'' which will be discussed in Chap. 20, can often be used to advantage.

Wire the receptacles so that one receptacle will be on the first circuit, and the next on the second circuit, then alternately first and second cir-

[2] An exception does permit a clock-hanger outlet or an outdoor receptacle to be connected to one of the two small appliance circuits; no lighting outlets may be connected to them.

cuit. This will somewhat reduce the likelihood of overloading one of the circuits if several appliances are used at the same time.

Three-wire circuits will be discussed in Chap. 20. One such circuit is equivalent to two 2-wire circuits, but permits using the same amount of power with less material and with less voltage drop. When installing receptacles on such a 3-wire circuit, it will be found most practical if you connect the upper halves of all the receptacles to the neutral and the black wire of the 3-wire circuit, and all the bottom halves to the neutral and the red wire. If protected by a GFCI, as is recommended but not required, that must be of the special type designed for 3-wire circuits, and with 20-amp rating. When "split-wired" receptacles are installed on a 3-wire circuit, be sure the branch-circuit disconnecting means opens both "hot" wires with a single operation, so that future maintenance at these outlets can be done safely. This will require a double-pole switch in a fused circuit, or a double-pole circuit breaker, or "handle ties" on two adjacent single-pole breakers.

Types of Branch Circuits. The Code recognizes two *types* of branch circuits. The first type is a circuit serving a single current-consuming appliance or similar load, such as a range or water heater. The second type is the ordinary circuit serving two or more outlets, consisting of permanently installed lighting fixtures or appliances, and receptacles for portable loads such as lamps, vacuum cleaners, and similar small appliances, or a combination of both.

Branch Circuits Serving Single Outlets. It is customary to provide a separate circuit for each of the following appliances:

1. Range (or two circuits for separate oven and counter unit).
2. Water heater.
3. Clothes dryer.
4. Garbage disposer.
5. Dishwasher.
6. Each permanently connected appliance rated at 1000 VA or more (for example, a bathroom heater).
7. Each permanently connected motor rated at $^1/_2$ hp or more, as, for example, on a water pump.

Some local codes require an oil burner and blower motor to be on a separate circuit regardless of the horsepower rating of the motor. Some even require the oil burner motor to have a suitable switch at the head of the basement stairs, identified by a *red* face plate.

According to the Code, it is not necessarily required that an individual circuit be installed for each of the items listed, but as a practical matter, this is common practice.

The circuit for the appliance may be either 120 or 240 volts, depending on the voltage rating of the appliance. The wire used must have sufficient ampacity for the current rating of the load it is to serve. The ampere rating of the overcurrent device in the circuit depends on the marked ampere rating of the specific appliance or motor served. If the appliance load consists of a motor only and is not marked with a maximum circuit size, see Chap. 15 for sizing the branch-circuit overcurrent device. If the appliance is expected to operate continuously (3 h or longer), the branch circuit must be 125% of the marked rating. The ampere rating of the plug on a UL Listed cord- and plug-connected appliance will establish the proper branch-circuit size.

Circuits Serving Two or More Outlets. There are five such circuits: 15, 20, 30, 40, and 50 amp, based on the ampacities of Nos. 14, 12, 10, 8, and 6 wire. Of course circuits larger than 50 amp may be installed but may serve only one outlet except on industrial premises.

Note that the ampere rating of a circuit depends on the rating of the breaker or fuse protecting the wire, not on the size of the wire in the circuit. While normally the rating of the breaker or fuse is matched to the ampacity of the wire in the circuit, this is not always the case. A 20-amp circuit is normally wired with No. 12 wire with ampacity of 20, but if you used a 20-amp breaker or fuse and used No. 10 wire (for example, for reduced voltage drop, or for mechanical strength in an overhead run), it would still be a 20-amp circuit.

Number of Outlets per Circuit. For wiring in houses or apartments, the Code places no limits on the number of outlets that may be connected on one branch circuit. But if you have provided the required number of circuits, you will have no need to overload any circuit supplying lighting and receptacle outlets. As already discussed, separate circuits must be provided for special equipment.

Fifteen-Ampere Branch Circuits. This is the size of circuit usually used for ordinary lighting circuits, which include general-use receptacle outlets other than those in cooking, dining, and laundry areas. It is wired with No. 14 wire and protected by a 15-amp breaker or fuse. The receptacles connected to it may be rated at no more than 15 amp, which means that only the ordinary household variety of receptacle may be used. Any type of socket for lighting may be connected to it. No cord- and plug-connected appliance used on the circuit may exceed 12 amp (1440 VA) in rating. If the current serves lighting outlets and/or portable appliances, as is usually the case, and also serves appliances fastened in place, the total load of the fixed appliances may not exceed 7½ amp (900 VA).

Twenty-Ampere Branch Circuits. The special small appliance and laundry circuits described earlier in this chapter are required to be 20-amp circuits. But 20-amp circuits are also permitted to be used for lighting and general-use receptacles, if No. 12 wire is used.

Any circuit wired with No. 12 wire and protected by a 20-amp overcurrent device is a 20-amp circuit as defined in the Code. It may serve lighting outlets, receptacles, permanently connected appliances, or an assortment of them. Any kind of socket for lighting may be connected to it. Receptacles may have either a 15- or a 20-amp rating. No single cord- and plug-connected appliance may exceed 16 amp (1920 VA) in rating. If appliances fastened in place are also on the circuit, the total of all such appliances may not exceed 10 amp (1200 VA).

Thirty-, Forty-, and Fifty-Ampere Branch Circuits. Circuits of these ratings, each serving a single outlet for a range, clothes dryer, or similar appliance, are commonly used and will be discussed in Chap. 21. However, such circuits serving *lighting* outlets are *not* permitted in dwellings; where serving *more than one outlet* for appliances, they are rarely used in dwellings, but such circuits are commonly used in nonresidential wiring and will be discussed in Chap. 27.

Balancing Circuits. In a house served by a 120/240-volt service (which of course contains a neutral wire), the white wire of each 120-volt branch circuit is connected to the incoming neutral, at the neutral busbar in the service equipment or panelboard. Great care must be used in connecting the hot wires of the branch circuits so that the loads will be divided approximately equally between the two incoming hot wires. If this is not done, practically all the load may be thrown on only one of the two hot wires; unbalanced conditions lead to frequent tripping of breakers or blowing of fuses.

Location of Branch-Circuit Overcurrent Protection. The service equipment, which consists of the service disconnecting means and service overcurrent devices [circuit breaker(s) or fuses on pull-out block(s)], is required by NEC Sec. 230-70(a) to be located near the point where the wires enter the building. In some houses, the cabinet of the service equipment, as just defined, also contains the branch-circuit breakers or fuses.

In many houses, most of the power is consumed in the kitchen, while others have large appliances such as a water heater and clothes dryer in the basement, but in either case there are only relatively small loads in the remainder of the house. For minimum cost, plan your layout so that the wires enter the house at a point from which the relatively large wires to the range, the water heater, the clothes dryer, and the kitchen appliances can be short; let the smaller, less expensive wires to the remainder of the building be the long ones.

In most houses today, the service equipment is located in the basement. Sometimes it is located in the kitchen, where it is more accessible. Of course, if the house is really adequately wired, fuses will rarely blow, so this point becomes less important. Be sure you locate the service equipment where there will be plenty of working space, headroom, and, if indoors, illumination.

Circuit Voltage. In dwellings and in guest rooms of hotels and motels the voltage *between conductors* on branch circuits supplying screw-shell lampholders, receptacles, or appliances per NEC Sec. 210-6(c) may not exceed 150 volts, except for permanently connected appliances or cord- and plug-connected appliances rated at more than 1380 watts or 1/4 hp or more. Since fluorescent lighting fixtures do not have lampholders of the screw-shell type, the 150-volt restriction does not apply to them.

Branch-Circuit Schemes. The most usual scheme of locating all branch-circuit fuses (or breakers) at one point is shown in Fig. 13-1. So far as fusing is concerned, the wires beyond the branch-circuit fuses may be as long as you wish, consistent with keeping voltage drop to a reasonable level. Note that 240-volt circuits may be run from any point by simply

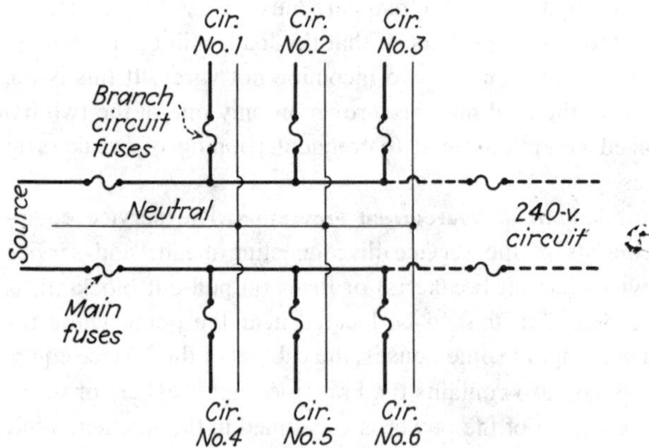

Fig. 13-1 The most common scheme: Locate all branch-circuit fuses in one location.

tapping off the two ungrounded wires. For example, the ungrounded wire of circuit no. 1, and the ungrounded wire of circuit no. 4, together would make one 240-volt circuit, the grounded wire of course being disregarded. One such 240-volt circuit is shown in dashed lines.

In very large houses, the owner may occasionally prefer to locate the branch-circuit fuses in various locations throughout the house, the fuses protecting the basement circuits, for example, being placed in the basement, those protecting the first floor being placed in the kitchen, and so on, as shown in Fig. 13-2. Where the service equipment and the branch-

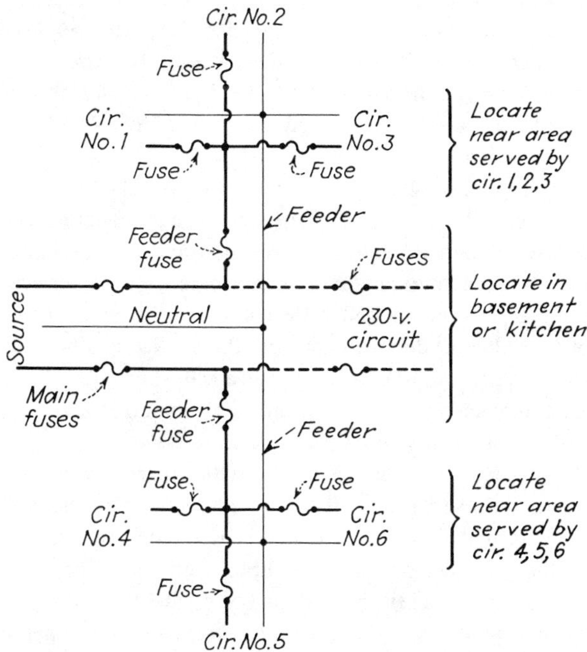

Fig. 13-2 In very large houses, this scheme is sometimes used. Groups of branch-circuit fuses are located in various parts of the house.

circuit fuses are in the same cabinet, as would be the case in an installation made in accordance with Fig. 13-1, the fuses (other than the service or main fuses) are branch-circuit fuses, and the wires beyond these fuses are branch-circuit wires. If the installation is made in accordance with Fig. 13-2, the same wires (but now larger and running to another fuse box called a "panelboard") become *feeders,* and the fuses (but now of a higher ampere rating) become feeder fuses.

In Fig. 13-2, the wires from the service equipment to each location where a fuse cabinet is installed are *feeders,* as already explained. The wires of any feeder must be big enough to carry the maximum load that will be imposed on the feeder by the branch circuits, at any one time. The ampere rating of the overcurrent device for the feeder must not be greater than the ampacity of the feeder. NEC Sec. 215-2(a) requires a minimum size of No. 10 wire for a feeder where the load it supplies consists of (a) two or more 2-wire branch circuits supplied by a 2-wire feeder, or (b) more than two 2-wire branch circuits supplied by a 3-wire feeder, or (c) two or more 3-wire branch circuits supplied by a 3-wire feeder. NEC Sec. 215-2(b) requires that the ampacity of the feeder must not be lower than the ampacity of the service-entrance wires, *if* the feeder carries all the current supplied by service-entrance conductors that are No. 6 or smaller.

If the scheme of Fig. 13-2 is used, one point is important. The fuse cabinets at the end of feeders downstream from the service equipment panelboard will of course contain a neutral busbar to which all the white grounded wires of the branch circuits will be connected. In the service equipment, this busbar *must* be bonded to the cabinet of the panelboard. In a fuse cabinet (panelboard) downstream from the service equipment, this not only is not required, it is prohibited. To repeat, the neutral busbar in the service equipment *must* be bonded to the cabinet; in a downstream panelboard it must be *insulated* from the cabinet.

If you are using nonmetallic-sheathed cable, with a bare equipment ground*ing* wire in addition to the insulated wires, the neutral busbar (which you have just insulated from the cabinet) is used only as a connection point for all the grounded wires of the branch circuits. The bare equipment ground*ing* wires of all the cables must be connected to a separate grounding busbar, and that busbar *must be bonded* to the cabinet.

Distribution of Outlets on Circuits. If the usual scheme of Fig. 13-1 is used, all the branch-circuit overcurrent devices will be in one location, and the question of what particular outlets to place on each circuit must be carefully studied. It is not wise to place all the first-floor lighting circuits on one circuit, all those on the second floor on another circuit, and so on, for then an entire floor would be in darkness if one fuse blows. It is best to have on each circuit lighting outlets of several different rooms and preferably different floors, for then when a fuse blows, there will still be at least some light on each floor.

Service Entrance. Every installation includes wires to bring the electric power into the building, and a good deal of related equipment. According to NEC Art. 100 "service *equipment*" includes only the *main* circuit breaker(s) or fuses to protect the entire installation, with a method of disconnecting all power from the building. But as a practical matter, in residential and farm wiring, the breakers or fuses to protect individual branch circuits are usually contained in the same cabinet with the service equipment as defined by the Code, as is also a neutral busbar to which the incoming neutral wire, the ground wire, and the white wires of all 120-volt branch circuits are connected. If nonmetallic-sheathed cables with bare equipment grounding wires are used, these wires must be connected to a separate grounding busbar, although some inspectors may allow their connection to the neutral busbar in the service equipment if there is a terminal for every wire.

In this book the words "service *equipment*" will be used in the broader sense outlined above. But the term "service *entrance*" (which is not defined in the Code) will include everything from the point where the service drop (the power supplier's wires) ends at the outside of the building, the service-entrance wires in conduit or cable from that point to the meter socket, and from there to the service equipment as broadly defined above, and also the ground wire from the service equipment to the ground. Some of these components have already been discussed in Chap. 9.

The details of these several components depend on many factors, such as the size of the building, the amount of power required, the number of branch circuits, and many others. In this chapter we shall discuss only the general details and general design; in Chap. 17 we shall discuss their actual installation.

Minimum Size of Service. This chapter will concern mostly houses and farm buildings, not industrial or commercial buildings, which will be discussed in Part 3 of this book. NEC Sec. 230-79 specifies the minimum ampere rating of the *disconnecting means,* and NEC Sec. 230-42 specifies the minimum ampacity or size of the *service-entrance conductors.* Both are required to be adequate for the load supplied in all cases, but absolute minimums are also specified as follows:

1. A 15-amp service equipment supplied by two No. 12 copper (or

two No. 10 aluminum) wires is acceptable for a limited load having only one 2-wire branch circuit, such as for a farm storage shed, or for a temporary service for a small power saw used during construction of a house.

2. A 30-amp service equipment supplied by two No. 8 copper (or two No. 6 aluminum) wires is acceptable for a load consisting of not more than two 2-wire branch circuits, such as a small farm building or a temporary construction service for small power tools. By special permission (by your inspector), this size of service may also be used where the load is otherwise limited so that the equipment and wires will not be overloaded, as in a service for a water pump in an isolated location.

3. A 100-amp 120/240-volt 3-wire service supplied by three wires is acceptable (and required as a minimum) for a single-family house that has an initial *computed* load of 10 kW or more, or that initially has six or more 2-wire branch circuits (one 3-wire branch circuit, which will be discussed in Chap. 20, is the equivalent of two 2-wire circuits). This will cover *all* single-family houses. Note that the *computed* load is usually less than the total *connected* load, as will be made clear later in this chapter. The two ungrounded wires for this service must have an ampacity of not less than 100 amp. The grounded neutral in the service often can be smaller than the hot wires, but not less than specified in NEC Sec. 220-22, as will be explained.

4. A 60-amp service is required for all services not covered by items 1, 2, or 3 above, supplied by the number of service wires required with ampacity ratings as follows: If two wires, both wires must have ampacity of 60 amp; if more than two wires, the ungrounded wires must again have an ampacity of 60 amp, but the grounded neutral may sometimes be smaller as will be explained shortly. This 60-amp service may sometimes be used for some apartment buildings where each individual apartment has its own service, and for many small buildings such as some farm buildings.

Thinking now of houses specifically, even if a 100-amp service may be permissible, do remember that the minimum specified by the Code is based on safety, not practicability or convenience. Look ahead: Instead of installing the acceptable 100-amp minimum, install 150- or 200-amp equipment. A few local codes already require a service larger than the 100-amp minimum.

Size of Wires for Various Services. As you already learned in Chap. 6, the ampacity of a given size of wire depends not only on its size but also on its insulation, and how the wire is installed. The ampacity of each size and kind of wire can be found in NEC Tables 310-16 through 310-19. These tables are shown in the Appendix of this book.

When using Type SE cable, sometimes you will find the type of wire used inside the cable marked on the outside of the cable. In this case, the temperature rating of the cable is that of the individual conductors. When the conductor type is not marked on the outside, the cable has a 75°C rating.

In NEC Table 310-16 the ampacity of No. 10 wire is shown to be 30 amp; therefore you might expect it to be suitable for a 30-amp service. As already stated, the minimum required is No. 8. Remember that the wire sizes discussed are for the service-entrance conductors, not those in the service drop, which consists of the power supplier's wires up to your building; the power supplier will determine the size of those wires. On farms,[3] however, wires from the meter pole to a building, while considered to be service-drop wires, must be based on the ampacities shown in NEC Table 310-17 or 310-19 if they are single copper wires in free air or in Table 310-16 if buried. Do remember that the smaller the wires, the greater the voltage drop; you may want to use a size larger than the minimum permissible.

When using fused service equipment of a size *larger than the minimum required*[4] (because a switch rated at the specific minimum is not available), it is not contrary to the Code to use service wires with a smaller ampacity rating than the ampere rating of the switch, provided the fuses in the switch do not have a higher ampere rating than the ampacity of the wires. But if you are using a wire that has an ampacity such that there is no standard fuse of that ampere rating, you may use a fuse of the next larger standard rating. Standard available ratings of fuses have already been discussed in Chap. 5.

Computing the Load. The service equipment and conductors do not need to have an ampere rating equal to the total ampere ratings of all the individual branch circuits. There never will be a time when every circuit in the house is loaded to its maximum capacity. Therefore "demand fac-

[3] Farm wiring will be covered in more detail in Chap. 23.

[4] Service switches are rated at 30, 60, 100, 200 and 400 amp (with of course higher ratings for larger buildings) with no in-between ratings.

tors" have been established, based on many tests and past experiences, that represent the maximum part of various types of loads that are likely to be in use at any one time. These demand factors and methods of computing loads are covered by NEC Art. 220. For residential occupancies, the loads are computed for specific types of loads: lighting, small appliance circuits, laundry circuit, special (usually heavy-duty) fixed or stationary appliances, heating, central air conditioning, and similar loads.

Demand Factor: Lighting. Compute the lighting loads on the basis of 3 VA per sq ft. This of course will include the receptacles for lamps, radios, TV, vacuum cleaners, and so on. It makes no difference whether the lighting circuits are 15- or 20-amp circuits—the total will be the same. *Temporarily* do not consider any demand factor; how to do so will be explained later.

Demand Factor: Small Appliance and Laundry Circuits. Even if each such 20-amp circuit can provide 2400 VA, allow only 1500 VA for each circuit in your computations, for all these branch circuits will not be fully loaded at the same time. Since a minimum of two small appliance circuits and one laundry circuit are required by the Code, allow 4500 VA in your computations. Remember that these circuits are for portable appliances such as toasters, coffee makers, irons, and refrigeration equipment in kitchen and dining areas and also for clothes washers and perhaps an iron in the laundry area. But these 20-amp circuits must *not* serve ranges, range tops, separate ovens, water heaters, and similar heavy-duty stationary and fixed appliances.

Demand Factor: Lighting, Small Appliance, and Laundry Circuits. Now add together the volt-amperes computed for lighting and the 4500 VA of the preceding paragraph. Count the first 3000 VA of the total at 100%, but the remainder above 3000 VA at 35%. Examples will be shown later in this chapter.

Special Appliances. As used here, "special appliances" includes all appliances not supplied by a lighting circuit, a small appliance circuit, or the laundry circuit. Examples are ranges, range tops, ovens, water heaters, clothes dryers, garbage disposers (and sometimes, but not always, room air conditioners, which will be discussed in Chap. 21). Each such appliance must be supplied by its own circuit, which serves no other loads. The load in watts stamped on the name-plate of the appliance must be used in your computations, at its full watts, or volt-ampere, rating, with several exceptions.

Demand Factor: Four or More Appliances. If there are four or more *fixed* appliances other than ranges, clothes dryers, space-heating equipment, or air-conditioning equipment *and* if all are supplied by the same service-entrance conductors or the same feeder (which is usually the case in a residence), then a demand factor of 75% may be applied to the total volt-ampere ratings of these items.

Demand Factor: Ranges. A range is actually a multiunit load assembled as a single appliance. The watts rating of the range is the total of the oven load, plus the total load of all burners turned to their highest setting. All these heating elements are not at all likely to be in use at any one time. Therefore a load of only 8000 watts (8 kW) may be used for any range rated at not over 12 000 watts (12 kW). If the range is rated at over 12 000 watts, start with 8000 watts; then for each kilowatt (or fraction thereof) above 12 000 watts, add 400 watts. For example, a range might have a rating of 16 kW, which exceeds 12 kW by 4 kW. Add 400 watts for each of the extra kilowatts, or 1600 watts altogether; add that to the first 8000 watts, for a total of 9600 watts, in your computation. For ranges and other cooking appliances, kilovolt-amperes and kilowatts are considered equivalent.

Demand Factor: Separate Ovens and Range Tops. If one range top and not more than two ovens are in the same room and supplied by *one* circuit, add the total of their separate ratings, and proceed as with self-contained ranges discussed in the preceding paragraph.

In all other cases, the name-plate rating of each item must be used separately, and no demand factor may be used. It is much simpler and much better to install a separate circuit for each range top and each oven; maintenance and replacement problems are greatly simplified if this is done. Voltage drop will be lower with separate circuits.

Note that ordinary portable hot plates and portable microwave ovens are *not* range tops or ovens. (The Code name for a range top is "counter-mounted cooking unit," and for a separate oven is "wall-mounted oven.")

Demand Factor: Dryers. The demand factors of Table 220-18 (see Appendix) may be applied to five or more clothes dryers. Dryers must be calculated at 5000 VA or the name-plate rating, whichever is larger.

Demand Factor: Other Loads. There are no demand factors for other loads; use their full rating in computing the total for the building. If there are loads that would never be used at the same time (for example, space heating and air conditioning), count only the larger load. For space heating, use the total volt-ampere rating of all the elements; if the circuits are so arranged that not all the elements can be on at the same time, count only the maximum load that can be used at one time.

Many times the equipment will consist of a motor or of a motor plus other equipment. If the equipment is rated in watts, use that figure. If it is rated only in amperes, multiply the amperes by the volts of the circuit (120 or 240); the figure will be a little high because of the power factor of the motor, but by using that figure you will be on the safe side. For motors as such, use a figure of 1000 watts for each horsepower; this is by no means precise but will serve the purpose.

Computing a Specific House. Computations for several different houses will follow. Note that the computations shown will determine the size of the *ungrounded* wires in the service equipment. Often the grounded neutral wire in the service may be smaller than the ungrounded wires, and that will be discussed later in the chapter.

Computing a 2000-Sq-Ft House. Assuming a house of 2000-sq-ft area (determined in the same way as outlined for determining the number of circuits) and assuming it has no major electrical appliances such as a water heater and range, the computations are as shown in the first table below.

	Gross computed volt-amperes	Demand factor, %	Net computed volt-amperes
Lighting, 2000 sq ft at 3 VA	6 000		
Small appliances (minimum)	3 000		
Special laundry circuit	1 500		
Total gross computed volt-amperes	10 500		
First 3000 VA		100	3000
Remaining 7500 VA		35	2625
Total net computed volt-amperes			5625

The ampere load is 5625/240, or 23.4 amp. Since No. 10 wire has an ampacity of 30, it would appear large enough for the service, but remember the Code requires a minimum of a 100-amp service if there are more than five branch circuits. Since this house will have a minimum of four lighting circuits, two small appliance circuits, and one laundry circuit, a minimum of seven altogether, a 100-amp service is required.

Assume now that an electric range rated at 12 kW, and also a permanently installed bathroom heater, are to be installed. The calculations then will be:

	Gross computed volt-amperes	Demand factor, %	Net computed volt-amperes
Lighting, 2000 sq ft at 3 VA	6 000		
Small appliance circuits	3 000		
Special laundry circuit	1 500		
Total gross computed volt-amperes	10 500		
First 3000 VA		100	3 000
Remaining 7500 VA		35	2 625
Bathroom heater	1 500	100	1 500
Range		Min.	8 000
Total net computed volt-amperes			15 125

The ampere load now is $15\,125/240$, or 63 amp. Therefore a 70-amp[5] service would appear to be big enough, but again we have more than five circuits, so a 100-amp service is required. The small extra cost will be a good investment.

Now assume a large suburban house with an oil-burning furnace and its own independent water system, since there is no city water. In addition to the range and bathroom heater of the previous example, let us add a water heater rated at 3500 VA, a clothes dryer rated at 5000 VA, two $1/4$-hp motors (one for the oil burner and one for the fan on the furnace), and one $1/2$-hp motor for the water pump. If the house has an area of 3000 sq ft, 9000 VA will have to be included for light circuits. The computation will then be as shown in the next table.

The ampere load, not including motors, is $24\,675/240$, or 102.8 amp. The ampere load of motors must be added separately. Use the name-plate ratings[6] if you have the motors. If you do not yet have the motors, use the figures from NEC Table 430-148 (see Appendix), which show that the $1/4$-

[5] If fused equipment were installed, a 100-amp switch would have to be used because there is no switch rated at more than 60 amp but less than 100 amp. However, 70-amp circuit breakers are available.

[6] If the motors are 115-volt (see Chap. 15 for explanation of why we suddenly use 115 volts instead of 120 volts), use one-half of the name-plate ratings to arrive at the equivalent 230-volt ratings; if using NEC Table 430-148, use the 230-volt column.

	Gross computed volt-amperes	Demand factor, %	Net computed volt-amperes
Lighting, 3000 sq ft at 3 VA	9 000		
Small appliance circuits	3 000		
Special laundry circuit	1 500		
Total gross computed volt-amperes	13 500		
First 3000 VA		100	3 000
Remaining 10 500 VA		35	3 675
Range		Min.	8 000
Fixed appliances:			
Bathroom heater	1 500	100	1 500
Water heater	3 500	100	3 500
Clothes dryer	5 000	100	5 000
Net computed volt-amperes, less motors			24 675

hp motors will consume 2.9 amp each, or 5.8 amp together, and the $^1/_2$-hp motor will consume 4.9 amp, all at 240 volts. But the Code requires that the largest motor must be included at 125% of its rated amperage, so for the $^1/_2$-hp we will use 125% of 4.9, or approximately 6.1 amp. So the total for the motors will be 2.9 + 2.9 + 6.1 or 11.9 amp. So add 11.9 amp to the 102.8 amp already determined for other loads, making a total of 114.7 amp. You should install a service of at least 125- and preferably 150-amp capacity; if you use a fused switch, it will have to be rated at 200 amp. In any case the service-entrance wires will have to have an ampacity of at least 115; larger ones would be desirable. Larger service-entrance wires will permit adding future loads.

Computing Houses: Optional Method. Instead of following the procedures already outlined, NEC Sec. 220-30 permits an optional method that many prefer. Include lighting at the usual 3 VA per sq ft, the small appliance circuits at 1500 VA each, and the laundry circuit at the usual 1500 VA, but *do not* apply a demand factor to the total of these three items. List the range (or oven and counter units) at their full name-plate ratings, instead of the arbitrary 8000 watts. Add all other loads at their actual ratings. Add all these loads together and then apply a demand factor: the first 10 000 VA at 100%, and the remainder at 40%. See the example below.

	Gross computed volt-amperes	Net computed volt-amperes
Lighting, 3000 sq ft at 3 VA	9 000	
Small appliance circuits	3 000	
Special laundry circuit	1 500	
Range, name-plate rating	12 000	
Fixed appliances:		
Bathroom heater	1 500	
Water heater	3 500	
Clothes dryer	5 000	
1/4-hp motor, oil burner	700	
1/4-hp motor, blower on furnace	700	
1/2-hp motor, water pump	1 175	
Total gross computed volt-amperes	38 075	
First 10 000 VA at 100% demand factor		10 000
Remaining 28 075 VA at 40% demand factor		11 230
Total net computed volt-amperes		21 230

Add the motors at their volt-ampere equivalent: the 1/4-hp at 2.9 × 240 or approximately 700 VA each, the 1/2-hp at approximately 1175.

The total computed by this optional method is 21 230 VA, which at 240 volts is 88 amp, compared with 115 amp as computed in the usual way, in the preceding examples. Obviously, by this optional method, the minimum 100-amp service would be acceptable. However, you would be wise to install a larger service.

If the house had been equipped with four or more separately controlled space-heating units, the total name-plate load in volt-amperes of all the units would have been added with the other loads such as water heater, etc. However, if a *central* electric *space-heating* system had been installed, then per NEC Sec. 220-30(c) only 65% of its rated load would have to be added to the 21 230 VA. Similarly, if a *central* air-conditioning system had been installed, 100% of its name-plate rating would have been added to the 21 230 VA. But if both a central space-heating unit and a central air-conditioning system had been installed, *only the larger* of the two loads (100% of the air-conditioning load, or 65% of the space-heating load) would have been added since both would never be used at the same time.

Recap Concerning Computations. These sample computations, regardless of whether you used the ordinary or the optional method, will serve merely as examples. It will be well for you to compute other houses, using both methods, for example: (1) the house in which you now live as it is now wired, (2) the same house as you would like it after adding more appliances, air conditioning, etc., and (3) the "ideal" house in which you would like to live. But always remember that the answers you reach by these Code methods will tell you only the minimum size of wire in your service that the Code considers necessary for *safety*. You may want to use wires and other equipment larger than the minimum to allow for future loads that you will probably add later.

Ampacity of Neutral in Service. First of all, there is no neutral in a 2-wire service or 2-wire branch circuit, and both the grounded and the ungrounded wires must be the same size. But in a 3-wire 120/240-volt *service,* as you have already learned, there is a neutral, grounded as explained. In most cases the grounded neutral service wire may be smaller than the two ungrounded wires.

If a load operates at 240 volts, the neutral wire of the service carries none of the current consumed at 240 volts. A range (or a separate oven plus a counter unit) is a combination appliance operating sometimes at 120 volts, sometimes at 240 volts, depending on where the burners are set between the highest and the lowest heats.

To determine the size of the grounded neutral in your service, to arrive at the figure required by NEC Sec. 220-22, proceed as follows:

a. Watts load for ungrounded wires as discussed in preceding paragraphs . . . . ______ VA
b. *Deduct* all 240-volt loads, and also the range (or separate oven and cooking units) and/or dryer . . . . ______ VA
c. *Add* 70% of the range (or separate oven and cooking units) and/or dryer loads . . . . ______ VA
d. Total of *a*, minus *b*, plus *c* . . . . ______ VA
e. Divide volt-amperes of *d* by 240, which produces ampacity required for the grounded neutral . . . . ______ amp

Should the total of *e* be over 200 amp, count only 70% of the amount over 200 amp. For example, should the total be 280 amp, to the first 200 amp add 70% of the additional 80 amp or 56 amp (80 × 0.70 = 56). The minimum ampacity of the grounded neutral is then 256 amp. However, the grounded neutral may never be smaller than the size of the ground wire of the installation.

The methods outlined above are also correct for calculating the size of the neutral wire of a feeder in a larger installation.

Disconnecting Means and Overcurrent Protection. A device must be provided at or near the point where the service wires enter the building, to disconnect the entire building from its source of supply. It is also necessary to have devices that disconnect each separate branch circuit. This is a safety factor, for there are times when it is necessary to disconnect parts or all of the wiring, as in case of fire or when working on parts of the installation.

Likewise it is necessary to provide overcurrent protection for the entire installation, or parts of it (individual branch circuits or feeders) against short circuits, ground faults, or overloads.

The equipment to disconnect the entire building and the overcurrent device to protect the entire installation, but not the branch circuits individually, are what the Code calls "service equipment." The equipment may contain one (or up to six) main circuit breakers, or one (or up to six) main fused switches, which both disconnect and protect the installation as a whole. As a practical matter, the devices to control and protect individual branch circuits are often (in the case of residential and farm installations) included in the same cabinet. The equipment used must bear the UL label and *must* be marked as suitable for service equipment.

Solid Neutral. Service equipment, as well as much other equipment, very often is described as "solid neutral" (abbreviated "SN"). This simply means that the grounded neutral conductor is not switched or protected by an overcurrent device. NEC Sec. 240-22 prohibits an overcurrent device, and Sec. 380-2(b) prohibits a switch or breaker, in the grounded wire, unless the device is such that it opens the ungrounded wires at the same time it opens the grounded wire. Such devices seldom if ever are used in residential or farm wiring.

Yet NEC Sec. 230-70 requires that means be provided for disconnecting *all* conductors, including the grounded neutral, from the service-entrance wires. Contradictory as this sounds, the solution is simple. The incoming grounded neutral wire is connected to the neutral busbar in the service equipment by using a solderless connector, which can in turn be disconnected by removing a nut or bolt or cap screw, and that satisfies the requirements.

Bearing these facts in mind, it is obvious that a *three*-wire 120/240-volt service equipment will contain only *two* fuses (or a 2-pole breaker) for each of the one to six disconnects.

Neutral Busbar. Service equipment as purchased includes a "neutral busbar," which is a piece of copper or brass containing usually all of the following: a solderless connector to which the incoming neutral wire is connected; one or more connectors for the neutral wires of 120/240-volt 3-wire branch circuits of large ampere rating, such as for a range; one connector for the ground wire; and numerous solderless connectors or terminal screws for the grounded wires of all 120-volt circuits.

But note that in equipment as purchased, this busbar is *insulated* from the cabinet, but if the equipment is used as *service* equipment, it *must* be bonded (grounded) to the cabinet. This is very easily done. The busbar in some brands will include a heavy screw that you must tighten securely, which bonds it to the cabinet. In other brands, you will find a short flexible metal strap already bonded to the cabinet; you must connect the free end to one of the connectors on the neutral busbar. But note that while the neutral busbar *must* be bonded to the cabinet when used as *service* equipment, it must *not* be bonded (must remain *insulated* from) the cabinet if the cabinet is used "downstream" from the service equipment.

Selecting Circuit-Breaker Service Equipment. An individual circuit breaker looks much like a toggle switch. On overload it opens itself; reset it as shown in Fig. 13-3 after you have corrected the trouble that caused it

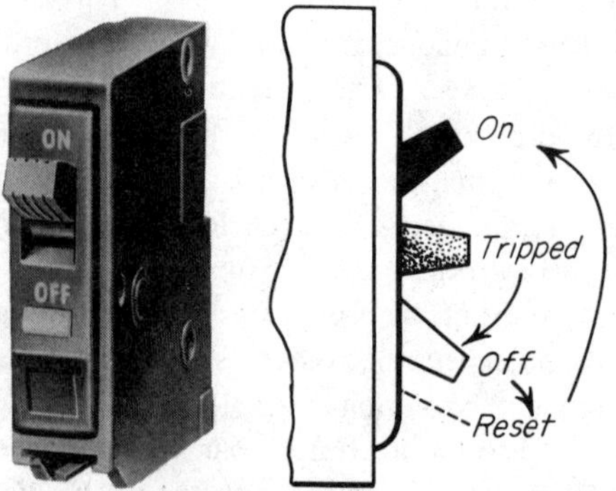

Fig. 13-3 A typical single-pole circuit breaker. If it trips because of overload, force the handle *beyond* the off position, then return it to on. (*Square D Co.*)

to trip. (In some brands it is not necessary to force the handle beyond the off position before resetting.)

Breaker-type service equipment may contain up to six separate main breakers, making it necessary to turn off up to six breakers to disconnect the entire building. Equipment containing only one *main* breaker is preferable. Service equipment as purchased usually includes the main breaker(s) plus an arrangement of busbars into which you plug as many breakers of the required ratings as needed to protect individual branch circuits. One single-pole breaker protects one 120-volt branch circuit; a 2-pole (double-pole) protects one 240-volt circuit (or a 3-wire 120/240-volt circuit such as for a range). A 2-pole breaker has a single handle and occupies twice the space of a single-pole breaker.[7]

A typical breaker-type service equipment is shown in Fig. 13-4, together with its internal wiring diagram. Note that it contains *one* main

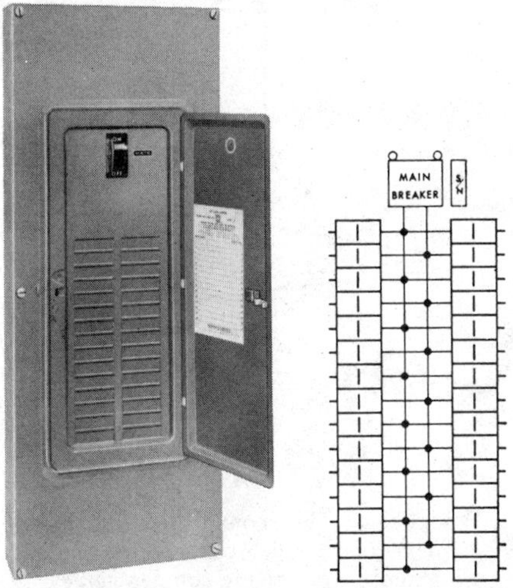

Fig. 13-4 Service equipment with a main breaker, and space for many branch-circuit breakers. The wiring diagram shows only single-pole breakers, but 2-pole breakers may also be used. (*Square D Co.*)

[7] So-called thin breakers are available that occupy only half the space of ordinary breakers, so that two of the thin or tandem variety can be plugged into the space for one ordinary breaker. This is not permitted unless the panelboard has an ampere rating high enough to accommodate the additional load.

breaker. Figure 13-5 shows breaker equipment of the "split-bus" type (prohibited in a few localities). Note that the top portion contains two main breakers, so that both breakers must be turned off to disconnect the entire installation. Code Sec. 384-16(a) requires that new service panelboards having more than 10% of their overcurrent devices rated at 30 amp or less, for which neutral connections are provided, must be protected on the supply side by *not more than two* main circuit breakers. Note, however, in the diagram of Fig. 13-5 that each of the main breakers in the upper portion of the equipment protects half of the breakers in the lower portion of the equipment. *These provisions apply equally to fused equipment.*

In a few areas it is common practice to use an outdoor cabinet containing the meter socket and six 2-pole main breakers, as shown in Fig. 13-6.

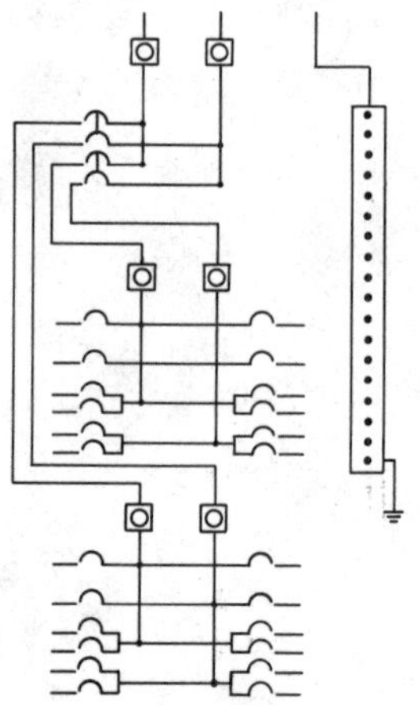

Fig. 13-5 Diagram for typical "split-bus" service equipment. Each of the 2-pole breakers in the top section becomes the *main* breaker for half of the branch-circuit breakers in the lower section.

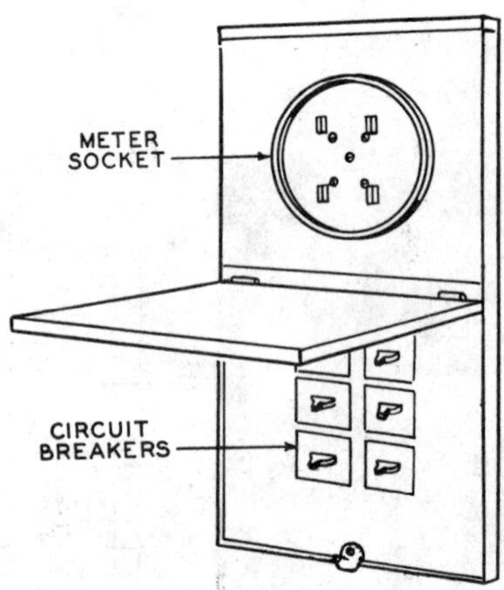

Fig. 13-6 This combination of meter socket and circuit breakers, in an outdoor cabinet, is favored in some localities.

One of them protects a panelboard inside the house containing all 15- and 20-amp branch-circuit breakers; the other five breakers protect 240-volt circuits in the house. Except in an existing installation, all the service-entrance breakers must be rated at larger than 30 amp.

Selecting Fused Equipment. In very old houses you will find service equipment that consists of a main switch with two hinged blades and an external handle, used to turn the entire load in the building on and off. The switch cabinet also contains two *main* fuses. Usually it also contains as many smaller fuses as are needed to protect all the individual branch circuits in the house; sometimes these fuses are in a separate cabinet.

Instead of a switch with an external handle, the type widely used in modern residential work has two main fuses mounted on a pull-out block as shown in Fig. 13-7. Insert the fuses into their clips while the pull-out is

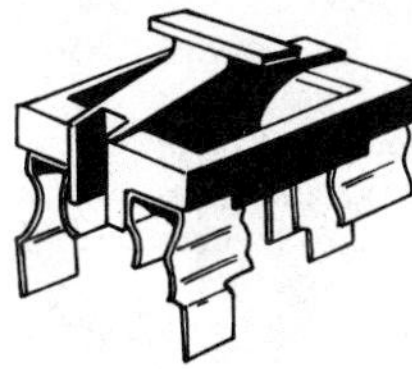

Fig. 13-7 In fused service equipment, cartridge fuses are installed in insulating blocks. When such a pull-out block is removed from its holder, it also functions as a switch.

in your hand. One side of each fuse clip has long prongs, so that the entire pull-out has four such prongs. The equipment in the cabinet contains four narrow open slots, but no exposed live parts; the live parts are behind the insulation. Plug the pull-out with its fuses into these slots; the four prongs on the pull-out make contact with the live parts, thus completing the circuit. Plugging the pull-out into its holder is the same as closing a switch with hinged blades; removing it is the same as opening such a switch. If you wish, you can insert the pull-out upside down, which leaves the power turned off.

Just as in the case of circuit breaker equipment, such service equipment usually contains one main pull-out with large fuses, protecting the entire load, plus additional pull-outs to protect 240-volt circuits with large loads, such as ranges, plus as many fuseholders for *plug* fuses as are needed to protect individual branch circuits: one for each 120-volt circuit, two for each 240-volt circuit. Equipment of this type is shown in Fig. 13-8. The wiring diagram is the same as shown in Fig. 13-4, except that fuses are used in place of breakers.

And again, as in the case of breaker equipment, the fused equipment *may* contain two *main* pull-outs, connected internally so that the service wires will run directly to each. To disconnect the load completely, you must pull out both the main pull-outs. The wiring diagram of connections within the equipment again is that of Fig. 13-5, except that fuses replace the breakers.

Fig. 13-8 Typical fused service equipment. It has a 100-amp *main* pull-out, plus additional cartridge fuses on pull-out blocks and fuseholders for plug fuses, to protect the branch circuits.

A word of explanation regarding fused switches is in order. If the fuses are of the cartridge type, the "60-amp" holder will accept fuses rated from 35 to 60 amp. The "100-amp" holder will accept fuses rated from 70 to 100 amp. The "200-amp" holder will accept fuses rated from 110 to 200 amp. Holders for plug fuses will of course accept fuses up to the 30-amp size, the largest made in the plug type. Thus you can always install the proper fuse to match the ampacity of the wires you are using.

Installation of Service Entrance. The equipment having been selected, it must be installed. The details are discussed in Chap. 17.

14
Good Lighting

In the very early days of electric lighting, even one lamp[1] hung in the center of a room was such an improvement over the ordinary kerosene lamp, or even the gas light then in use, that apparently little time was spent in considering whether the new illuminant provided sufficient light for good seeing. Even today too little thought is given to providing good lighting. As a result many homes and other buildings have insufficient or the wrong kind of illumination for good seeing.

This chapter will be devoted to the fundamentals of lighting, as well as to the selection, installation, and use of lighting fixtures and the lamps that go with them, in order to provide good lighting. Many volumes have been written on the subject, some covering only one aspect of the science, and on some points there is a good deal of disagreement among the authorities. The author does not flatter himself therefore that he can cover the subject in a chapter or two. He does propose, however, to set

[1] Is it a "lamp" or a "bulb?" The complete "light bulb" is properly called a lamp; the glass part of the lamp is the bulb. In this book they will be called lamps. Fluorescent lamps are often called "tubes."

forth some of the fundamentals involved, together with what are today considered standards, so as to give some degree of working knowledge to the reader.

Dozens of factors can be enumerated to make up a good lighting system. The more important ones are that the lighting system must provide (1) a sufficient quantity of light and (2) the right kind of light—light that is free from glare and free from objectionable shadows.

Good lighting is important in many ways. It contributes to personal comfort and reduces fatigue. It leads to greater efficiency in all activities and promotes safety by preventing accidents often caused by poor visibility.

Extent of Impaired Vision. Many people wear glasses, indicating that they have impaired vision. People with impaired vision need better lighting than those with normal vision. Since most families and other groups have at least one person with impaired vision, it follows that lighting should be designed not for those who have perfect vision, but for the majority who have impaired vision—usually older people.

Lumens and Footcandles. The total amount of light produced by a light source is measured in lumens.[2] When you buy a lamp, you will find the number of lumens it produces printed on the carton. The degree of illumination at any point or surface is measured in footcandles (abbreviated fc).

When one lumen of light falls on an area of exactly one square foot, that area is said to be lighted at one footcandle. Ten lumens falling on 1 sq ft produces 10 fc, and so on.

One point that often is misunderstood is that the illumination in footcandles remains the same, no matter what the distance from the light source, as long as the number of lumens of light falling on each square foot does not change. Assume a reflector so perfect that it condenses *all* the light produced by a lamp producing 100 lumens into a narrow beam so that it illuminates a spot *exactly 1 sq ft in area* on a sheet of paper 10 ft from the lamp. The illumination on the spot will then be 100 fc. If now the sheet of paper is moved to a point 20 ft away, the beam will illuminate a spot 4 sq ft in area and the illumination will be only 25 fc. If, however, a different reflector is then substituted, producing a much narrower beam, so that at the new 20-ft distance the entire 100 lumens will again illuminate a spot only 1 sq ft in area, the illumination on the spot will again be 100 fc. As long as *all* the light produced by a source delivering 1 lumen falls on an area of 1 sq ft, that area is illuminated to 1 fc no matter what the distance.

[2] This term is derived from the Latin word *lumen,* meaning "light."

It is impossible in practice to concentrate light to this degree, for reflectors are not perfect, absorbing some light and allowing some to spill in various directions. As a starting point, however, the relation can be considered correct. It is a most important rule, and you must remember this fundamental fact: *1 lumen of light on 1 sq ft of area produces 1 fc.*

To illustrate the utility of this rule, assume that an area 12 × 12 ft is to be lighted to 15 fc. Since the total area is 144 sq ft, it will require 144 lumens to provide 1 fc. Fifteen fc will require 144 × 15, or 2160 lumens.

The approximate lumens produced by general-purpose incandescent lamps are as follows:

Watts	Lumens	Watts	Lumens
15	125	150	2 880
25	235	200	4 010
40	455	300	6 360
60	870	500	10 850
75	1190	750	17 040
100	1750	1000	23 740

A 150-watt lamp delivering 2880 lumens in a room 12 × 12 ft should therefore produce about 20 fc if all the light produced by the lamp could be directed to the floor and none allowed to fall on the ceiling or walls. But that is a theoretical figure. Instead of 20 fc, you will actually attain a much lower figure, between 4 and 6 fc, depending on many factors, such as the reflector used with the lamp, the age of the lamp, the reflecting ability of the ceilings and walls, the type of fixture used, and many other factors that will be discussed in more detail in Chap. 29.

Do note, too, that fluorescent lamps produce more lumens per watt than ordinary incandescent lamps. Naturally, then, 150 watts of fluorescent lighting will produce more footcandles of illumination than 150 watts of ordinary lamps. This fact will be discussed in more detail later in this chapter.

Note that brightness and footcandles of illumination are not the same. A black sheet of paper illuminated to 10 fc will not seem so bright as a white one illuminated to 5 fc, because the white paper *reflects* most of the light falling upon it while the black *absorbs* most of it.

A Few Yardsticks. To provide some starting point of known values in footcandles, bright sunlight on a clear day varies from 6000 to 10 000 fc. In the shade of a tree on the same day there will be somewhere in the neighborhood of 1000 fc. On the same day the light coming into a window on the shady side of a building will be of the order of 100 fc, while 10 ft back it will have dropped to something between 7 and 15 fc. At a point 4 ft directly below a 100-watt lamp without a reflector, and backed by a black ceiling and walls, which have negligible reflecting power, there will be about 6 fc.

Law of Inverse Squares. The number of lumens of light produced by a light source is constant, other conditions remaining the same. A lamp produces the same number of lumens whether the observer is 1, 2, or 3 ft or any other distance away. Nevertheless it is easier to read, for example, a newspaper when you are close to a light source than when you are farther away.

See Fig. 14-1. Assume that the lamp produces enough lumens so that 10 lumens fall on an area of 1 sq ft when 1 ft away. But if the illuminated

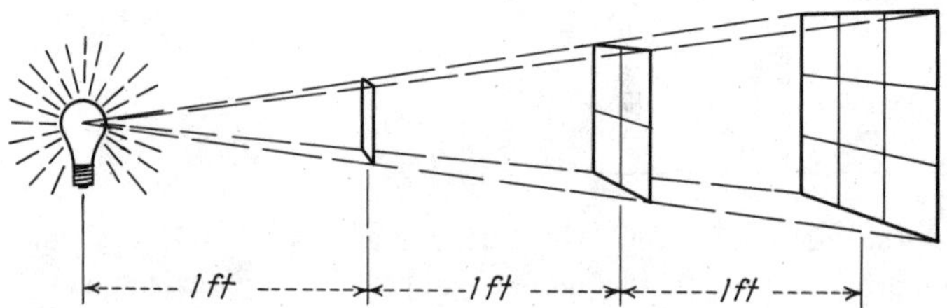

Fig. 14-1 The illumination on an object varies inversely as the square of the distance from the light source. (*General Electric Co.*)

area is moved to 2 ft away, those same 10 lumens now fall on an area of 4 sq ft, and at a distance of 3 ft they fall on 9 sq ft.

The area lighted was first 1 sq ft, then 4, then 9, yet the total amount of light remained the same, 10 lumens. Obviously it will be harder to read at the 3-ft distance than at the 1-ft distance, because there is only one-ninth the illumination—the light has been spread nine times as thin, if we may use that expression. *The illumination of a surface varies inversely as the square of the distance from the light source.* This is known as the "law of inverse squares." If a single light source gives satisfactory illumination for a given job when it is located at a distance of 5 ft, a light source giving four times as many lumens will be required if it is moved to a distance of 10 ft, other conditions remaining the same. To determine the relative amount of illumination, simply divide the square of one distance by the square of the other. For example, comparing the relative illumination of an object 7 ft from a light source with that of one 4 ft away,

$$\frac{4 \times 4}{7 \times 7} = \frac{16}{49} = 33\% \text{ (approx.)}$$

Just remember that the absolute degree of illumination at any given point, without regard to the power of the source from which the light comes, is measured in footcandles.

Law of Inverse Squares Is Treacherous. Later in this chapter there is described a simple instrument for measuring footcandles. Assume that such an instrument held 1 ft from a lighting fixture shows 160 fc. If it is held 2 ft away, according to the law of inverse squares, it should read 40 fc; 4 ft away it should read 10 fc; 10 ft away it should read 1.6 fc. It does not. Does that mean the law is no law at all, only a theory that does not work? Not at all.

The law is correct when all the light there is comes from one source and that source is dimensionally small (for example, a small lamp) and further that none of the light from the source hits a surface and is then reflected back into the lightmeter. The trouble is that such conditions are rarely found. The meter measures not only the light from the fixture that you are thinking about but also light from other fixtures. It measures not only the light that comes directly from the fixture but also light that is reflected from ceilings and walls. Generally speaking the law of inverse squares can be confirmed by a footcandle meter only if the distance from the light source to meter is at least five times the maximum dimension of the light source.

All that does not detract from the usefulness of the footcandle meter; it is a very convenient device and serves many useful purposes. More will be said about that later.

Measurement of Footcandles. Lumens are not measurable in a simple fashion. Fortunately footcandles can be measured as easily as reading a voltmeter. In Fig. 14-2 is shown a direct-reading footcandle meter, commonly known as a "lightmeter." Simply set the instrument at the point where the illumination is to be measured, and the footcandles are read directly on the scale. The device consists of a photoelectric cell, which is a device that generates electricity when light falls upon it. The indicating meter is simply a microammeter,[3] which measures the current generated, the scale being calibrated to read in footcandles. The use of this instrument is invaluable, especially in commercial work, as it takes the guesswork out of many lighting problems.

[3] A microampere is one-millionth of one ampere.

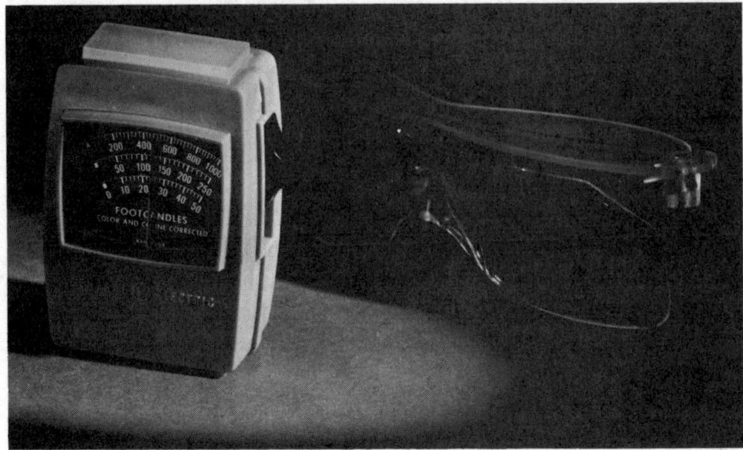

Fig 14-2 This instrument reads footcandles of light directly on its scale. (*General Electric Co., Nela Park.*)

Footcandles Required for Various Jobs. It must be remembered that there can be no absolute standard. First of all, the requirements for different individuals vary. Accordingly all that can be given are the commonly accepted standards of today. The more critical the task, the higher the level of illumination required. The more prolonged the task, the greater the amount of light needed; for example, you can quite easily read a paragraph of a newspaper in the relatively poor light of the dusk of evening, but it is almost impossible to read an entire page.

Before the energy crisis, recommendations for levels of lighting were considerably higher than those now considered acceptable. Especially in areas where no exacting work is being done, such as halls, stairways, and parts of living areas devoted primarily to conversation, relaxation, or entertainment, 10 fc is now considered adequate. But in parts of living rooms and other areas where reading is done, this should be supplemented by table or floor lamps to bring the level up to 30 or 40 fc. If sewing is being done, more light is needed, ranging from 50 fc for light fabrics to 150 fc for dark fabrics. In areas where studying is being done, levels of about 70 fc are considered proper.

In dining rooms 15 fc is sufficient, but in kitchens 50 fc in general (but up to 150 fc in areas where food is actually prepared) is recommended. In bathroom areas used for grooming, shaving, and makeup, the suggestion is 50 fc. In laundries, for areas where sorting and inspecting are done 50 fc is proper, while in washer and dryer areas 30 fc is sufficient. Do note that in all areas somewhat higher levels are proper for older people.

Important: More important than the exact levels of light you provide is forcing yourself to acquire the habit of turning off every light when you don't need it. *That will save much energy*—and reduce your electric power bill, too.

The need for energy conservation has focused attention on the fact that the *quality* of light is more important than the *quantity*. Proper selection of light sources and the selection, placement, and control of fixtures can deliver *more* light where it is needed while at the same time conserving energy.

Glare. Because so much space has been devoted to the amount or degree of illumination, do not think that this is the one all-important factor in good lighting. The *kind* of light is also important; this is dependent on three factors. One important factor is that the light must be free from glare.

Glare, generally speaking, is caused by a relatively bright area within a relatively dark or poorly lighted area. An exposed lamp in the lobby of a movie theater is not particularly noticeable when you enter during daylight, because your eyes are accustomed to the outer brightness; when you leave, after your eyes are accustomed to the relatively dark interior, the lamp will appear very bright and appear to create a glare that hurts your eyes. Similarly, a lighted automobile headlight is barely noticeable in the daylight but may be blinding at night. Glare may be defined as "any brightness within the field of vision of such a character as to cause annoyance, discomfort, interference with vision, or eye fatigue."

Glare is usually caused by exposed lamps so placed that they can be seen while looking at the object we want to see; the lamp may not be directly visible, but even if it is so placed that it can be seen by merely moving the eyes without moving the head, it still produces glare. Bright automobile headlights constitute a good example of this kind of glare.

Glare of an equally objectionable type may be caused by reflection, for example, from brightly polished and plated parts of typewriters and other office equipment. Modern equipment seldom has such parts, because manufacturers and users have learned that glare causes the evils recounted above, with resultant lower efficiency. For this reason, too, extremely glossy papers are being used less and less in printing.

How to Avoid Glare in Lighting. Everybody has tried reading in direct sunlight and found it difficult. On the other hand, it is not difficult to read in the shade of a tree on a bright day, even if the footcandles there are much lower than in direct sunlight. Therefore the answer cannot lie only in the footcandles of illumination prevailing at any given moment.

The answer does lie in the fact that direct sunlight comes essentially from a single point, the sun, and causes glare. In the shade of the tree the light comes from no point in particular but rather from every direction—north, south, east, west, and above. Not coming from one point, it does not cause harsh shadows or glare. It is *diffused* light.

Surface Brightness. When you look at an exposed 300-watt lamp of the *clear*-glass type, the filament appears as a concentrated spot of light less than an inch in diameter. The lamp itself is a little over 3 in. in diameter, but because the filament is so bright, the glass bulb itself is barely visible. That is why practically all lamps today are frosted—the filament is not seen, but rather the entire bulb. Since the bulb has a diameter of about 3 in, there is exposed to the eye an apparent area equivalent to the area of a circle of the same diameter, or about 7 sq in. The same total amount of light, which in the case of the clear-glass bulb was concentrated in an area of about 1 sq in, is now distributed over a much larger area of about 7 sq in. Obviously, then, while the total amount of light is the same, the brightness of the larger area is greatly reduced—the "surface brightness" is lower. It is still uncomfortably bright, however, if looked at directly.

Put the lamp inside a globe of translucent, nontransparent glass about 8 in. in diameter; this has an apparent area of about 50 sq in altogether instead of the 7 sq in observed before. The total amount of light is still the same, but it is far more comfortable to the eye because the surface brightness of the light source has been reduced still more. Consider the most common fluorescent lamp: It is the 40-watt type and is 48 in long. The surface brightness is very low because the total surface area of the tube is very large for the amount of light emitted and there are no bright spots. So, for comfortable lighting, use light sources, or fixtures with diffusers, that have low surface brightness.

Direct and Indirect Lighting. When light in a room is produced by fixtures that direct all the light on the area to be lighted, with none reflected from the ceiling or walls, the method is called "direct lighting." When the light is produced by fixtures that throw all the light onto the ceiling, which reflects it to the areas to be lighted, the method is called "indirect lighting."

An extreme but totally impractical example of *direct* lighting would be automobile headlights mounted on the ceiling, shining down toward the floor. Such lighting would produce extreme glare, sharp shadows, and uneven levels of light. It would be very poor lighting—direct lighting of the worst kind. Yet, direct lighting can be quite effective in some stores, particularly with recessed fixtures in the ceiling to light up particular items on counters or other displays.

An extreme and again totally impractical example of *indirect* lighting would again involve automobile headlights, now mounted high enough on the wall so that they could not shine directly into your eyes, and with their beams thrown against the ceiling, to be reflected downward. It would provide very bright spots on the ceiling and very uneven lighting in the area below—it would be indirect lighting of the worst kind. Indirect lighting of all kinds is very inefficient, but in establishments other than homes, it can be made quite effective by properly designed fixtures that distribute the light properly on the ceiling, which must be of a light color.

Good lighting in homes is neither completely direct nor completely indirect. Most light sources let some of their light fall on the ceiling and some on the areas below, which is what we look for in good lighting.

Figure 14-3 shows what happens in the case of a single-lamp fixture with a translucent shade. A goodly share of the light goes through the shade; it is diffused as it goes through, so that there are no dark spots below the fixture, or any extremely bright spots on the shade, to produce glare. Much of the light is reflected to the ceiling, and from there is reflected through the room, thus approaching the desired shade-of-a-tree type of light. In the illustration, note the paths of individual rays of light. The solid lines show what the paths would be if the ceiling and walls were absolutely smooth like a mirror. While a painted wall or ceiling may appear to be very smooth, if examined under a magnifying glass it will be

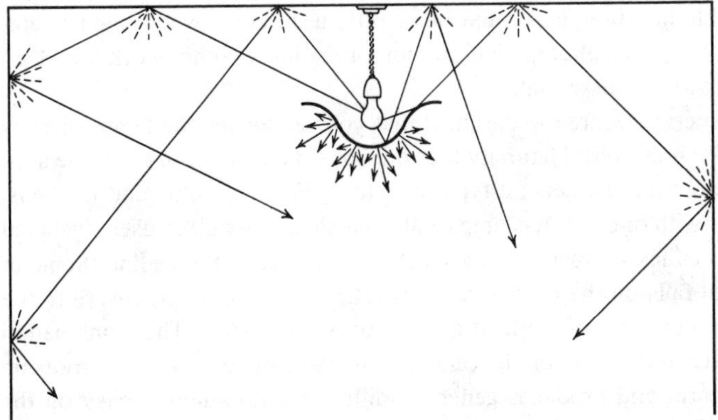

Fig. 14-3 This diagram shows how light becomes diffused as it is reflected from walls and ceilings.

found to be full of hills and dales. Therefore the rays as they strike any one point strike those hills and dales and are reflected and dispersed in every direction; what would otherwise be individual rays are therefore split into a multitude of rays (as shown in the dashed lines) giving diffused lighting. While fixtures of the exact type shown in the illustration are rarely used today, the picture does illustrate a principle.

In general, the entire area of a room (including walls and ceilings) must be lighted to a reasonable degree; smaller areas must be lighted to a higher degree for reading and other activities that require more light for comfort and efficiency. In some areas ceiling fixtures provide light for general seeing, and floor and table lamps provide light for reading and similar work. This gives good lighting. In other rooms only local lighting is provided by floor and table lamps. Such lamps, however, do not throw all their light downward; a considerable part goes up to the ceiling. Whatever the light source, a reasonably even distribution of light is important for comfort and for a room of pleasing appearance: no very dark areas and no areas of extreme brightness as compared with other nearby areas.

Fixtures for Homes. Where a room is lighted by fixtures using exposed lamps, most people think of it as direct lighting. It is a combination of direct and indirect, for a good deal of the light produced falls on the ceiling and is then reflected. But for most purposes it is not good lighting because the exposed lamps have high surface brightness, leading to great contrast with their immediate surroundings and much glare. Because the light comes from one point, such fixtures do not produce diffused shade-of-the-tree light that is so comfortable to the eyes.

A simple one-lamp fixture with an exposed lamp, on the other hand, is justifiable in attics, halls, and other little-used areas where light is employed for casual illumination and not for reading or other work for which good lighting is essential.

In general, fixtures so designed that you cannot see the lamps directly are to be preferred. Naturally the room in which the fixture is to be used will determine the general type of fixture. For bedrooms and kitchens, fixtures with one to three lamps and glass shades are often used, installed on the ceiling, or suspended a short distance below the ceiling. Some of the light falls on the ceiling, some is reflected from the glassware to the ceiling, and some of the light goes directly downward. The combination eliminates dark areas on the ceiling, eliminates extremely bright spots on the fixture, and produces generally diffused light which is easy on the eyes. Whether any given fixture will produce enough light in any given area will depend on the wattage of the lamps used, and the design of the fixture.

All that has been said refers to fixtures using ordinary incandescent lamps. In kitchens and other areas where their appearance is acceptable, fluorescent fixtures will provide up to three times as much light per watt. Be sure to select a type of fixture that permits part of the light to strike the ceiling, so that there will be no dark, dismal areas on the ceiling. Cove lighting is usually preferable to ceiling lighting fixtures for general lighting in bedrooms and living rooms; see Chap. 20.

What has been said about lighting fixtures has to do with their function as producers of good lighting. Fixtures also have a decorative function, to carry out a decorative scheme in a home. Indeed, most residential fixtures seen in showrooms seem to be designed with the aesthetic or decorative function in mind. Choose the kinds of fixtures that you like, but if they are largely decorative, be sure to supplement them with floor and table lamps so that an adequate level of illumination may be available where reading or similar activities are carried on.

Reflectance. If the ceiling and walls of the room were mirrors, obviously a great percentage of the light striking the ceiling and walls would be reflected downward. However, there would be direct reflections of the lamp, and the glare would be as bad as, or worse than, that from an exposed lamp. On the other hand, if the ceiling were black, most of the light would be absorbed and little reflected, and the lighting would be inefficient. Accordingly, for best results, the ceilings and walls should be a color that reflects as much light as possible; at the same time the finish should be dull or flat rather than glossy, to avoid the mirror effect of glare-producing spots.

Experience has shown that surfaces of varying colors reflect light to greatly varying degrees. The percentage of light that is reflected is called the "reflectance" of the surface. White, pale pink, pale yellow, or ivory colored surfaces reflect 85 to 80%. Cream, buff, gray, or light blue reflect 70 to 50%. Darker colors reflect still less, down to 5 to 2% for black. Woods in natural finish rarely reflect as much as 50%, and sometimes as little as 5%.

Shadows. Besides the annoying glare encountered when we read in direct sunlight, the contrast between bright light and sharp shadow is annoying and tiring. In the shade of a tree the footcandle level is much lower than in direct sunlight, but reading is more comfortable. The light is diffused; shadows are soft and not objectionable. Two sources of light, for example a floor lamp and an overhead fixture, lead to greater comfort in reading. Use lamps or fixtures that provide diffused light, light that comes from many points, as in the shade of a tree.

Color of Light. Sunlight has come to be accepted by most people as a standard, although the color and quantity may vary, depending on the hour of the day and the season. It may seem strange to speak of the "color" of sunlight, which appears colorless, yet sunlight is composed of a mixture of all colors. The rainbow is simply a breaking down of sunlight into its separate colors.

What makes one object red and another blue when this mixture of all colors that we call sunlight hits these objects? The explanation is simple. When "white" light (such as sunlight) strikes certain objects, the component colors are all reflected equally and we call such objects white. But when white light hits other objects and instead of being reflected it is absorbed, we see no light; we then say such objects are black.

Still other objects may absorb some colors of the spectrum and reflect others. For example, they may absorb colors except red, and reflect that. We then see only the red part, so we say that such objects are red. So it is with every other color: Whatever the color of the object, that is the color that object can reflect; all other colors are absorbed.

Everybody has observed that the color of an object appears different when observed in natural light as compared with artificial light. Light produced by an incandescent lamp does not have the same proportion of colors as exists in natural light; it has more orange and red and less of blue and green. It is not surprising, then, that efforts have been made to produce special lamps that more nearly duplicate the mixture and proportions of colors existing in natural light. The effort has led to lamps that approximate the light from a northern sky rather than direct sunlight. Such lamps are called "daylight-type" and tend to accent cool colors. Dawn pink tends to accent warm colors. Complexions are good under both colors, but such lamps do produce less light per watt.

Life of Lamps. The approximate average life of lamps, when burned at their rated voltage, varies a great deal with the size and type of lamp. The average life of any specific lamp will be found printed on its carton. For ordinary lamps of various sizes, the approximate average life will be found in Table 14-1. It is impossible to make lamps so that all will have exactly their rated life. Assume a large group of identical lamps with 1000-h rated life, all turned on at the same time and never turned off. At 800 h about 20% will have burned out. At 1000 h, 50% will have burned

TABLE 14-1 Characteristics of Incandescent Lamps

Size of lamp, watts	Average life, h	Total lumens	Lumens per watt
25	2500	235	9.5
40	1500	455	11.6*
60	1000	870	14.5
75	750	1 190	15.9
100	750	1 750	17.5
150	750	2 880	19.2
200	750	4 010	20.0
300	750	6 360	21.2
500	1000	10 850	21.7

* The first lamps made by Thomas Alva Edison in 1879 produced less than 1.4 lumens per watt.

out; at 1200 h, about 80%. At 1400 h some will be still burning. Their *average* life will be 1000 h.

It is a simple matter to make lamps that last longer, but in doing so their efficiency is reduced; the lamps will produce less light, fewer lumens *per watt.* A 100-watt lamp that costs 80 cents uses 75 kWh of energy during its 750-h life. At 6 cents per kWh, the cost of the energy used is therefore $4.50, as compared with the 80-cent cost of the lamp itself. If, then, to obtain longer life in an 80-cent lamp we reduce the efficiency, an extra $1 may be spent for electric power to produce the same amount of light. An added expense rather than economy results.

Where no great amount of light is needed and the lamp serves merely as a signal (such as a pilot light), the lamp may be designed for a life of 2000 h or more; on the other hand, where a great deal of light is needed and where it is important to limit the heat (such as lamps for a movie projector), the lamp may be designed for a relatively short life but with corresponding increase in efficiency, thus permitting smaller-wattage lamps producing less heat to be used. For example, a 1000-h 1000-watt lamp produces about 23 lumens per watt, but a 50-h lamp of the same wattage produces about 28 lumens per watt, and a photoflood lamp of the same approximate wattage, but only 10-h life, produces roughly 31 lumens per watt.

Efficiency of Various Sizes of Lamps. Larger lamps produce more light *per watt,* so that generally speaking, if there is a choice, it is better to use one large lamp than several smaller ones. Study Table 14-1, which is based on general-purpose lamps of current manufacture. It shows that one 150-watt lamp produces approximately as much light as twelve 25-watt lamps consuming 300 watts in total; one 300-watt lamp produces more light than seven 60-watt lamps consuming 420 watts.

Effect of Voltage on Lamp Life. If an incandescent lamp is operated at a voltage below that for which it was designed, its life is prolonged considerably, but the watts, the lumens, and the lumens per watt drop off rapidly. If it is operated at a voltage above normal, its life is greatly reduced, although the watts, the lumens, and the lumens per watt increase. For lowest over-all cost of illumination use lamps on a circuit of the voltage for which they were designed.

Careful study of Table 14-2 will confirm these statements. If a lamp is burned at the voltage for which it is designed, it will have characteristics shown in the 100% line. If it is burned at a voltage higher or lower than its design voltage, the various factors will change to a new value shown as a percentage of normal.

TABLE 14-2 Effect of Voltage on Incandescent Lamps

Circuit voltage as percentage of rated voltage of lamp	Total average life, h	Total output, lumens	Actual watts	Lumens per watt
85%	825%	58%	78%	74%
90%	400%	70%	85%	83%
95%	200%	84%	93%	92%
100%	100%	100%	100%	100%
105%	54%	119%	108%	109%
110%	29%	138%	116%	120%
115%	16%	160%	124%	130%

As an example, an ordinary 100-watt lamp designed for use on a 120-volt circuit will when burned at 120 volts have a life of about 750 h, produce 1750 lumens, and consume 100 watts, resulting in 17.5 lumens per watt. When burned at 108 volts (90% of 120 volts), it will have a life of about 400% or 3000 h and produce 70% of 1750 or about 1190 lumens; it will consume 85% of normal or about 85 watts, resulting in 83% of normal efficiency or about 14.5 lumens per watt.

A lamp burned at its rated voltage represents the lowest total cost of light, considering the cost of the lamp itself and the cost of the power consumed during its normal life. A 100-watt lamp during its normal 750-h life consumes 75 kWh of power, costing approximately six times as much as the lamp itself.

"Long-Life" Lamps. You will see advertised *for use in the home* lamps guaranteed for 5000 h or five years, or some other period far beyond the life of ordinary lamps. Naturally their price is *very high* compared with that of ordinary lamps. There is no secret or "gimmick" that makes such lamps last longer. They are simply lamps designed to burn at 130 or 140 volts but marked to pretend that they should be used on ordinary 120-volt circuits. Naturally they will last a long time, but they will deliver far less light *per watt* than ordinary lamps; sometimes they consume more watts than the size stamped on them, giving the illusion of producing as much light as ordinary lamps. You might as well use ordinary lamps of a smaller wattage. Remember that the cost of the power consumed during the life of a lamp is very much more than the cost of the lamp itself.

Nevertheless, there are locations such as factories and commercial establishments where it is quite proper to use lamps with a life much longer than that of ordinary lamps. Such lamps are called "extended-service" types and cost little more than ordinary lamps. They will be discussed in Chap. 29.

Lamp Bases. There are various standardized kinds and sizes of lamp bases, matched to the watts, the physical size, and the purpose of the lamp. In the screw-shell type the largest is the mogul (1.555 in. in diameter), used on 300-watt and larger lamps. Smaller bases are the medium (also called "Edison"), used on ordinary household lamps, and the intermediate and the candelabra. All four are shown in actual size in Fig. 14-4. Flashlights and similar lamps use a still smaller base.

Another base used on high-wattage lamps is the bipost of Fig. 14-5. The prefocus base of Fig. 14-6 is used mostly on lamps for projectors. It maintains the lamp in one particular position to give maximum light in one direction. Both these types are available in two sizes: medium and mogul. A three-light lamp is merely a lamp with two separate filaments, say, 100 and 200 watts, with a special base so arranged that either filament alone, or both at the same time, can be turned on, producing 100, 200, or 300 watts as desired.

Lamp Designations. The mechanical size and shape of an incandescent lamp are designated by standardized abbreviations such as A-19, PS-35, F-15. The letter designates the shape in accordance with the outlines shown in Fig. 14-7. The numeral designates the diameter in eighths of an

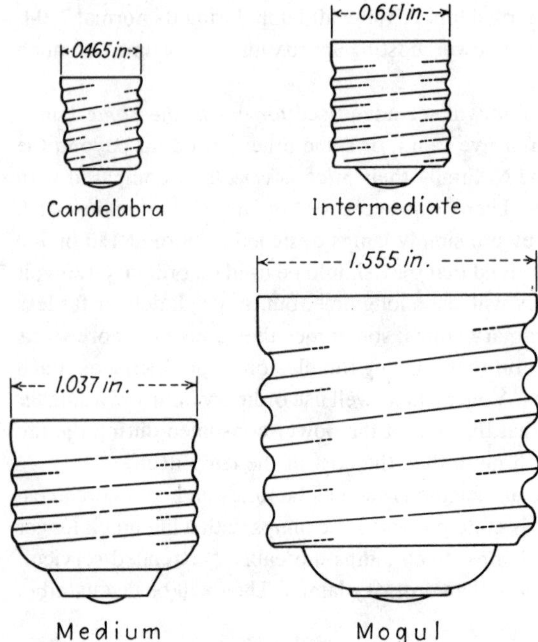

Fig. 14-4 The screw-shell bases used on lamps are standardized to the dimensions shown. The illustrations are actual-size.

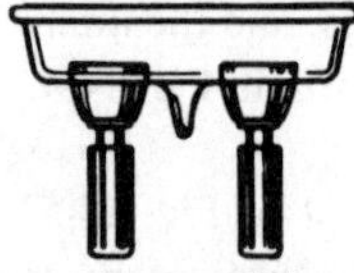

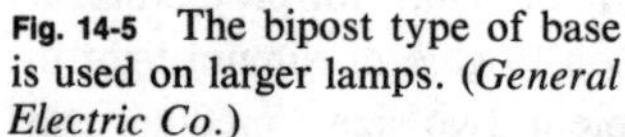

Fig. 14-5 The bipost type of base is used on larger lamps. (*General Electric Co.*)

Fig. 14-6 The prefocus type of base is used on lamps for projectors. (*General Electric Co.*)

inch. Thus an A-19 lamp has the simple A shape and is $^{19}/_{8}$ or $2^{3}/_{8}$ in. in diameter.

Lumiline Lamps. This is a tube-shaped incandescent lamp that has a special contact at each end. At one time it was in common use, but it is an inefficient lamp, producing less than 10 lumens per watt. For that reason it is seldom used in new installations but is available for replacements in 30-, 40-, and 60-watt sizes.

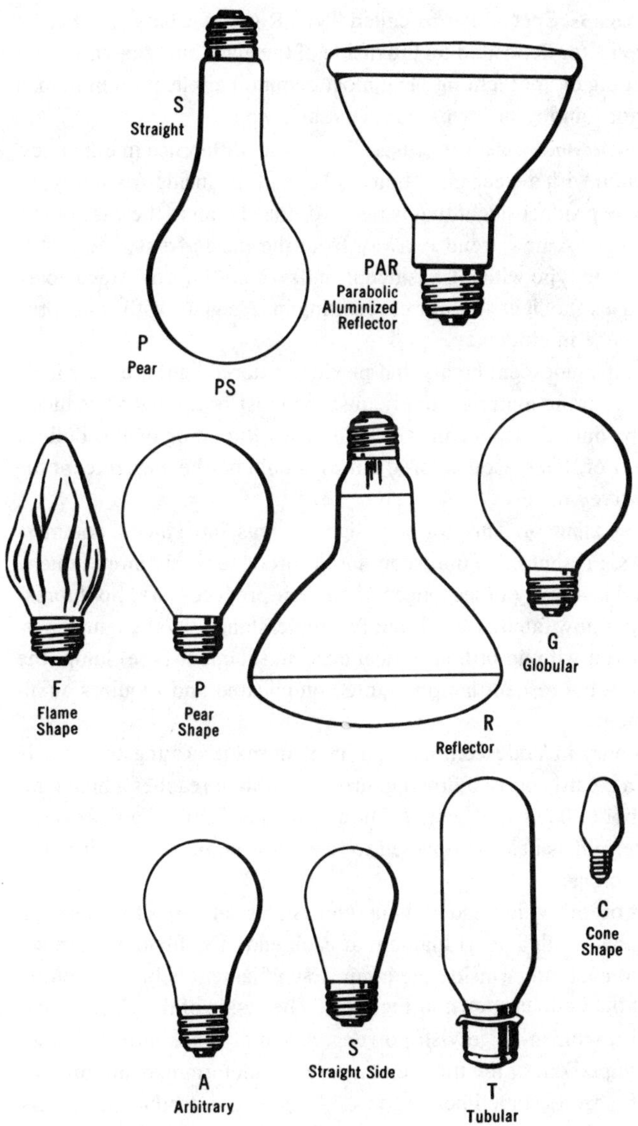

Fig. 14-7 The shape of lamps is well standardized. (*General Electric Co.*)

Reflector Lamps. Special lamps called Type R or PAR have a silver or aluminum reflector deposited on the inside of the glass bulb, and are used for floodlighting or spotlighting. The most common application in homes is for exterior lighting of lawns, gardens, and walks.

Efficiency of Various Colors of Lamps. There is no difference in efficiency between a lamp with a clear-glass bulb and one of the inside-frosted type. The latter type produces light that is better diffused than in the case of the clear-glass type. A new trend is away from the inside-frosted to a still newer soft-white type with a translucent, milky-color internal silica coating that diffuses the light still more, resulting in less glare with no appreciable difference in efficiency.

However, incandescent lamps that produce colored light are very inefficient. The coloring material simply absorbs most of the light produced by the lamp; only that portion which matches the color of the bulb is transmitted. For this reason colored lamps should not be used except for their decorative value.

Fluorescent Lighting. This form of lighting was introduced commercially in 1938. Present-day fluorescent lamps produce vastly more lumens per watt, and last many times longer, than their predecessors. So fluorescent lighting is now familiar to all, but few understand how the lamp operates. Compared with an ordinary incandescent (filament-type) lamp, the operation of a fluorescent lamp is quite complicated and requires auxiliary equipment.

In an ordinary incandescent lamp, a filament made of tungsten wire is heated by an electric current flowing through it until it reaches a high temperature (about 4000°F), at which time the filament emits light. It operates by its terminals being connected to two wires of an electric circuit of the proper voltage.

The most ordinary fluorescent lamp consists essentially of a glass cylinder or "tube" with a short filament at each end. The filaments are not connected to each other inside the lamp. Each filament is brought out to two pins on the end, as shown in Fig. 14-8. The inside of the glass tube is coated with a whitish or grayish powder, called a "phosphor." The air has been pumped out of the tube, and a carefully determined amount of a gas, usually argon, sometimes argon and neon, is introduced. A very small amount of mercury is also put into the tube. That is the basic machinery of a fluorescent lamp, but if it is connected directly to the ordinary 120-volt circuit, it will not operate.

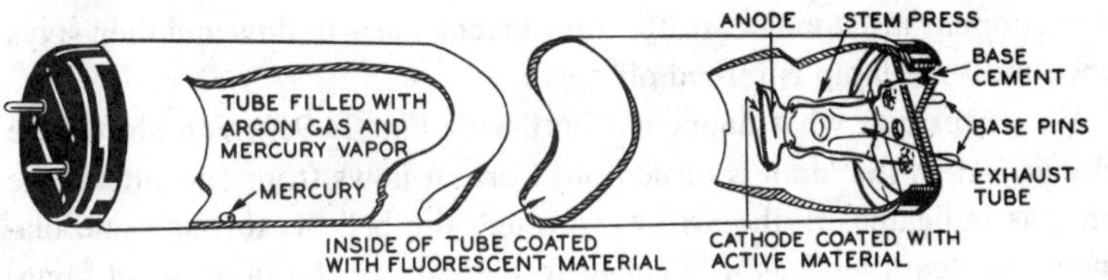

Fig. 14-8 Construction of a fluorescent lamp. (*General Electric Co.*)

If a fluorescent lamp is connected to a source of high-voltage current, the lamp will light up but will be quickly destroyed. If it is connected to a source of high-voltage current and just at the instant of starting the voltage is reduced to 120 volts, it will still be destroyed. Apparently, then, some special accessories are required to make the fluorescent lamp work.

Figure 14-9 shows the basic scheme. The auxiliary equipment consists of two devices: a ballast or choke coil and an automatic switch, called a

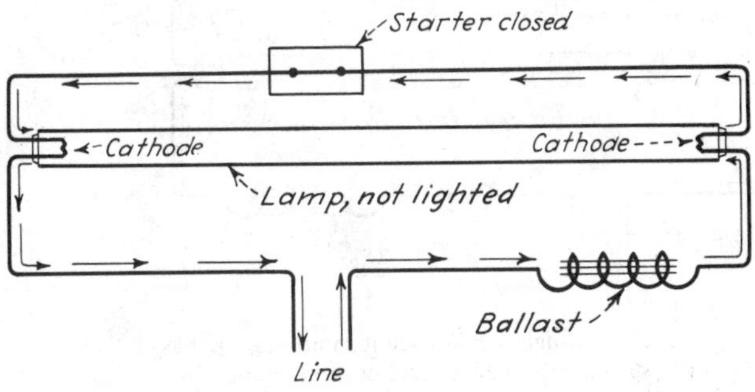

Fig. 14-9 Flow of current in a fluorescent lamp circuit at the instant when the lamp is turned on.

starter. Any coil of wire wound on an iron core has two peculiarities: (1) When connected to an ac circuit, it tends to resist any change of current flowing through it, and (2) when a current flowing through it is cut off, it delivers momentarily a voltage much higher than the voltage applied to it. The ballast for a fluorescent lamp is just such a coil. The automatic switch or starter (part of the lighting fixture) is so designed that it is ordinarily closed (while the lamp is turned off), but when the lamp is turned on, the starter opens a second or so after the current starts to flow and then stays open until the lamp is turned off again.

Visualize then what happens. Start with Fig. 14-9, which shows the circuit just as the lamp is turned on. Current flows from one side of the line, as indicated by the arrows, through the ballast, through one filament, or "cathode" as it is called in the case of the fluorescent lamp, through the automatic switch or "starter," through the other filament or cathode, and back to the other side of the line. During this period the lamp glows at each end but does not light. Then the automatic switch (the starter) opens, and the ballast does its trick—it delivers a high voltage as mentioned in the previous paragraph, a voltage considerably above 120 volts and high enough to start the lamp. The current can no longer flow through the starter because it is open; it then flows through the tube, jumping the gap and forming an arc inside the glass tube, following the arrows of Fig. 14-10 (in both Figs. 14-9 and 14-10 the current flows first

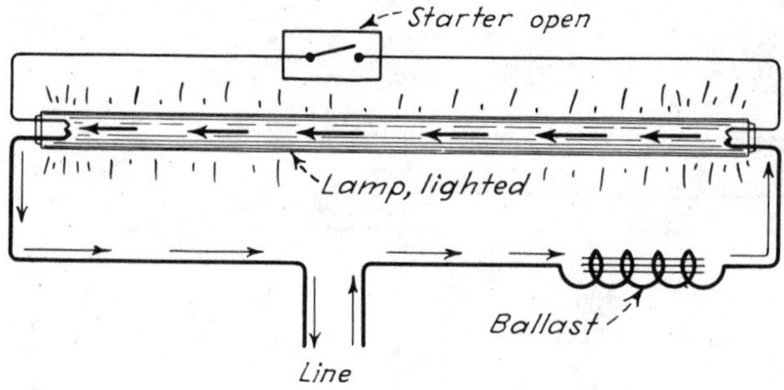

Fig. 14-10 Flow of current through a fluorescent lamp after it has started. The automatic starter has opened, and current cannot flow through it.

in the direction indicated by the arrows, then in the opposite direction, because the current involved is alternating current). The ballast then performs its other function: It limits the current flowing through the lamp to a predetermined proper value.

How does the current jump from one end of the tube to the other? The cathodes of coiled tungsten wire are coated with a chemical that when heated emits electrons, particles so small that billions of them laid side by side would still be invisible, being basically negative charges of electricity. They shoot out into space as popcorn does in a corn popper; and they travel more easily through argon gas than through a vacuum, which is why that gas is introduced. A stream of these particles constitutes an electric current, which heats the argon gas, which in turn heats the mercury, changing it to a vapor. The mercury vapor then becomes the path for the higher electric current of the lamp and its ballast.

The *preheat* circuit described, with a separate starter, is seldom used today. See Chap. 29 for descriptions of the *instant-start* and *rapid-start* circuits now in use.

If a fluorescent lamp as just described had a wall of clear glass, an insignificant amount of light would be visible, and yet the fluorescent lamp as installed produces a great deal of visible light. That is because the electric arc flowing through the mercury vapor produces only a very slight amount of visible light, but a great deal of invisible ultraviolet light. The glass "traps" the ultraviolet light, and very little passes through it.

However, the inside of the tube is covered with a layer of chemicals that become fluorescent or light-producing when exposed to the invisible ultraviolet light. In other words, invisible ultraviolet light striking the fluorescent chemicals makes the chemicals glow brightly, producing visible light. The particular chemical used determines the color of the light.

An explanation of the scientific principles that govern the emission of electrons from a heated cathode, the creation of the arc, the production of ultraviolet light by the arc, and the creation of visible light when invisible ultraviolet light strikes certain chemicals is beyond the scope of this book.

Advantages of Fluorescent Lighting. The greatest single advantage of a fluorescent lamp is its efficiency. Per watt of power used, it produces two to four[4] times as much light as an ordinary incandescent lamp. Its life is very much longer than that of an incandescent lamp. Being more efficient, it produces much less heat, which is important when a large amount of power is used for lighting, especially if the lighted area is air-conditioned.

Another major advantage is that the fluorescent lamp, being of relatively large size (in terms of square inches of surface compared with the total light output), has relatively low surface brightness, which in turn leads to less reflected glare, and less shadow, contributing to better seeing. Exposed fluorescent lamps, however, are bright enough to produce some glare discomfort, and enclosures are usually desirable.

[4] In the case of certain colors such as blue and green, the fluorescent lamp can produce over 100 times as much light per watt as an incandescent lamp.

Disadvantages of Fluorescent Lighting. Ordinary incandescent lamps operate in any temperature. Fluorescent lamps are somewhat sensitive to temperature. The ordinary type used in homes will operate in temperatures down to about 50°F. For industrial and commercial areas other types are available that will operate at lower temperatures, some as low as −20°F.

In the case of filament lamps, as the voltage is reduced, the light output drops off much faster than the voltage. However, even at greatly reduced voltages the filament still glows, producing some light. In the case of fluorescent lamps, as the voltage is reduced, the light output also drops off, but not as fast as in the case of ordinary filament lamps. But if the voltage is greatly reduced, the lamps go out completely, long before a filament lamp ceases to glow. Therefore fluorescent lamps cannot be used where there are frequent and violent fluctuations in voltage. Special voltage-regulating ballasts, however, are available for such locations, so that fluorescents need not be ruled out completely.

Life of Fluorescent Lamps. The life of an ordinary incandescent lamp varies from 750 to 2500 h depending on the size, type, and purpose of the lamp. It does not make any difference how many times the lamp is turned on or off. A *fluorescent* lamp that is turned on and never turned off will probably last over 25 000 h, or about three years. If it is turned on and off every 5 min, it may not last 1000 h. The reason lies in the fact that there is a limited amount of electron-emitting material on the cathodes; a specific amount of it is used up every time the lamp is turned on; when it is all gone, the lamp is inoperative. It is not possible to predict the exact number of starts the lamp will survive, and ordinary operation of the lamp also consumes some of the material, but the fact remains that the more often a fluorescent lamp is turned on, the shorter its life will be.

The published figures for approximate life are based on the assumption that the lamp will be lighted for 3 h every time it is turned on; for the most ordinary 40-watt lamp the average is 20 000 h. Under ordinary operation, fluorescent lamps will last five to fifteen times longer than ordinary incandescent lamps.

Rating of Fluorescent Lamps. An ordinary incandescent lamp marked "40 watts" will consume 40 watts when connected to a circuit of the proper voltage. A *fluorescent* lamp rated at 40 watts will also consume 40 watts within the lamp, but additional power is also consumed in the ballast; this additional power is from 10 to 20% of the watts consumed by the lamp itself, and must be added to the watts consumed by the lamp itself to obtain the total load in watts of the fixture.

Power Factor of Fluorescent Lamps. Ordinary *incandescent* lamps have a power factor (pf) of 100%. A single *fluorescent* lamp connected to a circuit has a power factor of somewhere between 50 and 60%. Assume one hundred 40-watt lamps connected singly to a 120-volt circuit. The ballast for each lamp can be expected to consume approximately 8 watts. The total for each combination is 48 watts, and the total for 100 such lamps is 4800 watts. The amperage consumed by these 100 lamps and their ballasts is, however, not $^{4800}/_{120}$ or 40 amp as might be expected (and is the case using ordinary incandescent lamps), but rather (assuming 60% pf) is $4800/(120 \times 0.60)$ or about 67 amp. Therefore the wiring serving this load must be capable of carrying 67 amp rather than a theoretical 40 amp.

Fortunately this is not as serious as it sounds. The common method is to have either two or four lamps per fixture and, in addition to the usual ballast, to use power-factor-correction devices built into the same case with the ballast, which bring the power factor up to about 90% or higher. Most two- and four-lamp fixtures on the market are the high-power-factor type, and especially so in the case of fixtures for commercial and industrial use. But if you buy "bargain" fixtures for the home, be sure they are of the high-power-factor type.

Sizes of Fluorescent Lamps. The cool-white lamps commonly used in homes (as well as in nonresidential occupancies) are listed in Table 14-3. Note that the watts consumed by the ballast are *not* included in the column headed "watts."

In addition to those listed in the table, circular fluorescent lamps are available in four sizes: 20 watts, $6^1/_2$ in. in outer diameter; 22 watts, $8^1/_4$ in. in diameter; 32 watts, 12 in. in diameter; and 40 watts, 16 in. in diameter. They are rather expensive and not very efficient as compared with

TABLE 14-3 Characteristics of Cool-White Fluorescent Lamps

Designation	Length, inches	Diameter, inches	Watts	Total lumens	Lumens per watt
T-8	18	1	15	870	58
T-12	24	$1\frac{1}{2}$	20	1250	62
T-8	36	1	30	2200	73
T-12	48	$1\frac{1}{2}$	40	3150	79

other fluorescent lamps, but much more efficient than incandescent lamps. Fixtures using two, three, or four different sizes are available.

Circular fluorescent lamps with ballast cases designed to screw into medium-base incandescent lampholders are also available as energy-saving replacements for incandescent lamps without the need to replace the fixture.

A very compact fluorescent lamp in a tightly folded U shape, in ratings of 8 to 15 watts, is being used primarily for porch, garden-pathway, and other exterior lighting, for energy savings.

Other Types of Fluorescent Lamps. For nonresidential use, there are many types of fluorescent lamps in addition to those already discussed, and these will be described in Chap. 29. Note that the circuit of Figs. 14-9 and 14-10 is that for a single lamp. More complicated circuits are used when the fixture has two or more lamps even if only one ballast is used.

Color of Light from Fluorescent Lamps. Everybody has noticed that the colors of flowers, clothing, and so on look different under ordinary incandescent lamps than in natural sunlight, because incandescent lamps produce light that is rich in red.

The color of the light produced by fluorescent lamps is determined entirely by the chemical and physical characteristics of the phosphor or powder that is deposited on the inside of the tube. The most ordinary color is "white" but there are many varieties of it, among them deluxe warm white, warm white, white, cool white and deluxe cool white. The differences lie in the proportion of red and blue in the light produced by the particular lamp under discussion. The "warm" varieties emphasize the red and yellow (like light from incandescent lamps), whereas the "cool" varieties emphasize the blue (more like natural sunlight). The kind of "white" to use depends on the effects desired.

If you look at a black-and-white printed page under each of the "white" colors in turn, you will see little difference. But if you look at a colored page, colored fabrics, or food under each kind in turn, you will see much difference.

Under deluxe cool white, people's complexions will appear much as in natural light, but the deluxe warm white will flatter them a bit, adding a ruddy or tanned appearance, much as is done by ordinary incandescent lamps. The cool white provides quite a good appearance also, but with a tendency toward paleness.

As to home furnishings, if your preference leans toward those having warm colors (red, orange, brown, tan), use deluxe warm white; if you prefer the cool colors (blue, green, yellow), use the deluxe cool white.

It must be mentioned that the "deluxe" varieties are not nearly so efficient as the others; they produce about 30% less light per watt of power used. If then you want maximum efficiency and are willing to sacrifice some color rendition, use the warm white in place of the deluxe warm white, and the cool white in place of the deluxe cool white. This difference in efficiency is probably the major reason why about 75% of all the fluorescent lamps in use are the cool-white variety. Yet, even the deluxe fluorescent lamps are far more efficient than incandescent lamps of the same rating in watts. So if color effects are important, use the deluxe varieties.

Some sizes of fluorescent lamps are also available in blue, green, pink, gold, and red. Their efficiency in colors is extraordinarily high compared with that of incandescent colored lamps. For example, a fluorescent green lamp produces about 100 times as much green light per watt as a green incandescent lamp. Colored lamps are used where spectacular color effects are needed, as in theater lobbies, lounges, stage lighting, and advertising.

15
Residential and Farm Motors

An average man can deliver no more than $^{1}/_{10}$ to $^{1}/_{8}$ hp continuously over a period of several hours. At $4 per h, a $^{1}/_{8}$-hp man would cost $32 per hp for each hour worked. At average rates, an electric motor will deliver a horsepower for an hour for about 6 cents. The motor costs little to begin with. It will operate efficiently on a hot day or a cold day, day after day, week after week. It never gets tired and costs nothing except while running. It uses electricity only in proportion to the power it is called upon to deliver. With reasonable care it will last for many years.

Horsepower. A horsepower is defined as the work required to lift 33 000 pounds one foot (33 000 foot-pounds) in one minute, which is equivalent to lifting 550 pounds one foot in one second. One horsepower is equal to 746 watts. Note that while the motor is *delivering* 1 hp or 746 watts, it is actually *consuming* more nearly 1000 watts from the power line, the difference of 254 watts being lost as heat in the motor, friction in the bearings, the power that it takes to run the motor even when it is idling, and similar factors.

General Information. In this chapter we will discuss some properties and characteristics of motors in general, then discuss various *kinds* of motors, and finally cover the wiring of the kinds of motors generally used in homes and farms; Chap. 30 will cover the wiring of motors used in larger industrial or commercial projects.

Starting Capacity. Some loads, such as compressors, require more starting torque (turning force supplied by the motor shaft) than do others such as fans. Motors deliver more torque while starting (some up to five times more) than when running. Naturally, the current consumed is also much higher while starting, and a motor will overheat quickly if it is delayed in reaching running speed by too great a load. Therefore the right kind of motor must be used for each load, depending on how hard it is to start the load.

Overload Capacity. Almost any good motor after reaching full speed will develop from $1^1/_2$ to 2 times its rated horsepower, or even more, for a short time. But no motor should be overloaded continuously, for overloading leads to overheating, and that in turn leads to greatly reduced life of the windings. Rewinding a burned-out motor is very expensive; in smaller sizes it usually is less expensive to replace the motor. Nevertheless this ability of a motor to deliver more than its rated horsepower is most convenient. For example, in sawing lumber, $^1/_2$ hp may be just right, but when a tough knot is fed to the saw blade, the motor will instantly deliver, if needed, $1^1/_2$ hp, dropping back to its normal $^1/_2$ hp after the knot has been sawed. A water-pressure system in a suburban home or on a farm may be properly equipped with a $^1/_2$-hp motor. When the motor is first turned on, with no pressure in the water tank, the pump may require a good deal less than $^1/_2$ hp. As the pressure in the tank builds up, the horsepower required also gradually increases. Just before the automatic pressure switch stops the motor, the motor will be delivering a good deal more horsepower than at the beginning of the running cycle. If the motor is undersized, and if an ''overload[1] protection device'' has been properly installed, the device will stop the motor before the pressure-activated switch on the pump will do so.

Gasoline Engines vs. Electric Motors. A gasoline engine is rated at the maximum horsepower it can deliver *continuously* (usually at a speed somewhat below its maximum speed); it has no overload capacity. Thus an engine rated at 5 hp can deliver 5 hp continuously when new, gradually diminishing with age, but unlike an electric motor, it cannot deliver more than 5 hp at any time. This explains why it is sometimes possible to re-

[1] An over*load* protection device is not the same as an over*current* device, as will be explained later.

place an engine with an electric motor of slightly lower horsepower rating —this is possible only if the engine never labors and slows down in normal operation. The motor that replaces the engine may at times deliver more than its rated horsepower, but don't overload the motor continuously—that will shorten its life.

Power Consumed by a Motor. The watts drawn from the power line by a motor are in proportion to the horsepower it is delivering. The approximate figures for a 2-hp motor are as follows:

While starting	4000 watts
While idling	400 watts
While delivering 1/2 hp	750 watts
While delivering 1 hp	1150 watts
While delivering 1 1/2 hp	1500 watts
While delivering 2 hp	2000 watts
While delivering 2 1/2 hp	2600 watts
While delivering 3 hp	3300 watts

Motors are designed to operate at greatest efficiency when delivering their rated horsepower; while they are delivering more or less power, the efficiency usually falls off. In other words, it costs a little more per hour to run a 1-hp motor at half load than it does to run a 1/2-hp motor at full load. So it costs less for power if you use a motor of the proper size for a machine than it does if you use a larger motor that you might have on hand. On the other hand, a motor that is continuously overloaded will not last long. Proper selection, application, and maintenance of a motor will extend its life substantially.

Speed of AC Motors. The most common 60-Hz ac motor runs at a theoretical 1800 r/min, but at an actual 1725 to 1750 r/min delivering its rated horsepower (the horsepower stamped on its name-plate). When the motor is overloaded, its speed drops. If the overload is increased too much, the motor will stall and burn out, if not properly protected by an overload device. Low voltage also reduces its speed and its horsepower output. A motor operating at a voltage 10% lower than normal will deliver only 81% of its normal power.

Motors of other speeds are available, running at theoretical speeds of 900, 1200, and 3600 r/min, with actual speeds a little lower. The slower-speed motors cost more, are larger, and will usually take longer to replace, although they do run a good deal quieter. In most cases the more readily available 1800 r/min will serve your purpose. If a very low speed is necessary, use a motor of the 1800 r/min type with built-in gears that reduce the shaft speed to a much lower figure, or use a separate reduction gear. However, be advised that for energy efficiency the closer the motor speed is to the required speed for the load, the better.

The speed of ordinary ac motors *cannot be controlled* by rheostats, switches, or similar devices, but the speed of shaded-pole and permanent-split-capacitor ac motors can be controlled by solid state (electronic) variable voltage controllers. Special-purpose variable-speed motors are obtainable, but they are very expensive and will not be discussed here.

Service Factors on Motors. At one time motors were rated only on the basis of a temperature *rise,* over and above the ambient temperature (the air temperature at the motor location when the motor is not running), while delivering their rated horsepower. Ordinary motors were based on a temperature rise of 40°C (72°F),[2] based on an ambient of 40°C (104°F). A 40°C-rise motor installed in a hot location, such as a pump house on a farm where the temperature might be 115°F (46°C), must have its temperature *rise* limited to 61°F (34°C) so as not to exceed the allowable rise *over a 40°C ambient.* For this application a 50°C-rise motor should be selected.

Over the years, the heat-resisting properties of insulations on wires that are used to wind motors, and the materials used to insulate these wires from the steel in a motor, have been vastly improved, so that they will not be harmed by temperatures that would have destroyed an old-time motor. This in turn has made it possible to reduce the physical size of motors a great deal—today's 10-hp 3-phase motor isn't much bigger than a 3-hp motor made, say, in 1945. Although these smaller motors run much hotter than the old larger size, they will not be burned out by temperatures that would have burned out the 1945 motor. They often run at temperatures that exceed 212°F (the boiling point of water) but are not damaged.

Motor ratings are based on a ''service factor'' stamped on their nameplates, ranging from 1.00 to 1.40. If for a specific motor the service factor is 1.00, it means that if the motor is installed in a location where the ambient temperature is not over 40°C or 104°F, it can deliver its rated horse-

[2] Do not confuse a *change* in the readings of two different thermometers with their *actual* readings. When a Celsius (formerly called centigrade) thermometer reads 40°, a Fahrenheit thermometer reads 104°. When a Celsius thermometer changes by 40° the Fahrenheit changes by 72°. Thus if the Celsius changes from 40 to 80° (a 40° rise), the Fahrenheit changes from 104 to 176° (a 72° rise). One degree on the Celsius scale is the equivalent of 1.8° on the Fahrenheit scale.

power continuously without harm. But if its service factor is, for example, 1.15, it means that it can be used at up to 1.15 times its rated horsepower under the same conditions. Multiply the rated horsepower by the service factor; a 5-hp motor, if its service factor is 1.15, can be used continuously as a 5.75-hp motor. But remember a motor will operate most efficiently and last longer if used at not over its *rated* horsepower. The service factor provides temporary extra power if needed.

Most open motors larger than 1 hp have a service factor of 1.15. Fractional-horsepower motors have a service factor of 1.25; some of them, a factor as high as 1.40. If the ambient temperature is above 104°F, the motor should not be operated at its full rated horsepower without a fan or blower in the area. Regardless of temperature, install motors where plenty of air is available for cooling.

Do note that some motors now being made are still rated on the basis of a 40°C temperature rise, and do not have a service factor.

Split-Phase Motors. A split-phase motor operates only on single-phase alternating current. It is the most simple type of single-phase motor made, which makes it relatively trouble-free; there are no brushes, no commutator. It is available only in sizes of 1/3 hp and less. It draws a very high current while starting. Once up to full speed this type of motor delivers just as much power as another type of motor, but it is not capable of *starting* heavy loads. Therefore it should never be used on a machine that is hard to start, such as a water pump or an air compressor that has to start against compression. It can be used on any machine that is easy to start or on one where the load is thrown on after the motor is up to full speed. It is suitable for grinders, saws, lathes, and similar equipment as used in home workshops.

Capacitor Motors. A capacitor motor also operates only on ordinary single-phase alternating current. It is similar to the split-phase type with the addition of a capacitor that enables it to start much-harder-starting loads. There are really two types of capacitor motors: (1) the capacitor-start type, in which the capacitor is cut out of the circuit after the motor has reached full speed, and (2) the capacitor-start, capacitor-run type, in which the capacitor remains in the circuit at all times.

There are several grades of such motors available, ranging from home-workshop types that can start loads from 1 1/2 to 2 times as heavy as the split-phase type will start, to the heavy-duty type, which will start almost any type of load. Capacitor motors are usually more efficient than split-phase motors, using fewer watts per horsepower. The current consumed *while starting* is very much less than that of the split-phase type. Capacitor motors are made in any size but are commonly used only in sizes up to $7^1/_2$ hp.

Repulsion-Start Induction-Run Motors. Often called an "RI" motor, this type again operates only on single-phase alternating current. It has very high starting ability and should be used for the heavier jobs; it will "break loose" almost any kind of hard-starting machine. The starting current is the lowest of all types of single-phase motors. It is available in sizes up to 10 hp.

Universal Motors. A universal motor operates either on direct current or single-phase alternating current. However, it does not run at a constant speed but varies over an extremely wide range. Idling, a universal motor may run as fast as 15 000 r/min, but under a heavy load may slow down to 500 r/min. This of course makes the motor totally unsuitable for general-purpose work. It is used only where built into a piece of machinery whose load is substantially constant, and predetermined. For example, you will find universal motors on vacuum cleaners, sewing machines, drills. Their speed can be controlled within limits by a rheostat as on a sewing machine.

Dual-Voltage Motors. Most motors rated at $^1/_2$ hp or more are constructed so that they can be operated on either of two different voltages such as 120 or 240 volts. Single-phase motors of this type have four leads. Connected one way, the motor will operate as a 115-volt motor; connected the other way, as a 230-volt motor. See Fig. 15-1. (Remember the standard voltage ratings of motors are still 115 and 230 volts, but entirely suitable for use on 120- and 240-volt circuits.)

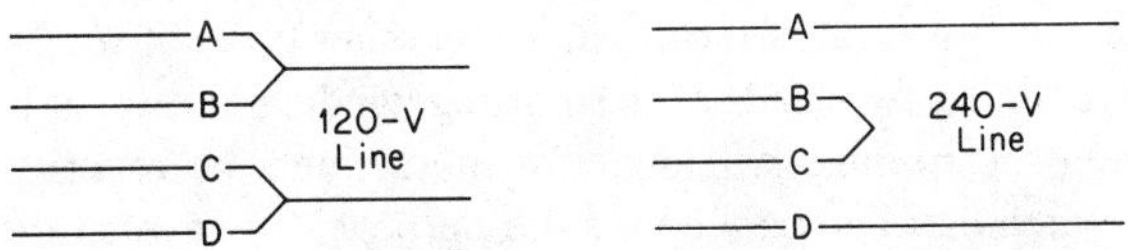

Fig. 15-1 Larger motors are made so that they can be operated on either 120- or 240-volt circuits, depending on the connections of the four leads from the motor, as shown in the diagram above. It is always wise to operate the motor at 240 volts if possible.

If a motor is of the dual-voltage type, two different voltage and two different current ratings will be shown on the name-plate. For example, it may be marked "Volts 115/230, Amps 24/12," which simply means that while delivering its rated horsepower, it will consume 24 amp if operated as a 115-volt motor, but only 12 amp if operated as a 230-volt motor.

A 3-phase dual-voltage motor has nine leads, and great care must be used to connect them properly for the voltage selected.

Three-Phase Motors. This type of motor, as the name implies, operates only on *3-phase* alternating current. Three-phase motors are the simplest of all kinds of motors and are available in any size. They cost less than any other type, so if you have 3-phase current available, by all means use them. But do not assume that because you have a *3-wire* service, you have *3-phase* current. In all likelihood you have 3-wire 120/240 volt *single-phase* current. If in doubt, see your power supplier.

Direct-Current Motors. Direct current (dc) is still found in the downtown sections of a few large cities, and perhaps in a few small towns. All battery-type small generating plants, once used on many farms and now used mostly in marine applications, are direct current. There are many types of dc motors, among them the series, the shunt, the compound types, and others. However, too few dc motors are used to warrant a full discussion here.

One important difference in operating characteristics between ac and dc motors is that the ac motor will run at a specific speed even if there are considerable changes in loads and voltages, and that speed cannot be controlled by a rheostat or similar device. A dc motor, on the other hand, even if designed for 1800 r/min, will run at that speed only with a specific combination of voltage and load. If either voltage or load change, the speed changes. Its speed can be controlled by a rheostat, although once adjusted, it will change with any change in voltage or load.

Reversing Motors. The direction of rotation of an RI motor can be reversed by changing the position of the brushes. On other types of *single*-phase motors, reverse the two leads of the starting winding in respect to the two leads from the running winding. If a motor must be reversed often, a special switch can be installed for the purpose. To reverse the rotation of a *three*-phase motor, simply reverse any two of the three wires running between the controller and the motor; do this in the controller. But if one of the wires is grounded, reverse the other two.

Problems with Large Motors. The size of farms is constantly increasing, leading to the use of larger sizes of farm machinery, requiring bigger motors: 10 hp, 25 hp, and even larger. In turn this reduces the number of hours of labor for a given output, contributing to the higher efficiency of labor on the farm, the food factory, that is being demanded to make the farm profitable.

But most farms have only a *single-phase* 3-wire 120/240-volt service. That requires only two high-voltage lines to the farm, and only one transformer. Single-phase motors are not usually available in a size larger than $7^1/_2$ hp, although a few larger ones are made. But before buying even a 5-hp single-phase motor, check with your power supplier to see whether the line and the transformer serving your farm are big enough to operate such a motor.

Single-phase motors 5 hp and larger require an unusually high number of amperes *while starting,* and the line and transformer often are too small to start such a motor. If you operate the motor only a comparatively few hours per year, your power supplier will object to installing a heavier line and transformer, just as the farmer would not buy a 10-ton truck to haul 10 tons a few times per year, while using it for much smaller loads most of the time.

In a few localities, at least some of the farms are served by a 3-phase line, requiring three (sometimes four) wires to the farm, and three transformers. If you are fortunate enough to have 3-phase service, your problems are solved. Simply use 3-phase motors, which cost considerably less than single-phase; being simpler in construction, they rarely pose a service problem. (Note: If you have 3-phase service, there will be available 3-phase power, usually at 240 volts, and also the usual 120/240-volt single-phase for lighting, appliances, and other small loads.)

But if you have only the usual single-phase 120/240-volt service, and still need larger motors, what to do? One solution is to use smaller machinery requiring motors not over 3 hp, but that is a step backward, for it increases the cost of labor, the number of expensive hours that must be expended in operating the farm. There is another solution. There are "phase converters" that permit *3-phase* motors to be operated on *single-phase* lines. The phase converter changes the single-phase power into a sort of modified 3-phase power that will operate ordinary 3-phase motors and at the same time greatly reduces the number of amperes required *while starting.* In other words, when a 3-phase motor is operated with the help of a phase converter, the same single-phase line and transformer that would barely start a 5-hp single-phase motor will start a $7^1/_2$-hp or possibly even a 10-hp 3-phase motor; a line and transformer that would handle a 10-hp single-phase motor (if such a motor could be found) would probably handle a 15-hp or 20-hp 3-phase motor.

Phase converters are not cheap, but their cost is partially offset by the lower cost of 3-phase motors, and in any event they make possible the operation of larger motors than would be possible without the converter.

There are two types of converters: static with no moving parts except relays, and the rotating type. The static type must be matched in size and type with the one particular motor to be used with it; generally there must be one such converter for each motor.

The rotating type of converter looks like a motor, but can't be used as a motor. Two single-phase wires run into the converter; three 3-phase wires run out of it. Usually several motors can be used on this at the same time; the total horsepower of all the motors in operation at the same time can be at least double the horsepower rating of the converter. Thus if you buy a converter rated at 15 hp, you can use any number of 3-phase motors totaling not over 30 to 40 hp, but the largest may not be more than 15 hp, the rate of the converter. The converter must be started first, then the motors started, the largest first, then the smaller ones.

But some words of caution are in order. A 3-phase motor of any given horsepower rating will not *start* as heavy a load when operated from a converter, as it will when operated from a true 3-phase line. For that reason it will often be necessary to use a motor one size larger than is necessary for the *running* load. This will not significantly increase the power required to run the motor, once it is started. Again, the converter must have a horsepower rating at least as large as that of the largest motor.

The voltage delivered by the converter varies with the load on it. If no motor is connected to the converter, the 3-phase voltage supplied by it is very considerably higher than the input voltage of 240 volts. Do not run the converter for significant periods of time, without operating motors at the same time, or it will be damaged by its own high voltage. Do not operate only a small motor from a converter rated at a much higher horsepower, for the high voltage will damage the motor or reduce its life. To be prudent, the total horsepower of all the motors operating at one time should be at least half the horsepower rating of the converter.

Last but not least, check with your power supplier; some do not favor or permit converters. If they do permit converters, the line and the transformer serving your farm must be big enough to handle all the motors you propose to use.

Care of Motors. Motors require very little care. The most important is proper oiling of the bearings. Use a very light machine oil, such as SAE No. 10, and use it sparingly—most motors are oiled too much. Never oil any part of the motor except the bearings; under no circumstances put oil on the brushes, if your motor has brushes.

If your motor has a commutator and brushes, occasionally while the motor is running, hold a very fine piece of sandpaper (never emery) against the commutator to remove the carbon that has worn off the brushes. *Be sure you are standing on something absolutely dry to avoid shock.*

Bearings. In most commonly used motors, there is a choice of sleeve bearings or ball bearings. If the motor is to be operated with the shaft in the ordinary horizontal position, there is no need for ball bearings. If the motor is to be operated with the shaft in an up-and-down position, ball bearings should be used because the usual sleeve-bearing construction lets the oil run out. Ball bearings are also better able than sleeve bearings to absorb the weight of the rotor. Ball bearings of some types are filled with grease and permanently sealed, thus doing away with the nuisance of greasing. If this type is purchased, be sure the bearings are double-sealed, that is, with a seal on each side of the balls. Some bearings are sealed on only one side, so that the grease can still get out the other side.

Wiring for Motors. The Code sections that govern the installation of motors are extremely complicated, for they cover all motors from the tiniest to those developing hundreds or thousands of horsepower. Such motors will be discussed in Chap. 30, but the wiring of motors *for homes and farms* can be covered by a few simple rules, covering five basic components of motor circuits, as follows:

Wire Size. First of all, extension cords made of the usual No. 18 or No. 16 wire should never be used, even with fractional-horsepower motors. A cord on a portable motor is necessary but if a longer extension cord is added, the voltage drop in the cord during the starting period might be so great that the motor will never reach full speed and switch off its starting winding; a damaged motor might easily result.

All wires in a motor branch circuit, all the way up to the motor, must be large enough to carry the starting current, and the length of the circuit must also be taken into consideration because of voltage drop. The Code requires that the wires have an ampacity of at least 125% of the name-plate ampere rating of the motor. So, multiply the motor ampere rating on its name-plate[3] by 1.25, which is 125%. Then use wire with an ampacity at least as great as the figure determined from NEC Tables 310-16 through 310-19, whichever is applicable (see Appendix).

Next consider the one-way distance to the motor to see whether the

[3] If you do not yet have the motor, determine the ampere rating from NEC Table 430-148 (see Appendix).

wire size you have selected will limit the voltage drop to a reasonable value. But if you calculate your answer based on the name-plate amperes, remember that the motor will consume many more amperes while starting, so that while starting, the drop will be much higher than while running. If the motor must start loads that are excessively hard-starting, too much drop might prevent the motor from reaching full speed, and that might lead to motor burnouts. To simplify the whole problem of correct wire sizes, use Table 15-1. If the machine that the motor drives is of the very-hard-starting type, use wire one size larger than shown.

Disconnecting Means. Every motor must be provided with a method of disconnecting it from the circuit. If your motor is portable, the plug and receptacle serve as the disconnecting means. If the motor is $^1/_8$ hp or less, no separate disconnecting means is required; the branch-circuit breaker or fuse will serve the purpose. All other motors must have a switch or circuit breaker as a disconnecting means, and it must open all ungrounded wires to the motor and its controller, if the motor has a separate controller.

A switch of the general type shown in Fig. 15-2 (or a larger one with cartridge fuses for a larger motor), or a circuit breaker in an enclosure, may be used as the disconnecting means. If the motor is *larger than 2 hp,*

Fig. 15-2 A switch of this type, or a larger one with cartridge fuses, may be used with small motors. If the switch has two fuses, it is for a motor operated at 240 volts; if it has only one fuse, it is for a motor operated at 120 volts. (*Square D Co.*)

the switch (or the plug and receptacle, for a portable motor) must be rated in horsepower not less than the horsepower rating of the motor; many switches are rated both in amperes and in horsepower. If the motor is *2 hp or smaller,* you may use the same kind of switch rated only in amperes, provided the ampere rating of the switch is at least double the ampere rating of the motor. Note that all circuit breakers are equivalent to switches rated in horsepower. You may also employ a general-use ac-

TABLE 15-1 Wire Sizes for Motors

Motor			Wire sizes								
Horse-power	Volts	Am-peres	14	12	10	8	6	4	2	1/0	2/0
1/4	115	5.8	55	90	140	225	360	575	900	1,500	1,800
1/3	115	7.2	45	75	115	180	300	450	725	1,200	1,500
1/2	115	9.8	35	55	85	140	220	350	550	850	1,100
3/4	115	13.8	25*	40	60	100	150	250	400	600	800
1	115	16.0	—	35	50	85	130	200	325	525	650
1½	115	20.0	—	25*	40	65	100	170	275	425	550
2	115	24.0	—	—	35	55	85	140	225	350	450
3	115	34.0	—	—	—	40*	60	90	160	250	325
1/4	230	2.9	220	360	560	900	1,450	2,300	3,600		
1/3	230	3.6	180	300	460	720	1,200	1,600	2,900		
1/2	230	4.9	140	220	340	560	875	1,400	2,200		
3/4	230	6.9	100	160	240	400	600	1,000	1,600	2,400	
1	230	8.0	85	140	200	340	525	800	1,300	2,100	
1½	230	10.0	70	110	160	280	400	675	1,100	1,700	2,200
2	230	12.0	60	90	140	230	350	550	900	1,400	1,800
3	230	17.0	—	65*	100	160	250	400	650	1,000	1,300
5	230	28.0	—	—	60*	100	160	250	400	650	800
7½	230	40.0	—	—	—	70*	110	175	275	450	550
10	230	50.0	—	—	—	—	90*	140	225	350	450

Note: Figures below the wire sizes indicate the *one-way* distance in feet (not the number of feet of wire in the circuit) that each size wire will carry the amperage for the size motor indicated in the left-hand column, with 1½% voltage drop. A dash indicates that the wire size in question is smaller than the minimum required by the Code for the horsepower involved, regardless of circumstances. Figures are based on single-phase ac motors.

* If you are using Type T or TW wire, the size of wire at the top of the column is too small for the amperage; a different type of wire may be suitable. See NEC Tables 310-16 through 310-19 in the Appendix.

type snap switch of the kind you use to control lights if its ampere rating is at least 125% of the motor ampere rating. In any case use a single-pole switch for motors on 120-volt circuits and a 2-pole switch for motors on 240-volt circuits. Always be sure that the disconnecting means is open (off) before working on a motor, a motor circuit, or the driven machinery.

Short-Circuit and Ground-Fault Protection. A motor and its circuit wires must be protected against short-circuit or ground-fault currents by a circuit breaker or fuse in each ungrounded wire. Since the motor's *starting* current for a short time is much greater than its full-load running current, the circuit breaker or the fuse must have a high enough ampere rating so that it will not trip or blow during the starting cycle. Its ampere rating must be at least 125% of the ampere rating of the motor, but it may be increased to as much as 175%[4] if time-delay fuses are used and 250% if circuit breakers are used. All this will be explained in considerable detail in Chap. 30.

Controller. Every motor must have a means of starting and stopping it: This is called a "controller." The controller may be an automatic device, part of a refrigerator, water pump, or any other equipment that starts and stops automatically. In that case you can safely assume that the proper controller comes with the equipment. The following discussion will be about manually operated controllers.

If the motor is portable and rated at $^1/_3$ hp or less the plug and receptacle serve as the controller. If, however, it is larger than $^1/_3$ hp, proceed as in the case of stationary or fixed motors, as discussed in following paragraphs.

If the motor is 2 hp or smaller, and used at not over 300 volts, you may use the disconnecting means (as already discussed under that heading) as the controller. If the motor is larger than 2 hp, a separate controller must be installed, rated in horsepower not less than the horsepower of the motor.

Regardless of the size of the motor, instead of using switches of the kind described, special motor starters of the type shown in Fig. 15-3 are generally used as controllers. They are rated in horsepower; use one rated at not less than the horsepower of the motor. They have built-in motor over*load* protection (which will be discussed in a later paragraph). For smaller (fractional-horsepower) motors use the kind shown in the left-hand part of the illustration; it isn't much larger than an ordinary tog-

[4] If such values seem to contradict everything said up to this point, that the branch-circuit overcurrent devices may not have an ampere rating greater than the ampacity of the wire in the circuit, bear in mind that each motor must be provided with an over*load* (not over*current*) device that protects the motor against overloads, or failure to start, as will be explained later in this chapter. The branch-circuit over*current* device therefore protects only against short circuits or ground faults.

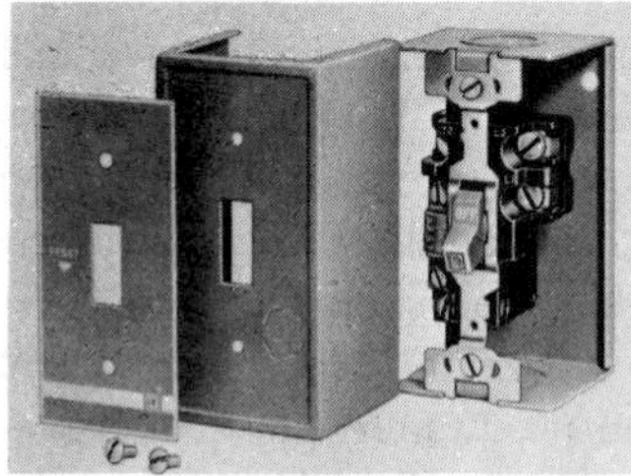

Fig. 15-3 Controls of these types are used to start and stop a motor; they also contain motor overload devices. (*Square D Co.*)

gle switch, and controls the motor just as an ordinary switch would. For larger motors, use the kind shown in the right-hand part of the illustration. It has push buttons in the cover for starting and stopping the motor. It is also available without the push buttons, which are then installed separately at a distance in some convenient location. Chapter 30 will discuss such starters and their method of operation in more detail.

"In Sight from." A motor *disconnecting means* must be in sight from the motor *controller;* there are no exceptions to this rule. The *controller* in turn must be in sight from the *motor and its driven machinery*. In the NEC "in sight from" is defined as being visible *and not over 50 ft* from the specified location. So if you can see one component from the other but they are more than 50 ft apart, they are *not* "in sight from" each other.

If the controller is not in sight from the motor and its driven machinery, you must do one of the following: (1) provide the type of disconnecting means that can be locked in the open position with a padlock, such as the type shown in Fig. 15-2, or (2) install an additional disconnecting means that is in sight from the motor and its driven machinery. This additional disconnecting means must meet the requirements for the disconnecting means already discussed, but if it is a switch, it need not have fuses.

Motor *Overload* Protection. The branch-circuit breaker or fuses, if of an ampere rating high enough to carry the *starting* current of the motor, will rarely protect the motor against damage caused by an overload that continues for some time, or by failure to start. Therefore a separate over*load* protective device must be installed to protect the motor against such damage.

The Code does permit manually controlled portable motors rated 1 hp or less to be plugged into an ordinary 15- or 20-amp 120-volt circuit. All others must be provided with separate overload protection.

Often, and especially in the case of smaller motors, the motor overload device is built into the motor; it is called a "thermal protector." A thermal protector has one or more heat-sensing elements that protect the motor from dangerous overheating due to overloads or to failure of the motor to start by opening the circuit when the motor temperature exceeds a safe value. Usually a thermal protector must be reset by hand when it trips; in some cases it resets automatically after a short time interval. Of course the overload condition should be corrected before a motor is put back into operation. Motors equipped with thermal protectors must be marked "Thermally Protected" or, for motors rated at 100 watts or less, "T.P." on their name-plates.

If a motor is permanently installed and not provided with built-in protection, separate overload protection must be provided. For the kinds of motors used in homes and on farms, the overload device must have an ampere rating not more than 25% above the ampere rating of the motor.

For this purpose, you will most likely use "heater coils" that are installed inside your motor controller, if it is the type shown in Fig. 15-3. The heater coils are rated in amperes and must be selected to match the ampere rating of the motor. If properly selected, the overload device(s) will carry the normal running current of the motor indefinitely, and will carry a small overload for some time, but will trip the controller and shut off the motor quickly in case of heavy overload or failure to start. Such a starter will then serve as the controller as well as the overload protection device. It must *never* serve as the disconnecting means. For a more detailed discussion of how such motor starters with their overload protection devices operate, see Chap. 30.

Do note that motor overload devices are quite capable of handling the normal motor current as well as the higher current that might result from a nominal overload or from failure to start. But they are entirely incapable of handling the very much higher currents that would develop in case of a short circuit or ground fault; such currents are interrupted by the motor branch-circuit short-circuit and ground-fault protection.

PART 2

Actual Wiring: Residential and Farm

Part 2 of this book explains the actual wiring of houses and farm buildings. The author feels that it will be much easier for you to study first these relatively simple installations than it would be if one chapter pertaining to one particular phase of the work included everything from a simple cottage up to an elaborate project.

If a building is wired while it is under construction, the electrical work is called "new work." If an existing building is wired or rewired, it is called "old work." Both are covered in Part 2.

All the fundamentals covered in Part 2 will in practice also be used in the bigger projects. They are the foundation for the more complicated methods to be covered by Part 3, which covers the wiring of larger projects.

16
Planning an Installation

The plans for an electrical installation consist of outline drawings of the rooms involved, with indications where the various outlets for fixtures, receptacles, and other equipment are to be located. Obviously a *picture* of a switch, a receptacle, or other device cannot be shown on the plans at each point where one is to be installed. Standard graphic symbols are used instead.

Graphic Symbols. In order to cover the wiring of all kinds of buildings from the smallest to the very largest, a great many symbols are required. Those likely to be encountered in the wiring of the kinds of buildings discussed in this book are shown in Figs. 16-1 and 16-2, excerpted by permission from the American National Standards Institute "Standard Y32.9-1972: Graphic Symbols for Electrical Wiring and Layout Diagrams Used in Architecture and Building Construction," copyright by the Institute of Electrical and Electronics Engineers, Inc.

Symbols of this kind are included in most wiring plans. You must understand them in order to be able to read and follow plans, so study the symbols of Figs. 16-1 and 16-2 until each one means as much to you as a picture would.

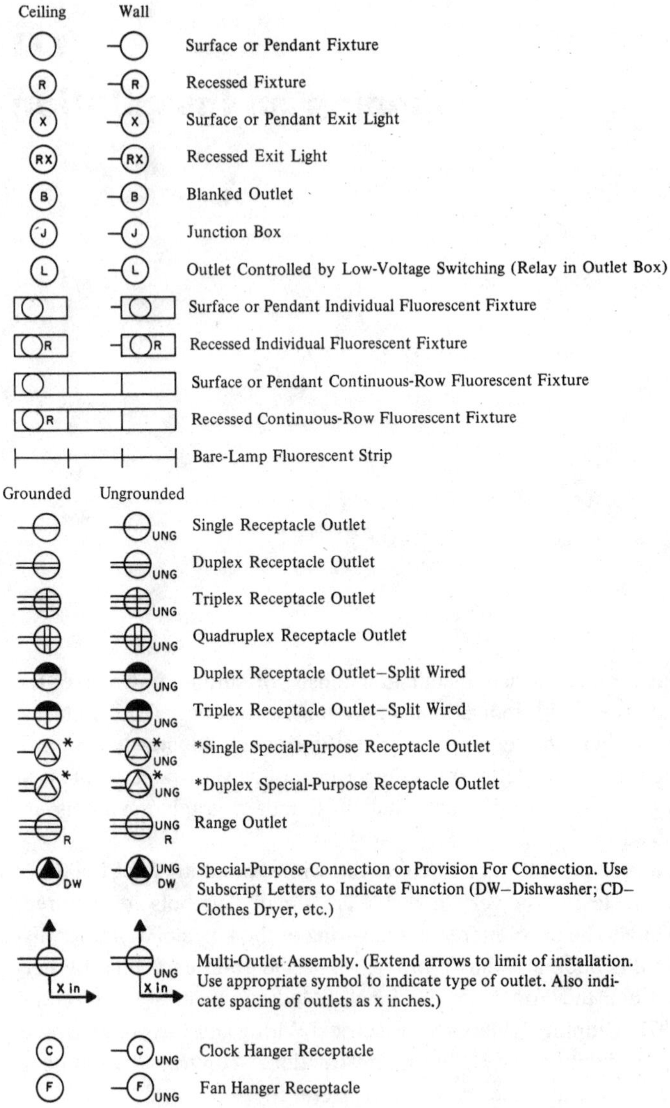

Ceiling	Wall	
		Surface or Pendant Fixture
R	R	Recessed Fixture
X	X	Surface or Pendant Exit Light
RX	RX	Recessed Exit Light
B	B	Blanked Outlet
J	J	Junction Box
L	L	Outlet Controlled by Low-Voltage Switching (Relay in Outlet Box)
		Surface or Pendant Individual Fluorescent Fixture
R	R	Recessed Individual Fluorescent Fixture
		Surface or Pendant Continuous-Row Fluorescent Fixture
R		Recessed Continuous-Row Fluorescent Fixture
		Bare-Lamp Fluorescent Strip

Grounded	Ungrounded	
	UNG	Single Receptacle Outlet
	UNG	Duplex Receptacle Outlet
	UNG	Triplex Receptacle Outlet
	UNG	Quadruplex Receptacle Outlet
	UNG	Duplex Receptacle Outlet–Split Wired
	UNG	Triplex Receptacle Outlet–Split Wired
*	* UNG	*Single Special-Purpose Receptacle Outlet
*	* UNG	*Duplex Special-Purpose Receptacle Outlet
R	UNG R	Range Outlet
DW	UNG DW	Special-Purpose Connection or Provision For Connection. Use Subscript Letters to Indicate Function (DW–Dishwasher; CD–Clothes Dryer, etc.)
X in	UNG X in	Multi-Outlet Assembly. (Extend arrows to limit of installation. Use appropriate symbol to indicate type of outlet. Also indicate spacing of outlets as x inches.)
C	C UNG	Clock Hanger Receptacle
F	F UNG	Fan Hanger Receptacle

* Use numeral or letter either within the symbol or as a subscript alongside the symbol keyed to explanation in the drawing list of symbols to indicate type of receptacle or usage.

Fig. 16-1 Graphic symbols used in architectural drawings to designate electric outlets.

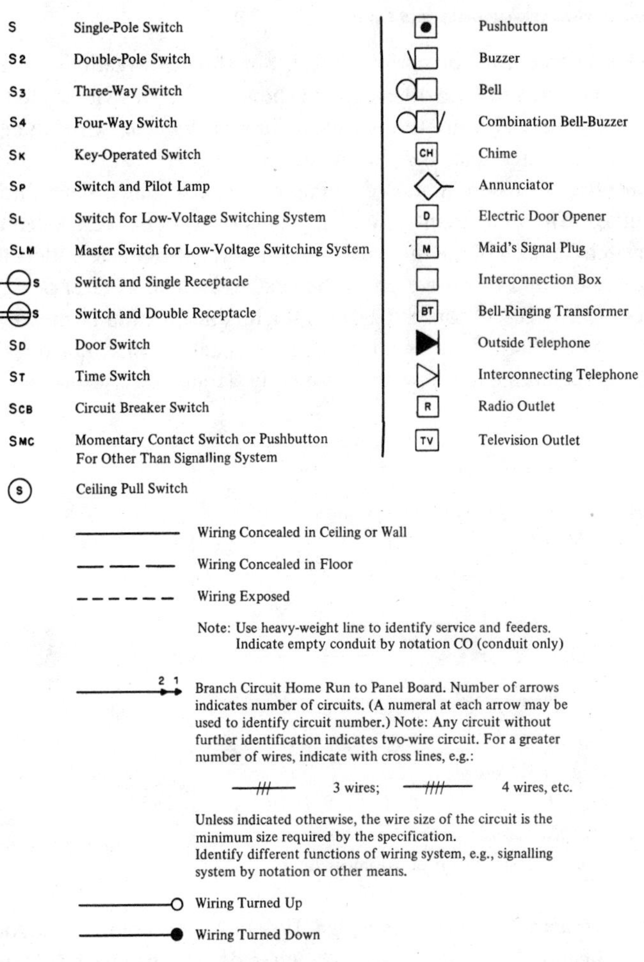

Fig. 16-2 More symbols. All are from American National Standards Institute (ANSI) "Standard Y-32.9-1972."

Sometimes a symbol is supplemented by additional letters near the symbol to more fully define the outlet. Examples:

WP	Weatherproof	G	Grounded
RT	Raintight	UNG	Ungrounded
DT	Dusttight	R	Recessed
EP	Explosionproof	DW	Dishwasher
PS	Pull switch	GFCI	Ground-Fault Circuit-Interrupter

Do note that the usual blueprint plans do *not* show how various outlets are to be connected to each other, but do show lines from switches to the particular outlet(s) they are to control. Solid lines indicate wires in ceilings or walls; dashed lines indicate wires under floors.

Typical Plans. Consider first a very simple plan, covering a small three-room cottage with two circuits, involving one ceiling outlet controlled by a wall switch for each of the three rooms, with three receptacle outlets for the larger room and one for each of the smaller rooms. The service entrance is 2-wire 120-volt only. The plan for this installation is shown in Fig. 16-3. Note that this does not provide adequate wiring, nor does it meet Code requirements; it is shown merely as an exercise in reading plans.

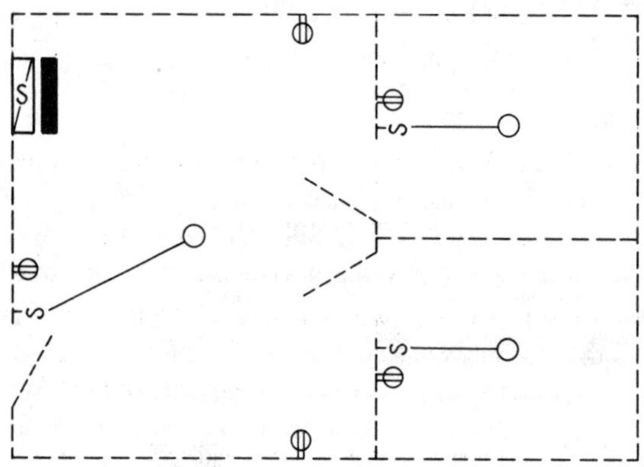

Fig. 16-3 Layout of a simple three-room project.

To make it easier to interpret this plan, Fig. 16-4 shows the same layout in pictorial fashion, with all the wires shown in detail. The grounded wire is shown as a light line; the "hot" wires, as heavy lines. Note how the grounded wire runs without interruption from the point where it enters the building to each device where current is to be used. The black wires run from their fuses direct to each receptacle outlet and to each switch; an additional length runs from each switch to the light it controls, and that completes the wiring.

A represents the main switch; *B* represents the main fuse. *C* and *D* represent the two fuses, one for each branch circuit. The first branch circuit comprises all the wiring served by the current that flows through fuse *C*;

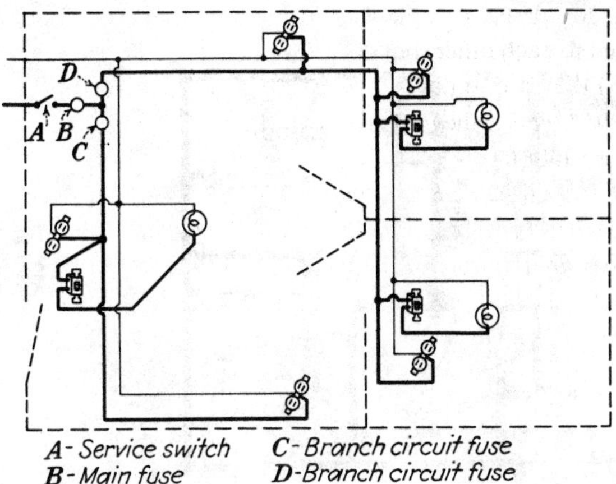

Fig. 16-4 The layout of Fig. 16-3, but here shown in pictorial fashion.

the second circuit comprises all wiring served by the current that flows through fuse *D*.

The wiring plans for a larger house are shown in Figs. 16-5 and 16-6. These diagrams[1] may at first sight seem rather formidable, but with study they become simple. Such plans are supplemented by detailed written specifications which give such information as size and type of service entrance, number of circuits, types of material to be used, and similar data. The plans in Figs. 16-5 and 16-6 show how symbols are used to indicate locations of outlets.

Make some plans of a similar nature for other installations—for example, (1) your own home as it is wired; (2) your own home as you would like to see it wired, probably with more appliances, such as a water heater, or a dishwasher, that you do not have now; and (3) an "ideal" home in which you would like to live. Do this until symbols are as clear to you as a picture would be.

Drawing Plans. Often the electrician may be called upon to make the plans for a job, instead of finding them ready-made. In that case include all those details found in plans of the general type shown in Figs. 16-5 and 16-6. Likewise include in the plans such things as the size and location of

[1] Reproduced by permission from the "Residential Wiring Design Guide."

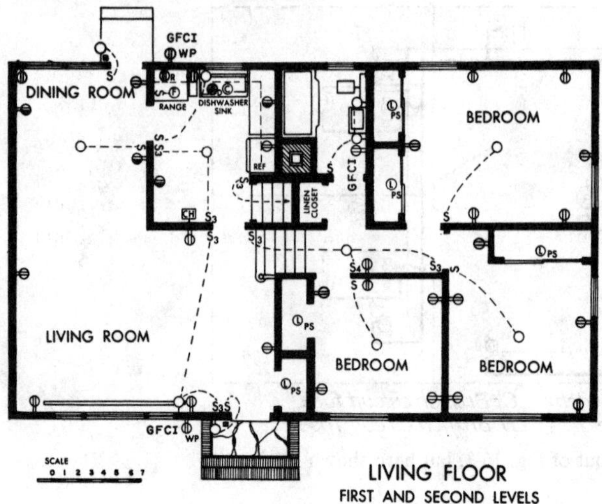

Fig. 16-5 Wiring diagram of first and second levels of a split-level house. ("Residential Wiring Design Guide.")

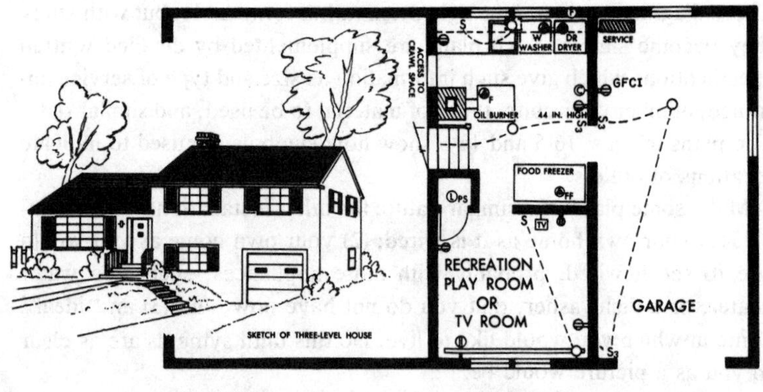

Fig. 16-6 Wiring diagram for basement of house shown in Fig. 16-5. ("Residential Wiring Design Guide.")

service-entrance wires and service equipment, the number of circuits, the materials to be used, and similar details.

In making such plans, it shouldn't be necessary to remind you to make sure the installation will meet the Code requirements. The average homeowners know little about such things, so their suggestions and ideas might result in an installation that is far from adequate, and possibly not even safe. Explain to your customers what advantages there are for them in an adequate installation, and it will mean a larger sale for you and better-satisfied customers.

17 Installation of Service Entrance and Ground

Chapter 13 covered the selection of the service-entrance wires, the rating of the service equipment, the number of branch circuits, and similar essentials. This chapter will cover the actual installation of the materials selected. Many variations are possible in the selection and arrangement of the different parts; the service-entrance wires may come in through conduit, or in the form of service-entrance cable; the supply may be overhead or underground; the meter may be inside or outside; the overcurrent devices may be circuit breakers or fuses, etc. Be sure to check with your power supplier for required location of the meter; usually the meter must be installed outdoors.

First we shall cover factors that apply to all services; then we'll discuss a service using conduit, next one using above-ground service-entrance cable, and finally one using underground service-entrance cable. We shall also discuss installation of what the Code calls the "grounding electrode conductor"; in this book it will be called just the "ground wire." It might be well for you at this point to review the part of Chap. 9 that defines the differences between the terms ground, ground*ed*, and ground*ing*.

Solderless Connectors. NEC Secs. 230-81, 250-113, and 250-115 prohibit the use of soldered connections in service wires, in the ground wire, or in a ground*ing* wire. The reasons are not difficult to understand. Service wires are ahead of ("upstream" from) the main overcurrent devices. A high fault current might melt the solder; a loose connection and heating would result, until this developed into a serious fault current, possibly leading to a blowup or virtual explosion in the service equipment. In the ground wire or a grounding wire, a melted soldered joint would result in a poor ground connection or no ground at all. But as a practical matter, you would use solderless (pressure) connectors in any event—soldered joints are a thing of the past.

Uninsulated Wire in Service Entrance. In the wiring of houses and farms, the neutral wire in the service entrance is grounded. NEC Sec 230-41 permits (but does not require) an uninsulated wire for the grounded neutral wire of the service entrance. If you use Type SE service-entrance cable, the neutral is uninsulated, but covered with an outer cable jacket.

Size of Neutral Wire. The neutral of the service entrance may often be smaller than the hot wires; this subject has already been covered in Chap. 13. Many people use a rule of thumb that says the neutral can be one size smaller than the hot wires if the hot wires are No. 6 or larger, but this rule of thumb is only an approximation, and while it is usually acceptable, it is a good idea to calculate the minimum size permitted for the neutral as outlined in Chap. 13.

The Meter. The power supplier decides whether the meter is to be located indoors or outdoors; in most cases it will be the outdoor type shown in Fig. 17-1 with its socket installed exposed to the weather. The power supplier furnishes the meter. The socket is sometimes furnished by the power supplier, sometimes by the owner; in any event it is installed by the contractor.

If the meter is to be the indoor type, install a plywood board large enough so that both the meter and the service equipment can be installed on it.

Service-Drop Wires. In Art. 100 of the NEC, service-drop wires are defined as "the overhead service conductors from the last pole or other aerial support to and including the splices, if any, connecting to the service-entrance conductors at the building or other structure."* If the wires are not overhead but underground, they are called not service-drop wires but

Fig. 17-1 A detachable outdoor meter and the socket in which it is installed. Meters of this type are exposed to the weather. (*General Electric Co.*)

service laterals, as discussed in the next paragraph. Service-drop wires are furnished and installed by the power supplier, although the owner or the contractor sometimes furnishes the insulators by which the wires are supported on the building. The power supplier also determines the size of the service-*drop* wires, which are often smaller than the service-*entrance* wires. If this leads you to wonder about too much voltage drop, remember the drop is ahead of the meter, so that the occupant of the building does not pay for wasted power. Moreover, the power supplier can adjust the voltage so that it is correct at the service equipment.

Service Laterals. When the power supplier's wires to a building are installed underground, they constitute a service lateral. The lateral ends where the wires enter the building and connect to the service-entrance wires, as shown later in Fig. 17-20. If they enter the building and continue without interruption directly to the service equipment, the continuous wires at the point where they enter the building become service-entrance wires, and the entire lateral must comply with Code requirements for the service-entrance wires.

Service-Entrance Wires. The wires from the point where the service drop ends, up to the service equipment, are service-entrance wires. They may be Types TW, RHW, THW, or any other type suitable for outdoor (wet) locations, in other words, any type that has a "W" in its type designation. They may be separate wires brought in through conduit, or wires made up into service-entrance cable approved for the purpose. If it is feasible, service-entrance wires should not be run inside the hollow spaces of frame buildings. In no case should they be run in such spaces for a greater distance than is absolutely necessary.

Openings into Buildings. If the building is of frame construction, it is a simple matter to bore a hole through the wall into the building. If the building is of brick or concrete construction, hard labor is involved. For an occasional masonry job, use a large star drill of the type that will be discussed in Chap. 22. If there are to be many openings, use an electric drill with carbide-tipped concrete drills (or hole saws).

Entrance Using Conduit. A typical installation is shown in Fig. 17-2. The size of the conduit and fittings will depend on the size of the service wires, as discussed in Chaps. 11 and 13.

Service Insulators. Insulators for supporting the service-drop wires where they reach the building (point *A* in Fig. 17-2) must be provided. These may be the simple screw-point insulators shown in Fig. 17-3; ac-

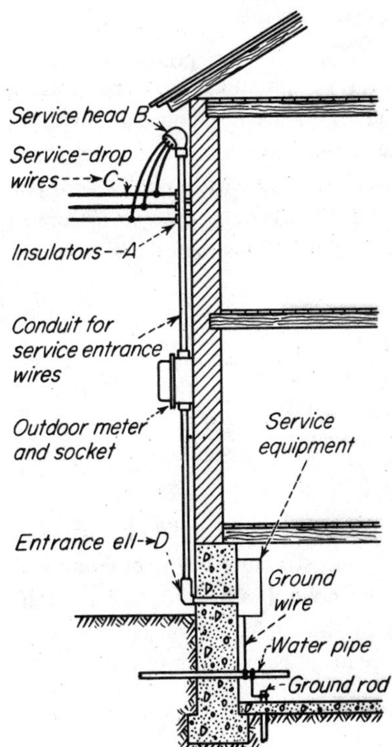

Fig. 17-2 Cross section of a typical service entrance, using wires in conduit. Locate the service head *B higher* than the insulators *A*. The ground rod may be inside or outside the wall, but it should reach permanently moist earth.

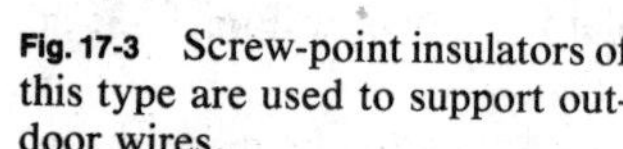

Fig. 17-3 Screw-point insulators of this type are used to support outdoor wires.

Fig. 17-4 A bracket with a number of insulators may be used instead of separate insulators.

cording to the Code they must be kept a minimum of 6 in apart. More often, however, the type shown in Fig. 17-4 is used. Choose one with insulators 8 in[1] apart. These are known as service brackets or secondary racks. Triplex cable consisting of two insulated wires spirally wrapped around a bare neutral, as shown in Fig. 17-5, is used more and more in service drops, and requires only one insulator. This material will be discussed in more detail in Chap. 23.

Fig. 17-5 Triplex cable of this kind is being used more and more by power suppliers in the service drop to the customer's building.

In ordinary residential construction, the insulators should be installed as high above the ground as the shape and structure of the building will permit, but *below* the level of the service head. If a building is very high, just observe the clearances required by the Code, as will be discussed shortly. NEC Sec. 230-54 requires the point of connection of the service-

[1] In some locations a separation of 6 in is acceptable.

entrance wires to the service-drop wires (point C in Fig. 17-2) to be *lower* than the service head (point B in Fig. 17-2). A difference of a foot or so is sufficient. If, however, because of the structure of the building this can't be done, the same NEC section permits locating the insulators and the service head at approximately the same level, but not more than 24 in apart. In that case drip loops as shown in Fig. 17-6 must be installed, and

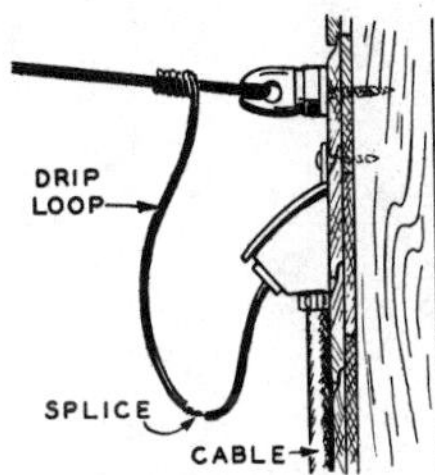

Fig. 17-6 If it is impossible to locate the service head higher than the insulators, install a drip loop in each wire. Be sure each splice is insulated with tape.

the splice between the drop and the entrance wires must be made at the *lowest* part of the loop. All this is required to minimize the danger of water following the wires into the conduit or cable. Of course, such splices must be carefully taped.

Clearance of Service-Drop Wires. NEC Sec. 230-24 specifies the clearances that service-drop wires must maintain from other objects. The clearances above the ground are as follows:

10 ft—from final grade, sidewalk, or other surface near the point of attachment, to the drop or drip loops, for service *cables* not over 150 volts to ground.

12 ft—over residential property and driveways, 300 volts or less to ground.

15 ft—over residential property and driveways and over commercial areas not subject to truck traffic, over 300 volts to 600 volts.

18 ft—over public ways, driveways other than residential, areas subject to truck traffic, and farmlands.

The same NEC section requires a minimum clearance of at least 3 ft from windows (unless installed *above* windows), doors, porches, fire escapes, and similar locations from which the wires could be reached. The clearance must be at least 8 ft from all points of roofs, except that when the voltage between conductors does not exceed 300 volts, a clearance of 3 ft is permitted where the slope of the roof is 4 in per ft or more, and a clearance of 18 in is permitted where not more than 4 ft of service drop passes over the overhanging portion of the roof and the drop terminates at a mast-type support.

Masts. In wiring the rambler or ranch-house type of residence, it is often impossible to maintain the prescribed minimum clearances. In such cases, use a mast as shown in Fig. 17-7. Such conduit masts must be able to support the service-drop wires and (unless your power supplier has

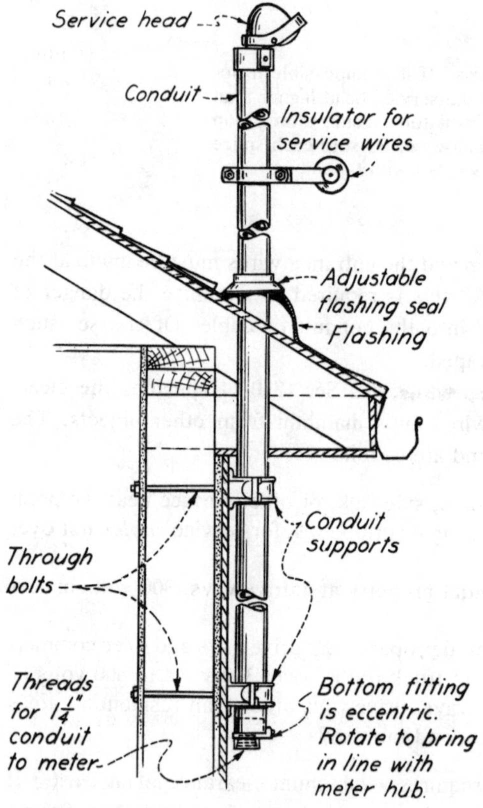

Fig. 17-7 Typical service mast installation. (*M. and W. Electric Mfg. Co.*)

other requirements) should not be smaller than 2-in rigid metal conduit, unless braced or guyed for extra-strain support. Some of the conduit fittings in mast kits will fit either 2- or $2^1/_2$-in conduit without threading. Install flashing to make a watertight opening through the roof.

Service Heads. At the top of the service conduit (*B* in Fig. 17-2), the Code requires a fitting that will prevent rain from entering the conduit. A fitting of this type is shown in Fig. 17-8; it is known by various names such as service head, entrance cap, or weather head. It consists of three parts: the body, which is attached to the conduit; an insulating block to keep the wires apart where they emerge; and the cover, which keeps the rain out and holds the parts together.

Entrance Ells. At the point where the conduit enters the building (point *D* in Fig. 17-2), it is customary to use an entrance ell of the type shown in Fig. 17-9. With the cover removed, it is a simple matter to pull wires around the right-angle corner. This fitting also must be raintight.

Fig. 17-8 A service head is installed at the top of the conduit through which wires enter the building. (*Killark Electric Mfg. Co.*)

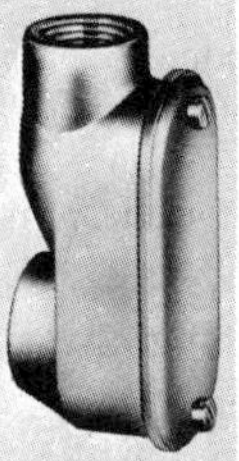

Fig. 17-9 An entrance ell is used at the bottom of the conduit, where it enters the building. (*Killark Electric Mfg. Co.*)

Service Conduit. We will assume that you have made an opening into the building for the service conduit to enter, and have marked the location of the meter, usually about 5 to 6 ft above ground level. (Some power suppliers have specified heights for meters, so check before deciding where to locate it.)

In some cases you can preassemble the service conduit, including the meter socket, before installing it on the wall. Cut a length of conduit to reach from the meter location to a point a foot or two *above* the service-drop insulators. It may take an entire 10-ft length of conduit, or more, for this run. Cut a length to reach from the service equipment inside the house, to a point just outside the wall, where you will install the entrance ell (point *D* of Fig. 17-2). Cut a third length to reach from the entrance ell to the meter socket. Ream the cut ends carefully and thread them as outlined in Chap. 11. Assemble the whole "stack" on the ground: service head at top, meter socket in the middle, entrance ell at the bottom, then the short length to the service equipment inside the house. Then install the whole assembly on the side of the house. Push the short end of the conduit in through the opening in the wall, and anchor the conduit on the side of the building, using pipe straps such as shown in Fig. 17-10, or similar ones.

Fig. 17-10 Use straps of this kind to support service-entrance conduit or cable on the building.

Inside the building, the conduit must be anchored to the service equipment, using the locknut-and-bushing procedure shown in Fig. 10-12, but instead of an ordinary bushing, use a *grounding* bushing that will be described later in this chapter.

Pulling Wires. The conduit will contain three wires. The neutral if insulated must be white. The other two wires should be red and black; larger sizes sometimes are available only in black. Examine the meter-socket neutral terminal. If the neutral can be stripped at that point and run through unbroken it will eliminate one place where a loose connection can occur. Otherwise, cut three lengths to reach from the meter socket to the top, plus about 3 ft extra. Pull these three wires into the conduit; connect to the meter socket as shown in Fig. 17-11; let the extra length project out of the top through the entrance cap. Since the conduit is very short you will probably be able to push the wires through it without any trouble; if you wish, use fish tape. Cut three more lengths to reach from the meter socket to the service equipment inside the house, allowing sufficient length so that they can be neatly placed along the walls of the cabinet to reach the most distant connector instead of being run directly to their connection points.

Electrical Metallic Tubing (EMT). Instead of rigid conduit, EMT may be used in the service entrance. The procedure is as outlined for rigid conduit, except that threadless couplings and connectors are used at the ends

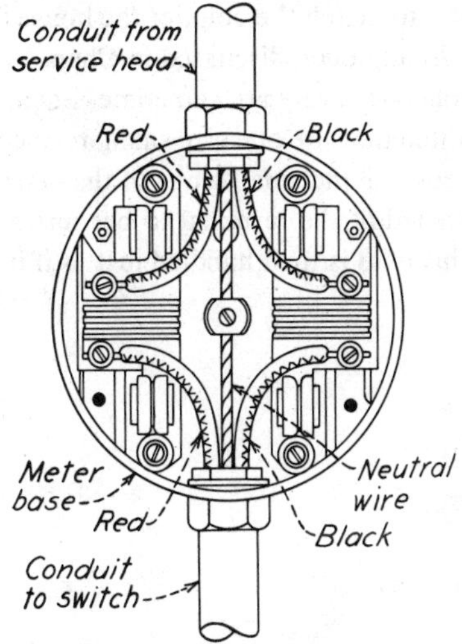

Fig. 17-11 How wires are connected to the meter socket. The neutral wire is always connected to the center terminal.

instead of the threaded fittings used with rigid conduit. Entrance caps and entrance ells of a slightly different design that clamp to the EMT are used. However, caps and ells designed for rigid conduit may also be used, for the external threads on the EMT connectors will fit threaded fittings designed for use with rigid conduit.

Service-Entrance Cable. Instead of separate wires in conduit, service-entrance cable is also commonly used. The most usual kind is NEC Type SE, shown in Fig. 17-12. It contains two insulated wires, black and red.

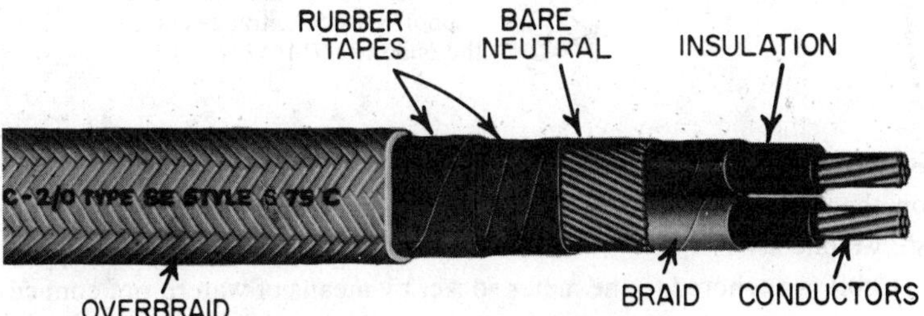

Fig. 17-12 Service-entrance cable, Type SE. The neutral consists of many strands of bare wire wrapped spirally over the insulated wires.

The bare uninsulated wire consists of many strands of small tinned wires wrapped spirally around the insulated wires. Over the neutral there is a layer of rubber tape for protection against moisture, with a final fabric jacket, usually gray. If you wish, paint it to match the color of the house. The ampacity rating of the cable has already been discussed in Chap. 7.

All the small wires in the bare neutral collectively are sometimes equal to the size of each insulated wire, but often they are one size smaller, as is usually permitted for neutrals of services. All the small wires of the neutral must be bunched together and twisted at the terminal to become a single wire, as shown in Fig. 17-13. This wire is then handled just as if it were an ordinary stranded wire.

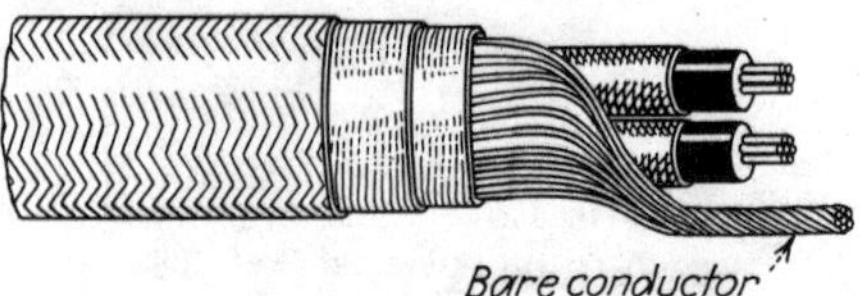

Fig. 17-13 The small wires of the neutral are bunched together into a single wire for connection to a terminal.

Fittings for Service-Entrance Cable. Secure the cable to the building within 12 in of the meter socket, entrance cap, entrance ell, service equipment cabinet, etc., and at additional intervals not over $4^1/_2$ ft. Use straps of the general type shown in Fig. 17-14 or 17-10. The service head

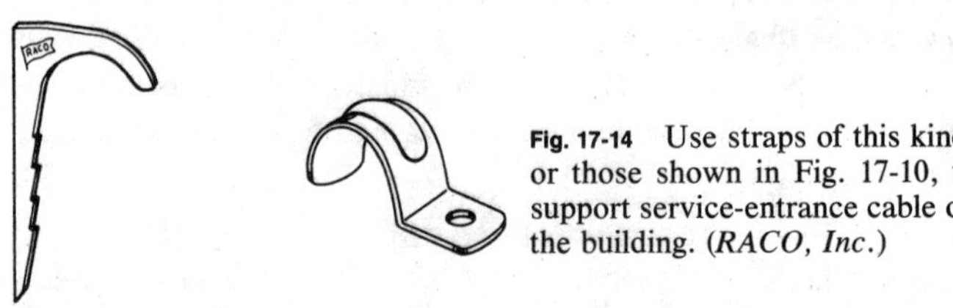

Fig. 17-14 Use straps of this kind, or those shown in Fig. 17-10, to support service-entrance cable on the building. (*RACO, Inc.*)

is slightly different from the type used with conduit, in that it is supported on the building instead of being supported by the conduit; Fig. 17-15 shows one style.

Cable is anchored to the meter socket by means of waterproof connectors; two types are shown in Fig. 17-16. Such connectors incorporate soft rubber glands; as the locking nut or screws are tightened, the rubber is compressed, making a watertight seal around the cable. In use, the connectors are screwed into the threaded openings of the meter socket (the threads are usually treated with a waterproofing compound). Next the cable is slipped through the rubber gland, and the locking nut or screws

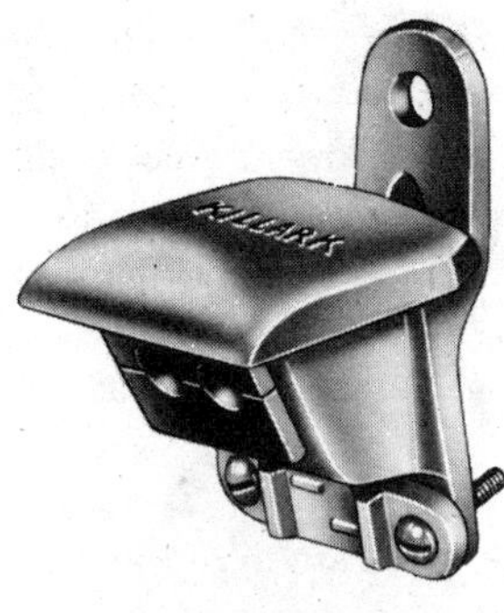

Fig. 17-15 A typical service head for service-entrance cable. (*Killark Electric Mfg. Co.*)

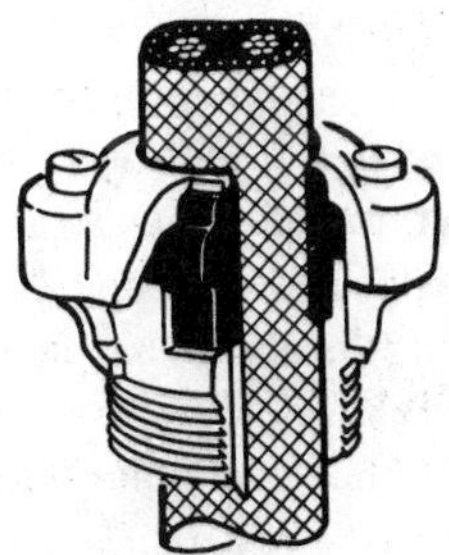

Fig. 17-16 Watertight connectors for service-entrance cable. The rubber gland in the connector squeezes against the cable, making the installation watertight. (*Killark Electric Mfg. Co.*)

are tightened, making a watertight connection. Indoors, waterproof connectors are not required; ordinary connectors of appropriate size are used.

At the point where the cable enters the building, the Code requires a seal to prevent water from following the cable into the building. The simplest way to seal this point is to use a sill plate such as shown in Fig. 17-17. Soft waterproofing compound that comes with the plate is used to seal any opening that may exist.

Service Wires on Side of Building. In most cases it is simple to let the service-entrance conduit or cable run straight down from the top, through the meter socket, and then straight down some more to the point where it enters the building, as shown in Fig. 17-2. Sometimes that is impossible, in which case do a neat job. Never run conduit or cable *at an angle* to the sides of the building. Always make the runs either vertical or horizontal as required.

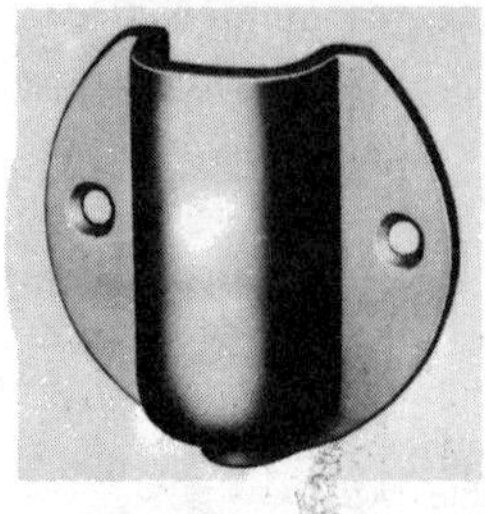

Fig. 17-17 Install a sill plate where cable enters the building. Soft sealing compound seals the opening to keep out water. (*Killark Electric Mfg. Co.*)

Underground Service Entrance. Where service wires are to run underground, two basically different methods may be used. Any kind of wire suitable for use in a wet location (any wire with a "W" in its type designation) may be installed in an underground raceway. But in most installations of the kind covered in this part of the book, a special cable known as Type USE (standing for *U*nderground *S*ervice *E*ntrance) is often used. It may be buried directly in the ground. The single-conductor type is shown in Fig. 17-18, but it is also available in multiconductor type. The individual conductors are insulated with an especially moisture-resistant compound, with another final layer of insulation that is mechanically very tough besides being moisture-resistant.

Fig. 17-18 Type USE cable. It is generally used in the multiconductor variety. (*General Cable Company.*)

In NEC Sec. 300-5 you will find the requirements that must be observed in installing *any* underground wires, including service-entrance wires, as follows: (1) If installed *under* a building, all conductors (including those normally acceptable for direct burial) must be installed in a raceway extending beyond the outside wall of the building. (2) If installed underground but not under a building, cable buried directly in the ground must be buried at least 24 in except in residential occupancies, where if the cable serves as a branch circuit and not as a service and if the circuit has overcurrent protection of 30 amp or less, the required depth is reduced to 12 in. (3) When direct-burial cable is used, the cable must be protected against possible physical damage. Use only clean earth. If the backfill contains rock, slate, or similar materials, the cable must first be covered by sand or clean soil. (4) Where a direct-burial cable enters a building, it must be protected by a raceway from underground to a point inside the building; at the pole end, to a height of at least 8 ft above ground level. These raceways for the physical protection of direct-burial cables must extend not less than 18 in below grade and be bushed at the ends.

Extra Precautions. If there is any likelihood of future disturbance by digging, it is wise to place a board or similar obstruction over the cable to serve as a warning. If instead of a board you pour a concrete pad at least 2 in thick, NEC Sec. 300-5 then permits a reduced depth of 18 in for cables and 12 in for nonmetallic conduit. Under an area subject to heavy vehicular traffic, the minimum depth for any cable or raceway is 24 in.

If you use single-conductor cables, bunch them instead of having them spread apart. Do not pull any kind of cable tight from one end to the other; "snake" it a bit to permit expansion and contraction under the action of settlement, temperature, or frost.

Underground service conductors usually begin at the power supplier's pole or, where the distribution system is underground, at a subsurface or pad-mounted transformer.[2] Even if you have used Type USE cable, suitable for direct burial, the Code requires mechanical protection such as conduit for the cable above ground. Although the Code requires rigid metal conduit or PVC Schedule 80 to extend only 8 ft up the pole, it is sometimes run all the way to the top of the pole, with a service head at the top. At that point all the wires of the cable can be spliced to the power supplier's wires. Some power suppliers require that the conduit stop at about 8 or 10 ft above ground level, in which case continue the cable to the top of the pole, but make a good watertight seal at the top end of the conduit where the wires (individual wires or multiconductor cable) emerge from the conduit. Above the conduit, the wires must be enclosed in a wooden trough, $1^1/_2$ in thick, or the equivalent, to protect them from climbing hooks (sharply pointed leg irons used for climbing poles). The power supplier usually does the work on its poles above the 8-ft level.

The bottom end of the protective conduit should end between the bottom of the trench and the ground level, not less than 18 in below grade; it should end pointing straight down rather than in a horizontal sweep. Install a bushing on the end of the conduit. Be sure to provide a sort of S loop in the cable, which will tend to prevent damage to the

[2] Subsurface and pad-mounted distribution transformers are generally installed and maintained by the power supplier. Should you have occasion to make connections at such a transformer, first consult your power supplier.

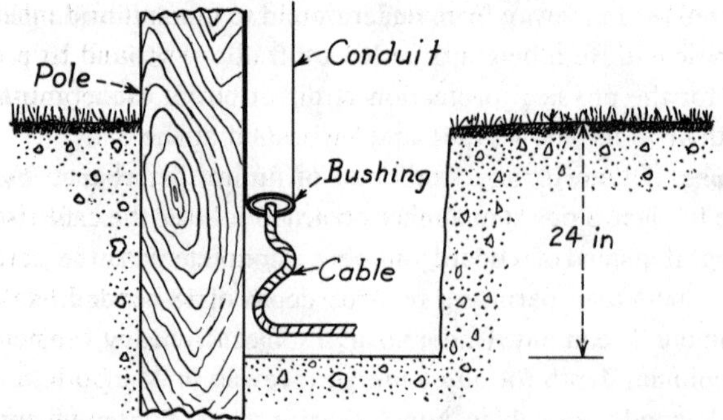

Fig. 17-19 Provide an "S" loop where an underground run of cable ends, to protect the cable against damage from "frost-heaving" or other earth movement.

cable (especially in northern climates) during movement of the earth (frost-heaving) in the change from winter to spring. See Fig. 17-19.

At the house end, if the meter is the outdoor type, follow the same procedure as at the pole, using a short length of conduit from the meter socket down into the trench. But if the cable is to run through the foundation to an indoor meter, follow the procedure in Fig. 17-20. After the

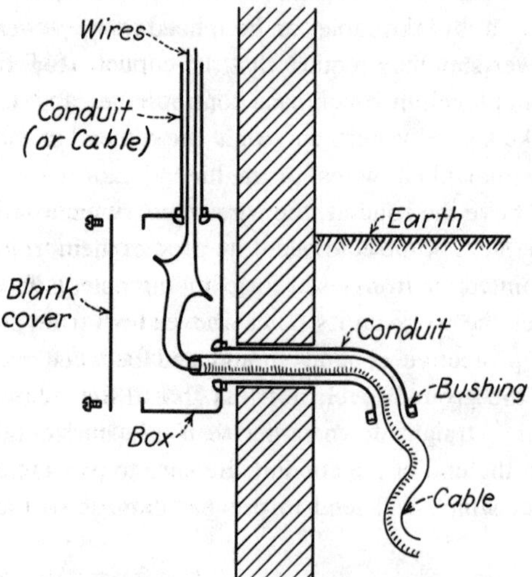

Fig. 17-20 If underground conductors cannot be run directly to the service equipment, use the construction shown, in changing from underground to interior-type wiring.

cable has been installed, fill the openings around the conduit and inside the conduit with commercially available "duct seal" compound to prevent rainwater, melted snow, or other moisture from underground sources from following the cable and entering the building.

One more detail: Some power suppliers will accept the use of PVC extra-heavy-wall (Schedule 80) conduit as a pole riser, but if the pole riser is of metal, it must be grounded; the fact that it projects into the earth is not effective grounding. The metal pole riser must be bonded to the service equipment at the building or to the neutral conductor at the pole. The simplest method is to install a grounding bushing on the bottom of the conduit at the pole and at the house and to run a grounding wire from one to the other along with the entrance cable. This bonding conductor is sized in accordance with the size of the service-entrance conductors; see Table 250-94 in your copy of the Code. (The conduit at the house end is bonded to the service equipment cabinet and to its grounding busbar, which is grounded as was explained in Chap. 13.)

Underground Bare Neutral Wire. As covered by NEC Sec. 230-41, if the underground service is made using individual wires in conduit, the neutral may be bare if copper, but not if aluminum. If individual Type USE conductors are used, buried directly in the earth, the neutral if copper may be bare "where bare copper is judged to be suitable for the ground conditions," which means "if the inspector agrees." There are a few locations where soil conditions are not suitable for bare copper. If all the wires are in the form of a multiconductor Type USE cable, which has a moisture- and fungus-resistant outer cover, the neutral may be bare whether it is copper or aluminum.

Service Equipment. Most of the details of the service equipment have already been explained in Chap. 13. It might be well for you to review that material now. Run the hot incoming wires to the terminals marked "Mains" and the neutral to the grounded busbar. Spend a few extra minutes to run the wires *neatly* within the cabinet: Instead of running them this way and that at random, bend them to run parallel to the sides or ends of the cabinet. You can then take pride in your workmanship.

Note that in residential and farm wiring the overcurrent devices to protect *branch* circuits are often in the same cabinet or enclosure with the service equipment.

Grounding. As already discussed in Chap. 9, the Code differentiates between ground*ed* conductors (which carry current in normal operation) and equipment ground*ing* conductors (which carry current only under abnormal fault conditions). The ground*ed* conductor of a circuit is always white or natural gray. The ground*ing* conductor may be bare, or insulated if green; if armored cable is used, or a metal raceway (other than *flexible* metal conduit), no separate grounding conductor is required. At the service equipment a *single* ground wire (which the Code calls the "grounding electrode conductor") is used to connect both the ground*ed* and the ground*ing* wires to the earth. This is made possible by the main bonding jumper that bonds the neutral busbar of the service equipment to its cabinet, by means of a screw or flexible strap as was already explained in Chap. 13.

Grounding Electrode. In the city the usual method of grounding for many years has been to run the ground wire to a metal pipe of an underground water system, and Codes prior to the 1978 Code required that the water pipe be used wherever available. However, because of the increasing use of nonmetallic piping materials, and of insulating fittings in runs of metal pipe, experience has shown that the water pipe is no longer the reliable grounding electrode it once was. The Code now requires that all of the following which exist at the building be bonded together to form a "grounding electrode system": 10 ft or more of metal underground water pipe; the effectively grounded metal frame of the building; a concrete-encased electrode (20 ft or more of 1/2-in steel reinforcing rod or No. 4 or larger solid copper wire encased by at least 2 in of concrete and located near the bottom of a concrete foundation or footing that is in direct contact with the earth); or a ground ring of No. 2 or larger bare copper wire not less than 20 ft long buried at least 2 1/2 ft deep and encircling the building. If only one of these exists it may serve as the grounding electrode, but water piping alone *must* be supplemented by an additional electrode. If none of the electrodes above exists, then an electrically continuous underground gas piping system (subject to approval of *both* the gas supplier *and* your local inspector) or other underground metal piping or tanks may be used. If none of these alternatives is available use a "made" electrode as discussed later in this chapter. For a water-pipe electrode, use the cold, not the hot, water pipe because the former usually runs more directly to the earth. Make the connection to the pipe on the street side of the water meter, if possible; on the other hand, make the ground wire as short and direct as possible. Other things being equal, run the ground wire to the nearest cold-water pipe. On farms a metal well *casing* (but never a drop pipe in a dug well) may be used.

A word of caution: Your home, if wired under pre-1978 Codes, probably relies on the incoming water service piping as the grounding electrode. Plumbing changes within the building, or changes in the water supply system (metal water-service lateral replaced with plastic pipe; water meter electrically isolated by insulating unions; metal water mains replaced by nonmetallic pipe, or by cast-iron pipe having no electrical continuity due to neoprene seals at every joint), can leave an existing building without a grounding electrode for the electrical system. These are some of the reasons the Code now requires the water pipe to be supplemented by an additional electrode, and why you should be aware that changes in the water supply system can affect the safety of your electrical system.

Ground Wire. The Code calls this the "grounding electrode conductor" and explains the method of installing it in NEC Sec. 250-92, and the minimum size in NEC Sec. 250-94. The ground wire is usually bare, but there is no objection to using insulated wire. It may never be smaller than No. 8, but there is no objection to using a size larger than the minimum required. When using only a made electrode (ground rod or plate), the ground wire never needs to be larger than No. 6, regardless of the size of the service wires. Regardless of size, the ground wire must be securely stapled or otherwise fastened to the surface over which it runs.

If the service wires are No. 2 or smaller, the Code permits a No. 8 ground wire. *Don't use it; use No. 6.* Number 8 must be enclosed in conduit or armor regardless of circumstances, but for No. 6 this is not required. The cost of material when using No. 6 is lower than when using No. 8; it also takes less time to install the No. 6.

If the service wires are No. 1 or 1/0, the ground wire may not be smaller than No. 6. If it is installed free from exposure to physical damage and if it is stapled to the surface over which it runs, neither conduit nor armor is required.

If the service wires are No. 2/0 or 3/0, use No. 4 ground wire. The ground wire must be fastened to the surface over which it runs. Conduit or armor is never required, but if installed where physical damage is likely, it must be otherwise protected. For example, if installed outdoors in an alley or driveway where it might be damaged by vehicles, protect it by installing a sturdy post or steel beam near the wire.

Ground Clamps. The type of clamp must be carefully selected. The ordinary sheet-metal strap type is prohibited by NEC Sec. 250-115. If the grounding electrode (the water pipe, ground rod, etc.) is iron or steel, the ground clamp must be made of galvanized iron or steel, per NEC Sec. 250-115. If the clamp is to be installed on copper or brass pipe or rod, the clamp must be made of copper, brass, or bronze. Unless this point is observed, electrolytic action can take place, resulting in a high-resistance ground, which in severe cases might be no better than no ground at all.

Ground-Wire Conduit or Armor. If you install the ground wire without protective armor or conduit, use the ground clamp shown in Fig. 17-21.

If you use ground wire with armor similar to that of armored cable, use the clamp shown in Fig. 17-22. This clamps the armor of the wire, and the wire itself is connected to a terminal on the clamp. At the service equipment cabinet, use a connector of appropriate size to bond the armor securely to the cabinet; the wire itself is connected to the grounded busbar.

If you use conduit, use a clamp of the type shown in Fig. 17-23. Thread

Fig. 17-21 Ground clamp for bare ground wire. (*RACO, Inc.*)

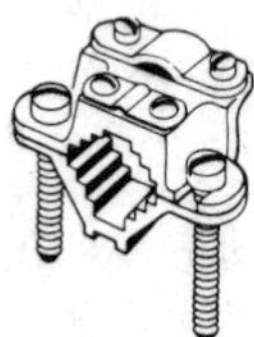

Fig. 17-22 Ground clamp for armored ground wire. (*RACO, Inc.*)

Fig. 17-23 Ground clamp for ground wire run through rigid conduit. (*RACO, Inc.*)

the conduit into the clamp. At the other end anchor it securely to the service equipment cabinet with the usual locknut and a grounding bushing.

Most important: If the conduit or the armor is poorly clamped, or clamped only at one end, or not clamped at all, the resultant ground is *very* much less effective than it would be had a wire without conduit or armor been used. This is because the metal armor or conduit forms a "reactive choke" that causes high impedance and heat.

Water Meters. It is not unusual for a water meter to be removed from a building, at least temporarily for testing. If the ground connection is not made on the street side of the meter, removing the meter then leaves the system ungrounded, which causes a hazardous condition. In some cases the joints between the water pipes and the water meter are purposely insulated. Therefore NEC Sec. 250-81(a) requires that unless the ground wire is connected on the street side of the meter, a bonding jumper must be installed around the meter, as shown in Fig. 17-24. The bonding jumper is also required around any other type of a fitting, such as a union, that is likely to be disconnected at one time or another. Use two ground clamps and a jumper of the same size as the ground wire.

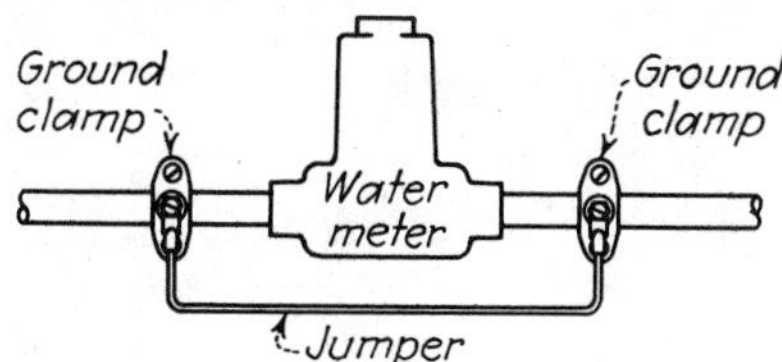

Fig. 17-24 A bonding jumper must be installed around the water meter.

A similar jumper is required around any equipment in the incoming water pipe if there is likelihood that the electrical continuity of the pipe will be or could be disturbed. A water softener could be temporarily disconnected for servicing. A jumper is required around the softener if the ground wire is connected to a pipe running to the softener, rather than to the pipe on the street side of the softener.

Neutral Busbar. The service equipment cabinet contains a busbar called the "neutral busbar," to which all grounded wires must be connected as already explained in Chap. 13; at this point you might do well to review that portion of the chapter.

In the finished installation, the service conduit (and all conduit or cable armor in the branch circuits) is bonded to the service equipment cabinet. Therefore the main bonding jumper (screw or flexible strap) effectively bonds the neutral busbar in the cabinet, and all conduit and cable armor, together. However, the NEC in Sec. 250-72 requires a lower resistance between the service conduit and the service equipment than is provided by ordinary locknuts and bushings. The Code in that section requires the use of grounding bushings and bonding jumpers, or the equivalent, for service raceways (armored cable is not permitted in services).

Concentric Knockouts. The cabinets of service switches and breakers are too small to permit installing knockouts of all the different sizes that may be required in various installations. Such cabinets are therefore provided with "concentric" knockouts as shown in Fig. 17-25. Remove as many parts of this complicated knockout as required to provide the size you need. This is a tricky operation, so be especially careful to remove only as many sections as are required to provide the size you need. Remove the center section only to provide the smallest size; remove two sections for the next larger size, and so on. Should a knockout be larger than the raceway or connector to be installed, use a pair of reducing washers—as shown in Fig. 17-26, one inside and one outside the enclo-

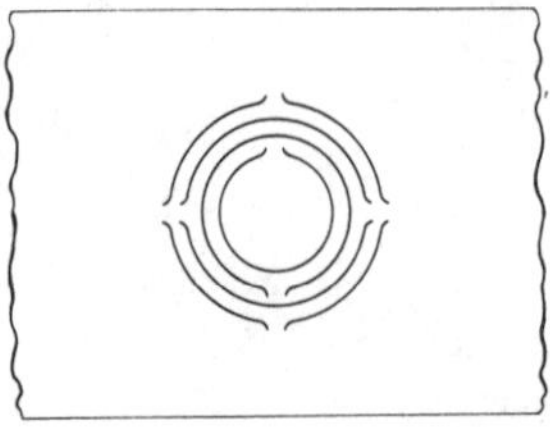

Fig. 17-25 A concentric knockout. Remove as many sections as necessary to provide the size of opening required.

Fig. 17-26 Reducing washers are available in any combination of sizes from 1/2 in through 4 in. (*Efcor products manufactured by Gould, Inc.*)

sure wall—of the proper size. When the voltage is over 250 volts to ground and entry is through concentric, eccentric, or oversized knockouts, one of the bonding methods for services must be used to bond raceways or metal cable sheaths to enclosures.

Grounding Bushings. Where the service conduit enters the service equipment cabinet, use a grounding bushing of the general type shown in Fig. 17-27 instead of an ordinary bushing. The bushing shown has a set-

Fig. 17-27 A grounding bushing. (*Union Insulating Co.*)

screw in its side that bites into the conduit; other brands have a setscrew that bites into the metal of the cabinet. Either type prevents the properly installed bushing from turning, and the setscrew bites into the metal for the good ground continuity that is so important for safety. The bushing shown has a connector on the side for a bonding jumper to the neutral busbar. The bushing shown has an insulated throat, but all-metal bushings may be used if the service wires are No. 6 or smaller. If the wires are No. 4 or larger, bushings with an insulated throat must be used. Bushings made entirely of insulating material may be used if a grounding locknut is installed between the bushing and the service cabinet.

If the conduit entry to the service equipment enclosure is through a threaded hub bolted to the enclosure or is of the two-piece type shown in Fig. 17-28, additional bonding is not required. The knockout for the two-piece hub must be of the same size as the hub, since the outer rings of

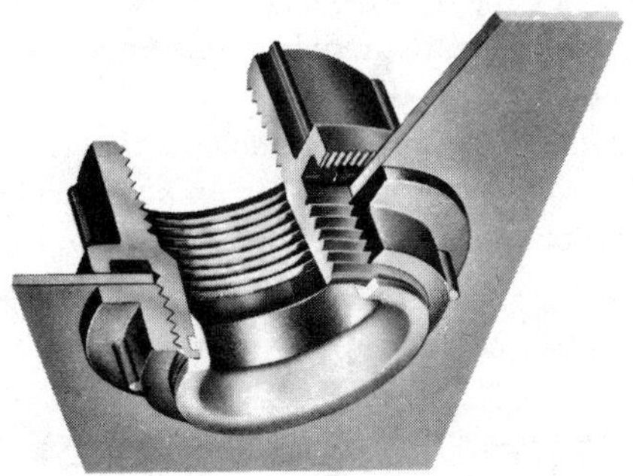

Fig. 17-28 Two-piece threaded hub. Serrations on both pieces bite into enclosure metal, making a bonding jumper unnecessary. The O ring in the outside piece forms a raintight seal. (*Myers Electric Products, Inc.*)

a concentric knockout cannot be depended on for service-conduit bonding.

If you have used EMT, use a grounding-type locknut such as shown in Fig. 17-29 instead of the regular locknut on the EMT connector. You may also use a grounding wedge, shown in Fig. 17-30; install it on the inside of the cabinet, between the wall of the cabinet and the locknut of the connector.

Bonding Jumpers. Your service equipment cabinet probably has concentric knockouts. If the conduit is large enough so that *all* the parts of the knockout have been removed, a grounding locknut or wedge (shown in Figs. 17-29 and 17-30) is all that is required. For a smaller conduit,

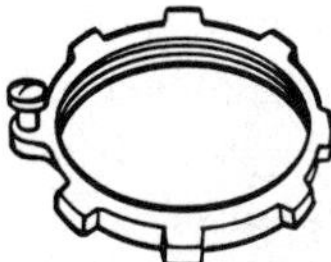

Fig. 17-29 A grounding locknut.

Fig. 17-30 A grounding wedge.

some part of the knockout remains in the cabinet, and in that case you must install a grounding bushing and connect a bonding jumper from the grounding bushing to the neutral busbar, as shown in *A* in Fig. 17-31, which shows only the neutral wire of the service and the ground wire. The jumper wire must be of the same size as the ground wire.

If you have used service-entrance cable with a bare concentric neutral, the grounding bushing and the jumper are not required where the cable enters the cabinet, as shown in *B* of Fig. 17-31.

It is not necessary to use grounding bushings, grounding locknuts, or wedges where conduit or cables supplying *feeders* or *branch* circuits enter the cabinet (except in hazardous locations, for which see Chap. 33).

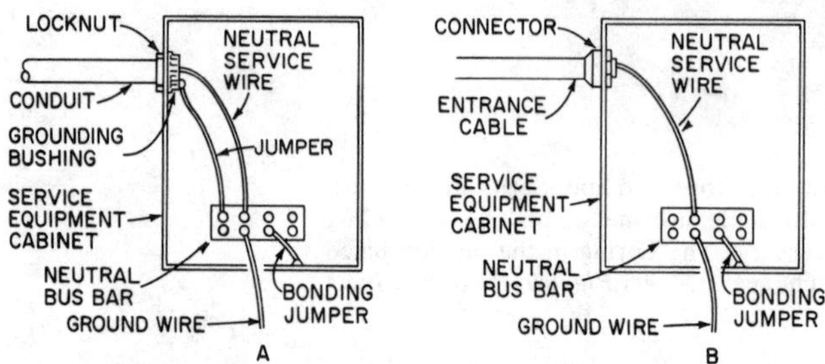

Fig. 17-31 Typical ground connections: at left, where conduit is used for the service entrance; at right, where Type SE or Type USE service-entrance cable is used.

Made Electrodes. Underground metal piping systems are the usual form of grounding electrode. The metal casing of a driven well, if located so that the ground wire to it is not too long, makes an excellent ground (but do not use drop pipes in a dug well; such wells can go dry, so the drop pipe would not provide any ground at all).

In the absence of existing electrodes, you must use what the Code calls a "made electrode," which may be a driven rod or pipe, or a metal plate exposing 2 sq ft of surface to the soil. In this book the ground rod will be referred to, as it is the most commonly used made electrode. The usual ground rod is a pipe or rod that must be driven at least 8 ft into the earth. Galvanized pipe may be used if it is at least 3/4-in trade size. Solid rods may be used, and if of iron or steel they must be at least 5/8 in. in diameter. The most common ground rod is copper, or an approved substitute, and need be only 1/2 in. in diameter. The most widely used ground rod that UL has Listed is the Copperweld type—a rod of steel with a layer of copper fused to its outside surface. For more information about how to install such ground rods, see Chap. 23.

NEC Sec. 250-86 prohibits a lightning rod conductor or ground rod from also being used as the ground rod for the electric system. However, the two electrodes (rods) for the two systems may, and should be, bonded together. If *not* bonded together, they must be kept at least 6 ft apart.

The question often arises as to what to do if the only underground metal water pipe to a building is *less* than 10 ft long. NEC Sec. 250-81 requires that the water pipe be used as the ground, but you *must* also use another available electrode or ground rod, and bond the two together. This is an important safety measure; don't overlook it. In any case, the *interior* water piping must *always* be bonded to the service equipment enclosure, the neutral conductor at the service, or to the grounding electrode system, so that whether the water pipe outside the building qualifies as an electrode or not, the usual residential service ground will be to the water pipe and a driven rod, as well as to any other existing electrodes.

18
Installation of Specific Outlets

In previous chapters installations of electrical devices were considered in rather general fashion; in this chapter the exact method of installing a variety of outlets using assorted materials will be discussed in detail. Only methods used in *new work* (buildings wired while under construction) will be explained. *Old work* (the wiring of buildings after their completion) will be discussed in Chap. 22.

When a building is wired with any kind of conduit, the conduit is installed and the switch or outlet boxes are mounted while the construction is in its early stages. This is called "roughing-in." The wires are not usually pulled into the conduit until after the lath and plaster, or the wallboard, have been installed; the switches, receptacles, fixtures, and so on naturally can't be installed until the wall finishes have been completed. However, to avoid repeating later, the pulling in of wires will be included in this chapter.

Each type of outlet will be treated separately. In the illustrations, each outlet will be shown in four separate ways: (1) as it would appear on a blueprint, (2) as it would appear in diagrammatic or schematic fashion, (3) in pictorial fashion using rigid metal conduit or EMT, and (4) in pictorial fashion using nonmetallic-sheathed cable. The primary purpose of this chapter is to show you how to connect the *circuit* wires (the wires carrying current in normal use) so that lighting fixtures, switches, receptacles, etc., will work properly. The equipment ground*ing* wire of nonmetallic-sheathed cable, and the bonding jumper from the green terminal of a receptacle to its box, will be omitted in the illustrations to keep them reasonably simple. This means that receptacles *without* the green grounding terminal screws will be shown, even though the grounding type *must* be used. The diagrams will *not* show the grounding wire of nonmetallic-sheathed cable. Review Chap. 11, which explains how to connect the green grounding terminals and the grounding wire of the cable.

The details shown in Fig. 18-1 will be used in other illustrations, and since these will be smaller, the individual parts will not be named, but you will easily recognize them. Where, in conduit wiring, a wire passes through a box without tap or termination, the drawings show them short (see Fig. 18-7, *D*), but you must always leave a loop. Fold the loop back in the box for easy future access.

Note that grounded circuit wires are shown as a light line in the diagrams, while ungrounded or hot wires are shown as a heavy line. This does not mean that one wire is larger than the other. Light and heavy lines are used only to make it easier for you to trace grounded and ungrounded wires.

In each case the point from which the current comes is labeled SOURCE. The white grounded wire always runs from SOURCE *without interruption by a switch, fuse, or breaker* to each point where current is to be consumed at 120 volts. Splices are permitted if made in boxes, as in the case of the hot wires. When wiring with conduit a black wire should be connected only to another black; a third wire should be red and connected only to red. This is not required but is common practice. In a 3-wire circuit,[1] it is best to use red and black for the two hot wires. Green wire must never be connected to a wire of any other color.

Ceiling Outlet, Pull-Chain Control. This is the type of lighting outlet sometimes used in closets, basements, attics, and similar locations, with the wires ending at the outlet. It is the simplest possible outlet to wire; Fig. 18-2 shows it. At *A* is shown the designation as found on blueprint layouts; the wires running up to it are not shown, for in blueprints lines are used only to show which switch controls which outlet(s). At *B* is a

[1] Three-wire circuits will be explained in Chap. 20.

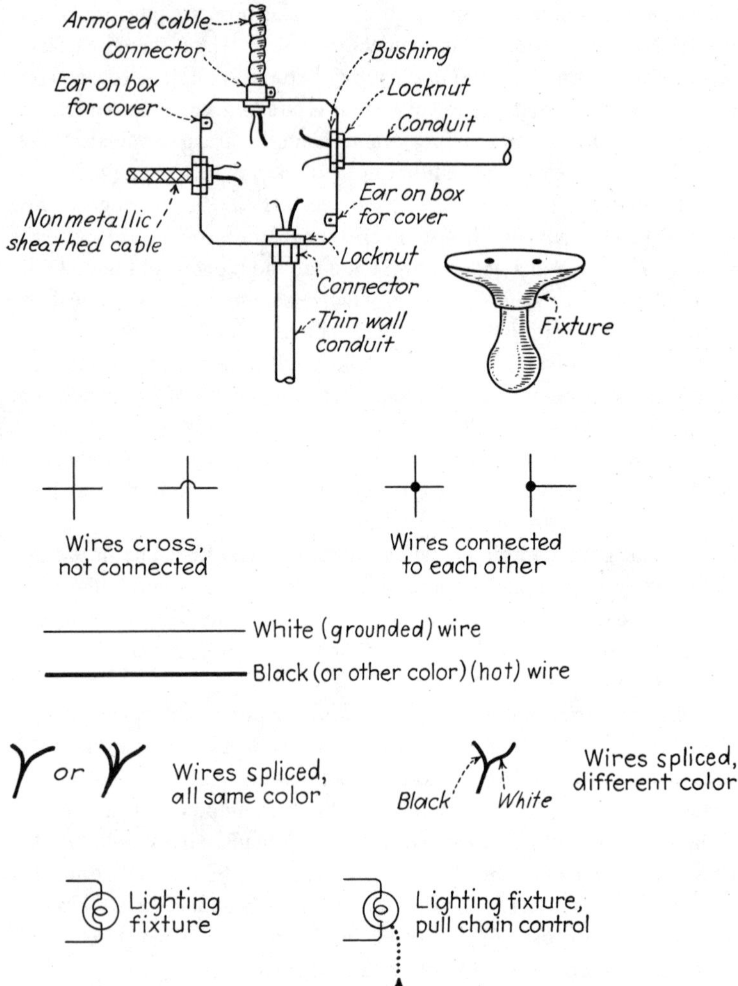

Fig. 18-1 Study these details well, so that other diagrams in this chapter will be clear to you.

wiring diagram for the same outlet. Assume you have already installed the outlet box for the fixture by one of the methods shown in Chap. 10. Installing this outlet by using conduit as in Fig. 18-2, *C*, assume that you have properly reamed the cut end of the conduit, that you have properly anchored it to the box using a locknut and bushing, and that you have

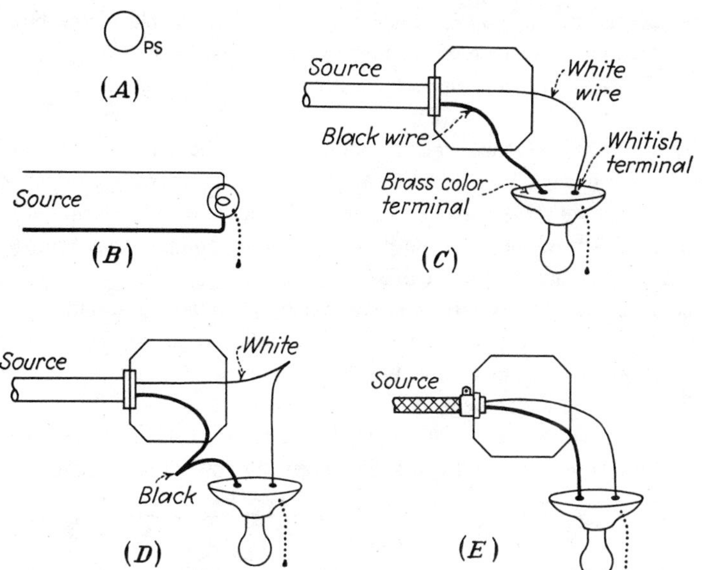

Fig. 18-2 A simple hookup of a pull-chain outlet, with wires ending at that outlet.

pulled two wires, white and black, into the conduit with about 10 in sticking out of the box. To complete the outlet all you have to do is to connect the wires to the fixture and mount the fixture in one of the ways discussed in the next chapter.

If the fixture is one with two leads (two wires instead of two terminals to which the wires from the box can be connected), the wiring is still the same except that a couple of splices must be made in the box, as shown in Fig. 18-2, *D*. If you are using EMT instead of rigid metal conduit, follow the same procedure as with rigid conduit, but use threadless fittings such as were shown in Figs. 11-16 and 11-17 instead of the locknut and bushing used with rigid conduit.

How to install this outlet using nonmetallic-sheathed cable is shown in *E* of Fig. 18-2. Connect the ground*ing* wire of the cable to the box as discussed in Chap. 11. The cable must be supported within 12 in of the box and at intervals not over $4^1/_2$ ft.

If you are using armored cable, again follow *E* of Fig. 18-2. You have inserted an insulating bushing between the wires and the armor; you have bent the grounding strip back under the connector that you have installed on the cable. You have solidly anchored the cable to the box. There is nothing further to do except to connect and install the fixture. Again, the cable must be supported within 12 in of the box and at intervals not over $4^1/_2$ ft.

If you are using flexible metal conduit, follow the procedure for armored cable except that the wires are pulled into it after the conduit is installed. You must also install a separate grounding wire as discussed in Chap. 11; it serves the same purpose as the bare equipment ground*ing* wire in nonmetallic-sheathed cable.

Lighting Outlet, Wires Continued to Next Outlet. From the first outlet, the wires continue to the next; it makes no difference what may be used at that next outlet. The only problem is how to connect at the first outlet the wires running on to the next.

This combination is pictured in Fig. 18-3, and again at *A* is its designation as found on blueprint layouts. Comparing this with Fig. 18-2, *A*, you

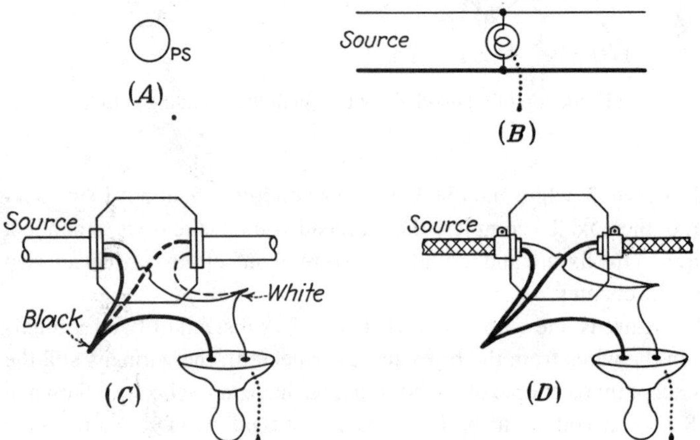

Fig. 18-3 The hookup shown in Fig. 18-2, but with the wires running on to another outlet.

will find no difference. This may be confusing, but on blueprints, the wires between different outlets are not indicated [except, as already mentioned, lines between switches and the outlet(s) they are to control]. It is up to the installer to use his or her own judgment as to the exact fashion of hooking together the various outlets in a proper manner. At *B* is shown the wiring diagram for the combination.

At *C* is shown the outlet using conduit. Compare it carefully with Fig. 18-2, *D*. The new outlet, Fig. 18-3, *C*, is the same as the former except for the addition of two wires running on to the second outlet, at the right, and these two wires have been shown in dashed lines to distinguish them from the first wires. Simply splice all the white wires together, and all the black wires, with connectors such as wire nuts.

Receptacle Outlets. In Fig. 18-4 at *A* is shown the designation for the outlet as found on a blueprint; at *B* is shown the wiring diagram. To wire

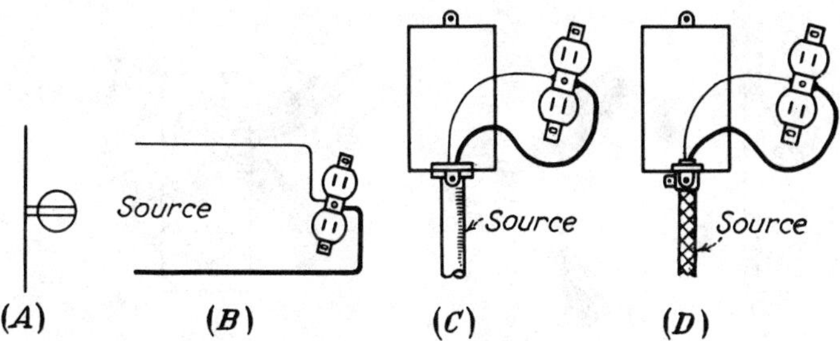

Fig. 18-4 Installation of a baseboard receptacle outlet.

this outlet using conduit as at *C*, merely connect the wires to the receptacle, *and* connect a bonding jumper from the green terminal of the receptacle to the box, as explained in Chap. 11, unless either the box is surface mounted, or the receptacle is of the type that does not require the jumper, as also explained in Chap. 11.

At *D* is shown the same receptacle wired using nonmetallic-sheathed cable. The grounding wire is not shown but must be used. If you are using armored cable, proceed as with nonmetallic-sheathed cable. A bonding jumper must be connected from the green terminal to the box. Review Chap. 11.

Receptacle Outlet, Wires Continued to Next Outlet. In Fig. 18-5, *A*, is shown the wiring diagram of the circuit under discussion. At *B* and *C* are

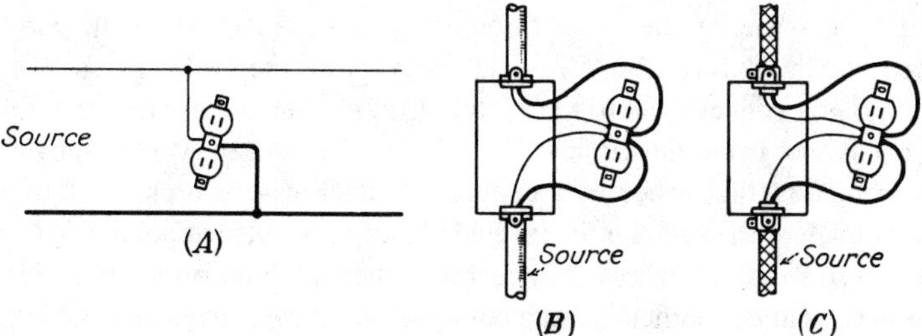

Fig. 18-5 Receptacle outlet, with wires running on to another outlet.

shown the actual installations using conduit and cable. Run the incoming white wire to the white terminal of the receptacle, and continue with another white to the next outlet. Do the same with black wire, using the brass terminals of the receptacle.

Because it is frequently necessary to do this, receptacles are provided with double terminal screws so that two different wires can be connected to the same terminal strap, as shown in *A* of Fig. 18-6. If conduit is used,

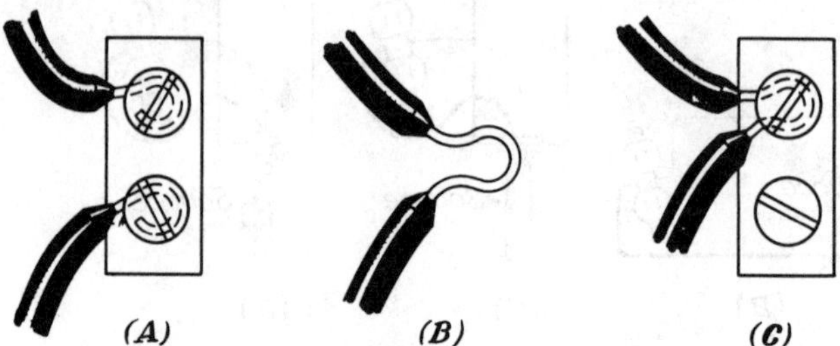

Fig. 18-6 Duplex receptacles have double terminal screws on each side, for convenience in running wires on to another outlet.

you might find it more convenient to pull a continuous wire from SOURCE to box 1, to box 2, and to box 3 than to pull one wire from SOURCE to box 1, another from box 1 to box 2, and another from box 2 to box 3. If you use a continuous length, allow a loop of wire to project about 6 in at each box, as shown in Fig. 18-6, *B*. Remove the insulation as shown at *B* in the same illustration, and connect the wire under one of the terminal screws, as shown in *C*. Where the receptacle is installed on one side of a three-wire circuit, the white wire must be connected as shown in *C*, or pigtailed as for the lighting fixture in Fig. 18-3, so that removal of the receptacle will not interrupt the continuity of the white grounded wire.

Light Controlled by Wall Switch. This combination is very simple to wire. Figure 18-7, *A*, shows the blueprint symbol, and *B* shows the wiring diagram. At *C* is shown the outlet wired using conduit. Run the white wire directly to the fixture. Run two black wires from the outlet box to the switch, and connect them to the switch. Connect the upper ends, one to the black wire from the SOURCE, the other to the fixture. At *D* is shown the optional method whereby a continuous black wire is brought all the way through to the switch box, instead of having a splice at point *X* in *C*.

When this outlet is wired using cable, either armored or nonmetallic-sheathed, as at *E*, a difficulty becomes apparent. According to everything

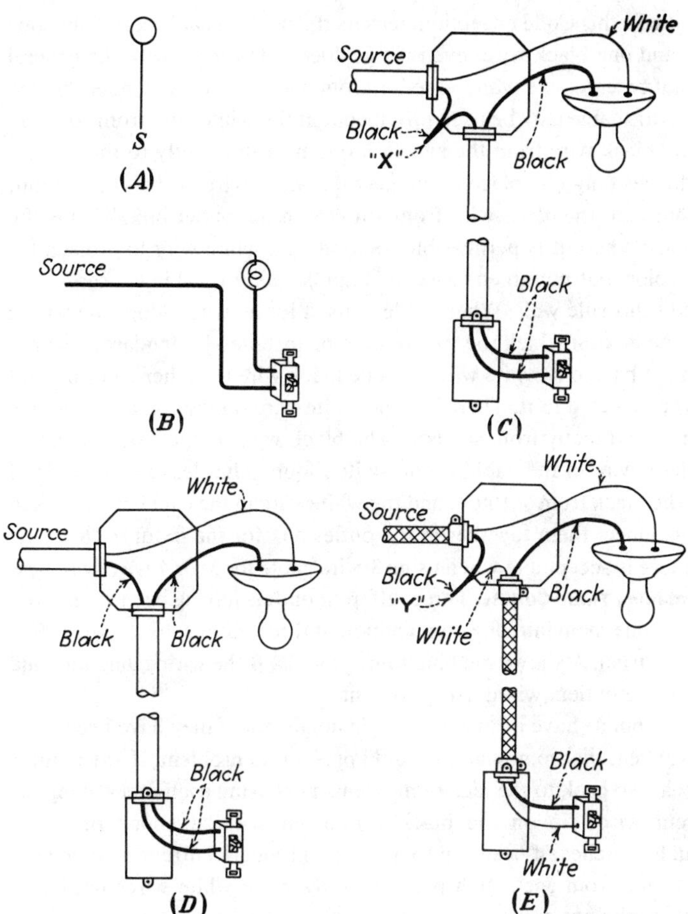

Fig. 18-7 Fixture outlet controlled by wall switch (see also Fig. 18-10).

you have learned up to this point, not only the wire from the outlet box to the switch, but also the wire from the switch to the fixture, should be a *black* wire, since neither one is grounded. But all 2-wire cable contains one black wire and one white. Should manufacturers, distributors, and contractors then be forced to stock a special cable containing two black wires? That would be impractical.

The Code in Sec. 200-7 permits an exception for this. In the case of a switch loop (as the wiring between an outlet and the switch that controls it is called), this Code exception permits the use of a cable containing one white and one black wire, even if this does not comply with the general rule that reserves the white wire for grounded wires only. Under this exception the cable may be used provided that the white wire from SOURCE, and the black wire from the switch loop, are run directly to the fixture. That leaves only one place to connect the *white* wire of the switch loop, and that is to the *black* wire from SOURCE in the outlet box. This is the only case where it is permissible to connect a white wire to a black (or other color, but not green) wire, and applies *only* if cable is used.

Study this rule well: When cable is used for a switch loop, the wiring up to the box on which the fixture is to be installed is standard. The fixture must have one white wire and one black wire (or other color but not green) connected to it. Therefore the white wire on the fixture is the one that comes directly from SOURCE. The black wire on the fixture must be the black wire in the cable in the switch loop. That leaves two ends of wire: the black from SOURCE, and the white wire in the cable of the switch loop. Connect them together in the outlet box for the fixture. The same procedure is acceptable when using 3-wire cable for 3- or 4-way switches.

Combining Three Outlets. Three different outlets having been wired, you can combine them into one combination of three outlets, as shown in Fig. 18-8. As usual, *A* shows the blueprint symbols, *B* the wiring diagram, and *C* the three outlets wired using conduit.

All the points have been gone over in detail, and if they have been carefully studied, this combination should present no problem. If any point is not clear, go back to the idea of messengers chasing each other along the different wires, in on the black wire from SOURCE, along that wire (through switches, if used) up to every point where current is to be consumed, and from each such point back over the white wires until they emerge at the SOURCE.

Taps. At times an outlet box is used merely to house a splice or tap where one wire branches off from another. Where a box is installed only for splices, it is always closed with a blank cover. Per NEC Sec. 370-19 such boxes must be installed in locations permanently accessible without removing any part of the structure of the building.

Figure 18-9, at *A*, shows how taps or splices are shown on blueprints; at *B*, as shown in wiring diagrams. At *C* is shown the method using conduit; merely connect together all the black wires and all the whites. At *D* is shown an optional method in which, instead of three ends of each color, one loop of each color is used. The loop is formed by pulling a con-

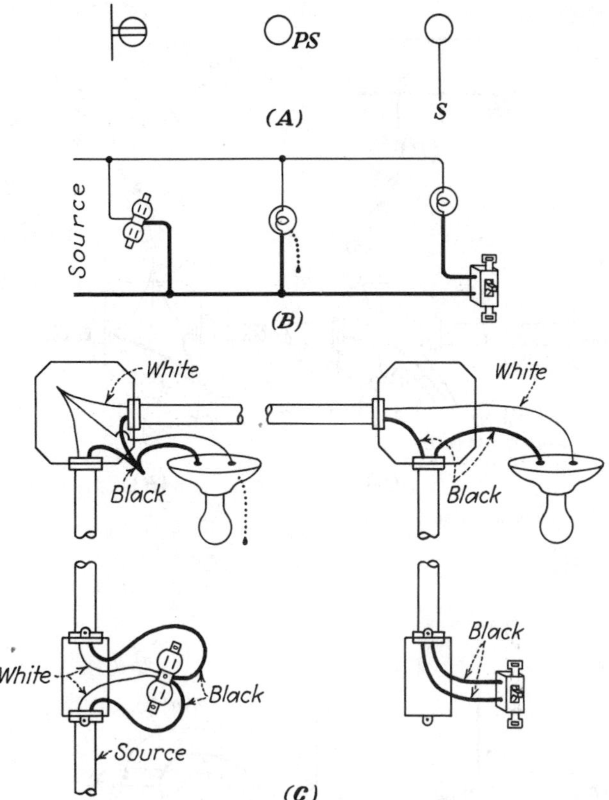

Fig. 18-8 Combining the outlets of Figs. 18-2, 18-4, and 18-7 into one three-outlet combination.

tinuous wire through the box, the ends of the loop being skinned and joined to the remaining ends of the wires to the next outlet. *E* shows the same outlet using cable.

Outlet with Switch, Feed through Switch. Switches are connected in different ways to the outlets they control. The SOURCE wires do not always run through the ceiling to the outlet box on which the fixture is installed and from there on to the switch, as in Fig. 18-7, already discussed. Sometimes the SOURCE wires come in from below, and run through the switch box and then on to the ceiling outlet, where they end, as in Fig. 18-10. At *A* is shown the usual blueprint symbol, and *B* shows the wiring diagram. At *C* is shown the outlet using the conduit system, and little explanation

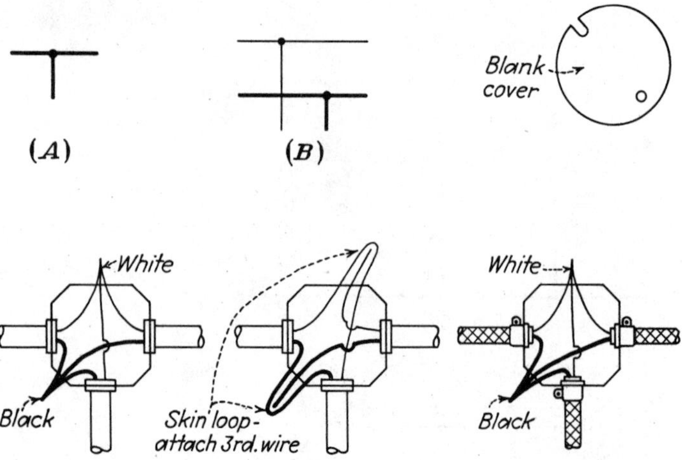

Fig. 18-9 A splice made in an outlet box.

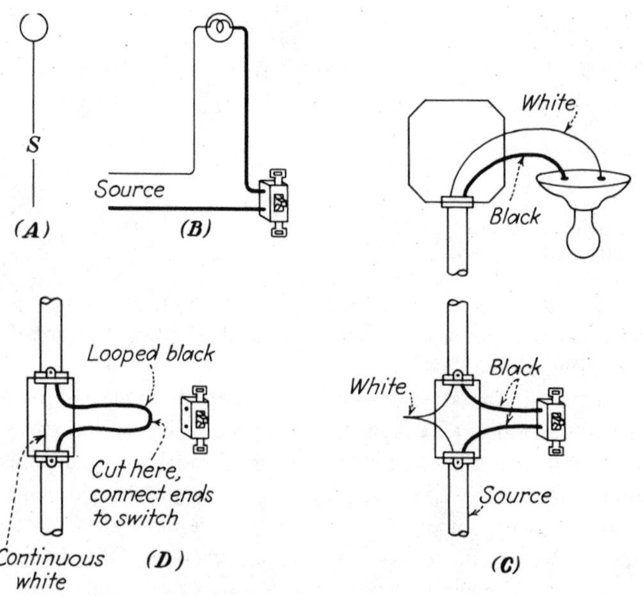

Fig. 18-10 Same as Fig. 18-7, except that the wires from SOURCE enter the switch box instead of the outlet box on which the fixture is mounted.

is needed. At *D* is shown an optional method of handling the wires at the switch box by pulling the white wire straight through. The black wire is also pulled straight through, but with a loop, which is later cut, the two ends being connected to the switch.

The diagram for this outlet when wired with cable is not shown because there is no new problem. When the cable feeds through the switch box, there is no difficulty with the colors of the wires in the cable as in a previous example. The wires can be run through to completion of the outlet without the need of joining black to white, exactly as when using conduit (*C* of Fig. 18-10).

Outlet with Switch, with Another Outlet. This is simply a combination of two outlets that have already been discussed separately. The wiring of an outlet with wall-switch control was covered in connection with Fig. 18-7. The wiring of an outlet with pull-chain control was covered in connection with Fig. 18-2. The two have been combined in Fig. 18-11, the left-hand

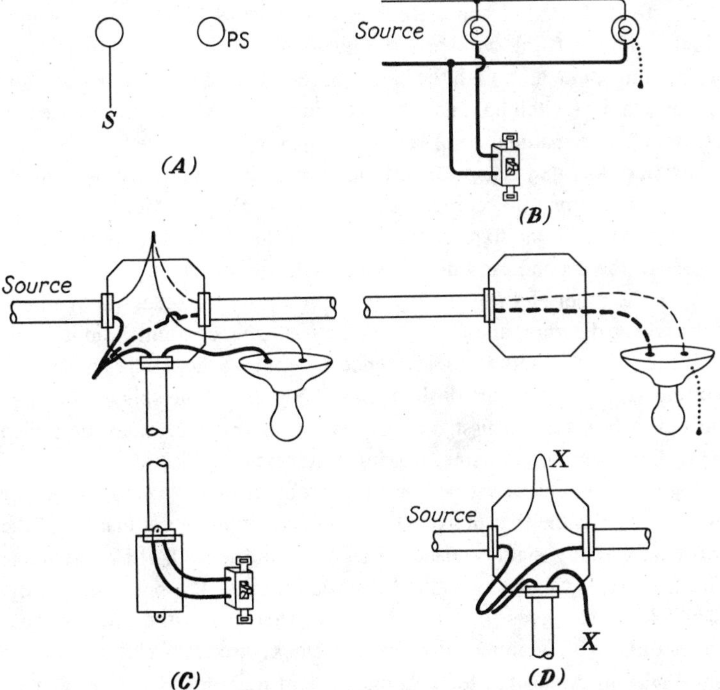

Fig. 18-11 The switch controls the first fixture; wires run on to a second fixture that is controlled by a pull chain.

portion of which is identical with Fig. 18-7; to it have been added in dashed lines the wires of Fig. 18-2, making the new combination as shown. At *D* is shown an optional method of handling the wires through the outlet box, the wires from the fixture being connected at the points marked *X*. No new problems are involved in the cable methods; for that reason they are not shown.

If the wires from SOURCE (instead of coming in through the ceiling outlet) come in from below through the switch box, the problem is different in that three wires instead of two must run from the switch box to the first outlet box, as shown in Fig. 18-12, *B*. A good way to analyze this combination is to consider first the right-hand outlet with the pull chain. Both a white and a black wire *must* run to this from SOURCE, uninterrupted by any switch, so that the light can always be controlled by the pull chain. Run the third wire from the switch to the left-hand outlet to control that.

Analysis of Fig. 18-12, *C*, shows that it consists of a combination of Figs. 18-10, *C* (wires in solid lines), and 18-2 (wires in dashed lines).

If instead of conduit you use cable, as in Fig. 18-12, *D*, the problem is equally simple. If the wires are fed through the switch box, there is no problem in connection with the colors of the wires. Merely connect white to white in the switch box, and then continue the white to each of the two fixtures. The remaining colors are as shown in the picture.

Switch Controlling Two Outlets. When a switch is to control two outlets at the same time, the connections are most simple. Merely wire the switch to control one fixture, then continue the white wire from the first fixture to the second, and do the same with the black.

As in the other pictures, in Fig. 18-13, *A* shows the blueprint symbols, *B* the wiring diagram, and *C* the combination using conduit. Compare this with Fig. 18-7, *C*; there is no difference except that these wires have been continued as shown in the dashed lines. To avoid all the splices shown in the outlet box for the first fixture, several of the wires may be pulled through as continuous wires, making a neater job.

If you use cable, there is again the problem of having to use in the switch loop a cable that has one black and one white wire, instead of the two blacks that should be used. Handle it as in Fig. 18-7, *E*. The white wire from SOURCE goes to each of the two fixtures. The black wire in the switch loop also goes to the first fixture, then on to the second. That leaves only two unconnected wires, the black wire from SOURCE and the white wire in the switch loop; connect them together as before, as permitted in the Code's exception.

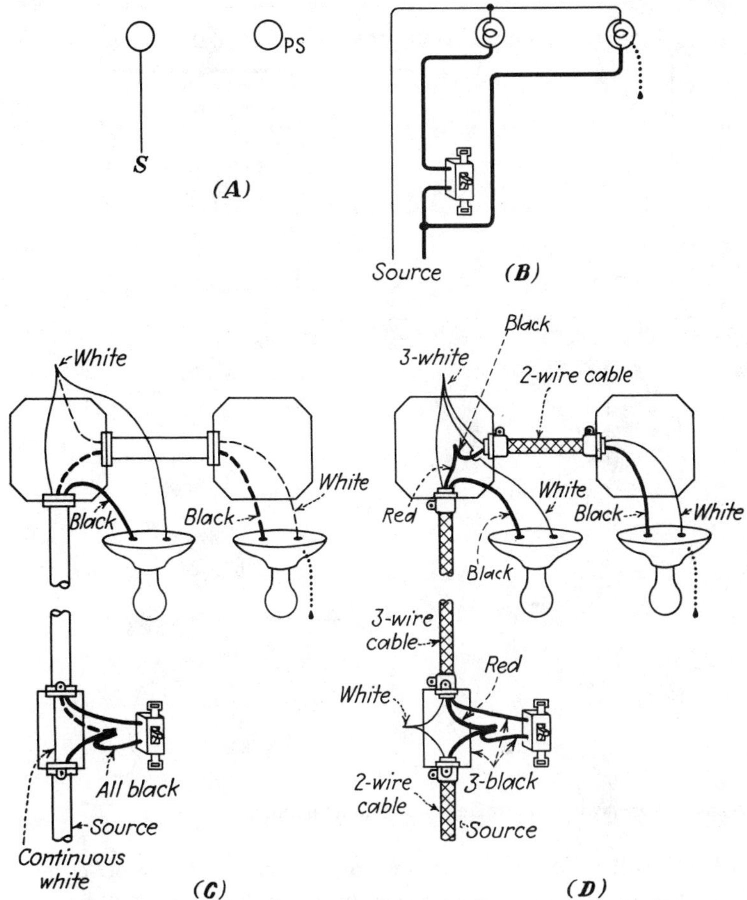

Fig. 18-12 This is the same as Fig. 18-11, except that the wires from SOURCE enter through the switch box.

Three-Way Switches. When a pair of 3-way switches controls an outlet, there are many possible combinations or sequences in which the SOURCE, the two switches, and the outlet may be arranged; a great deal depends on where the SOURCE wires come in. The three most common are

SOURCE—Switch—Switch—Outlet.

SOURCE—Outlet—Switch—Switch.

SOURCE—Switch—Outlet—Switch.

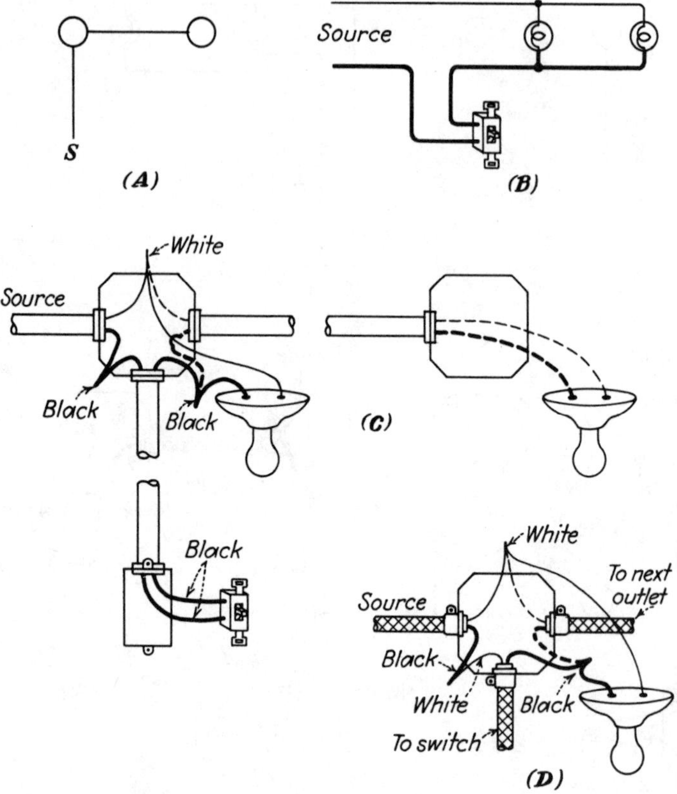

Fig. 18-13 One switch controlling two separate fixtures.

A fourth is the sequence where the SOURCE comes into the outlet box, from which point two runs are made, one to each switch.

These sequences are shown in Fig. 18-14, parts *A*1 to *A*4; the wiring diagrams are shown in *B*1 to *B*4. Comparing *B*1, *B*2, *B*3, and *B*4, you will see little difference except the exact location of the light that the switches control.

Reviewing the subject of 3-way switches, remember that one of the three terminals on such a switch is a ''common'' terminal, corresponding to the middle terminal of an ordinary porcelain-base single-pole double-throw switch shown in Fig. 4-19; this terminal is usually identified by being of a different color from the other two. The exact location of this common terminal with relation to the other two varies with different brands; for the purposes of this chapter, where 3-way switches are shown in pictorial fashion, the terminal that is alone on one *side* of the switch will always be the common terminal.

Reviewing the subject a bit further, run the incoming black wire from SOURCE directly to the common or marked terminal of either switch. From the corresponding common or marked terminal of the other switch, run a black wire directly to the proper terminal on the fixture. From the remaining two terminals on one switch, run wires to the corresponding terminals of the other switch, which are the only two terminals on that switch to which wires have not already been connected. To complete the circuit, connect the incoming white wire from SOURCE to the fixture.

With these facts in mind, the wiring of any combination of 3-way switches with conduit becomes quite simple. Assuming that the boxes and conduit have been properly installed, as shown in Fig. 18-14, *C*1, which covers the sequence of *A*1, pull the white wire from SOURCE through the switch boxes up to the outlet where the fixture is to be used; run the black wire from SOURCE to the common terminal of the nearest 3-way switch. From the two remaining terminals of this switch, run two black wires to the corresponding two terminals of the second switch. From the common terminal of this switch, run a black wire to the fixture. All wires are black, but sometimes one red wire is used for identification purposes; any color may be used except white or green.

The wiring of other combinations or sequences with conduit is equally simple if you remember the points of the two previous paragraphs, and so no diagrams are shown. It will be good practice to draw the circuits in a fashion similar to that shown in Fig. 18-14, *C*1.

This same combination wired with cable is shown in Fig. 18-14, $C1^1/_2$. Note how the white wire from SOURCE is continued from box to box until it reaches the fixture. The black wire from SOURCE, as in the case of conduit, goes to the common terminal of the nearest switch. The 3-wire cable between the switches contains wires of three different colors, of which the white has already been used, leaving the black and the red. Therefore run these two wires from the two remaining terminals of the first switch to the corresponding terminals of the second; the red and the black may be reversed at either end, being completely interchangeable. This leaves only one connection to make, and that is the black wire from the common terminal of the second switch to the fixture.

When some of the other sequences, such as *A*2, *A*3, or *A*4 in Fig. 18-14, are wired with cable, the usual difficulty in connection with the colors of the wires in standard 2- and 3-wire cables is met. The red wire of 3-wire cable is interchangeable with the black. Many times, however, you must take advantage of the Code's exception permitting, in switch loops, white wire to be connected to black.

*C*2 shows the sequence *A*2 and *B*2 of Fig. 18-14 wired with cable. The incoming cable from SOURCE contains one black and one white wire; run the white directly to the fixture. The Code requires that the other wire on the fixture may not be white; consequently the black wire of the 2-wire cable that runs to the first 3-way switch is connected to the fixture; connect the opposite end of it to the "common" terminal of the first 3-way

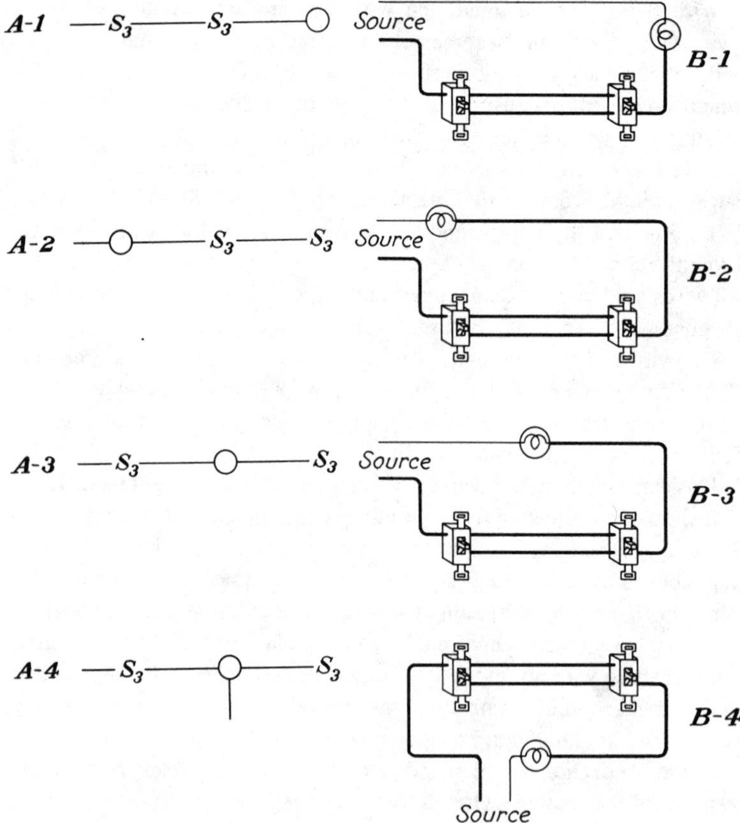

Fig. 18-14 Part 1. Four different sequences of parts in a circuit consisting of one fixture and a pair of 3-way switches.

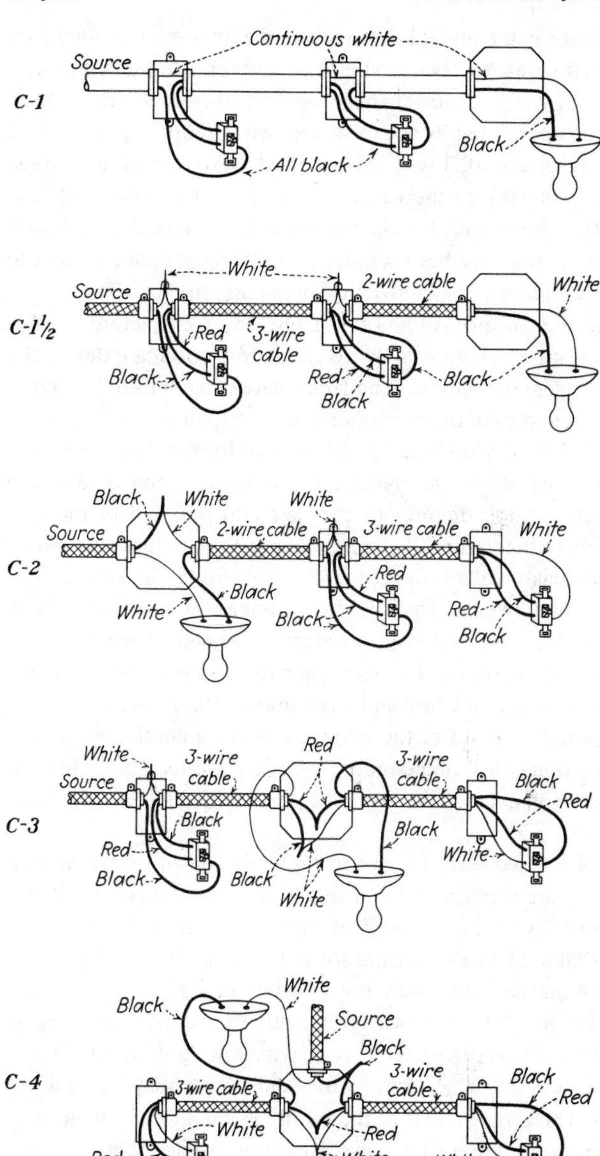

Fig. 18-14 Part 2. Wiring methods of the combinations shown in Part 1 of this figure.

switch. That leaves in the outlet box on which the fixture is mounted only two unconnected wires: the black wire from SOURCE and the white wire of the next run of cable. Connect them together; this is contrary to the general rule but is permitted by the Code for switch loops. From the first 3-way switch to the second, 3-wire cable is used. Since one of these three wires is white, connect it to the white wire of the 2-wire cable and continue it on to the ''common'' terminal of the second switch. That leaves one black and one red wire between the two switches; connect them to the remaining terminals of each switch, completing the installation.

In wiring the sequence of *A*3 and *B*3 of Fig. 18-14, as pictured in *C*3, similar problems arise. There is a 2-wire cable from SOURCE entering the first switch box; from that point a length of 3-wire cable runs to the outlet box in the center, and from there another length of 3-wire cable runs on to the second switch box. Continue the white wire from SOURCE from the first switch box directly to the fixture, as the Code requires. Run the black wire from SOURCE directly to the common terminal of the first switch. Going on to the fixture, the second wire on the fixture may not be white; therefore make it black and continue it onward to the common terminal of the second switch. That leaves unconnected two terminals of each of the two switches, and your problem is to connect them to each other. There are yet two unused wires in each run of 3-wire cable, a black and a red in the one, and a white and a red in the other. Therefore make one continuous red wire out of the two reds, make a continuous white-black out of the other two, and connect the extreme ends to the two remaining terminals on each of the switches, respectively. This completes the connections.

In the case of the sequence of *A*4 and *B*4 of Fig. 18-14, as pictured in *C*4, the problems are similar, and you should have no difficulty in determining for yourself why each wire is of the color indicated. The fundamental rule is that the white wire from SOURCE must go to the fixture, and the second wire on the fixture may not be white or green.

Pilot Lights. In Fig. 18-15 is shown a frequently used combination toggle switch and pilot light, together with its internal wiring diagram. This is nothing more or less than a separate switch and a separate pilot light in a single housing. The arrangement of the terminals on another brand may be totally different from that shown in the picture. A pilot light is used at a switch that controls a light not visible from the switch, for example, at a switch in the house controlling the light in a garage; it is used simply as a reminder that the light is on.

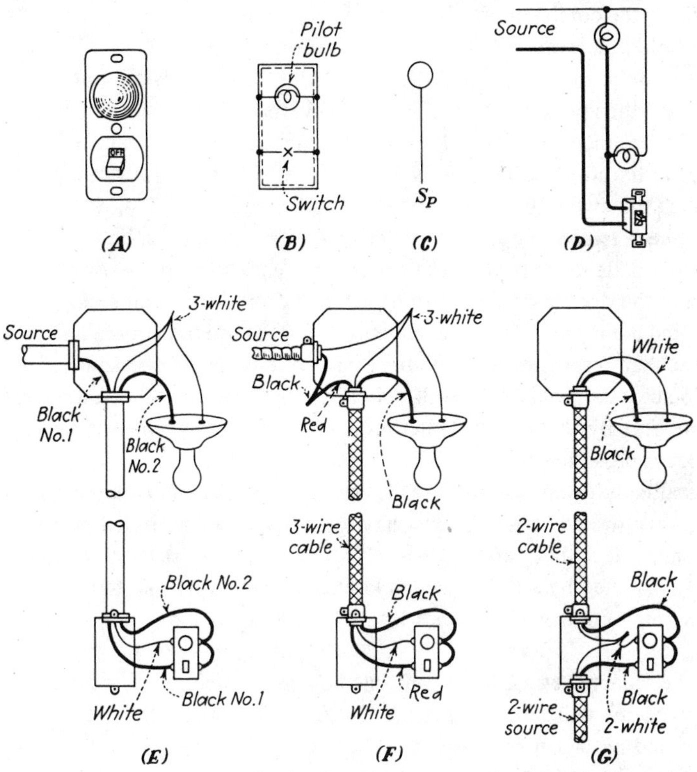

Fig. 18-15 A switch with a pilot light.

In the same figure at *C* is shown the blueprint symbol for this combination and at *D* the usual wiring diagram. Compare this with Fig. 18-13, *B*; it is merely a case of one switch controlling two different lights, one of which happens to be located at the same point as the switch.

Using the conduit system, the connections are shown at *E*. Run the incoming white wire from SOURCE directly to the first fixture and extend it from there to the second fixture, which in this case happens to be the lamp in the combination device. In the outlet box on which the fixture is mounted, run the black wire from SOURCE to the switch. From the other side of the switch, run a black wire to the first lamp (in the combination device) and then on to the fixture, finishing the job.

Wiring the combination with cable, as shown in Fig. 18-15, *F*, presents no problem. It would not be a Code violation to run the black wire from SOURCE on to the switch, and then a red wire from the switch to the first lamp and then on to the fixture, but then the fixture would have the usual white wire and a *red* wire. It is much better to let the second wire on lamps or fixtures be consistently *black*. If the wires from SOURCE come in through the switch box, the problem is still simpler, as *G* shows.

Switched Receptacles. When entering a dark room it can be very hard to find a table or floor lamp to turn on. It is easy to stumble over something in the dark and thus cause an accident, and lamps can be knocked over and damaged. It is also a real nuisance to go around a room and turn off each lamp separately at bedtime. One solution is to have some of the receptacles controlled by a wall switch, so that they can all be turned off at one time by the switch. Other receptacles can be left unswitched for clocks, radios, and the like.

As already discussed in Chap. 12, the Code in Sec. 210-70(a) requires that every room of a house must have some lighting controlled by a wall switch; review that part of Chap. 12. The switch must control permanently installed lighting fixtures in kitchens and bathrooms, but in other rooms may control receptacles into which floor or table lamps can be plugged.

If the receptacles are to be controlled by one switch, no particular wiring problems will be encountered. But in some rooms, they should be controlled by a pair of 3-way switches. Then the wiring becomes a bit more complicated. There can be a great many different sequences of outlets and switches; that shown in Fig. 18-16 at *A* and *B* is perhaps as common as any. When conduit is used, the problem is simple indeed. Run the white wire from SOURCE to each of the receptacles, connect the remaining terminal of each of the three receptacles together with black wire, and continue to the common terminal of one of the 3-way switches. From there run two wires over to the other 3-way switch. To the common terminal of this second switch, connect the black wire from SOURCE, and the job is finished. It is pictured in Fig. 18-16, *C*. All the necessary wires in the boxes housing the receptacles will badly crowd the ordinary switch boxes, so the preferable method is to use 4-in square boxes with raised single-gang covers designed to take a duplex receptacle (See Fig. 10-5).

To wire this same combination using cable is practically impossible with the particular sequence shown, because it requires four wires at some points and 4-wire cable is not stocked by dealers. Therefore, when

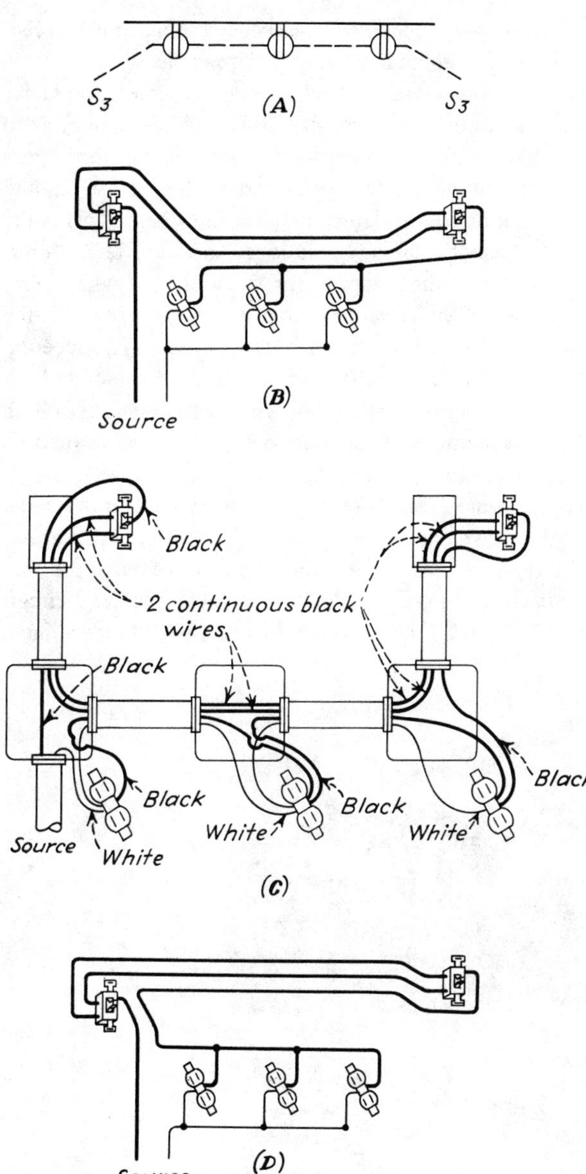

Fig. 18-16 Switched receptacle outlets.

using cable, it is best simply to modify the sequence to that shown in Fig. 18-16, *D*, which requires nothing more than 3-wire cable.

Two-Circuit Duplex Receptacles. Instead of having some receptacles switched for lamps, and others permanently live, a better solution is to use 2-circuit receptacles. These are so designed that half of each receptacle can be permanently connected for clocks, radios, TV, and so on, and the other half switched to control lamps plugged into them. This is the 2-circuit or "split" receptacle already mentioned in Chap. 12. An *ordinary* receptacle has two terminal screws for the grounded white wire, common to both halves of the receptacle, and two more screws for the hot wire, common to both halves of the receptacle. The *2-circuit* receptacle also has two screws for the white or grounded wire, common to both halves, but two separate screws for the hot wires, *not* common to both halves. Each of the two screws for the hot wires serves *one* of the two halves of the receptacle.

Instead of buying special 2-circuit receptacles, you will find that most better-quality receptacles have a removable link between the two brass-colored terminal screws for the hot wires that can be pried out at the time of installation, thus converting the ordinary receptacle into the 2-circuit type in a moment. How to do this is shown in Fig. 18-17. In installing a

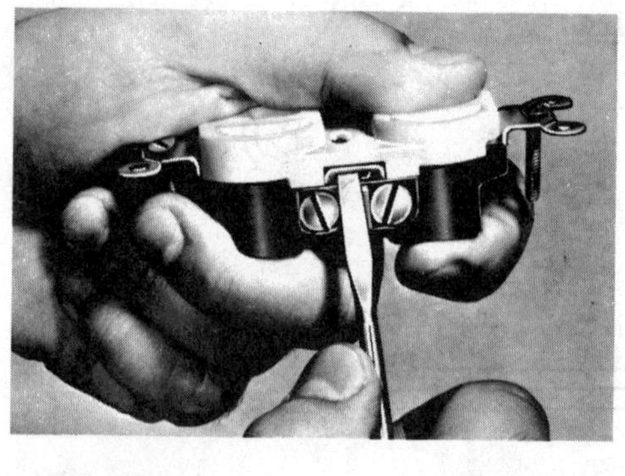

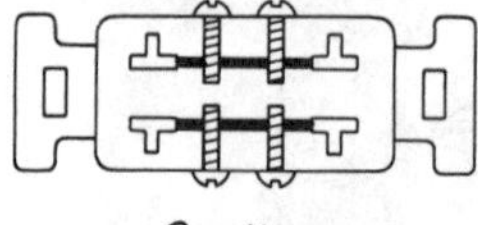

Fig. 18-17 Two-circuit receptacles have many advantages. In better-quality receptacles, the ordinary receptacle can be changed to the 2-circuit type by breaking out a small metal strip as shown. (*General Electric Co.*)

2-circuit receptacle, install it so that the switched half is at the bottom, leaving the unswitched half at the top, more readily available and easy to get at for plugging in a vacuum cleaner and other things that are often plugged in and later unplugged. Figure 18-18 shows the blueprint symbol for a 2-circuit or split-wired receptacle at *A*.

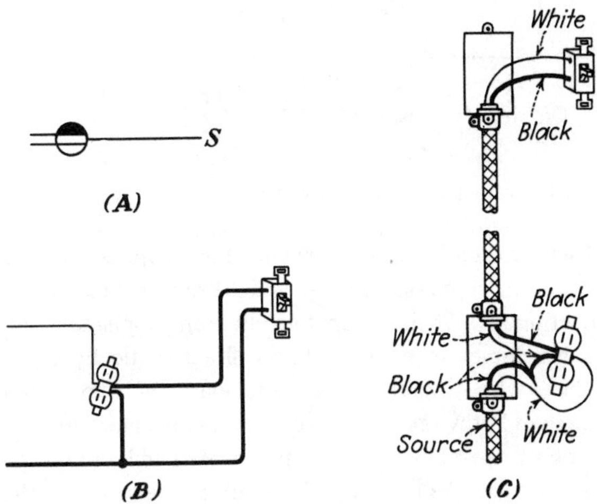

Fig. 18-18 Wiring a 2-circuit receptacle, using cable.

The wiring presents no great problem. Run the white grounded wire from SOURCE to one of the terminal screws on the metal strap on the receptacle that is not broken in the middle. Run the black wire from SOURCE to one of the terminal screws on the opposite side of the receptacle, and continue it to the switch. From the switch run another black wire back to the remaining terminal screw on the receptacle, on the side where the center of the strap is broken out. All this is shown in Fig. 18-18, but for more detail see the additional illustration of Fig. 18-19.

If you use cable to the switch, again you will have the problem that one of the wires is white where it should be black; take advantage of the Code's special dispensation and connect the white wire in the switch loop to the incoming black wire from SOURCE, as shown in *C* of Fig. 18-18.

If you use 3-way switches, the problem of getting the colors is more difficult, but may be solved exactly as in the case of the lighting outlets shown in Fig. 18-14.

Another common application for the 2-circuit receptacle is in the two 20-amp small appliance circuits in the kitchen. If these two circuits are

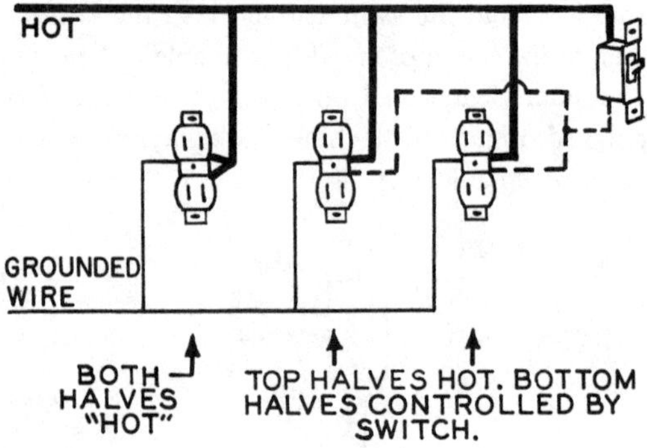

Fig. 18-19 More detail for wiring 2-circuit receptacles.

merged into one 3-wire circuit (as will be explained in Chap. 20) run the white grounded wire to the common terminal, and the red and black wires to the other two terminals. Then, when two different appliances are plugged into the same duplex receptacle, they will automatically be on opposite legs of the 3-wire circuit. At the panelboard the two circuit breakers feeding the two hot wires must have their handles mechanically tied together so that both operate with one motion of the hand, or you can use a 2-pole breaker. If fuses are used, install a double-pole switch in the circuit where it leaves the panelboard. Thus one half of the circuit cannot inadvertently be left on should you need to replace one of these split-wired receptacles at some future time.

Combining Outlets. Just as in the early part of this chapter the outlets of Figs. 18-2, 18-4, and 18-7 were combined into one three-outlet combination of Fig. 18-8, so outlets may be combined into any desired combination with any desired total number of outlets, which then forms a circuit running back to the service equipment. To add an outlet at any point, connect the white wire of that new outlet to the white wire of the previous wiring (provided it is not a white wire in a switch loop, when using cable), and the black wire of the new outlet to any previous wiring where the black wire can be traced back to the original SOURCE without being interrupted by a switch. Simply connect the new outlet at any point on the previous wiring where a lamp connected to the two points in question would be permanently lighted. Review Chap. 4 if this point is not entirely clear.

Installing Recessed Fixtures. The temperature inside recessed (flush) fixtures is higher than in fixtures in free air. The terminals may reach temperatures higher than the temperature limits of the kinds of wire normally used in ordinary wiring. Assuming the fixture is UL Listed (and in *recessed* fixtures it is more important than ever that you use only UL Listed fixtures), it will be marked with the minimum temperature rating of the wire supplying the fixture.

If a UL Listed fixture is unmarked or marked "60°C," it may be used with any kind of wire normally used in wiring. Some recessed fixtures are marked for a temperature considerably above 60°C. In that case you have a choice of two methods. The first is to wire the entire circuit up to the fixture with a wire having a temperature limit[2] at least as high as the temperature shown on the fixture. (Fixture wire *must not* be used for the branch circuit supplying a fixture, or in *any* branch circuit.) Such wire is likely to be an expensive wire but may be feasible if the circuit to the fixture is very short. In most cases, the other permitted method is better.

Using the second method, install a junction box, which may be an ordinary outlet box, not less than 12 in from the fixture, and run a length of flexible metal conduit not less than 4 ft nor more than 6 ft long from that junction box to the fixture. Loop the conduit as shown in Fig. 18-20 if

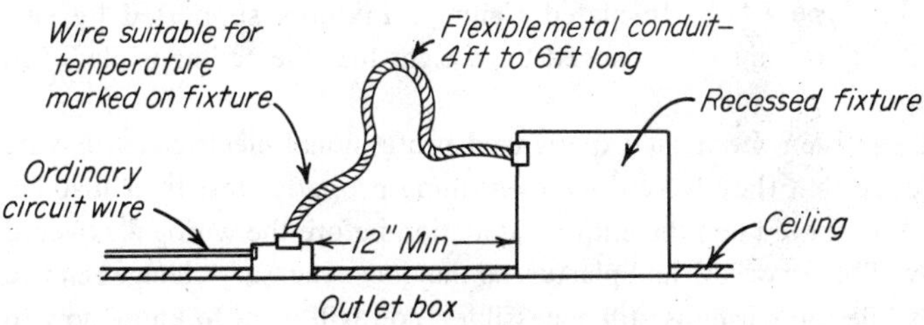

Fig. 18-20 Schematic outline for installing a recessed fixture.

necessary to consume the entire length. In that length of flexible conduit, install wire of the proper temperature rating. Most inspectors will accept fixture wire if it is of the proper temperature rating.[3]

It is usually more convenient to use recessed fixtures designed so that the terminal connections operate within the 60°C limits of ordinary wire. A fixture so constructed is shown in Fig. 18-21.

[2] The temperature limits of any kind of building wire can be found in NEC Tables 310-16 to 310-19, all in the Appendix of this book.

[3] Fixture wires Types TF and TFF have a 60°C rating; Types RFH and FFH have a 75°C rating; all other types have a rating of 90°C or higher. For detailed information, see Table 402-3 in your copy of the Code.

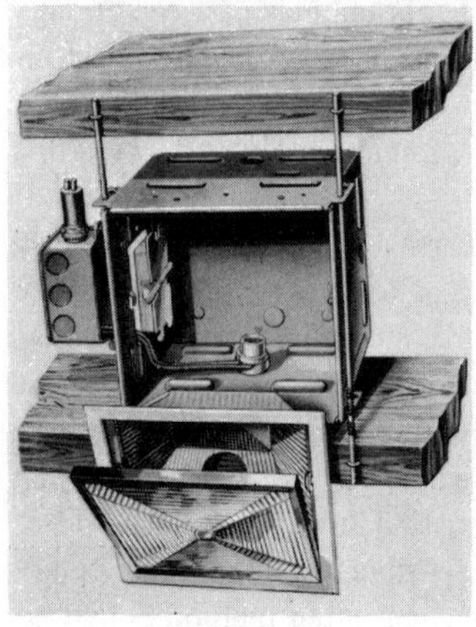

Fig. 18-21 A well-designed and properly installed recessed fixture. (*The Kirlin Co.*)

Note that NEC Sec. 410-66 requires a clearance of at least $^1/_2$ in between a recessed fixture and all combustible material, as, for example, wooden joists. Thermal insulation must not trap heat above the fixture, and it must be kept at least 3 in from a recessed fixture unless the fixture is marked "Type I.C." (Insulated Ceiling). Fixtures supported by suspended ceilings must not exceed the weight that the ceiling is classified to support.

Testing. Even the most experienced professional electricians, having confidence that they have done everything properly, test their installations. You should test the entire installation before the wiring is covered up by wallboard or lath and plaster, so that any necessary changes can be made while the wiring is still accessible. You will want to know how to make the tests anyway, for troubleshooting if the need arises. For this purpose, you can use a doorbell or buzzer and two dry-cell batteries connected in series, as shown in Fig. 18-22. Tape the bell to the dry cells; also tape the test leads to the cells, so that you can lift the entire unit by the leads without putting strain on the bell or the dry-cell terminals.

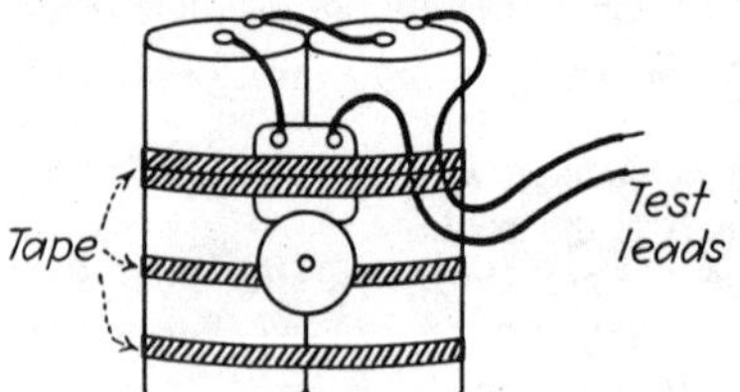

Fig. 18-22 Test outfit consisting of two dry cells and a doorbell.

It should not be necessary to say so, but the testing must be done *before* the power supplier has connected its wires to the service-entrance wires.

Before doing any testing, do the following at each outlet. If it contains wires that are to be permanently connected to each other, connect them now, using wire nuts or similar connectors. Where receptacles or fixtures are to be installed, let the wires project freely from the outlet. At all points where a switch is to be installed, be sure that all wires that will be connected to one switch are *temporarily* connected together; the quickest way of doing that is to use a wire nut that will later be removed. Be sure that a transformer for doorbell or chimes has *not* been connected to a circuit. Be sure that no exposed bare wire is allowed to touch any box, or conduit, or the armor of cable.

Testing Metal Wiring Systems. We must assume that you have connected the white grounded wires of all 120-volt circuits to the neutral busbar in the service equipment. Turn all circuit breakers off, and remove all fuses. Then proceed as follows: At the service equipment, connect one lead of the battery-and-bell tester to the neutral busbar. Then touch the other lead of your tester in turn to the black (or red) wire of each 120-volt circuit. The bell must *not* ring. If it does, there is a short circuit, probably at a box where you have inadvertently permitted a bare wire to touch the box. Clear the fault up until there is no ring.

Next, make another test for continuity. Connect a *temporary* wire jumper across the two hot service wires where they enter the service equipment. (After you have completed the tests, remove this jumper, or you will have a total short circuit across your 240-volt service wires.) If you have used circuit breakers, turn them all to the on position; if you have used fused equipment, make sure all the fuses are in place. Then remove the bell from the batteries, leaving just the leads from the batteries. Connect one of them to the terminals in the service equipment across which you have placed the jumper; connect the other to the grounded neutral busbar in the service equipment. You will then have temporary 3-volt current across each 120-volt circuit you have installed. But to repeat a caution, *remove* that temporary jumper after you have completed the tests.

Then take the bell to each outlet where a fixture or a receptacle is to be installed. Touch it across the black and white wires; the bell should ring, just as lamps in your fixtures or floor and table lamps will later light when connected to these same wires. If the bell does not ring, check to make sure all circuit breakers are in the on position, or that all fuses have been installed. After the check across the black and white wires of each outlet, touch the bell across the black wire and the box itself; the bell should again ring because through the conduit or cable armor all the boxes are connected together and grounded, as is the white wire. If the bell rings feebly, a poor job has been done somewhere; probably the locknut at one or more boxes has not been run down tightly enough. If the bell rings at each point, the wiring is all right. *Remove that temporary jumper in the service equipment.*

Do note that the procedure just outlined is for 120-volt circuits only. For 240-volt circuits, touch the bell across one of the two wires at the outlet, and the box. The bell should ring. But if you touch it to the two wires of the circuit, it will *not* ring. To further test the 240-volt circuit, remove that jumper in your service equipment. Connect one of the battery leads to one of the two hot incoming service wires, the other lead to the other of the two hot service wires. Then connect the bell across the two wires at the 240-volt outlet; it should ring.

Testing Nonmetallic-Sheathed Cable Installations. If you are using *metal* boxes and cable with the bare equipment ground*ing* wire, properly installed, proceed as in the case of metal conduit or armored cable installation, but be sure you have connected the bare ground*ing* wires to the grounding busbar in your service equipment or panelboard, and be sure they have been connected to each metal box.

If you have used nonmetallic boxes, test by touching the bell not only between the black and white wires, but also between the black wire and the bare ground*ing* wire. The remainder of the test is as with the metal conduit system.

If all the tests check properly, *remove that jumper* and proceed to finish the installation as outlined in the next chapter.

Circuit Testers. Depending on how much work you do, it may well be worth your while to buy a commercial tester. Many types are available. The one shown in Fig. 18-23 can be used only to test the continuity of a circuit. It operates on a couple of penlight batteries and of course can be used only while the power is turned off. It can also be used, for example, to test cartridge fuses, removed from their holders, to see whether they are blown or not. Cartridge fuses look the same whether blown or not.

Another handy device is the tester shown in Fig. 18-24. It contains a tiny neon bulb that glows when energized; it can be used on either 120- or 240-volt circuits to determine whether the circuit is "hot" or not. It can

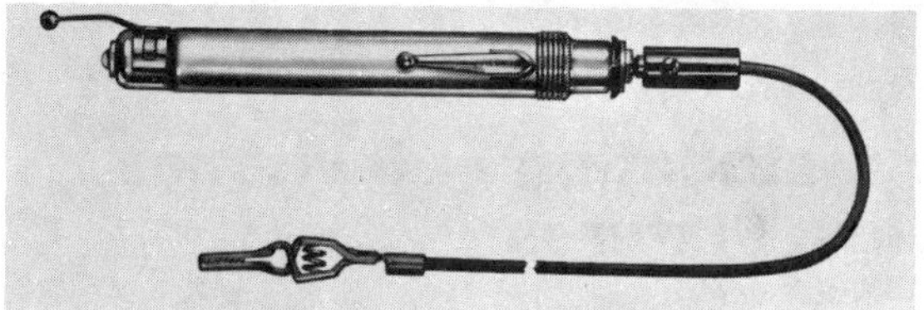

Fig. 18-23 A test light of this kind is convenient in tracing circuits, testing fuses, and so on. (*Ideal Industries, Inc.*)

Fig. 18-24 This test light is handy for determining whether a circuit is live or not.

also be used to test cartridge fuses while they are still in their pull-out blocks, still in the switch. On most pull-out blocks you will find tiny holes at the top and the bottom of each fuse. Insert the leads of the tester into both top and bottom holes. If the power is on, the neon light will glow if connected across a blown fuse, but it will not glow if connected across a fuse in good condition.

After the wiring is completed and the power turned on, the tester shown in Fig. 18-25 is used only to determine whether 15- and 20-amp

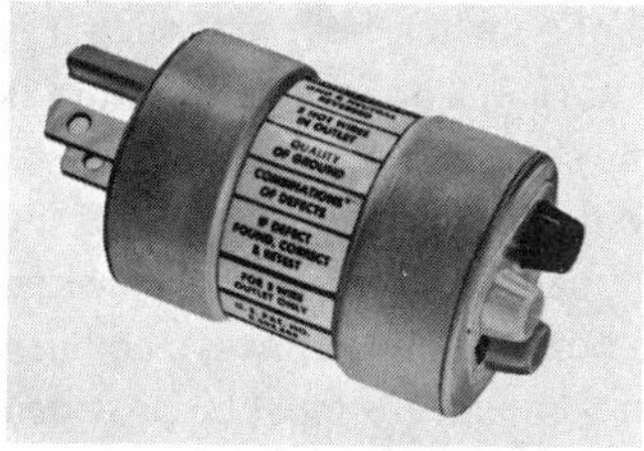

Fig. 18-25 This tester tells you whether 15- and 20-amp receptacles have been properly connected. (*Cirfico Holdings Corp.*)

receptacles have been properly connected. The tester has three neon lights; when plugged into a receptacle, it indicates whether the receptacle has been properly connected. If the receptacle has been wrongly connected, the combination of lights will show the nature of the wrong connection.

19

Finishing: Installation of Switches, Receptacles, Fixtures

In the preceding chapter the roughing-in of outlets was discussed. You are now ready to install the receptacles, switches, face plates, and fixtures.

Installing Switches, Receptacles, Etc. Every device of this kind is provided with a metal mounting strap or yoke that has holes in its ends, so spaced as to fit over the holes in the ears of a switch box, on which it is mounted using screws that come with the device. The face plate in turn is anchored to the device, not to the box, as shown in Fig. 19-1.

For a neat installation, the yoke of the device must be flush with the plaster or wallboard. Since plaster or wallboard is added after the switch boxes are installed, the front edges of the boxes may not always be flush with the finished surface of the wall. Then some means must be provided to bring the device flush with the surface. Most devices have plaster ears on the ends of the yokes, as was shown in Figs. 4-6 and 4-16. These ears lie on top of the wall surface, thus bringing the device flush with the surface. The metal is scored near the end of the yoke so that the ears can easily be broken off if not needed. If your device does not have plaster ears, you can insert small washers between the box and the yoke.

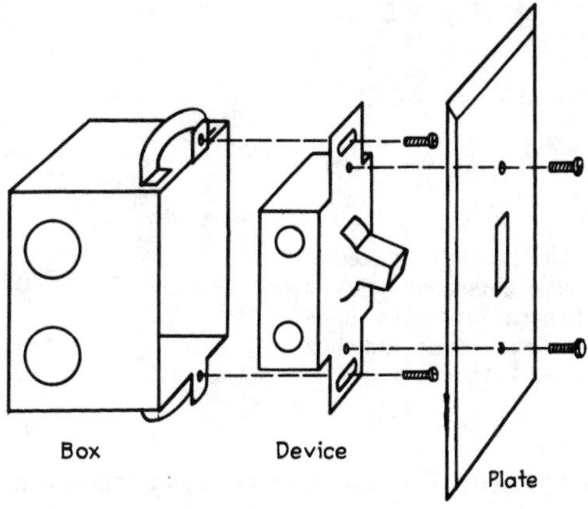

Fig. 19-1 The switch or other device is installed on the box, and the face plate is installed on the device.

Where the wall finish is combustible, such as wood paneling, the Code in Sec. 370-10 requires that the box face not be back of the finished surface. Any space between the box face and the wall surface can be protected by the use of a switch box extension similar to that shown in Fig. 19-2.

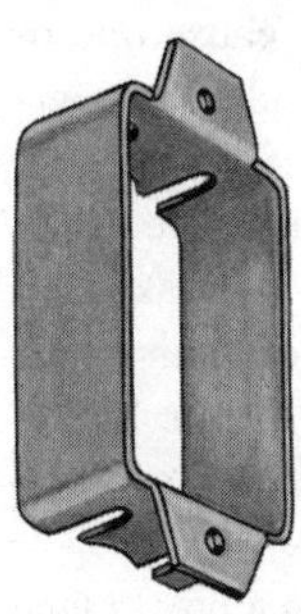

Fig. 19-2 Switch box extension, for use when the box is not flush with combustible wall finish. (*Appleton Electric Co.*)

Neatly installed devices must be absolutely straight up and down; often the boxes are not entirely straight. For this reason the mounting holes in the ends of the yoke are not round but elongated, so that the device itself can be mounted straight even if the box is not straight. A glance at Fig. 19-3, which exaggerates the usual condition, should make this clear.

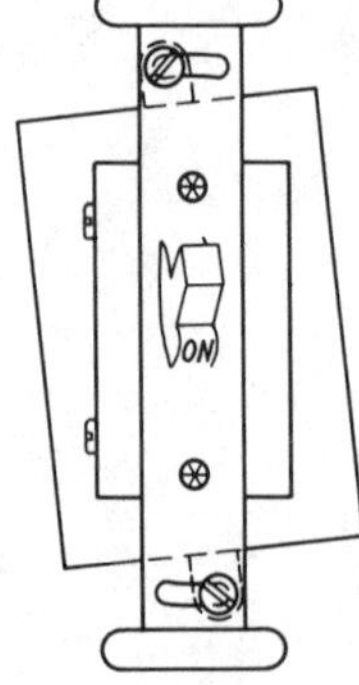

Fig. 19-3 The mounting yokes of devices have elongated holes in the ends to permit vertical installation of the device, even if the box is not mounted straight.

You should have left about 10 in of wire at each box, as explained in Chap. 8. About 4 in of this will be used up in making the connection to the terminal of the device, as was also explained in Chap. 8. Don't leave the wires *too* long, or the box will be overcrowded. The 6 in that is left can be neatly folded in the box so that there is plenty of space for the device to be installed in the box, if the right size of box was installed. The insulation of the wire must extend up to the terminal. *There must be no bare wire exposed between the terminal and the end of the insulation.* Review Chap. 8 regarding this.

Here is an important safety tip: when you place devices with screw terminals in a box, the stiffness of the wire can sometimes cause one or more terminals to loosen slightly. Loose terminals have resulted in overheating, arcing, and even fires. To avoid this possibility, carefully fold the wires around the back of the device before placing the device in the box so that any mechanical resistance offered by the wires is taken by the device body rather than by the terminals. Be careful, of course, not to damage the wire insulation. Fold the wires into place in the box—don't just push. If you have used a box of the proper size, there will be room for everything.

Installing Face Plates. After a device has been installed in a box, a face plate must be installed, using screws that come with the plate. If the device has plaster ears, it will automatically be at the right height as compared with the wall surface, so that the plate will fit snugly against both the device and the wall surface. But if the device does not have plaster ears, install spacing washers between the box and the mounting yoke, to bring it as nearly as possible flush with the wall surface. But if the device is slightly below the wall surface, one caution is in order. Do not pull up too tightly on the screws holding the plate to the device. It is not unusual to crack the small bridge between the two openings of a plastic plate for a duplex receptacle, ruining the plate.

If the box is of the 2- or 3-gang type, the mounting of the plate will not be entirely simple because of the elongated mounting holes in the yokes of the individual devices. These elongated holes are a great advantage when mounting devices in a single-gang box, but they require extreme care when ganged boxes are used, for if several devices are mounted not *exactly* parallel with each other, they will not match up with the absolutely parallel openings (and mounting holes) in a 2- or 3-gang plate. So several adjustments of the individual yokes may have to be made before the screws in a multigang plate will match up with the tapped holes in the yokes of the individual devices.

If you install multiganged devices, it often will be to your advantage to make a line-up tool as shown in Fig. 19-4. Start with a metal plate having

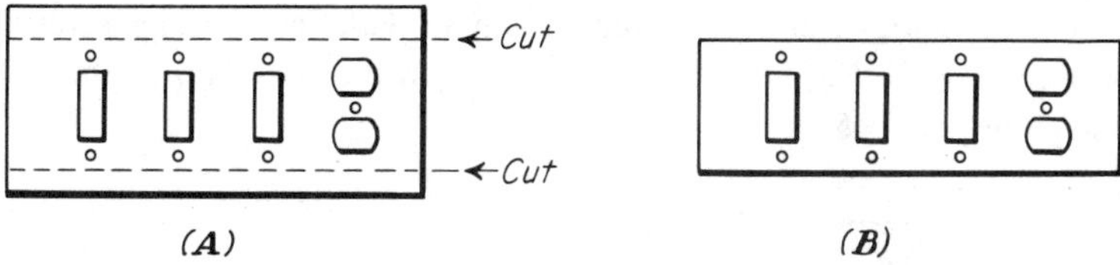

Fig. 19-4 This gadget saves time when you install several devices in a multigang box.

the necessary arrangement of devices, as shown at *A*. Saw off the edges close to the mounting holes that normally are used to install the plate on the switches, all as shown at *B* of the illustration. After the switches (and the receptacle if there is one) have been properly connected, install them in the box, but install the mounting screws loosely; do not tighten them. Then temporarily install the sawed-off cover on the switches, tightening the mounting screws of the cover, so that all the devices are solidly anchored to the temporary cover. Then all the devices will be parallel and the right distance apart, in spite of the oval holes in their mounting straps. Only then tighten the screws in all the individual straps, solidly mounting the devices in the box; then remove the temporary sawed-off cover. The holes in the final face plate will now fit perfectly over the various switches and receptacles and can easily be installed.

Wallpaper is combustible, and therefore no loose edges should be exposed at an outlet, but when metallic-foil wallpaper is used, carefully trim the paper away from the *entire* opening of the outlet before installing the face plate. There have been cases of shock which were caused by foil wallpaper being energized through contact with live terminals at a switch or receptacle.

Wiring Lighting Fixtures. The fixture may have two terminals for connecting the wires, in which case connect the white wire to the whitish terminal, the black wire to the other terminal. More often the fixture has two wire leads. If these are black and white there is no problem. Usually they consist of fixture wires as described in Chap. 6: one a solid color, the other of the same color but identified by a colored tracer thread in the outer cover. The wire with the colored tracer is for connection to the white wire in the box. In case the identification is not clear, trace the wires into the fixture; the one that connects to the screw shells of the individual sockets is for connection to the white wire.

Hanging Fixtures. There are many ways of installing lighting fixtures, all dependent on the style, shape, and weight of the fixture, the particular box involved, and the method of mounting the box in the ceiling or wall.

Simple fixtures can be mounted directly to outlet boxes using screws that come with the fixture and fit into ears on the box. This method is shown in Fig. 19-5.

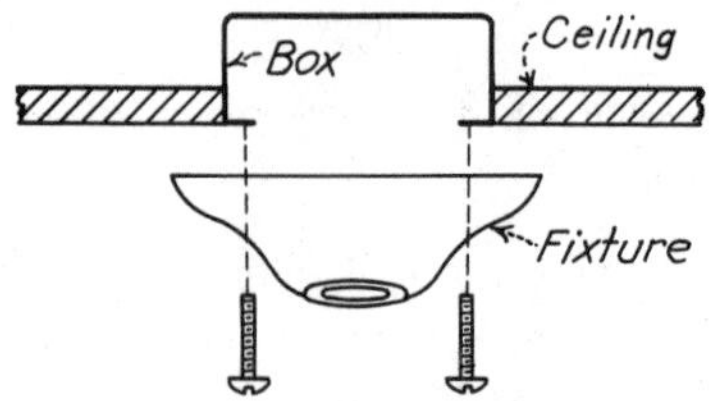

Fig. 19-5 Small, simple fixtures are mounted directly to the box.

Other fixtures are too large to permit mounting in this direct fashion, in which case a strap is used as shown in Fig. 19-6. The strap is first installed on the box, then the fixture is mounted on the strap.

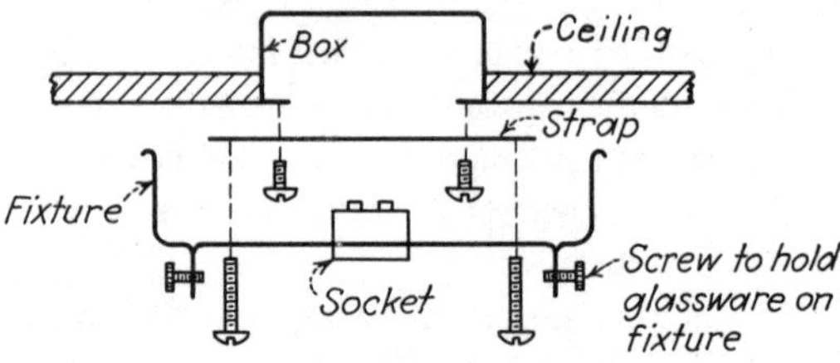

Fig. 19-6 Larger fixtures are installed on a strap that has been mounted on the box.

Often a fixture stud such as shown in Fig. 19-7 is used in mounting the fixture. Mount the stud in the back of the outlet box by means of screws through holes provided for the purpose. Some boxes have the stud as an integral part of the box. If you use a hanger of the type that was shown in the upper part of Fig. 10-16, the stud that is part of the hanger goes through the center knockout of the box, anchors the box to the hanger, and at the same time permits the stud to be used to support the fixture.

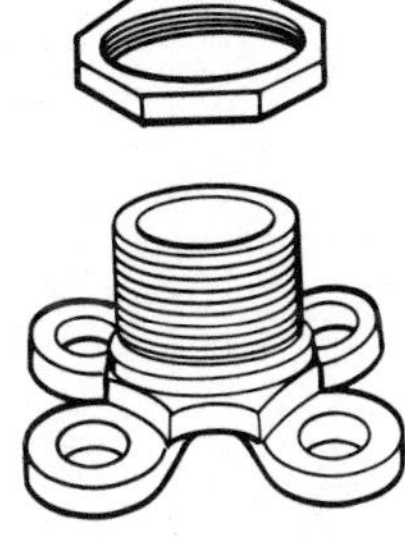

Fig. 19-7 Fixture studs are installed in boxes to support fixtures.

The outside of the stud is tapped to fit 3/8-in-trade-size pipe; sometimes there is an inner female thread fitting 1/8-in-trade-size pipe.

Figure 19-8 shows the fittings generally used, called a lockup unit. It

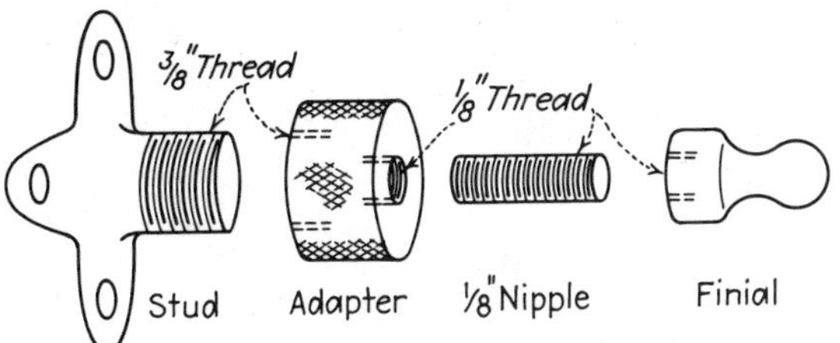

Fig. 19-8 A typical lockup unit for installing small fixtures.

consists of an adapter fitting over the fixture stud, a short length of 1/8-in running-thread pipe, and a nut to hold the assembly together, the nut usually being of an ornamental type and then called a finial. Figure 19-9 shows these parts used to hold up a simple fixture. It is a simple matter to drop down the fixture while making connections, then to mount it on the ceiling.

If the fixture is a larger unit, the mounting is similar, and Fig. 19-10 should make it clear. The top of the fixture usually consists of a hollow stem, with an opening in the side through which the wires of the fixture

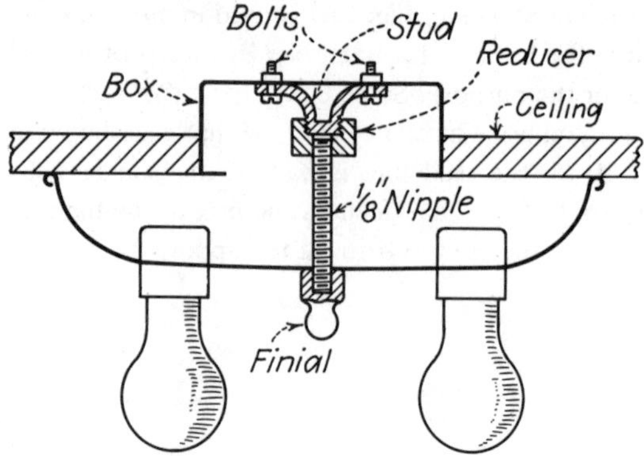

Fig. 19-9 This shows the parts of Fig. 19-8, used to install a small fixture.

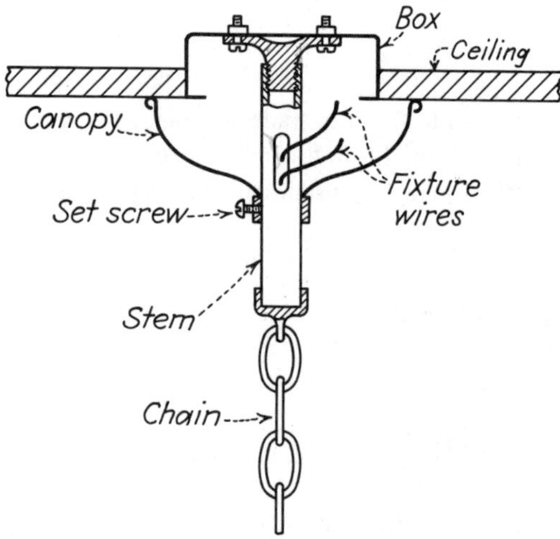

Fig. 19-10 The usual method of installing larger fixtures.

emerge. The top is threaded to fit on the fixture stud, and the mounting is as shown in Fig. 19-10. Drop the canopy down while the connections are being made, and then slip it back to be flush with the ceiling.

Sometimes the wires from the fixture come out of the end of the stem instead of through an opening in its side. In that case use a hickey shown in Fig. 19-11 between the end of the stem and the fixture stud. Sometimes the stud is too short or the box too deep, in which case use an extension

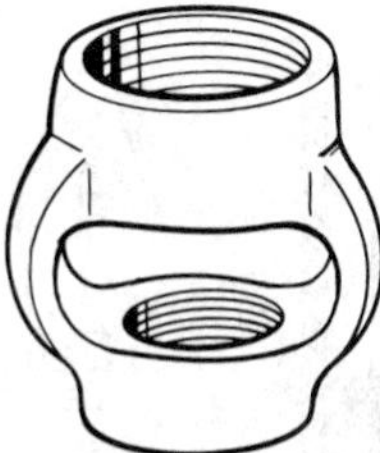

Fig. 19-11 A hickey, sometimes used between the end of the stem on the fixture and the fixture stud.

Fig. 19-12 In deep boxes it is sometimes necessary to use an extension piece between the stud and the fixture stem.

piece as shown in Fig. 19-12. Fixtures weighing more than 50 lb must be supported to the building structure, independently of the outlet box (NEC Sec. 410-16).

If the ceiling or wall material is combustible (such as wood or low-density wallboard) and there is an open space between the box and the edge of the fixture or canopy, NEC Sec. 410-13 requires the space to be covered with noncombustible material such as metal, asbestos, or spun glass, sometimes provided with the fixture.

Installing Wall Brackets. Wall brackets are more decorative than useful. The method of installing a bracket depends to a large degree on the type of box used. Some brackets are too narrow to completely cover a 4-in box, so there is little choice except to use a switch box. Sometimes a fixture stud is mounted in the back of the box; in that case the bracket is mounted using the lockup unit that was shown in Fig. 19-8, the completed installation being as shown in Fig. 19-14. More usually a fixture strap as shown in Fig. 19-13 is first mounted on the switch box, the fixture in turn being mounted on the strap, as shown in Fig. 19-15.

Fig. 19-13 Fixture straps are mounted on boxes, and the fixture is then supported by the strap.

If the wall surface is combustible, be sure to cover any part exposed between the outlet box and the edge of the fixture canopy with noncombustible material.

Adjusting Height of Fixture. A fixture with a chain is adjustable as to the height above the floor, to compensate for ceilings of different heights.

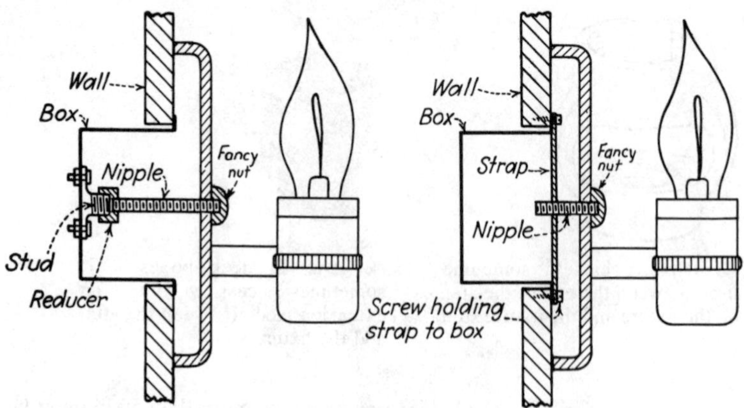

Fig. 19-14 Wall bracket installed on box by means of a fixture stud in the box.

Fig. 19-15 Wall bracket installed on box by means of a fixture strap mounted on the box.

The height is simply controlled by removing as many links of the chain as necessary. The actual height will be governed by personal preference and location. If the fixture is located above the dining-room table, hang it at least 24 in and preferably more above the table. Again, this will be dependent on the type of fixture, the ceiling height, and similar factors.

20
Miscellaneous Wiring

In the wiring of any house, there are problems and niceties of detail that could not readily be included in the discussions of previous chapters, so they will be grouped here in a separate chapter. Study this chapter well because from it you should get ideas that will help you make your installations better than just average.

Cove Lighting. When a room is lighted only by floor and table lamps, large parts of the ceiling remain relatively dark. This can be prevented by installing cove lighting, giving the room a far more cheerful appearance. Cove lighting consists of a series of single-tube fluorescent fixtures, end to end, in a kind of trough all around the room, near the ceiling. See Fig. 20-1. The trough should be at least 12 in below the ceiling, and in that case the center of the fluorescent tube should be about $2^1/_2$ in from the wall. As the distance below the ceiling increases, the distance from the wall should be increased somewhat, up to about $3^1/_2$ in for a distance of 24 in below the ceiling. A reflector behind the tube, aimed at about 20 deg above the horizontal plane, produces a more even level of light on the ceiling. Note the baffle: This must be of a height so that the fluorescent tube cannot be seen from anywhere in the room.

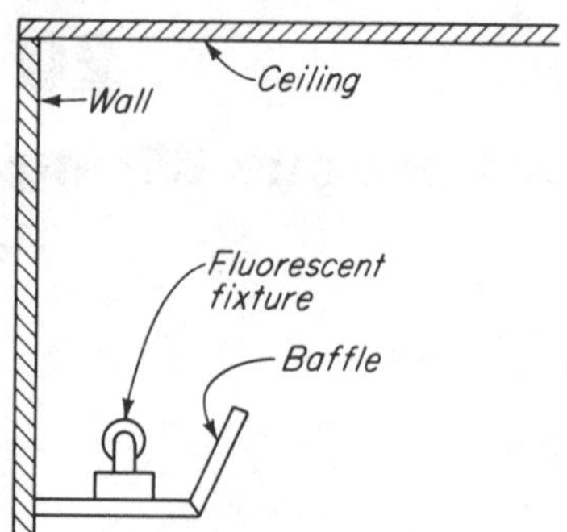

Fig. 20-1 Cove lighting does away with dark areas on the ceiling.

Track Lighting. Ceiling-mounted tracks, either surface or flush, accommodate decorative fixtures at any point along their length and are used effectively to wash walls with light, highlight art objects, or illuminate wall-hung paintings. A large assortment of fixtures are available for a variety of lighting effects. The track is fed from a ceiling outlet, usually placed at one end. See Fig. 20-2.

Quiet Switches. Ordinary switches have an annoying click as they are

Fig. 20-2 Lampholders may be located at any point along the open-bottom track. Two on the left "wash" the mantel wall, while the one on the right floods the ceramics on the shelves. (*Halo Lighting Division, McGraw-Edison Co.*)

turned on or off. Usually this click may not even be noticed, but in sickrooms, children's rooms, and similar locations the click can be decidedly annoying. In the dead of night, a switch turned on or off in the hall on the *outside* of a bedroom wall can sound very loud inside that room.

Two kinds of noiseless switches are available and are in common use. The original noiseless switch was the mercury type, in which a pool of mercury inside a glass tube takes the place of mechanical contacts. Mercury switches must be installed in a vertical position, and right end up, or they will not operate. The "general-use ac-only" switches described in Chap. 4 are much less expensive, and are very quiet.

Lighted Switches. Often it is quite a nuisance to find a switch in the dark. Use switches with a lighted handle. Such switches have a small neon lamp in the handle, which lights up when the switch is turned off. The lamp consumes only a fraction of a watt and will last almost indefinitely. Similar switches have the handle lighted when the switch is on, serving as a pilot light for loads that are out of sight from the switch, or loads such as heaters that can easily be left on in an unoccupied room.

Dimmers. If you would like to control the brightness of the lights in part of your home, such as the dining room or an amusement room, use special dimming switches, usually called just dimmers. They are available for either incandescent or fluorescent lighting, but be sure to buy the right kind: They are not interchangeable. In any case, remove the ordinary switch and replace it with a dimmer. See Fig. 20-3.

One type of dimmer for incandescent lights is a quite inexpensive device that has only high-off-low positions, controls only up to 300 watts, and is not available in a 3-way type. Others control up to 1000 watts continuously from off to full brightness, and are available in both ordinary

Fig. 20-3 A dimmer controls the brightness of lights; it also serves as a switch. (*Pass & Seymour, Inc.*)

and 3-way types. One type of dimmer for fluorescent lighting can control up to 600 watts, from off to full bright, but requires special ballasts and is available only in the single-pole type.

Some of the less expensive types may cause radio or TV interference. Better ones have noise-eliminating devices built into them, and should so state on their cartons.

Interchangeable Devices. Sometimes there is not enough space at the desired location to install a box for more than one device. Or perhaps after a single-gang box has been installed, it becomes desirable to use two or more devices. In either case, there are available various types of devices small enough in physical size so that two or three can be installed in a single-gang box. The basic devices, such as receptacles, switches, and pilot lights, are stocked separately; typical pieces are shown in Fig. 20-4.

Fig. 20-4 Separate devices of this kind are assembled on the job into any desired combination. Three of them occupy no more space than one ordinary device. (*Pass & Seymour, Inc.*)

They are mounted on a skeleton strap or yoke (which comes with the face plate) as shown in successive steps in Fig. 20-5.

Although a single yoke with three devices can be installed in a single-

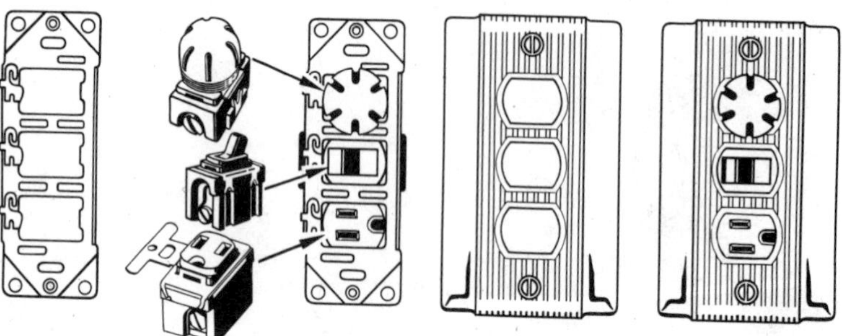

Fig. 20-5 The devices of Fig. 20-4 are assembled on a skeleton strap, as shown in the steps above. (*Pass & Seymour, Inc.*)

gang box or two yokes with six devices in a 2-gang box, this is likely to result in overcrowding of boxes (more wires than permitted in a box of any given size). So, use *deep* switch boxes, or better yet, use 4-in square boxes with raised covers, as were shown in Figs. 10-2, 10-4, and 10-5, wherever this is possible.

Note that face plates with only one or two openings are available to preserve continuity of appearance throughout an installation.

Plug-In Strip. No matter how many receptacles are provided in a home, there always seems to be need for more. At least for better homes, consider installing one or more multioutlet assemblies, defined in NEC Art. 100 as "a type of surface or flush raceway designed to hold conductors and receptacles, assembled in the field or at the factory"* and usually called "plug-in strips." One is shown in Fig. 28-17. It consists of a steel channel with a cover providing receptacles at regular intervals. Receptacle spacing varies, but for homes the 18-in interval is popular. In some plug-in strips the receptacles are wired by the installer; in others the receptacles have already been connected to continuous wires at the factory.

The material is sometimes installed on top of a baseboard, producing the effect of seeming to be part of it. In kitchens it is installed some dischtance above the counter. Connections are made to the back of the channel with conduit or cable, as to an outlet box. This material will be discussed further in Chap. 28.

Surface Wiring. Where the wiring is to be permanently exposed, as in some basements, attics, or garages, you have a choice of several methods. You may use cable or conduit with surface-type boxes and covers that were shown in Figs. 10-6 and 10-21. Where appearance is more important, you can use surface raceways, which will be described in Chap. 28.

Where it is not subject to physical damage, you can use nonmetallic-sheathed cable with special surface wiring devices, which are combinations of a nonmetallic box and a device such as a switch or receptacle. Since such devices are frequently used in providing additional outlets in buildings where appearance is not too important, such as in farm buildings, they will be described in Chap. 23.

Telephones. In ordinary residential work, too frequently no attention is paid to telephones, the problem of installation being left strictly up to the

*Reprinted with permission from NFPA 70-1984, National Electrical Code®, Copyright© 1983, National Fire Protection Association, Quincy, Massachusetts 02269. This reprinted material is not the complete and official position of the NFPA on the referenced subject, which is represented only by the standard in its entirety.

telephone people. They do a splendid job, but in many cases they have to install exposed cable. To avoid this, install EMT or ENT from a suitable, easily accessible location to all locations where telephones may be wanted. Then install a switch box for each phone. Cover it with a special face plate with a single bushed opening for the telephone cord.

Drop Cords. In making up a drop cord, tie an Underwriters knot at the top so that the weight is supported not by the copper conductors where they are connected to terminals, but by the knot. This knot is easy to make, as shown in Fig. 20-6. The drop will be supported from the ceiling

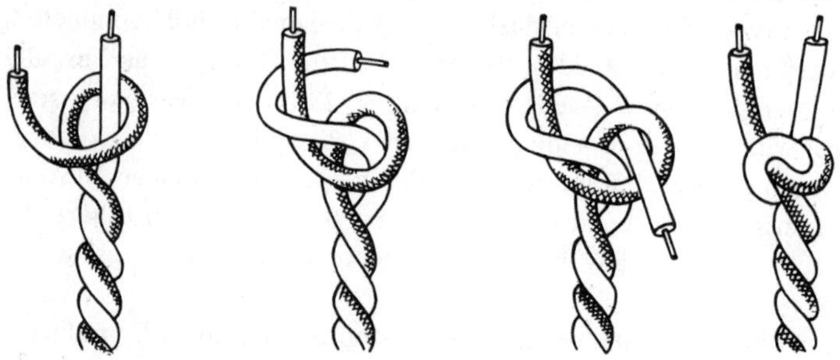

Fig. 20-6 Drop cords are not often used, but if one is installed, provide an Underwriters knot at each end.

by a blank cover with a bushed hole in the center, mounted on an outlet box. The knot goes above the bushed hole. Tape wrapped tightly around the cord to form a ball, above the hole, will also serve to support the weight. In general, the use of drop cords (pendant lights) can and should be avoided.

Garages. If the garage is attached to the house, treat it as another room of the house. Three-way switches, one at the outer door of the garage, and another at the door between the garage and the living quarters of the house, are most desirable. The Code requires at least one GFCI-protected receptacle outlet (in addition to any provided for laundry equipment) for tools and appliances such as a battery charger or a "trouble light" of the type shown in Fig. 20-7. Although not required, receptacles in a detached garage should also be GFCI-protected, for exposure to potential hazards is the same as in an attached garage.

If the garage is a separate structure some distance from the house, it can be handled in several ways. If there is only a light and nothing else installed in the garage, and if it is to be controlled only by a switch in the garage, *or* only by a switch in the house, only two wires are necessary.

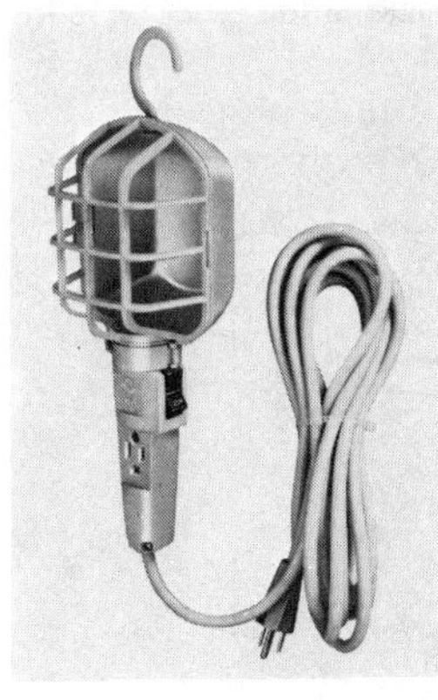

Fig. 20-7 A "trouble light" is practically a necessity in any garage. Provide at least one receptacle for it. (*General Electric Co.*)

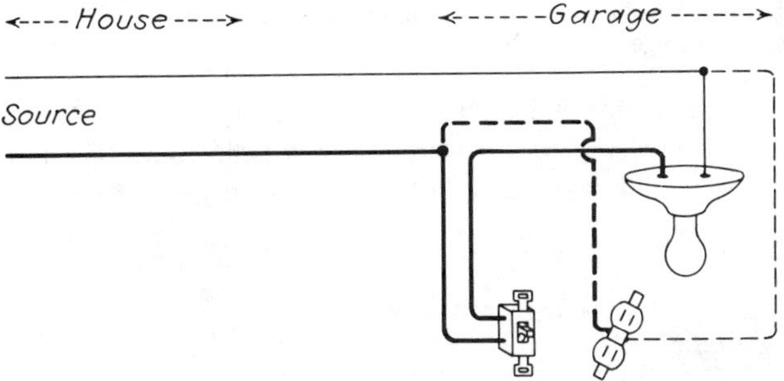

Fig. 20-8 Simple garage circuit. The receptacle is always on.

Using only two wires, as in Fig. 20-8, if a receptacle is to be installed in addition to the light, and the light is to be controlled only by a switch in the garage, the receptacle can be continuously live, but if the light is controlled only by a switch at the house, the receptacle will of course be dead whenever the light is turned off.

If there is to be no receptacle outlet in the garage but its light is to be controlled by 3-way switches at both house and garage, three wires must be run as shown in Fig. 20-9. If the wires shown in dashed lines are disregarded, this becomes identical with Fig. 4-22, the basic diagram for 3-way switches. If, however, a receptacle is installed as shown in Fig. 20-8, the receptacle will be turned off when the light is turned off by either 3-way switch. This is undesirable because, for example, frequently it is most convenient to have a charger running all night to charge your car battery, without having the light burn all night. Therefore a fourth circuit wire as shown in dashed lines in Fig. 20-9 is necessary, making the receptacle independent of the switches.

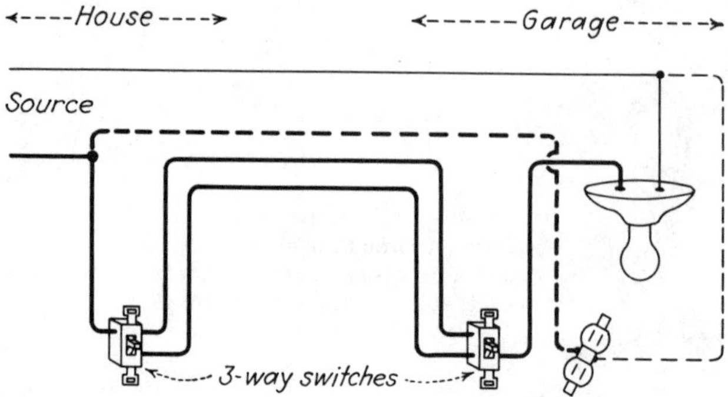

Fig. 20-9 The garage light is now controlled from either house or garage. The receptacle is always on. This requires four circuit wires between house and garage.

A pilot light at the switch in the house is also desirable. See Fig. 20-10. If the dashed lines are disregarded the circuit becomes the same as that of Fig. 20-9. To add a pilot light, run a fifth circuit conductor as shown in dashed lines.

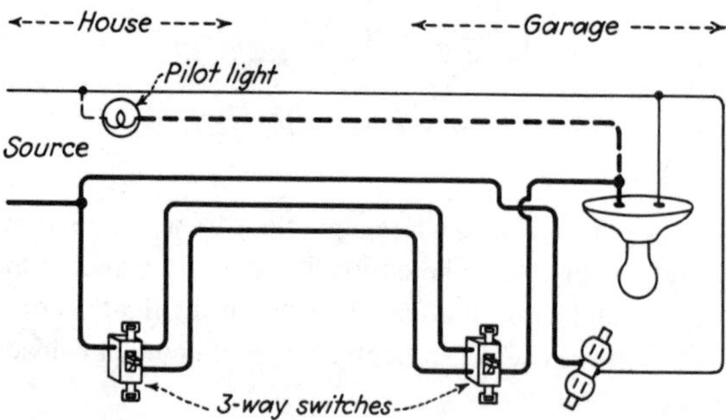

Fig. 20-10 The circuit of Fig. 20-9, with the addition of a pilot light in the house, to indicate whether the light in the garage is on or off. This requires five circuit wires.

Although the circuit shown in Fig. 20-10 is the usual one where 3-way switches are used with a pilot light at the house end, and a permanently live receptacle in the garage, there is another circuit arrangement that requires only four wires instead of five. It nevertheless meets Code require-

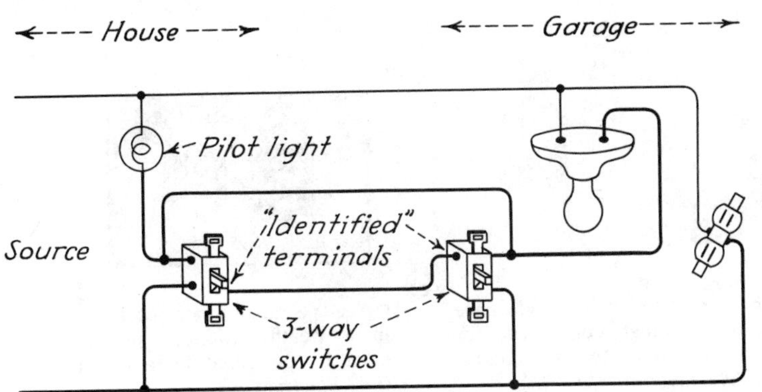

Fig. 20-11 This circuit, using only four circuit wires, serves the same purpose as the circuit of Fig. 20-10, which requires five circuit wires.

ments and therefore may be used. It is shown in Fig. 20-11. It requires more care in installation to make sure that all connections are correct.

Note that all receptacles must be of the grounding type (even if the diagrams do not show them, to keep the diagrams simple), and in all the diagrams in Figs. 20-8 to 20-11, an equipment grounding wire must be run in addition to the circuit wires in the diagrams, and be connected to the green terminals of the receptacles, and to all metallic wiring enclosures.

If the wires to the garage are to run overhead, they must be securely anchored at each end. Either of the insulators of Figs. 17-3 or 17-4 may be used. Where the wires enter or leave a building, either of the methods shown in Fig. 20-12 is suitable. A convenient fitting is shown in Fig. 20-13

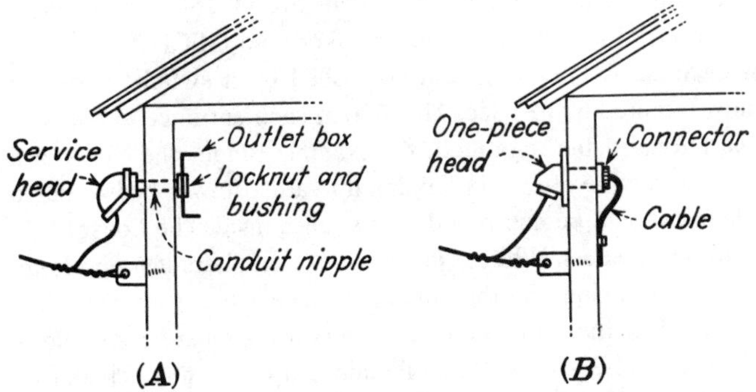

Fig. 20-12 Two ways of having wires enter the garage. The same methods can be used for other wiring, for example, farm wiring.

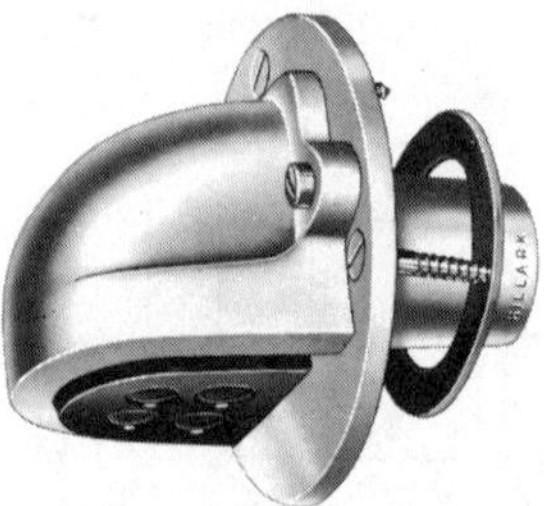

Fig. 20-13 This wall-type entrance fitting is most convenient for bringing wires into outbuildings. (*Killark Electric Mfg. Co.*)

Fig. 20-14 An outdoor receptacle in a weatherproof housing. It must be protected by a GFCI. (*Bell Electric Co.*)

and shown installed in *B* of Fig. 20-12. Be sure the insulators are installed at a point lower than the entrance of the wires into the building.

Outdoor Receptacles. One outdoor receptacle is required in every one- or two-family dwelling by NEC Sec. 210-52(d); two receptacles, one in front and one in back, would be better. In turn, NEC Sec. 210-8(a) requires that each 15- and 20-amp 120-volt outdoor receptacle directly accessible from grade level must be protected by a GFCI. Thus, a receptacle serving a grade-level patio or located on a porch reached by a few steps must be GFCI-protected; but one on the roof for convenience in servicing air-conditioning equipment or one at a second-story balcony need not be. The GFCI protection may of course be provided for the entire branch circuit or at each individual receptacle. Figure 20-14 shows a cover for an outdoor receptacle, to be installed on a cast box (such as a Type FS that will be shown in Fig. 28-2), while Fig. 20-15 shows a combination weatherproof receptacle, circuit breaker, and GFCI.

Outdoor Lighting. An outdoor light controlled by a switch inside the house is now required by NEC Sec. 210-70(a) at each entrance to a house. Optional additional lights, mounted, for example, under the eaves of a roof, provide attractive effects in lighting up yards, flower beds, and so on. Such lights should be controlled by switches inside the house.

Lighted House Numbers. While ordinary house numbers if properly installed will be well lighted by the outside light near the front door, it is unfortunately a fact that most house numbers are almost impossible to read at night. Figure 20-16 shows an illuminated house number that requires no special wiring. It operates from the transformer for your doorbell or chimes, and is always on; it goes out only while the push button is pushed; it consumes very little power.

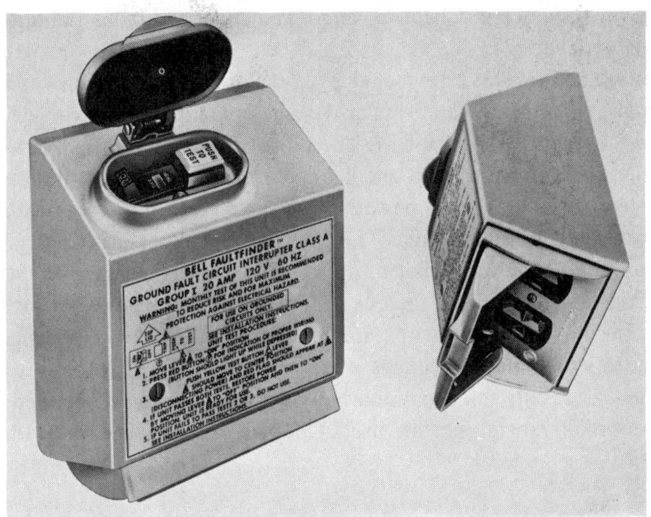

Fig. 20-15 The receptacle of Fig. 20-14, with a built-in GFCI. (*Bell Electric Co.*)

Outdoor Switches. If a switch must be installed exposed to the weather, use a cast Type FS box (Fig. 28-2). Install an ordinary toggle switch in it, with a weatherproof cover as shown in Fig. 20-17. Outdoor wiring for

Fig. 20-16 Your friends will appreciate a lighted house number. (*General Electric Co.*)

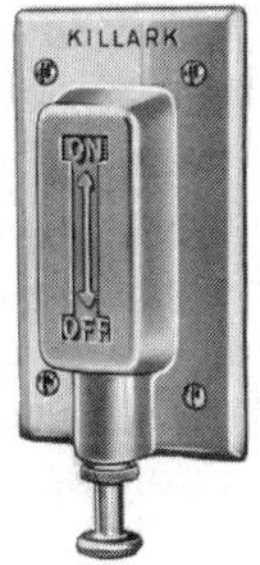

Fig. 20-17 Install an ordinary toggle switch in a cast weatherproof box, and use this waterproof cover to control it. (*Killark Electric Mfg. Co.*)

farms will be discussed in Chap. 23, and commercial and industrial outdoor wiring in Chap. 28.

Low-Voltage Wiring. This term includes wiring for doorbells, chimes and other signals, thermostats, and similar devices operating at low voltage. Usually this means 30 volts or less. The power for operating such low-voltage circuits is derived from small transformers. Under no circumstances may low-voltage wires be run in the same conduit or cable with other wires carrying full voltage. They must never enter an outlet or switch box containing full-voltage wires unless a metal barrier of the same thickness as the walls of the box separates the two types of wiring, or unless the power supply wires are introduced solely for energizing equipment to which the low-voltage wires are connected.

Transformers. If only the usual doorbell or chimes are to be operated, ordinary doorbell transformers are used. One type is shown in Fig. 20-18.

Fig. 20-18 A transformer for operating doorbells or chimes. For bells, select one delivering about 8 volts. For chimes, select one delivering about 18 volts. (*General Electric Co.*)

Mount the transformer near an outlet box. Connect the flexible (primary) leads to the 120-volt wires in the box. Similar transformers are available that are mounted on an outlet box cover, which is installed directly on the box. The screw terminals on the transformer deliver the secondary or low-voltage output of the transformer. Most such transformers have a maximum capacity of about 5 watts, and usually deliver from 6 to 10 volts if intended for doorbells, but more nearly 20 volts if intended for chimes. Some provide a choice of voltage, such as 6, 12, or 18 volts; others are available in larger wattage ratings. Such transformers consume a very insignificant amount of power from the 120-volt line, except while they are operating a bell or chimes.

Transformers of this type are so designed that, even if the secondary is short-circuited, the power flowing will be limited to the rating of the transformer. Listed Class 2 transformers are rated at not over 100 VA, and the type used for residential wiring is usually from 5 to 10 VA. Because of this limited current there is little danger of fire, and because of the low voltage there is no danger of shock. Therefore the Code requires only that the wires have insulation that is suitable. Therefore inexpensive single-wire or 2- or 3-wire cables insulated only with a thin layer of plastic are used for bells, chimes, and thermostats if both the voltage and the current are limited to 100 VA.[1] The usual size of the cable is No. 18, though sizes may range from No. 16 to No. 22. The size must be chosen so as to be suitable for the load, the length of the run, and the voltage available. In use the cable is merely fished through walls without further protection; if run on the surface, it is fastened with insulated staples.

Low-Voltage Circuits. Circuits for low-voltage bells, chimes, and similar items are very simple. Merely consider the secondary of the transformer as the SOURCE for the circuit, and consider the push buttons as switches, which they are. The basic circuit is shown in Fig. 20-19, covering the installation of one bell. Push buttons can be installed in as many locations as desired, as shown by the dashed lines.

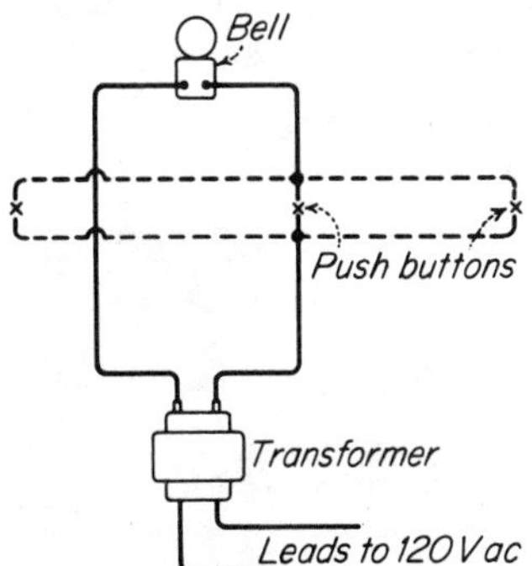

Fig. 20-19 The basic diagram for a doorbell is very simple.

Figure 20-20 shows the diagram for a combination bell and buzzer, which is a device with three terminals, one of them connected directly to the frame of the device, and not insulated from it. This is the terminal from which a wire must be run to the transformer; the wires from the other terminals run to the push buttons and then to the other terminal of the transformer.

[1] For some types of low-voltage (30 volts or less) circuits where the power is limited to not over 1000 VA, overcurrent protection must be provided and cables approved for the purpose must be used, as explained in NEC Sec. 725-11, but such circuits are a rarity in ordinary residential and farm wiring.

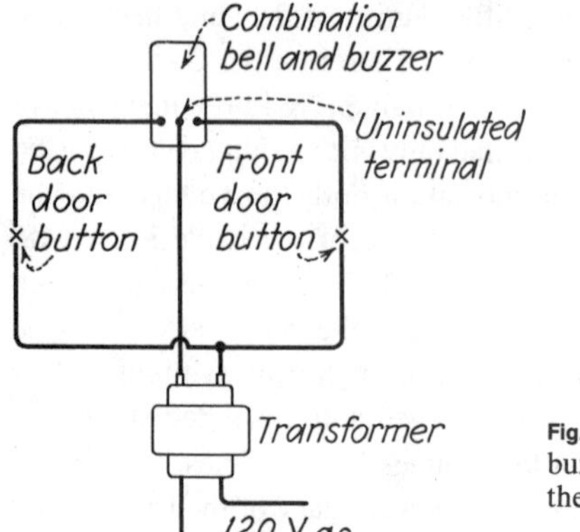

Fig. 20-20 A combination bell and buzzer has been substituted for the bell in Fig. 20-19.

In new construction, chimes rather than bells are used just about all the time. Several styles are shown in Fig. 20-21. Connect them as just described for bells. But remember that most chimes require a transformer delivering from 15 to 20 volts, as compared with 6 to 8 volts for bells. If you are replacing a bell with chimes, you will have to change the transformer for satisfactory operation.

Fig. 20-21 Chimes with musical notes have replaced bells in new construction.

Low-Voltage Switching. If a light is to be controlled from three or more points, the wiring becomes expensive because 3- and 4-way switches, long runs of cable or conduit, and very much labor are involved. Yet control from many points is often desirable, and the low-voltage switching system makes it possible to control a light from any number of points at low cost. A motor-operated master switch which will turn all the lights in the house on or off with the touch of one switch can be installed in the bedroom.

In this system the 120-volt wires end at the outlet boxes for fixtures that are to be switched; they are *not* run to switch locations. An electrically operated switch called a relay is installed *in the outlet box*[2] on which the fixture to be controlled is installed. A relay is shown in Fig. 20-22; it

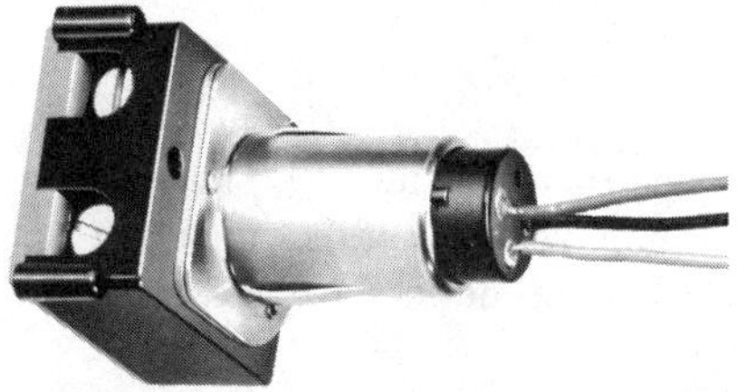

Fig. 20-22 The remote-control relay is an electrically operated switch. It is installed in the outlet box on which the fixture is mounted. (*General Electric Co.*)

will control a load up to 20 amp at any voltage up to 277 volts. This is a latching relay which remains in either the on or the off position in response to the momentary closing of either the on or the off contacts in any of the switches connected to it. From each relay, three small wires are run to low-voltage switches located in as many places as you wish.

The power for operating all the relays comes from a single transformer (Fig. 20-23) installed in the basement or some other convenient location. It steps the voltage down from 120 volts to about 24 volts. Because the voltage is so low, there is no danger of shock; even if the wires on the secondary side of the transformer are short-circuited, the transformer delivers so little power that there is little danger of fire.

[2] The relay does not necessarily have to be installed in that outlet box; it could be installed near the load, or several could be installed in a centrally located cabinet. As a practical matter, however, in residential work they are usually installed in the outlet box involved.

Fig. 20-23 This transformer delivers about 24 volts. One transformer furnishes the power for all the relays in an installation. (*General Electric Co.*)

The relay is made so that its round shank fits into a ½-in knockout. Push the relay through a knockout from the inside of the outlet box. The 120-volt terminals on the relay are then inside the box; the three low-voltage leads, on the outside.

Figure 20-24 shows one of the switches generally used. It resembles an ordinary toggle switch but requires special face plates to fit. It is a single-

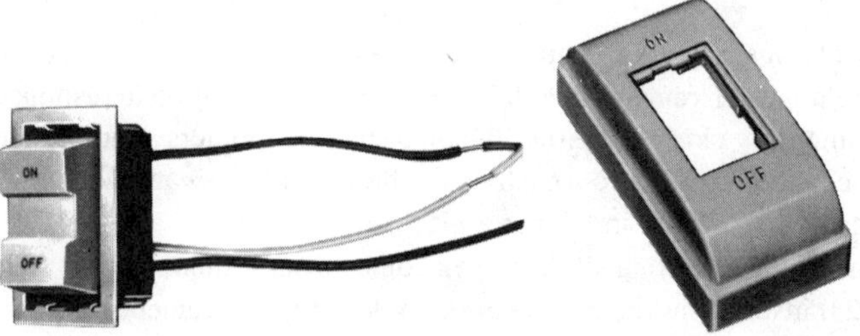

Fig. 20-24 The switch used in remote-control switching systems is equivalent to two push buttons. If it is to be surface-mounted, use the cover shown. (*General Electric Co.*)

pole, double-throw momentary-contact switch. Such switches have a neutral position for the handle. Push the top end momentarily to turn a light on; the handle then returns to the neutral position when released. Push the lower end momentarily to turn the light off; again the handle returns to the neutral position when you remove your finger. Because the voltage involved is only 24 volts, supplied by a special transformer with little total power, the switches need not be installed in switch boxes, but they usually are for the sake of neatness. If you want to install them without boxes, mounting frames are available. Some will fit the mounting strap for interchangeable devices shown in Fig. 20-5.

Because of the low voltage and limited power, wires do not need to be the relatively expensive kind used in 120-volt wiring, nor do they need to run inside a raceway. Almost any kind of wire No. 20 or larger may be used. Two kinds of cable used for the purpose are shown in Fig. 20-25,

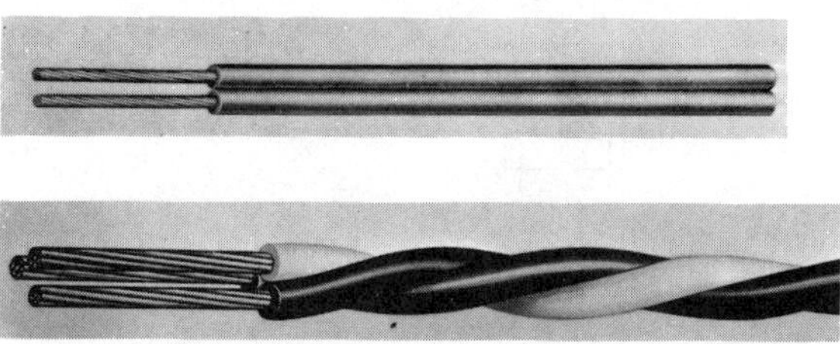

Fig. 20-25 Wire used to connect relays to switches does not need much insulation because of the low voltage. The special cable shown is convenient. (*General Electric Co.*)

available in 2-, 3-, or 4-conductor types. The 4-wire type is used when a pilot light is needed. Run the cable any way you find convenient. Staple it to the surface over which it runs; in old work fish it through walls or run it behind baseboards, or run it exposed. Just be sure that it is not exposed to physical damage.

Figure 20-26 shows an installation of one light controlled by any of four switches. Install as many switches as you wish.

Merely connect the white wires of all the switches together, all the red wires together, and all the black wires together. The Code does not require any particular color scheme for these *low-voltage* wires; the colors shown in Fig. 20-26 are those used by one manufacturer of this equipment. If you wish to control two or more outlets at the same time from any switch, merely connect the several outlets together with the usual 120-volt conductors just as if ordinary wiring were being used and install the relay in the most convenient box.

Numerous diagrams for wiring many different combinations are usually supplied by the manufacturer of the relays and related equipment.

Installation Hints. In new work, two methods of installation are possible. In the first method, all the work is done before the lath and plaster or

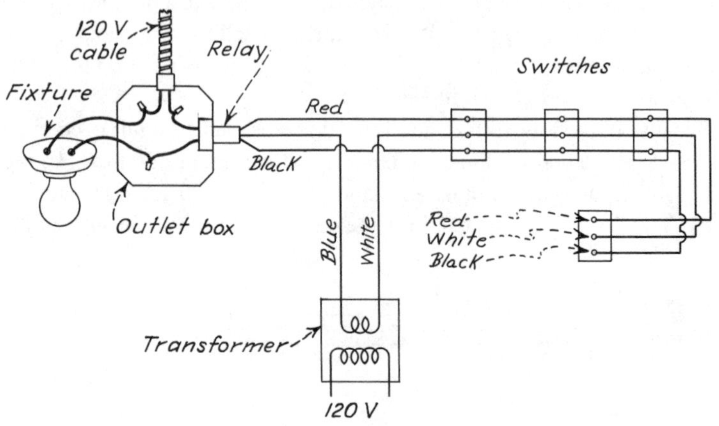

Fig. 20-26 One outlet controlled by any of four switches.

the wallboard are installed. The relays are installed in the boxes, and the low-voltage wires from the switches and transformer are connected to the relays. In that case, leave at least 6 or 8 in of slack in the low-voltage wires at the relay, so that if at a future date a relay should prove defective, it can be removed from the inside of the box, which would be impossible if you did not provide the slack. The alternate method is to let from 6 to 8 in of the low-voltage wires project into the outlet box through the knockout in which the relay will later be installed. At that time, connect the low-voltage wires to the proper leads on the relay, push them out through the knockout, and push the relay into the knockout.

Use of a low-voltage remote-control switching system will make the electrical system much more flexible than when using ordinary switches to control outlets. A switch can be installed in an additional location as an afterthought, at small cost. Using ordinary full-voltage 3- and 4-way switches for basement or garage lights, you never know whether the light is on or off, unless you can see the light, or use pilot lights. With a remote-control system, simply push the off button, and if the light was on, it is now off; if it already was off, nothing happens. Use this system to avoid long runs of expensive 120-volt wiring.

Three-Wire Circuits. Any building that has a 3-wire 120/240-volt service can have 3-wire circuits, which reduce voltage drop. Let us analyze what a 3-wire circuit is. See Fig. 20-27, which shows an ordinary 2-wire 120-

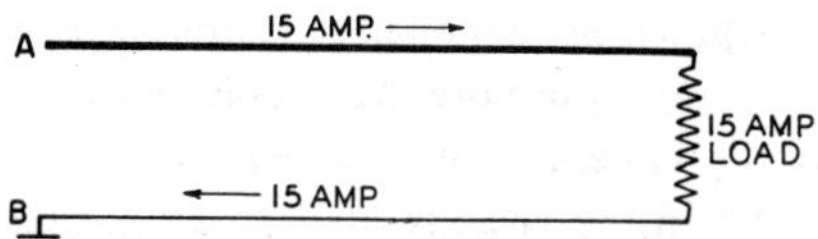

Fig. 20-27 One 2-wire circuit carrying 15 amp. One wire is grounded at the service equipment.

volt circuit. Assume that it is wired with No. 14 wire and is 50 ft long, which means that the current flows through 100 ft of wire. Assume the load is 15 amp, the same as the ampacity of the wire. The voltage drop is then about 3.86 volts, or about 3.2%.

Now see Fig. 20-28, which shows two such circuits, one on each leg of the 3-wire service. The voltage drop on each circuit will still be 3.86 volts. Note, however, that the two grounded wires are *connected to each other* at the service equipment, and that they run parallel to each other. Then, why use *two* grounded wires? Why not use just one wire as in Fig. 20-29? You might easily jump to the conclusion that one grounded wire serving two circuits would have to be twice as big as before to carry 2 × 15, or 30, amp. That is a wrong conclusion. In Fig. 20-28, each of the grounded wires *B* and *C* does indeed carry 15 amp, but note the direction of the arrows in the illustration. The flow of current in wire *B* at any instant is in a direction opposite to that in *C*. Therefore in Fig. 20-29, at any given instant, the single wire *BC* can be said to carry 15 amp in one direction and also 15 amp in the opposite direction, so the two cancel each other. In other words wire *BC* has now become a neutral conductor.

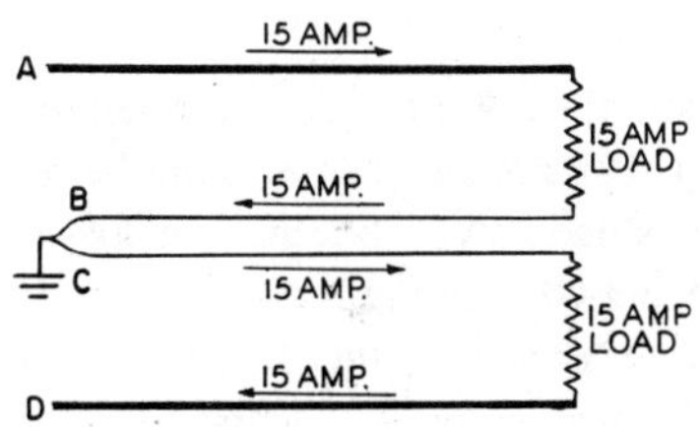

Fig. 20-28 Two 2-wire 120-volt circuits, each carrying 15 amp, and fed by opposite legs of a 120/240-volt 3-wire service. Wires *B* and *C* are grounded at the service equipment.

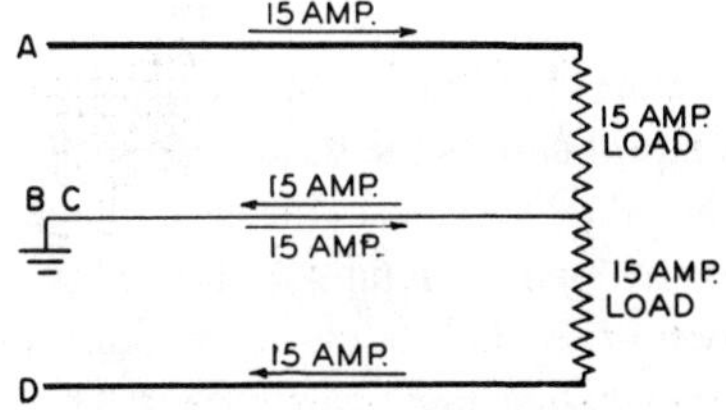

Fig. 20-29 If the two circuits of Fig. 20-28 run to the same location, you can use only one grounded wire *BC* to make a 3-wire circuit. Here each half of the 3-wire circuit carries 15 amp. The wire *BC* has become a neutral wire.

If the currents in the two circuits are precisely identical, the circuit would work just as well if wire *BC* were missing. But note that this is true only if the current in wire *A* is exactly the same as that in wire *D*.

But what about the voltage drop? In the circuits of Figs. 20-27 and 20-28, the voltage drop is 3.86 volts, based on 15 amp flowing through 50 ft of wire *A* plus 50 ft of wire *B*, a total of 100 ft. In Fig. 20-29, however, the current in each circuit flows through only 50 ft of wire, for wire *BC* carries no current. Therefore the voltage drop in the circuit involving wires *A* and *BC* is only half as great, or 1.93 volts. In the entire 3-wire circuit there is only 150 ft of wire as compared with 200 ft in two separate 2-wire circuits. Therefore by using one 3-wire circuit we have saved 25% of the wire and halved the voltage drop.

If the two halves of a 3-wire circuit are not equally loaded, as, for example, in Fig. 20-30, there is still an advantage. Suppose, as shown in that

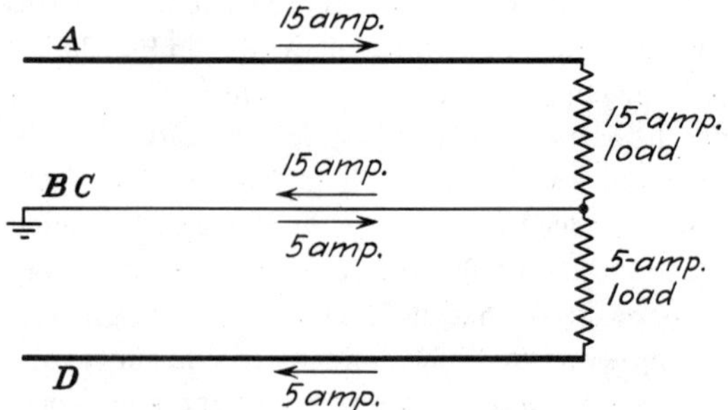

Fig. 20-30 The 3-wire circuit of Fig. 20-29, one half carrying 15 amp and the other half 5 amp. The neutral *BC* carries 10 amp.

picture, half the circuit carries 15 amp and the other only 5 amp; then the neutral carries the difference or 10 amp. The voltage drop will not be reduced by 50%, as in the case of equally loaded halves, but the total losses in the 3-wire circuit will still be less than in two separate 2-wire circuits. If one of the halves carries no current at all, then the other half functions exactly like any 2-wire circuit. This is also what happens if a fuse blows in one of the two hot wires; what is left is an ordinary 2-wire circuit. Therefore the neutral wire in a 3-wire branch circuit must be the same size as each of the two hot wires.

While Figs. 20-29 and 20-30 show 3-wire circuits with a single load at the end of each line, 3-wire circuits are not limited to such applications.

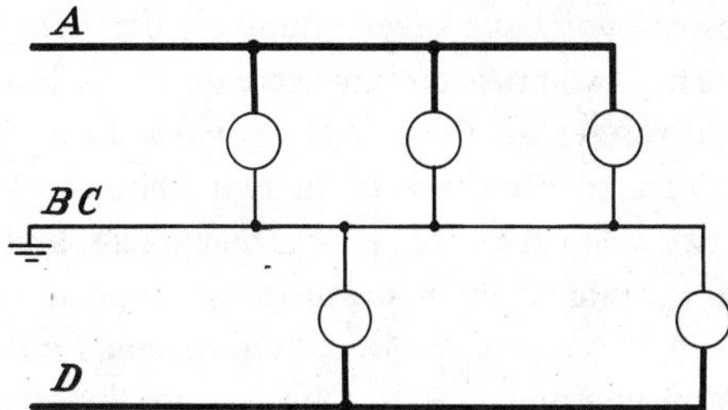

Fig. 20-31 Three-wire circuits need not carry single loads. Install loads where required.

Figure 20-31 shows a 3-wire branch circuit with loads connected at various points. No matter how these loads are spaced, the total losses in a 3-wire[3] circuit are always lower than in two separate 2-wire circuits.

Note that if a combination GFCI/breaker is installed on a 3-wire circuit, it must be of a type designed specifically for a 3-wire circuit.

Application of Three-Wire Circuits. In ordinary wiring (whether house, barn, or other building), two separate 2-wire 120-volt circuits are often required to supply two loads that are near each other. In such cases it is often more economical to install one 3-wire circuit. Where a 3-wire circuit is installed using cable, the cable will have one white wire for the grounded neutral, and one red and one black for the hot wires. When using conduit, choose the same colors. Having one hot wire black and the other red makes it easy to know which leg, or half of the circuit, you are working on. Be sure that the white wire runs to each outlet, for if you were to connect a receptacle to the red and black wires, the outlet would operate at 240 instead of 120 volts. Be sure to divide the outlets on the circuit as evenly as possible between the two legs. Outlets for two known heavy or long-time loads should be on opposite legs of the circuit where practical.

Neutrals in Three-Wire Branch Circuits. The Code in Sec. 300-13(b) states that the continuity of the grounded neutral conductor of a multiwire circuit "shall not be dependent upon device connections, such as lampholders, receptacles, etc., where the removal of such devices would interrupt the continuity"* of the conductor. This means that the neutral

*Reprinted with permission from NFPA 70-1984, National Electrical Code®, Copyright© 1983, National Fire Protection Association, Quincy, Massachusetts 02269. This reprinted material is not the complete and official position of the NFPA on the referenced subject, which is represented only by the standard in its entirety.

[3]A 3-wire branch circuit is called a "multiwire circuit" in NEC Art. 100. A 3-phase multiwire circuit would have three hot wires plus a grounded neutral.

must be specially watched. Assume you have several duplex receptacles on a 3-wire circuit. A receptacle has two brass terminal screws for the hot wires, and two whitish terminal screws for the grounded wires. In a *2-wire* circuit it is quite all right to connect the ends of the two white wires of two lengths of cable to the two whitish terminals of a receptacle, and then run the cable on to another receptacle where the process is repeated, then on to a third receptacle, and so on. *In a 3-wire circuit this must not be done*. If using wires in conduit you may use the loop in a continuous length of wire, as was shown in *B* and *C* of Fig. 18-6, but where using cable there is no way of doing this. You must follow the procedure that was shown in Fig. 8-4. Connect the ends of the white wires of two lengths of cable to each other *and* to a short length of white wire, and then connect the other end of that short wire to the whitish terminal of the receptacle. Unless this precaution is observed, removing a receptacle for replacement would (while there is no connection to the receptacle) result in a break in the neutral wire. In turn this would place all the receptacles connected to one leg (beyond the receptacle that is temporarily removed) in series with all the receptacles on the other legs, *all at 240 volts*. If for example you have a 1000-watt appliance plugged into a receptacle on one leg, and a 100-watt appliance on the other leg, the 1000-watt appliance would have a voltage of about 20 volts across it, and the 100-watt appliance about 220 volts. Regardless of what you plugged in, everything plugged into receptacles on one leg would operate at more than 120 volts, and everything on the other leg at less than 120 volts, depending on the total wattage plugged into the receptacles of one leg as compared with the wattage plugged into receptacles on the other leg. Anything subjected to more than 120 volts would be quickly damaged. Therefore, be most careful to connect the white neutral so that temporary removal of a receptacle does not interrupt the neutral to the other receptacles.

Safety tip: When working on an existing circuit, be sure *all* the branch-circuit disconnecting means (circuit breakers or fused switches at the panelboard) are *open*. In Chapter 9 you were told that there is no hazard in touching the grounded wire, but on a 3-wire circuit if only one of the hot legs is disconnected and you open the neutral wire which is part of the circuit supplying load on the other hot leg, you can get severely shocked because the neutral is a current-carrying wire and that current could flow through you if you become part of the neutral circuit with one side of the circuit energized.

"Safety" Receptacles. Children have a normal tendency to explore. What is more natural than for a child to push part of a toy, a loose hairpin, or any other loose metal part into one of the slots of a receptacle? If the child is on a completely dry floor, probably no harm will be done. But if the child is also touching a grounded object such as a radiator, a water pipe, or the framework of a defective lamp, he or she may be subject to a violent shock. Every year the deaths of many children can be traced to this cause.

For that reason "safety" receptacles were developed. One type has a spring-loaded cover over the slots of the receptacle; unless the receptacle is in use with a plug in it, the slots of the receptacle are concealed. Another type is designed so that inserting anything into just *one* of the two slots will not make electrical contact; both prongs of a plug must be inserted before contact is made. Using safety receptacles is sensible where small children are members of the household.

Smoke Detectors. Self-contained smoke detectors which will sound an alarm when visible or invisible products of combustion are sensed have been credited with saving many lives in dwellings by waking up the occupants early enough for them to escape from a fire. UL lists two basic types of smoke detectors.

1. The ionization detector, in which a radioactive source emits particles that ionize the air between two electrodes, resulting in a flow of current. Smoke particles entering the chamber combine with the ionized air, slow down their action, and reduce the current flowing between the electrodes, initiating the alarm signal.

2. The photoelectric detector, which has a light source and a photocell so arranged in a chamber that normally no light falls on the photocell. When smoke particles enter the chamber, however, they scatter or reflect part of the light to the photocell, initiating the alarm signal.

Units combining both ionization and photoelectric detection are also Listed.

Most building codes require a smoke detector in the hallway outside bedrooms, or above the stairway leading to bedrooms on an upper floor. Plug-in and direct-wired models, as shown in Fig. 20-32, are available, and the ionization type is also available battery-operated. The required location will be on the ceiling, or on the side wall not more than 12 in from the ceiling. Be sure not to place the detector in the path of ventilation that will move air past the detector faster than in other parts of the room. If

Fig. 20-32 Direct-wired 120-volt ionization-type residential smoke detector, with monitor light to provide a visual check that the unit is "on." (*Square D Co.*)

the detector is to be direct-wired, choose a circuit with often-used lights on it, such as the bathroom light, to ensure that the circuit cannot be dead without it being noticed; but wire the detector directly across the two circuit wires, unswitched.

Swimming Pools. The Code covers this subject in Art. 680. For portable above-ground pools, the motor of the portable filter pump must have double insulation as described in Chap. 9, and *in addition* must have all its internal, nonaccessible non-current-carrying metal parts *grounded,* using a 3-wire cord with a 3-prong plug. The circuit feeding the pump must be protected by a GFCI, also described in Chap. 9. No receptacle may be installed within 10 ft of the pool. Beyond that, use good common sense. Install lighting so that it, and especially the switches, cannot possibly be touched by anybody in the pool, and so located that the wires if damaged couldn't possibly fall into the pool.

For permanently installed pools, the Code requirements are rigid and severe, as they should be, for nowhere is an individual more subject to shock, injury, or even death than when in a pool in which the electrical equipment was improperly installed. You should study NEC Art. 680 and not attempt an installation until thoroughly familiar with all the Code requirements; work with others familiar with the requirements, before attempting your own installation.

Television Antennas. A television antenna mast or other metal support is required by NEC Sec. 810-15 to be grounded, and the lead-in conductors must also be provided with antenna discharge units, in accordance with Sec. 810-20.

Many fatal accidents have occurred when antennas being erected or moved have contacted overhead high-voltage lines. Before erecting an antenna look around to be certain that there is adequate clearance from overhead lines. In most states it is illegal to bring any conducting object within 10 ft of an energized high-voltage overhead line.

21
Wiring for Special Appliances

The Code does not define a "special appliance." As used in this chapter, that term will generally mean an appliance of relatively high ampere rating that requires a branch circuit of its own. The 1975 and earlier Codes classified appliances into three groups: portable, stationary, and fixed. Since all portable, most stationary, and some fixed appliances could be connected by means of a cord, plug, and receptacle, the classifications were vague in identifying the type of electrical connection required, and the terms were not used consistently throughout the Code. The Code now uses two classifications for the *wiring* of appliances: permanently connected and cord- and plug-connected. Classifications used to indicate the *location* of the appliance are (1) fastened in place and (2) located to be on a specific circuit.

Cord- and Plug-Connected. This group includes appliances which in use can be and normally are moved about: toasters, irons, coffee makers, vacuum cleaners, and similar appliances. Each is equipped with a cord and plug.

Permanently Connected. This group includes all appliances not cord- and plug-connected.

Fastened in Place. This group comprises appliances that once installed cannot readily be moved because they have plumbing connections, are built into the structure of the building, or are otherwise permanently secured in place. Examples are water heaters, garbage disposers, furnace motors, central air conditioners, and wall-mounted ovens and counter-mounted cooking units. Some of these might occasionally be installed by using a cord and plug for convenience in servicing.

Located to Be on a Specific Circuit. This group comprises appliances that once installed are left in their original location but can nevertheless be moved fairly easily. Self-contained ranges, window air conditioners, and clothes dryers are good examples. They are connected by using a cord and plug, but if the owner moves from one location to another, he or she can move them with other household goods.

Receptacle Ratings. Receptacles all carry voltage ratings, indicating the maximum voltage at which they may be used: 125, 250, or 125/250 volts for appliances such as ranges. For nonresidential use, others are rated at 277, 480, and 600 volts. Each also carries a maximum ampere rating: 15, 20, 30, 50, or 60 amp.

The illustrations of Figs. 21-1 and 21-2 show both the small and some larger types of receptacles and their most usual applications. Note that they are identified as 2-pole 2-wire and also 2-pole 3-wire; as 3-pole 3-wire and also 3-pole 4-wire; and so on. The number of *poles* indicates the number of circuit conductors carrying the current in normal use. If the number of *wires* is one greater than the number of *poles,* it means that the receptacle has an extra opening, and the plug an extra prong, for connection of an equipment grounding conductor, provided only for safety, and never carrying current under normal conditions. Thus a 2-pole 2-wire receptacle is used for an appliance with two current-carrying wires but not an equipment grounding wire; the 2-pole 3-wire is used for an appliance having two current-carrying wires plus an equipment grounding wire.

The small figures on the side of each receptacle in the illustration show the diameter of the receptacle in inches. On each receptacle the opening marked "G" is for the equipment grounding wire; the opening marked "W" is for the white grounded wire of the circuit; those marked "X," "Y," or "Z" are for other current-carrying wires.

Note that a 2-*pole* 125-volt receptacle may be used only for appliances operating at 125 volts or less; a similar 2-*pole* receptacle but rated at 250 volts may be used only to supply a load operating at strictly 240 volts; a 3-*pole* 125/250-volt receptacle may be used only for loads operating partially or at various times at either 120 or 240 volts (an electric range is a typical example). (Of course, any of the receptacles mentioned may have an additional opening for the equipment grounding wire.)

The illustrations do not show all available configurations, only the more common types. The last column shows the NEMA (National Elec-

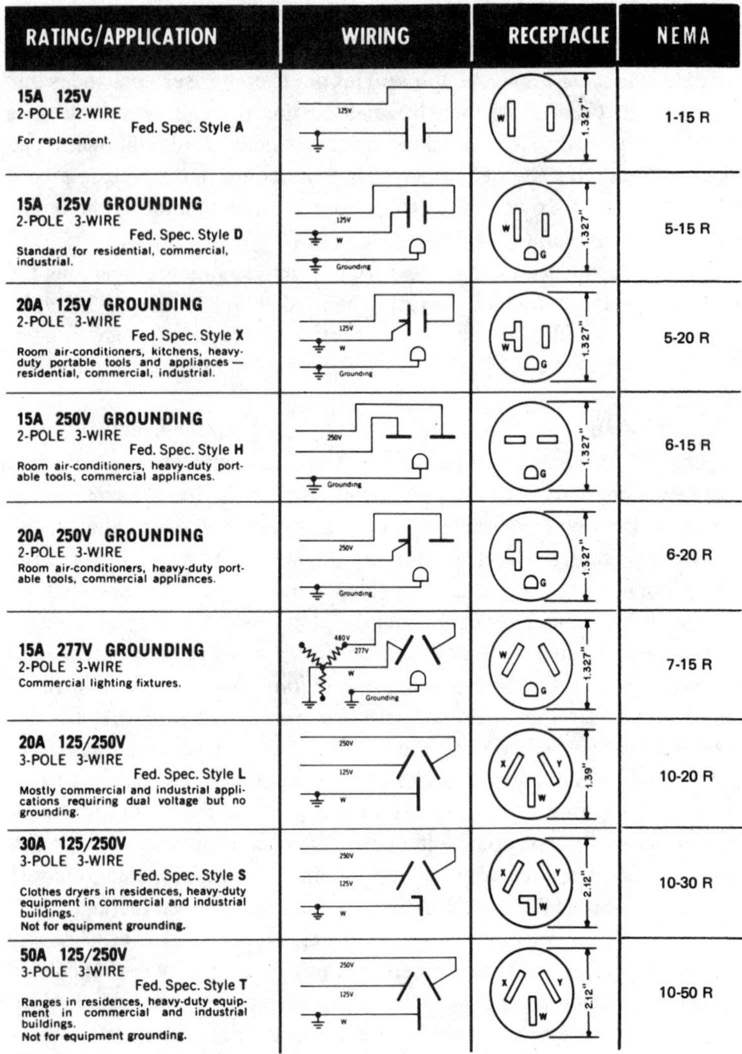

RATING/APPLICATION	WIRING	RECEPTACLE	NEMA
15A 125V 2-POLE 2-WIRE Fed. Spec. Style **A** For replacement.	125V	W; 1.327"	1-15 R
15A 125V GROUNDING 2-POLE 3-WIRE Fed. Spec. Style **D** Standard for residential, commercial, industrial.	125V; W; Grounding	W; G; 1.327"	5-15 R
20A 125V GROUNDING 2-POLE 3-WIRE Fed. Spec. Style **X** Room air-conditioners, kitchens, heavy-duty portable tools and appliances — residential, commercial, industrial.	125V; W; Grounding	W; G; 1.327"	5-20 R
15A 250V GROUNDING 2-POLE 3-WIRE Fed. Spec. Style **H** Room air-conditioners, heavy-duty portable tools, commercial appliances.	250V; Grounding	G; 1.327"	6-15 R
20A 250V GROUNDING 2-POLE 3-WIRE Room air-conditioners, heavy-duty portable tools, commercial appliances.	250V; Grounding	G; 1.327"	6-20 R
15A 277V GROUNDING 2-POLE 3-WIRE Commercial lighting fixtures.	480V; 277V; W; Grounding	W; G; 1.327"	7-15 R
20A 125/250V 3-POLE 3-WIRE Fed. Spec. Style **L** Mostly commercial and industrial applications requiring dual voltage but no grounding.	250V; 125V; W	X; Y; W; 1.39"	10-20 R
30A 125/250V 3-POLE 3-WIRE Fed. Spec. Style **S** Clothes dryers in residences, heavy-duty equipment in commercial and industrial buildings. **Not for equipment grounding.**	250V; 125V; W	X; Y; W; 2.12"	10-30 R
50A 125/250V 3-POLE 3-WIRE Fed. Spec. Style **T** Ranges in residences, heavy-duty equipment in commercial and industrial buildings. **Not for equipment grounding.**	250V; 125V; W	X; Y; W; 2.12"	10-50 R

All spec. styles shown refer to Federal Specification W-C-596a.

Fig. 21-1 This shows receptacle configurations for many different ampere and voltage ratings. A plug that fits one receptacle will not fit another. (*General Electric Co.*)

trical Manufacturers Association) number for each configuration, which can be used as a universal identification independent of catalog numbers. For corresponding plugs, the suffix "R" is simply changed to "P." Re-

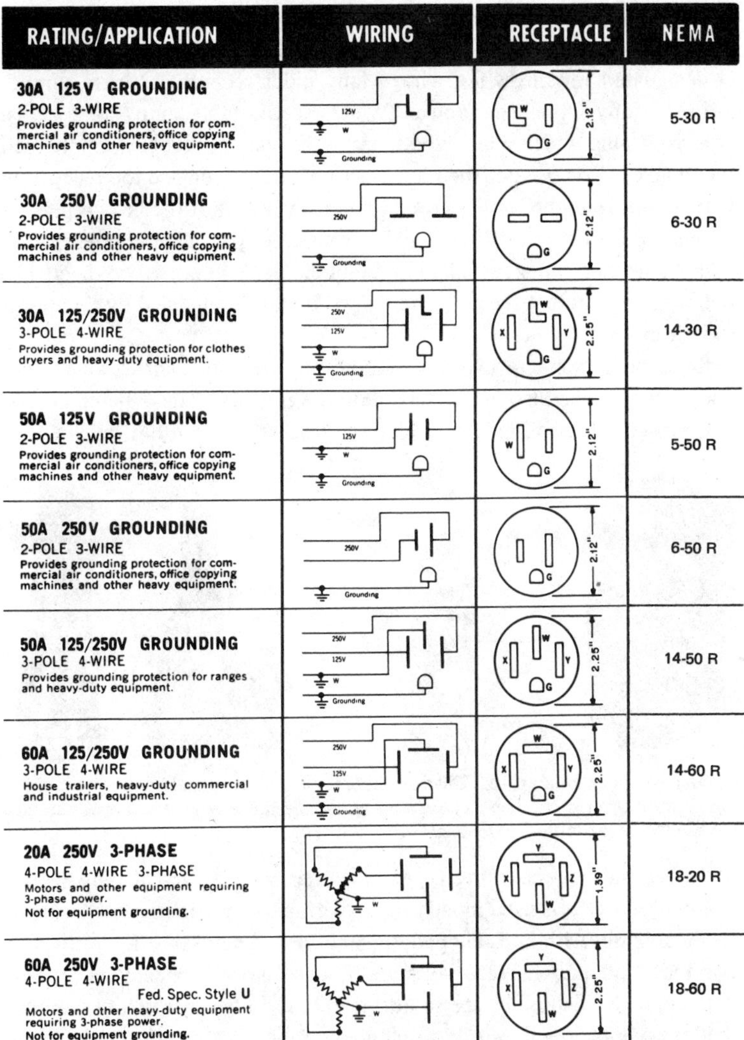

RATING/APPLICATION	WIRING	RECEPTACLE	NEMA
30A 125V GROUNDING 2-POLE 3-WIRE Provides grounding protection for commercial air conditioners, office copying machines and other heavy equipment.	125V W Grounding	W, G 2.12"	5-30 R
30A 250V GROUNDING 2-POLE 3-WIRE Provides grounding protection for commercial air conditioners, office copying machines and other heavy equipment.	250V Grounding	G 2.12"	6-30 R
30A 125/250V GROUNDING 3-POLE 4-WIRE Provides grounding protection for clothes dryers and heavy-duty equipment.	250V 125V W Grounding	W, X, Y, G 2.25"	14-30 R
50A 125V GROUNDING 2-POLE 3-WIRE Provides grounding protection for commercial air conditioners, office copying machines and other heavy equipment.	125V W Grounding	W, G 2.12"	5-50 R
50A 250V GROUNDING 2-POLE 3-WIRE Provides grounding protection for commercial air conditioners, office copying machines and other heavy equipment.	250V Grounding	G 2.12"	6-50 R
50A 125/250V GROUNDING 3-POLE 4-WIRE Provides grounding protection for ranges and heavy-duty equipment.	250V 125V W Grounding	W, X, Y, G 2.25"	14-50 R
60A 125/250V GROUNDING 3-POLE 4-WIRE House trailers, heavy-duty commercial and industrial equipment.	250V 125V W Grounding	W, X, Y, G 2.25"	14-60 R
20A 250V 3-PHASE 4-POLE 4-WIRE 3-PHASE Motors and other equipment requiring 3-phase power. Not for equipment grounding.	W	Y, X, Z, W 1.39"	18-20 R
60A 250V 3-PHASE 4-POLE 4-WIRE Fed. Spec. Style **U** Motors and other heavy-duty equipment requiring 3-phase power. Not for equipment grounding.	W	Y, X, Z, W 2.25"	18-60 R

Fig. 21-2 Similar to Fig. 21-1, except that still more types are presented. Note the application data at the left of each receptacle and the wiring diagram. (*General Electric Co.*)

ceptacles used for higher voltages are not shown. Besides the types shown, there are others so designed that a plug can be locked into the receptacle by twisting clockwise after insertion, and cannot be removed without first twisting counterclockwise to unlock.

Note the difference in the lengths of the two openings in the face of the NEMA 1-15R receptacle. This is a "polarized" receptacle: the wider slot is designated for the white wire. Many older receptacles were nonpolarized. Many appliances such as TV sets and portable lamps come with a polarized plug which cannot be inserted in one of the old nonpolarized receptacles. *Do not* alter the plug to make it fit but replace the receptacle with a polarized one, or if a grounding means exists at the box, replace it with a 2-pole 3-wire (NEMA 5-15R) grounding type. Proper polarization will ensure that the screw shell of a lampholder will not be connected to the hot wire and that the hot wire will be the one controlled by the on-off switch in your TV set.

Receptacles come in a variety of mounting methods to fit various boxes and face plates, also in surface-mounting types. The 50-amp 3-wire 125/250-volt type in Fig. 21-1 is also shown pictorially in Fig. 21-3 in both

Fig. 21-3 Typical 50-amp 125/250-volt receptacles, surface and flush types, and a pigtail cord connector. This is a typical receptacle for a range when grounding to the neutral is permitted. (*General Electric Co.*)

flush and surface-mounting types, together with a typical plug with a "pigtail" cord. It is used mostly in the wiring of electric ranges. A similar receptacle but with an L-shaped opening for the neutral (NEMA 10-30R) is rated at 30 amp 125/250 volts and is used mostly for clothes dryers.

Receptacles and plugs are so designed that a plug that will fit one particular receptacle will not fit a receptacle rated at a higher or lower ampere or voltage rating, with one exception: A plug that will fit a 15-amp 125-volt receptacle will also fit a 20-amp 125-volt receptacle, but a plug made specifically for the 20-amp 125-volt receptacle will *not* fit the 15-amp 125-volt receptacle.

Some receptacle diagrams show the configurations upside down as compared with Figs. 21-1 and 21-2, but that is of no consequence. They may be installed either way, although the preferred method is to install them with the opening for the equipment grounding blade at the top.

Individual Circuits for Appliances. The Code rules as to when an appliance requires an individual branch circuit are not too well defined. In general you will not only meet Code requirements but provide a better installation if you will provide a separate circuit for each of the following:

1. Range (or two separate circuits for oven and counter units).
2. Water heater.
3. Clothes dryer.
4. Dishwasher.
5. Garbage disposer.
6. Any 120-volt fixed or stationary appliance rated at 12 amp (1440 watts) or more. This includes motors.
7. Any fixed or stationary 240-volt appliance.
8. Any automatically started motor, such as a fan or furnace motor.

Appliance Circuits. The Code in Arts. 210 and 422 (and for motor-driven appliances in Arts. 430 and 440) outlines conditions for branch circuits serving appliances. If a circuit supplies *one non-motor-operated appliance and nothing else,* Sec. 422-27(e) provides that the overcurrent protection in the circuit may not exceed 150% of the appliance ampere rating if the appliance is rated at 16.7 amp or more. The branch-circuit wires must have an ampacity of at least 125% of the ampere rating of an appliance that is continuously loaded. If you know the rating of the appliance only in watts, it is of course easy to determine the ampere rating: Just divide the watts rating by the voltage at which it is to operate.

In the case of branch circuits serving other loads *in addition* to appliances, no cord- and plug-connected appliance may exceed 80% of the ampere rating of the circuit. In the case of 15- or 20-amp circuits, the total rating of *fixed* appliances may not exceed 50% of the rating of the circuit, *if* that circuit also serves lighting outlets or receptacles for *portable* appliances.

Disconnecting Means. Every appliance must be provided with a means of disconnecting it from the circuit, for safety when making repairs, or while inspecting or cleaning it.

Disconnecting Portable Appliances. A plug-and-receptacle arrangement is all that is needed. Of course both must have a rating in amperes and volts at least as great as the rating of the appliance.

Disconnecting Stationary and Fixed Appliances. If the appliance is rated at not over 300 watts or $^{1}/_{8}$ hp, no special disconnecting means is required. For larger appliances, see the following paragraphs.

If the branch circuit to which the appliance is connected is protected by a circuit breaker, or by a fused switch, or by fuses *in a pull-out block* in the service equipment, no further protection is required. But if the circuit is protected by *plug* fuses, a separate switch of the general type shown in Fig. 15-2 must be installed. If a stationary appliance such as a range or clothes dryer is connected by plug and receptacle, that is sufficient in place of a switch.

One further exception: If the appliance has built-in switches so that the entire appliance can be disconnected by the switch or switches, then in a *single-family* dwelling no separate disconnecting switch is required, for then the service-entrance disconnecting means will serve the purpose. In *two*-family dwellings this is also true if each family has access to its own service-entrance disconnecting means. If the building is an apartment house with three or more separate apartments, the disconnecting means must be within each apartment, or at least on the same floor as the apartment.

Wiring 240-Volt Appliances with Cable. You have already learned that the white wire may be used only as a grounded wire. But the white wire does not run to a 240-volt load. Therefore no wire to a 240-volt load may be white, but when you are using cable, it contains one black and one white wire. What to do? NEC Sec. 200-7 Exc. 1 allows you to paint the white wire black (or any other color except green) at each terminal, and at all points where the wire is accessible and visible after installation. Colored tape may be used instead of paint. Remember, an equipment grounding wire must also be installed unless you are using a suitable metal wiring system such as rigid conduit or armored cable.

Bear in mind that an electric range does not operate at strictly 240 volts. When any burner is turned to ''high'' heat, it operates at 240 volts; when turned to ''low'' heat it operates at 120 volts. In other words, it is a combination 120/240-volt appliance; therefore all three wires including the neutral must run to it.

Wiring Methods: Ranges and Other Special Appliances. The Code does not restrict the wiring methods that may be used. Use conduit or cable, as you choose. However, NEC Sec. 338-3(b) establishes one important point. For ranges, ovens, cooking units, and clothes dryers, you may use service-entrance cable with a bare neutral; that wire serves as the neutral *and* the equipment grounding conductor.[1] There is one restriction: The

[1] Service-entrance cable with a bare neutral may be used in any circuit provided the bare neutral is used only as the equipment grounding wire; it must not be used as a current-carrying wire in normal use, except in the case of the appliances mentioned.

cable must run directly from the service equipment. This is almost always the case in ordinary residential installations, but in apartment buildings, there is often a feeder from the service equipment to other panelboards, from which branch-circuit wires run to various locations, as required. Such panelboards are *not* part of the service equipment, and you may not use cable with a bare neutral beginning at such a panelboard.

If the appliance is portable or stationary, run your conduit or cable up to the receptacle, which may be either surface- or flush-mounted.

Wiring Ranges. It is not likely that all the burners and the oven of a range will ever be turned on to their maximum capacity at the same time. For that reason, NEC Sec. 220-19 and Table 220-19 establish a demand factor that permits wires in branch circuits feeding ranges (also ovens and cooking units if both are served by one circuit) to be smaller than the watts rating the appliance would otherwise require. If the total rating is not over 12 000 watts, determine wire size based on 8000 watts. On that basis No. 6 wires are usually used; for smaller ranges No. 8 is sometimes used. If the rating is over 12 000 watts, add 400 watts for each additional kilowatt or fraction thereof. Thus for a range rated at 13 800 watts, use 8000 + 400 + 400 or 8800 watts in determining wire size. But for a circuit serving one oven, or one cooking unit, you must use the full rating of the appliance in determining wire size.

Because of the way the heating elements are connected within a range, the neutral cannot be made to carry as many amperes as the hot wires. For that reason, per NEC Sec. 210-19(b) the neutral may be smaller, but never less than 70% of the ampacity of the hot wires, and in no case smaller than No. 10. The "rule of thumb" is to use a neutral one size smaller than the hot wires: No. 8 neutral with No. 6 hot wires, and so on.

Run the wires up to a range receptacle of the type shown in Fig. 21-3, flush- or surface-mounted, as preferred. The range will be connected to the receptacle using a pigtail cord shown in the same illustration. The plug and receptacle constitute the disconnecting means. The Code requires that the frames of ranges (and separate ovens and cooking units) be grounded, but does not require a separate equipment grounding wire.[2] Instead, it may be grounded to the neutral wire, as already explained.

Sectional Ranges. In many cases, the oven is a separate unit installed in a wall or on a counter, wherever the owner prefers. Groups of burners in a single unit are installed in the kitchen counter. This makes for a flexible

[2] A few local codes do not permit grounding to the neutral. Three insulated wires must be used, plus an equipment grounding wire.

arrangement and lets you use much imagination in laying out a modern, custom-designed kitchen.

The Code calls such separate ovens "wall-mounted ovens" and the burners "counter-mounted cooking units." Here they will be referred to merely as ovens and cooking units. Self-contained ranges are considered stationary appliances, but ovens and cooking units are considered fixed appliances.

You have a choice of two basic methods in wiring ovens and cooking units. The simplest way, and the way preferred by many inspectors, is to provide one circuit for the oven, another for the cooking unit. An alternate way is to provide one 50-amp circuit to supply both appliances.

Assuming that you install a separate circuit for the oven, proceed as already outlined for fixed appliances in general. Use wire of the ampacity required for the load. For ovens and cooking units, the full name-plate rating must be used. No demand factor is permitted, for the entire maximum load of either circuit is often used at the same time. The oven will probably be rated at 4600 watts, which at 240 volts is nearly 20 amp, so No. 12 wire would appear suitable, but the minimum is No. 10 when grounding to the neutral. In installing a cooking unit, proceed exactly as in the case of the oven, again using a minimum of No. 10 wire, which with an ampacity of 30 amp will provide a maximum of 7200 watts. This will take care of most cooking units, but be sure to verify the actual size needed from the watts rating of the cooking unit. Circuits installed before the exact appliance rating is known should not be smaller than No. 10 for ovens and No. 8 for counter-mounted cooking units to be on the safe side.

If you install a single circuit for the oven and cooking unit combined, follow the circuit of Fig. 21-4. To determine the wire size required, add together the ratings of the oven and the cooking unit, then proceed as if you had a self-contained range of the same rating. The same wire size must be used from the service equipment or panelboard, up to the oven and up to the cooking unit, or to the receptacles for these components. The receptacles shown are only for convenience and are not considered as the disconnecting means for these fixed appliances; and they must be suitable for the temperature of the space in which they are installed.

NEC Sec. 210-19(b) Exc. 2 is often misinterpreted to mean that the wire size may be reduced at the point where the circuit splits, with one set of wires to the oven and another to the cooking unit. Not so! It does permit the *leads* from the oven or cooking unit (as in a pigtail cord) to the receptacles, or to the splices with the circuit wires, to be smaller if large

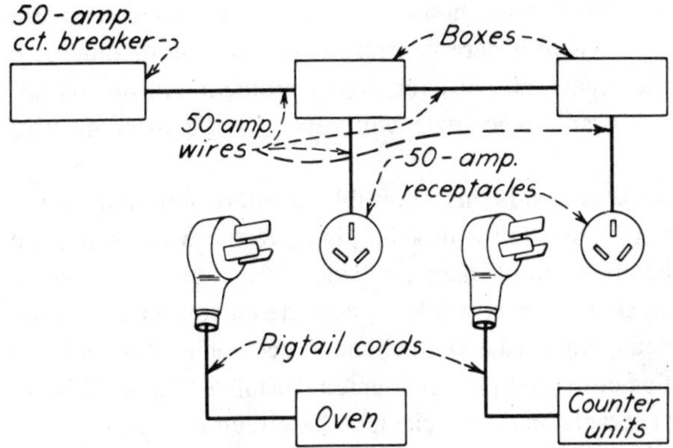

Fig. 21-4 It is best to provide one circuit for the oven and another for the cooking unit, but both *may* be connected to a 50-amp circuit. The receptacles shown are not required.

enough to carry the load, and if not longer than necessary to service the appliance, but with a minimum ampacity of 20.

As already mentioned, the receptacles in Fig. 21-4 are not required, but may be convenient in installing and servicing. The oven and the cooking unit in the diagram may be connected directly to the circuit wires in the junction boxes. Receptacles if used must be rated not less than the rating of the circuit. If you have used No. 6 wire, the receptacles must be rated at 50 amp.

The neutral of any circuit to a range, an oven, or a cooking unit may be uninsulated only if service-entrance cable from the service location is used. In using any other method, the neutral must be insulated.

Clothes Dryers. Dryers are basically 240-volt appliances, although most have 120-volt motors in them. Their frames must be grounded, but may be grounded to the neutral of the circuit if it is No. 10 or larger.

Wire the dryer as you would a range; NEC Sec. 220-18 requires a circuit with a minimum 5000-watt capacity, or a capacity based on the rating of the dryer, whichever is higher. Dryers are usually installed using cord and plug, and a receptacle, which must be a 30-amp 3-wire 125/250-volt type shown in Fig. 21-1, NEMA 10-30R (similar to the 50-amp range receptacle but with an L-shaped opening for the neutral instead of the straight opening of the range type). Use a pigtail cord similar to the one shown in Fig. 21-3 for a range, but with smaller wires though not smaller than No. 10.

Use any wiring method you choose. The Code permits you to use service-entrance cable with a bare neutral; as in the case of ranges, it must start from the branch-circuit overcurrent protection *in the service equipment*. The plug and receptacle will serve as the disconnecting means.

Automatic Clothes Washers. Is this a portable, fixed, or stationary appliance? If the washer is installed with solid piping to the water supply, it certainly should be considered a fixed appliance. If it is connected to the water supply by flexible hose, it would be considered stationary. In any case, it is connected using cord and plug, and a receptacle, and the Code simply refers to it as cord- and plug-connected. Install a 20-amp 125-volt grounding-type receptacle on the special laundry circuit that the Code requires in the laundry area, as has already been discussed.

Water Heaters. Water heaters vary greatly in size, the average domestic type being rated at about 4500 watts. A 4500-watt load consumes 18.75 amp at 240 volts, so a 20-amp circuit with No. 12 wire would appear adequate. However, the Code in Sec. 422-14(b) requires that a water-heater branch circuit must be sized 25% larger than the name-plate rating. Since 18.75 × 125% = 23.44 amp, provide a 2-pole 25-amp circuit with No. 10 wire for a 4500-watt water heater.

Be sure to check for markings at the terminal box to indicate the required temperature rating of the branch-circuit wires. If unmarked, your circuit wiring can be rated 60°C (T, TW, etc.), but if marked 75°C or higher, then your wires must be rated at least as high. A tap from a junction box near the heater is the convenient way to satisfy this requirement, with the circuit wire up to the tap being rated 60°C.

Room Air Conditioners. An air conditioner is considered a *room* air conditioner if it is installed in the room it cools (in a window or in an opening through a wall), if it is single-phase, and if it operates at not over 250 volts. The unit may have provisions for heating as well as cooling. Installation requirements are outlined in NEC Art. 440. The air conditioner may be connected by cord and plug. A unit switch and overload protection are built into the unit. The disconnecting means may be the plug on the cord, or the manual control on the unit if it is readily accessible and not more than 6 ft from the floor, or a manually controlled switch installed where readily accessible and in sight from the unit. If the unit is installed on a circuit supplying no other load, the ampere rating on the name-plate of the unit must not exceed 80% of the circuit rating. If it is installed on a circuit also supplying lighting or other loads, it may not exceed 50% of the circuit rating. Cords must not be longer than 10 ft if the unit operates at 120 volts, and not over 6 ft if it operates at 240 volts.

Efficiency of a machine, as you know from Chap. 2, is generally expressed as a percentage, obtained by dividing the output by the input. For room air conditioners a number is used, instead of a percentage, because the output and input are expressed in different terms. Energy Efficiency Ratios (EER) for room air conditioners may range from $4^1/_2$ to 12 (the higher the number the greater the efficiency) and are derived from the output in Btu/h (British thermal units per hour) divided by the input in watts. For example, an output of 6000 Btu/h and an input of 750 watts gives us an EER of $6000/750 = 8.0$. For room air conditioners an EER of 7.5 or higher is recommended. While the initial cost may be more, the higher-EER-rated models produce more cooling for a smaller amount of electricity used, thus costing less in the long run because of less power used—and at the same time they contribute to the conservation of energy.

There are many things you can do to conserve energy in an air-conditioned room. Remember that extra heat in the room (from lighting fixtures, from solar radiation through unshaded windows, or warm air introduced from other rooms or from outside through open doors or windows) places an additional load on your air conditioner, so conserve energy by eliminating sources of additional heat whenever and wherever possible.

If an air conditioner has a 3-phase motor, or operates at more than 250 volts, it is not a *room* air conditioner. Install it on a circuit of its own. Such a unit must meet the requirements discussed in Chap. 30 for sealed (hermetic-type) motor-compressors.

Electric Heat. As gas and oil become more scarce, heating of buildings by means of electricity [much of which is generated using plentiful energy sources such as nuclear or coal or nondepleting sources such as hydro (falling water) or geothermal (natural underground heat)] should come into wider use.

The many means by which electric space heating can be accomplished, including ceiling and wall panels, embedded cables, electrically heated boilers, central furnaces, baseboard resistance heaters, quartz and infrared lamps, and individual unit heaters, make it impractical to cover all the installation details here. In every case the manufacturer's instructions should be carefully followed with regard to circuit size and overcurrent protection (both of which must be at least 125% of a resistance heater rating, according to the Code), separation from combustible materials, circuit voltage, protection from physical damage, and means of control. Article 424 of the Code should be consulted for the requirements for the type of heating equipment you are installing.

A few important points to consider are: when the heating elements within the equipment are subdivided into loads not exceeding 48 amp each, the wires supplying the several supplementary overcurrent devices for the subdivided loads shall be considered not as a feeder, but as a branch circuit, and a disconnecting means must be installed within sight and on the supply side of these supplementary overcurrent devices; receptacle outlets may not be installed in walls above baseboard heating units because cords draped against the heaters could be damaged. However, the Code's receptacle-spacing requirements for dwellings can be met by using receptacle outlets furnished or installed by the heater manufacturer in filler sections between heaters, provided the outlets are not supplied by the heater circuit.

The kilowatts of heating capacity to install will usually be determined by a heating or other engineer, as the calculations involve consideration of the expected outside temperature; desired inside temperature; heat transmission (heat loss) characteristics of the floors, walls, ceilings, and windows; and infiltration losses. In the interest of energy conservation and of keeping operating costs down, electrically heated buildings must be well insulated thermally.

For dwellings and apartment houses see Chaps. 13 and 26, respectively, for information on branch circuit, feeder, and service calculations for heating loads.

Grounding. The grounding of ranges, ovens, counter units, and dryers has already been discussed. As for any other fixed or stationary appliance, if it is equipped with a 3-wire cord and plug, plugged into a properly installed receptacle, the grounding has been taken care of. If it is permanently wired using metal conduit or armored cable, anchored solidly to the terminal box on the appliance, it is properly grounded. If, however, you use nonmetallic-sheathed cable, it must be the type with the bare equipment grounding wire, which must be connected to the frame of the appliance, and grounded to the box nearest the appliance.

Since a water heater is connected to water pipes, you might think that the heater is automatically grounded. But the pipe can be disconnected in servicing the heater, so that it is no longer grounded. Therefore the heater must be grounded as already discussed.

22
Old Work; Modernizing

In old work, or the wiring of buildings completed *before* the wiring is started, there are few *electrical* problems that have not already been covered. Most difficulties can be resolved into problems of carpentry, in other words, how to get wires and cables from one point to another with the least effort and minimum tearing up of the structure of the building.

In new work it is a simple matter to run wires and cables from one point to another in the shortest way possible; in old work considerably more material is used because often it is necessary to lead the cable the long way around through channels that are available, rather than to tear up walls, ceilings, or floors in order to run it the shortest distance.

No book can give all the answers as to how to proceed in old work; here the common problems will be covered, but considerable ingenuity must be exercised in solving actual problems in the field. A study of buildings while they are under construction will help in understanding what is inside the wall in a finished building.

Wiring Methods in Old Work. It is impossible to use rigid or thin-wall conduit in old work without practically wrecking the building. It would be used only when a major rebuilding operation is in process, and installation then would be as in new work. The usual method is to use nonmetallic-sheathed cable or armored cable. The material is easily fished through empty wall spaces. It is sufficiently flexible so that it will go around corners without much difficulty. In some localities flexible conduit (greenfield) is used. Install it as you would cable, except that the empty conduit is installed first and the wire pulled into place later.

Concealed Knob-and-Tube. You may have occasion in old work where the existing wiring is concealed knob-and-tube to relocate an outlet, extend an existing circuit, or feed an existing circuit from a new panelboard. Make the change of wiring method in an outlet box or an accessible junction box if possible, or use an end fitting on the cable, called an "A head," as shown in Fig. 22-1. These fittings have an insulating part with a

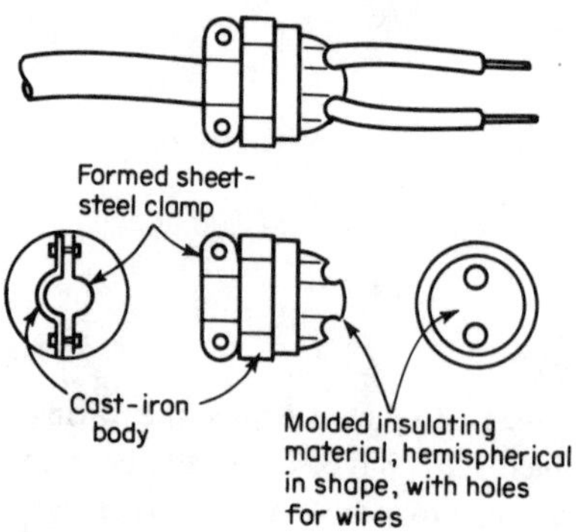

Fig. 22-1 End fitting for armored or nonmetallic-sheathed cable.

separate hole for each wire. Each wire between the fitting and the splice to the knob-and-tube wiring must be completely encased in a piece of flexible nonmetallic tubing called "loom." A porcelain split-knob must support the wires within 6 in of the splice. As knob-and-tube materials are getting scarce, you may have to salvage knobs, tubes, and loom from parts of the wiring you are to replace. A change from knob-and-tube in a wall to Type NM cable is shown in Fig. 22-2. As the splices will be concealed in the wall, they must be very carefully made and well insulated. Soldered joints are traditionally used in knob-and-tube work, but if you lack the equipment and skill required to do a good soldering job, use a splicing device, such as a wire nut, as shown, being careful that all parts of the finished job are maintained at least $1^1/_2$ in from the stud or joist.

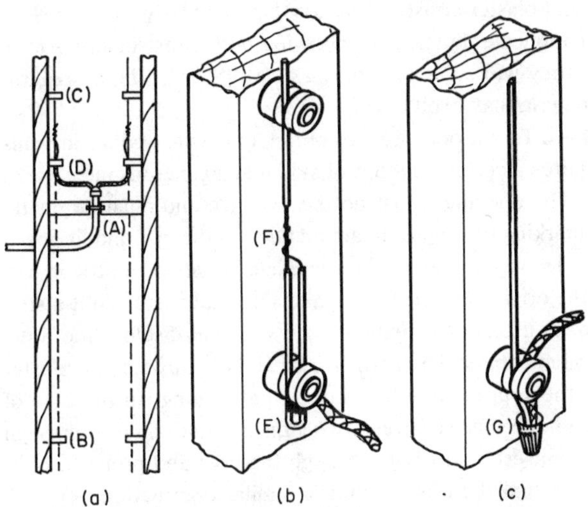

Fig. 22-2 (*a*) Install wooden support for Type NM cable at A. Relocate knobs from abandoned portion of knob-and-tube run at B to within 6 in of splice at C. Encase cable conductors in loom D. (*b*) Fold back and tape end of knob-and-tube wire E. Splice, solder, and tape joint F. (*c*) Alternate method, using pressure-splicing device, G, instead of solder.

Code Requirements for Old Work. For new work, boxes are usually $1^1/_4$ in or more in depth, but for old work boxes as shallow as $^1/_2$ in are sometimes necessary at lighting outlets. Figure 22-3 shows such shallow boxes. Cable is simply pulled through empty spaces in walls and ceilings; each piece must be a continuous length. Naturally it cannot be supported inside the walls or ceilings, but must be anchored to boxes with connectors or built-in clamps. Any box enclosing a flush device (switch, receptacle) must have an internal depth not less than $^{15}/_{16}$ in.

Illustrations. The walls and ceilings of the building under consideration

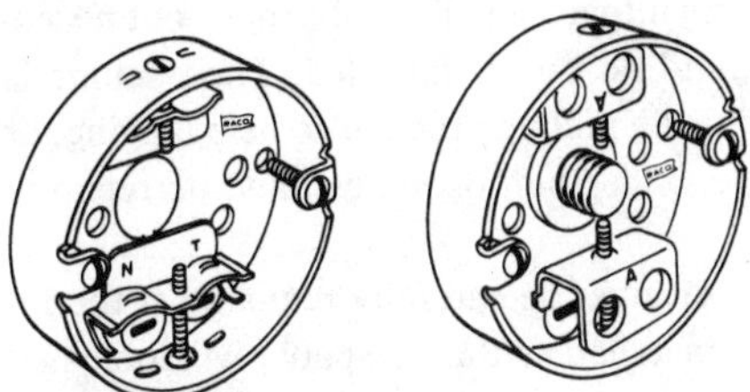

Fig. 22-3 In old work, shallow boxes are permitted for lighting outlets. (*RACO, Inc.*)

may be of lath-and-plaster construction, or they may be wallboard. It is not practical to duplicate illustrations showing both constructions, so in the illustrations they will be labeled only as "wall" or "ceiling" regardless of which material has been used.

Cutting Openings. To cut openings for outlet and switch boxes in walls and ceilings requires a certain amount of skill and a generous measure of common sense. The openings must not be oversize and must be neatly made. Start by marking the approximate location of the box and, if possible, allow a little leeway so that the opening can be moved a little in any direction from the original point. First make sure that there is not a stud or joist in the way; usually thumping on the wall will disclose the presence of structural members. Then punch a small hole through the plaster or wallboard at the approximate location of the opening. In the case of plaster, probe until the space between two laths is found; then go through completely. Then insert a stiff wire through the hole and probe to right and left to confirm that there is no stud or similar obstruction.

The sawing in a lath-and-plaster job can be done using a keyhole saw: Proceed gently so as not to loosen the bond between the plaster and the lath. Many prefer to use a hack-saw blade, heavily taped at one end to serve as a handle. Have the teeth of the blade lie backward, the opposite of the usual position, so that the sawing is done as you pull the blade out of the wall, not as you push the blade away from yourself as in usual sawing. This will tend to leave a firm bond between lath and plaster, especially if you hold your hand against the plaster as you do the sawing. Unless you watch this carefully, you may end up with a considerable area of plaster unsupported by laths that have become separated during the sawing operation. When the opening is in wallboard, the sawing may of course be done with the blade in the usual position.

Temporary Openings. It is sometimes necessary to cut temporary openings in odd places to make it possible to pull cable from one point to another, for example, from the ceiling into a wall. The cable does not necessarily run through the opening, and no box is installed. The opening is used only to get at the cable during the pulling process, to help it along, or to get around obstructions in the wall. Such openings must be repaired when the job is finished.

If the room is papered, the paper must be carefully removed at the location of the opening, and then reinstalled so that the paper will look as it did before the wiring job. This is easily done. With a razor blade cut the sides and bottom of a square, but not the top. Apply moisture with a rag or sponge, soak the cut portion until the paste has softened, and then lift the cut portion, using the uncut top as a hinge. Fold it upward, and fasten it to the wall with thumbtacks or masking tape. These steps are shown in Fig. 22-4. Do not cut the paper at the top and fold it *downward;* plaster or wallboard particles would adhere to its wet surface; then the paper would not fit smoothly when pasted back.

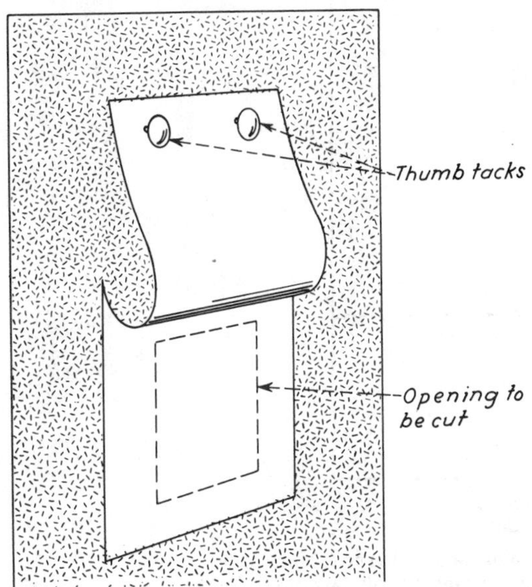

Fig. 22-4 Sections of wallpaper are easily removed temporarily. Use the top of the cut section as a hinge; this makes it easy to restore the wallpaper to its original condition after the job is finished.

If no box is to be installed (where the opening was made for access only), the opening is easily patched using a ready-mixed plaster mixed with water. The same mixture may be used to fill the openings around the boxes, for the Code does not permit large open spaces around boxes. If openings are carefully made, no patching around the boxes need be done.

Openings for Switch Boxes. The opening for a switch box must be about $3^1/_4 \times 2$ in. Make yourself a template as shown in actual size in Fig. 22-5: Simply lay a piece of thin cardboard under the page, and with carbon paper trace the outline on the cardboard, then cut it to size. This will save you much time, especially if many openings must be cut. Lay the tem-

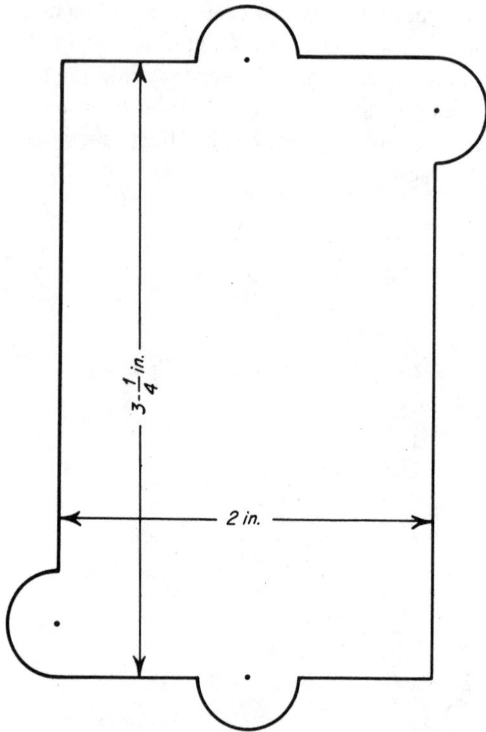

Fig. 22-5 Actual size of opening for a switch box. Make yourself a template of this drawing; it will save you much time.

plate against the wall where the opening is to be, and trace around it to mark the outline of the opening. Note two 1/2-in holes bored at opposite corners, to start your hack-saw cuts; there are two more 1/2-in holes at top and bottom, to provide clearance for mounting screws of switches or other devices installed in the box. The *centers* of the holes must be on the lines of the outline, so that *not over half* of each hole will be outside the rectangle. Unless you watch this carefully, part of the hole may later not be covered by the face plate for the receptacle or switch.

If the wall is lath-and-plaster, remember that the ordinary switch box is 3 in long, while two laths plus three spaces between laths measure more than 3 in. If you cut away two full laths, it will be difficult to anchor the box by its ears on the next two laths, for the mounting holes in the ears will then come very close to the edges of the laths, which will split when the screws are driven in. Cut one lath completely, and remove part of each adjoining lath, as shown in Fig. 22-6. The boxes do not necessarily have to be mounted using screws; the methods outlined in the next paragraphs for drywall construction can also be used.

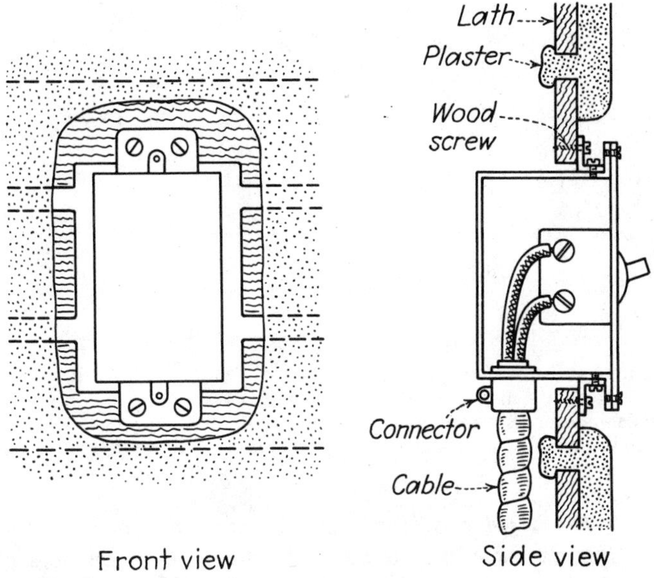

Fig. 22-6 In cutting an opening for a switch box in a lath-and-plaster wall, do not cut away two complete laths. Cut away one, and part of each of two others.

Ears on Boxes. The mounting ears on the ends of boxes are adjustable to compensate for various thicknesses of walls. They are also completely reversible as shown in Fig. 22-7. In the position shown at *A* they are used for installing on lath-and-plaster walls. Reverse them as at *B* when installing the box on wallboard.

Installing Switch Boxes in Drywall Construction. Wallboard is not sturdy enough to accept screws as in lath-and-plaster construction, so other methods must be used.[1] In each case adjust the brackets on the ends of the box so that when finally installed, it will be flush with the wall sur-

[1] The methods described for drywall can, of course, also be used for lath-and-plaster construction.

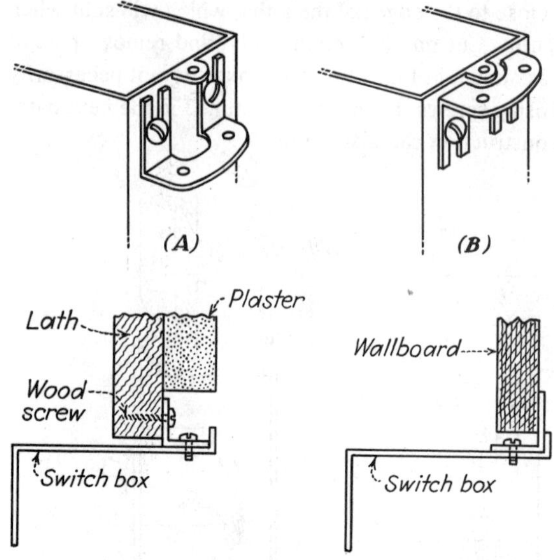

Fig. 22-7 The external mounting ears of switch boxes are reversible. In position *A* they are used when mounting boxes on lath under plaster. Reverse them as shown at *B* for mounting on wallboard.

face. Boxes with beveled corners and cable clamps are preferable but not essential. Bring the cable into the box and tighten the clamps, letting about 10 in project out of the box. If you are using boxes without clamps, install the connector on the cable, let it project into the box through a bottom knockout, and install the locknut only after the box is installed.

One way to deal with wallboard is to use the special box of Fig. 22-8. It has special clamps on the outsides of the box. After installing the cable, push the box into the opening, then tighten the screws on the external clamps. This makes the clamps collapse, anchoring the box in the wall.

Another way is to use an ordinary box plus the U-shaped clamp of Fig. 22-9. Install the U-clamp with the screw holding it in place, unscrewed about as far as it will go. Slip the box into its opening; the ends of the clamp will expand outwards, and when you tighten the screw holding the clamp, it will anchor the box ears firmly against the wall.

A third method is to use an ordinary box, plus a pair of special straps shown in Fig. 22-10. Insert one strap on each side of the wall opening, and push the box into the opening, being most careful not to lose one of

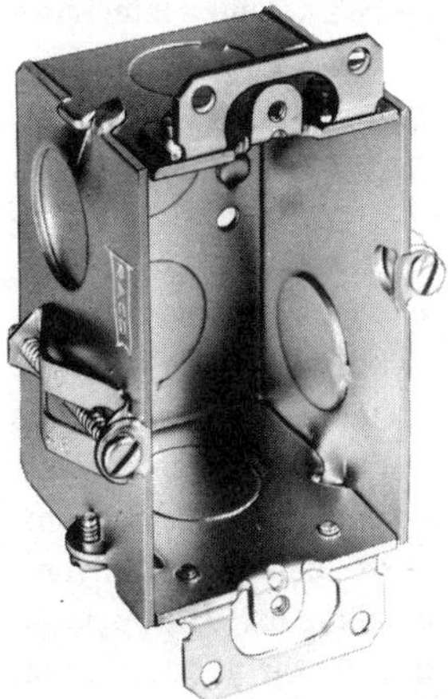

Fig. 22-8 This box has exterior collapsible clamps on each side. Push box into opening, tighten screws on sides, and the box will be anchored in the wall. (*RACO, Inc.*)

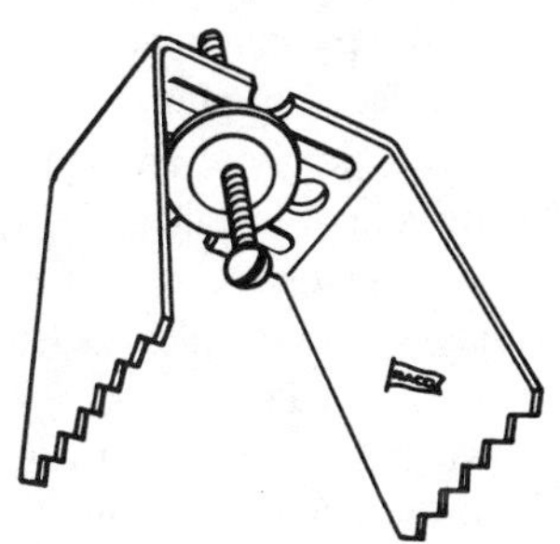

Fig. 22-9 This clamp can be used with an ordinary box, to anchor it in the wall.

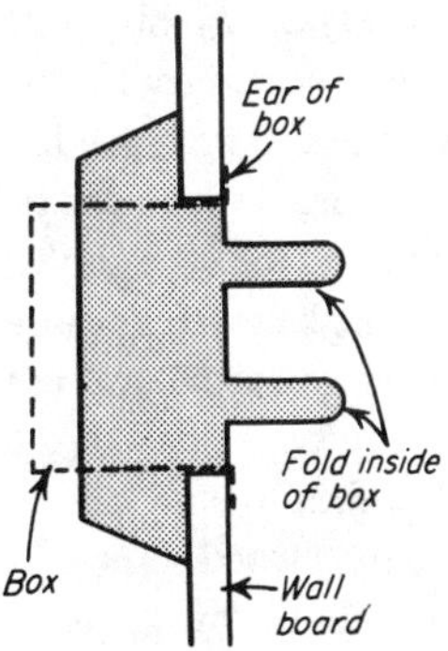

Fig. 22-10 A pair of these hanger straps is also convenient in installing a switch box in a wallboard wall.

the straps inside the wall. Then bend the short ends of each strap down into the inside of the box. *Be sure* they are bent back sharply over the edge of the box, and lie *tightly* against the inside walls of the box, so that they cannot touch the terminals of a switch or receptacle installed in the box, which would lead to grounds or short circuits.

Mounting Outlet Boxes. If there is open space above the ceiling on which the box is to be installed, and if there is no floor above (or if there is a floor in which a board can easily be lifted as will be explained later) proceed as in new work using a box hanger and the usual 1 1/2-in-deep box. The only difference is that you will be working from above.

If all the work must be done from below, the procedure is different. Let's assume you are going to install a ceiling fixture. Use a 1/2-in-deep box of the general type shown in Fig. 22-3, with cable clamps.

If your ceiling is lath-and-plaster, use a box that does not have a fixture stud in it. Make a hole about an inch smaller in diameter than the box, where the fixture is to be installed. Remove the center knockout from the box, and one additional knockout for each cable that is to enter the box. Run your cable(s) to this location and anchor the cable(s) to the box, letting about 8 in project from the box. Then push a straight bar hanger similar to that shown in Fig. 10-16, but without the nailing ears, through the opening in the ceiling, first removing the locknut from the fixture stud on the hanger. Let the fixture stud hang down through the opening in the ceiling, into the center knockout of the box, then install the locknut on the fixture stud. Of course the hanger can't be nailed to the joists, so be sure to turn it to lie *crosswise* across several laths, to distribute the weight over a wider area.

If your ceiling is made of wallboard, the same method may be used; or, the box may be supported by toggle bolts, shown in Fig. 22-24, but great care must be taken in making the holes no larger than necessary for the two toggles and the cable entry, lest there be too little ceiling finish left for support. There are also round boxes with ears or flanges which can be recessed into the ceiling using a snap-in bracket similar to the type shown in Fig. 22-9. If the lighting fixture to be supported by the box weighs more than a few pounds, the box should be supported from the structure (not the finish) of the building. If a ceiling joist happens to be located in the center of the room, a 1/2-in-deep round box with a fixture stud can be mounted on the surface and secured to the joist with long wood screws. If none of the above methods is used, and if the space above the ceiling is not accessible, it may be necessary to remove a section of either the wallboard or the floor above, in order to provide support for the box and fixture.

Lifting Floor Boards. In many cases the outlet and switch boxes may be so located with respect to wall or ceiling obstructions that it is necessary to lift floor boards in the floor above. This should be avoided if possible, but where it is necessary, use extreme care so that when the boards are replaced there will be a minimum of visible damage to the floor. Ordinary attic flooring is easily lifted, but tongue-and-groove construction presents a problem.

It is necessary to chisel off the tongue on at least one of the boards. The thinner the chisel, the less the damage that will be done to the floor. A putty knife with the blade cut off short and sharpened to a chisel edge makes an excellent chisel for the purpose, if a thin chisel is not available. Drive it down between two boards as shown in Fig. 22-11. This should be

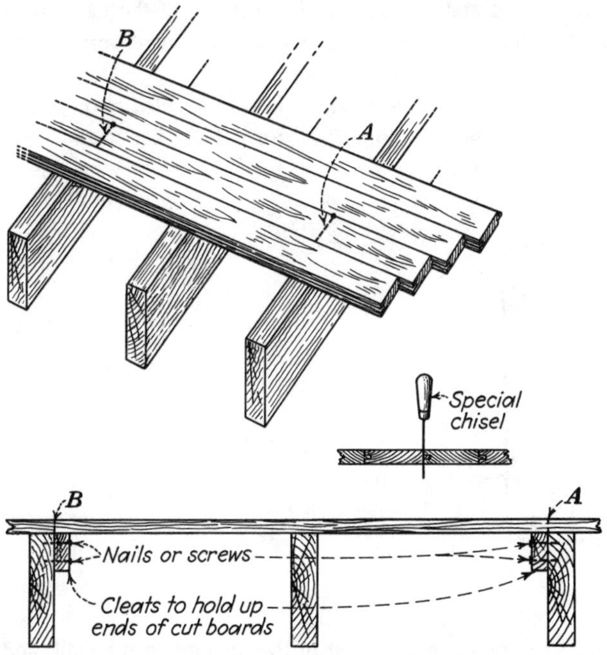

Fig. 22-11 Steps in lifting and replacing floor boards.

done on the entire length between three joists. Having the cut section extend over a joist gives the added advantage of a better footing when the board is reinstalled. In cutting the tongue, the exact location of the joists can be determined and in this way points *A* and *B* in the illustration can be located. Bore a small hole at these two points next to the joists, and with a keyhole saw cut across the boards as shown.

The board can then be lifted, the wiring done, and the board later replaced. It will be necessary to attach cleats to the joists for the floor board to rest on, at each end. Anchor these cleats securely to give the cut board a really solid footing. The bored holes can later be filled with wooden plugs.

Another and perhaps simpler way is to use an electric circular saw set to cut a depth exactly the thickness of the flooring. Then cut the board at the exact center of the two joists. The saw will of course also cut part of the two adjoining boards. *Save the sawdust.* When the wiring has been done, replace the board and nail each end to the top of the joists with finishing nails. Then take the sawdust that you have saved, mix it with glue, and patch the saw cuts with the paste.

Installing Switch Box in Wall. Use a *deep* box with *beveled* corners and internal cable clamps as shown in Fig. 22-12. Strip the jacket off the cable

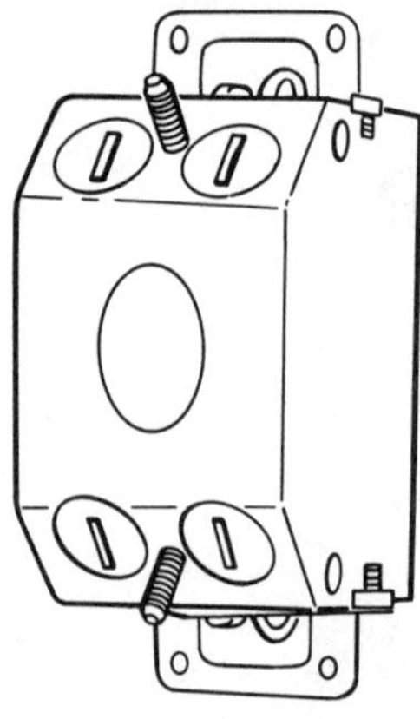

Fig. 22-12 Boxes with beveled corners, and internal cable clamps, simplify the procedure of Fig. 22-13.

for about 10 in, let the ends project out of the opening in the wall, and push them into the knockout of the box, which is still outside the wall. Then push the box into the opening; to hold it in place use any of the methods shown in Figs. 22-8 through 22-10. Pull the cable into the internal clamps of the box, tighten the clamps, and you are ready to install switch or receptacle. All this is shown in Fig. 22-13.

Nonmetallic boxes are available for old work, with supports similar to those shown in Figs. 22-8 and 22-9, and with cable clamps for nonmetallic-sheathed cable. Cable clamps must be used, as it is not possible to secure the cable within 8 in of the box, as required for new work.

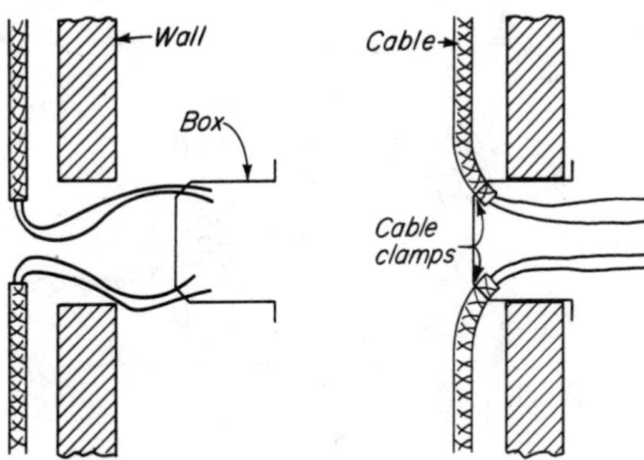

Fig. 22-13 Installing switch box in opening. The cable is loosely secured to the box by internal clamps before the box is inserted into the opening. The clamps are tightened only after box has been inserted into opening.

All that has been said assumes you are using nonmetallic-sheathed cable. If you are using armored cable, your problem becomes a bit more complicated because boxes with *beveled* corners are not available with built-in clamps for *armored* cable. You have two choices. Use a deep box with *square* corners and built-in clamps for armored cable, which will require the cable to enter at the top or bottom end of the box. After stripping off about 10 in of the armor, let the wires project out of the opening in the wall, and let the wires (but not the armor) enter through the knockout in the box, projecting into the box through the clamps. Anchor the box in the wall in any of the ways described. Then jiggle the wires in the box until the *armor* enters the clamp; tighten the clamp. Don't forget to install the insulating bushing to protect the wires from the cut end of the armor. See Fig. 11-36.

An alternative is to use an ordinary box without clamps. After removing the armor, install a connector on the cable in the usual way, but remove the locknut. Let the ends of the wires project into the box, but leave the connector outside. Install the box, then pull the connector into the knockout by pulling on the wires. Install and tighten the locknut.

Cable behind Baseboard. Assume that the bracket light on the wall at *A* in Fig. 22-14 is controlled by a switch on the fixture, but now you want to install a wall switch at the point *D* on the same wall. You can do that fairly easily by following this procedure: Cut an opening for the switch box at point *D*. Then remove the baseboard at the floor and cut two holes

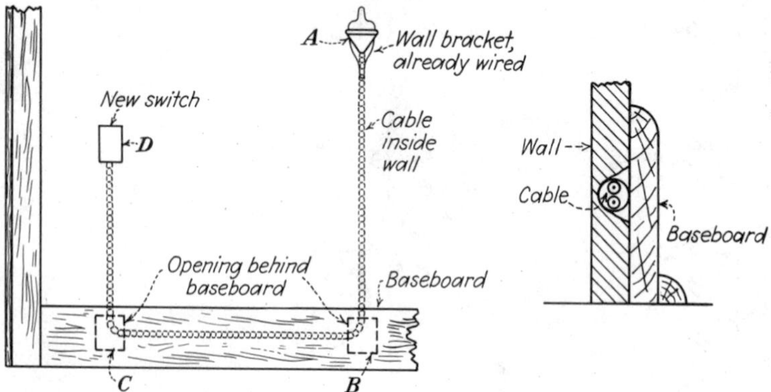

Fig. 22-14 Often cable can be concealed in a trough cut into the wall, behind a baseboard.

at *B* and *C* behind the baseboard. Then, cut a groove or trough (called a "chase") from *B* to *C*. If the wall is lath-and-plaster, it may be necessary to slice away parts of the laths. If wallboard is used and it is quite thin, you may have to slightly notch the studs.

When this has been done, prepare a length of cable long enough to reach from *A* to *B* to *C* to *D*, plus about 10 in of wire beyond the armor or jacket at each end. Install a connector at each end. Remove the locknuts. Temporarily remove the fixture at *A*; remove a knockout in the bottom of the box on which the fixture is installed. Considering the very short lengths involved, push a short piece of stiff but flexible steel wire (or fish tape) into the knockout at *A*, down toward *B*, where it can be fairly easily located. Attach the end of the cable to it, pull the cable inside the wall up to *A*, jiggle the connector into the knockout, install the locknut, and connect the wires properly at the fixture. Repeat the process at *D*, pull the cable up to *D*, jiggle the connector into the knockout in the box at *D*, install the locknut, and connect the switch. In the meantime lay the cable into the trough behind the baseboard; any slight excess length will lie inside the wall. Wherever the cable crosses a stud, cover it with a $^1/_{16}$-in steel plate before replacing the baseboard. This protects the cable from future penetration by nails.

Cable through Attic. For single-story houses, or for outlets on the second floor of a two-story house, it is generally practical to run the cable through the attic. It is a simple matter to lift a few boards of the usual rough attic flooring and lead the cable around, in that way avoiding the need for temporary openings in the living quarters. No baseboards then need to be lifted. It may require a few more feet of cable, but the saving in labor offsets the cost of a little extra cable, many times over. Always explore this possibility before considering a more difficult method. For example, in Fig. 22-14, the cable would run from outlet *A*, up to the attic, under the attic floor to a point directly above outlet *D*, and then drop down to *D*.

Cable through Basement. If the outlet shown in Fig. 22-14 is on the first floor, you can often run the cable through the basement—dropping straight down from *B*, into the basement, then over to the left, then upward again at point *C*. Often this will require boring holes through a sill plate, as shown in Figs. 22-15 and 22-16. Sometimes there will be obstruc-

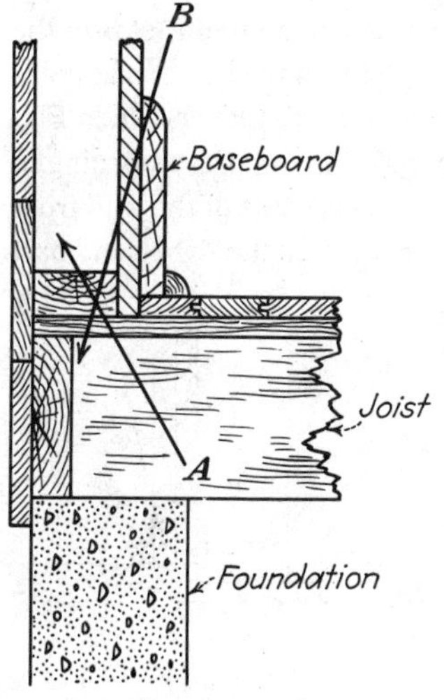

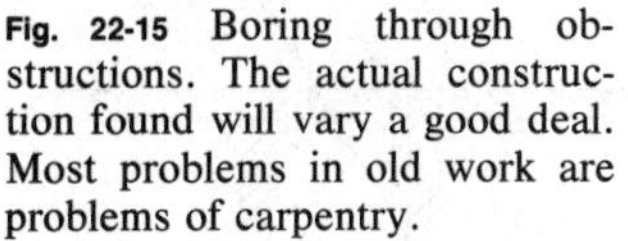

Fig. 22-15 Boring through obstructions. The actual construction found will vary a good deal. Most problems in old work are problems of carpentry.

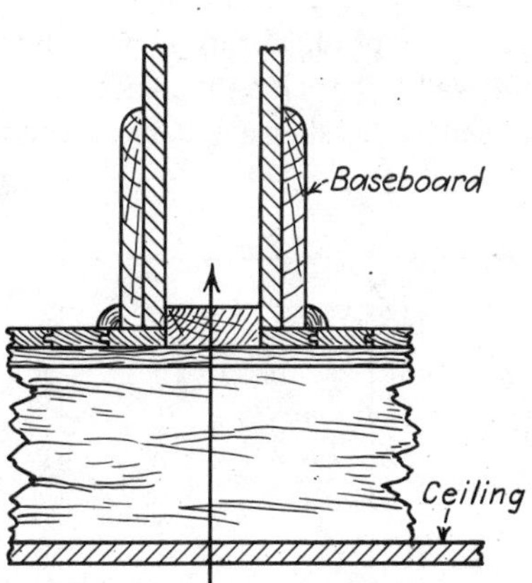

Fig. 22-16 Sometimes obstructions can be cleared by boring straight up.

tions in the walls, such as cross bracing, whether you go through the basement or use the method shown in Fig. 22-14. In such cases you will usually have to make an incision in the wall (to be patched later) in order to notch the cross braces.

If the point where the cable is to run down into the basement is in an outer wall, the construction is likely to be as shown in Fig. 22-15. In that case bore a hole either upward as indicated by arrow A, or downward from a point behind the baseboard as indicated by arrow B. If the cable is to enter the basement from an interior wall, it is often possible to bore upward from the basement as in Fig. 22-16. The long-shank electrician's bit shown with an extension in Fig. 22-17 makes the job easier.

Fig. 22-17 Electrician's bit and extension. (*Greenlee Tool Division, Ex-Cell-O Corporation.*)

Cable around Corner Where Wall Meets Ceiling. Figure 22-18 shows this problem: how to lead cable from outlet A in the ceiling to outlet B in the wall, around the corner at C. Usually there will be an obstruction at point C. Any of a dozen types of construction may be used; that shown in Fig. 22-18 is typical. The usual procedure is to make a temporary opening in the wall at the ceiling at point C, but on the opposite side of the wall from opening B, as shown in the enlarged view of point C in the right-hand part

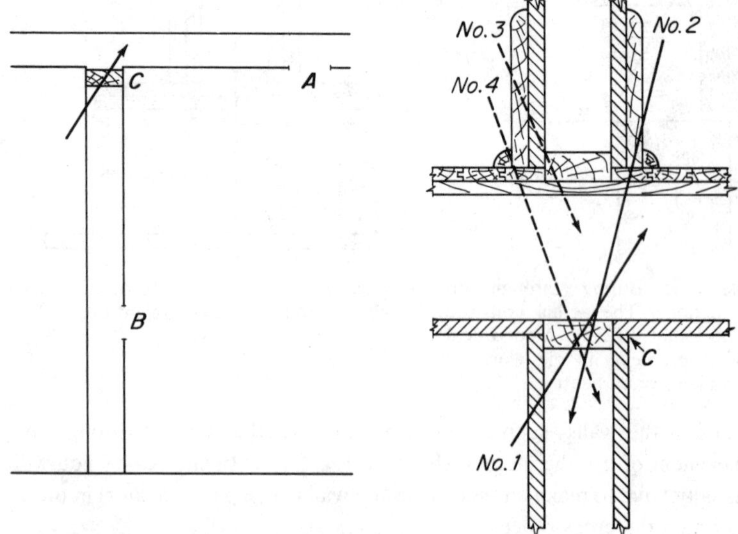

Fig. 22-18 To get cable from A in the ceiling to B in a wall, a temporary opening must often be made at C. Point C is shown enlarged at the right.

of Fig. 22-18. Bore upward as shown by arrow 1. Push a length of stiff wire into the hole until the end emerges at *A*. If the opening at *C* is large enough, push the opposite end of the wire downward to *B*, and pull the loop that is formed at *C* into the wall, pulling at either *A* or *B*; you will then have a continuous wire from *A* to *B* with which to pull in the cable. More usually the hole at *C* will be small; use two lengths of wire to do the fishing. Push one length from *C* to *A*, leaving a hook at *C* just outside the opening. Push another end from *C* to *B*, again leaving a small hook just outside the opening at *C*. Hook the ends together, pull at *B*, and it is a simple matter then to pull the longer wire from *A* through *C* to *B*, and, with this, to pull in the cable.

If there happens to be another wall directly above point *C*, it may be better to bore downward from a point behind the baseboard as shown by arrow 2. In that case one length of wire is pushed down from above through the bored hole to *B*, and another length from *A* toward *C*. When the hooks on the ends engage, pull down at *B* until a continuous length of wire extends from *A* through *C* to *B*. Attach cable to one end of the wire and pull into place.

Cable from Second Floor to First. If the first-floor partition is directly below a second-floor partition, it is usually simple to bring the cable through by boring as indicated by arrows 2 and 3 (or 3 and 4) in the enlarged view of Fig. 22-18. Use good judgment so that the two holes will, so far as is possible, lie in the same plane, thus simplifying the fishing problem. An opening behind the baseboard is usually necessary.

If the first-floor partition is not directly below the second-floor partition, handling it as shown in Fig. 22-19 will usually solve the problem. Bore holes as indicated by the two arrows.

In really old homes you may find a molding around the room on the walls, just below the ceiling; then it is usually better to remove the molding and chisel a hole in the corner, chiseling away a part of the obstruction to provide a channel for the cable, as indicated by the arrow in Fig. 22-20. If the cable projects a bit, it will be concealed when the molding is replaced.

Extension Rings. In old work it is often desirable to extend an outlet beyond an existing outlet. If the wiring must be entirely flush, it might involve a good deal of carpentry if the new outlet also must be flush. But at least in certain locations (such as basements) it will be acceptable to install the new outlet on the surface. In that case use an extension ring of the type shown in Fig. 22-21. Extension rings are outlet boxes without

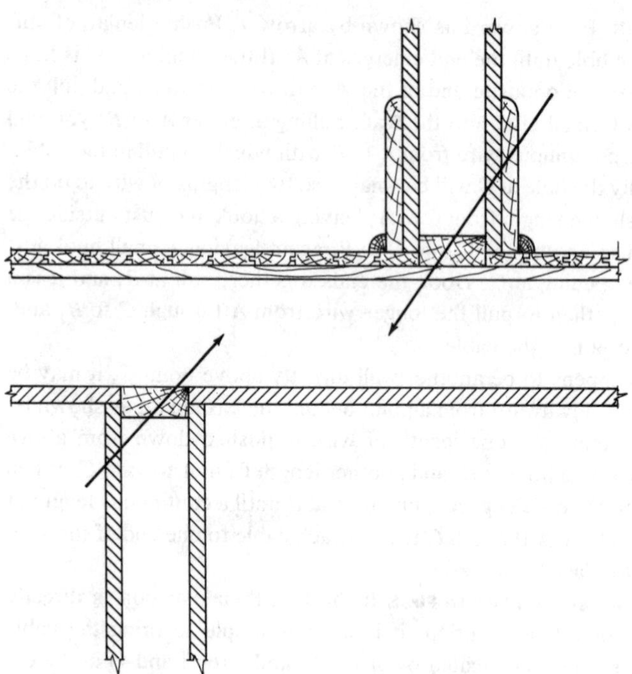

Fig. 22-19 Problem in bringing cable from a second-floor wall into a first-floor wall.

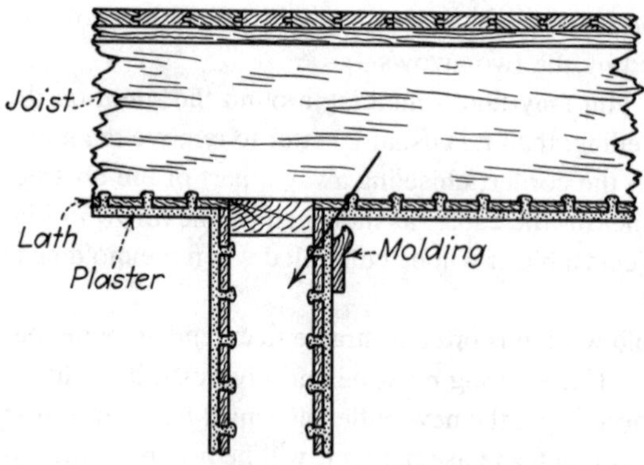

Fig. 22-20 In very old houses, temporary openings can sometimes be made to advantage behind a picture molding.

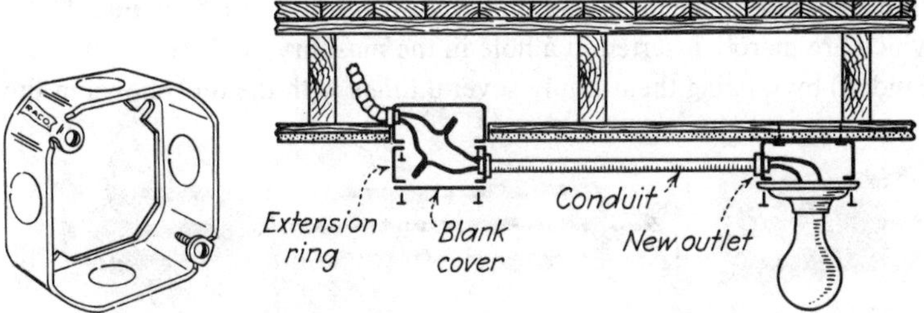

Fig. 22-21 If surface wiring is acceptable, an extension ring is handy in installing a new outlet, starting from an existing outlet.

backs; they are available to fit all kinds of outlet boxes. Simply mount the ring on top of an existing outlet box, and from there proceed as in any surface wiring. The extension ring must be covered with a blank cover or with the fixture that may have been mounted on the original box.

Boxes on Masonry Walls. When a house has been built using masonry (brick, concrete, etc.), an extraordinary amount of work is involved in providing space for a box, plus a channel for a raceway or cable, and serious consideration should be given to installing surface wiring. But if the decision is to install concealed wiring, space must be chiseled into the masonry to receive the box. Boxes cannot be secured directly to the masonry with screws, so it is necessary to use one of the many types of plugs or anchors available for the purpose. In any case it will be necessary to drill holes into the masonry, using a star drill of the general type shown in Fig. 22-22.

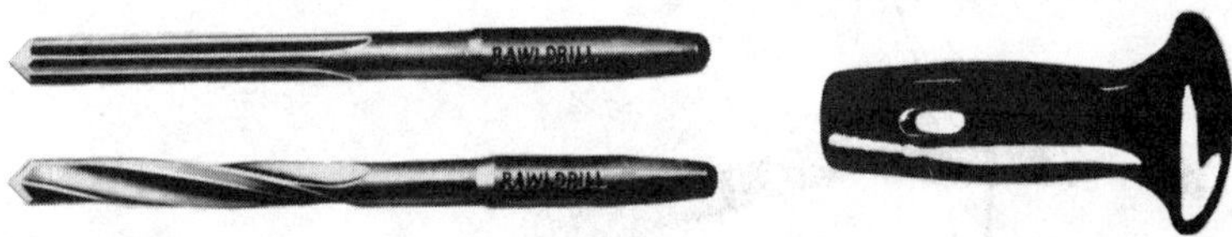

Fig. 22-22 To drill holes in masonry, use a star drill and a hammer or carbide-tipped drills with an electric drill. (*Rawlplug Co.*)

The drill is used by simply pounding on its head with a hammer, rotating the drill a bit after each blow. Be sure to use the drill holder shown; it prevents mashed fingers. Similar drills can be used with power hammers, and a different type is used in electric drills.

A very common mounting method is that using the well-known lead anchor expansion shields of the type shown in the top left of Fig. 22-23, which are merely inserted in a hole in the masonry, and then "set" (expanded) by tapping them firmly several times with the tool shown in the

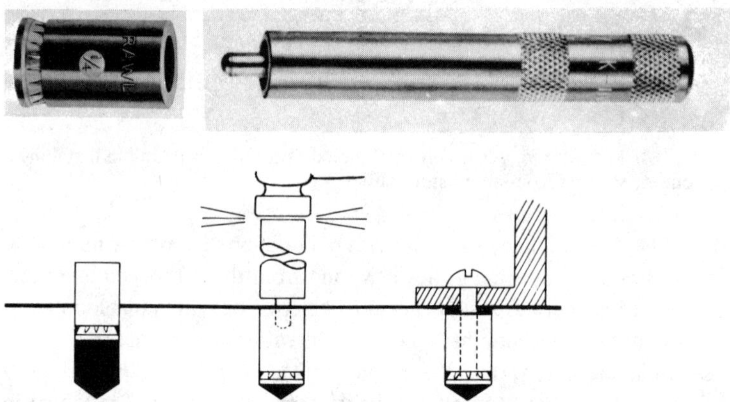

Fig. 22-23 One of many available varieties of screw anchors. (*Rawlplug Co.*)

upper right of the illustration. Ordinary machine screws are used to secure the box (or conduit strap, etc.) to the anchor, as shown in the bottom part of the same illustration. The Code prohibits using wooden plugs in holes.

For a hollow area, such as a hollow tile wall, use toggle bolts as shown in Fig. 22-24, which also shows the installation method. Merely slip the

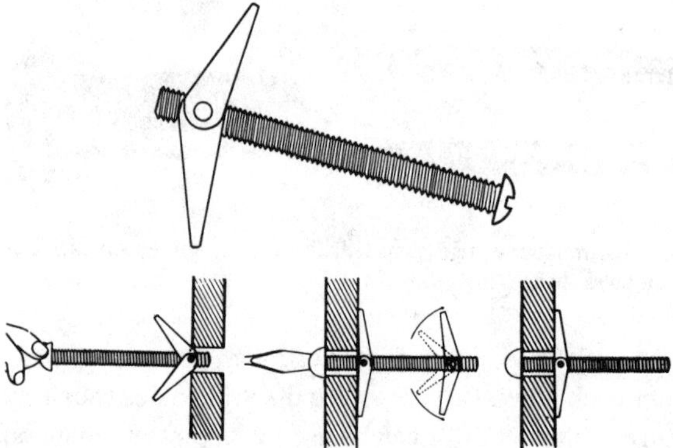

Fig. 22-24 Toggle bolts are convenient and practical in mounting equipment on hollow walls. (*Rawlplug Co.*)

collapsible wings through the opening in the wall; a spring opens the wings, which then provide anchorage when the bolt is tightened.

Surface Wiring. An excellent method of adding additional receptacles is to use multioutlet assemblies such as the plug-in strip shown in Fig. 28-17. Armored cable or similar wiring to supply the multioutlet assembly can usually be installed in a concealed area. This and other surface raceways will be discussed in Chap. 28.

Don't overlook the combinations of nonmetallic box and wiring devices shown in Fig. 23-9. They are convenient and practical wherever surface wiring is acceptable and appearance is not a major factor. They can be used *only* with nonmetallic-sheathed cable.

Use Common Sense and Patience. No book can outline all the problems in carpentry that will be encountered in old work. The method of construction in houses varies widely with the age of the house, the skill and integrity of the builder, the geographical area, and many other factors. Plenty of patience, coupled with a generous measure of "horse sense," is the greatest asset in old work.

Modernizing an Installation. The wiring in many old houses, and some that are not so old, needs modernizing. Some—usually the oldest—need to be completely rewired. Others, especially the not so old, need only additional circuits and outlets. In most cases, a larger service is needed. You must use good judgment when deciding what is best to do in such cases, for it is ridiculous to waste money replacing safe wiring when only additional wiring is needed. On the other hand, it is not only ridiculous but downright hazardous to allow unsafe wiring to remain in use. So examine the existing wiring before you do anything else.

If the insulation on the accessible ends of the old wires is brittle, if cables have been improperly spliced to old knob-and-tube wiring, or if improper splicing (such as without boxes) has been made in existing cables, the wiring should be replaced. If panelboard terminals have been overloaded for long periods of time and are in a deteriorated condition, the insulation on the concealed wiring might also be deteriorated and unsafe.

If the existing wiring has good insulation and has not been improperly spliced, the chances are that it can be left in service, especially if recently added *appliances* can be disconnected from the existing circuits and connected to new circuits installed for the purpose. Additional circuits should be added for both 240- and 120-volt appliances for which no wiring was originally provided. However, additional receptacles can be safely added to some existing circuits without overloading the circuits, where the additional outlets are for convenience only—where no additional load will be imposed, but where the same lamps or portable appliances as before are used; the new receptacles only add convenience: easier-to-get-at receptacles. An adequate number of receptacles eliminates the need for extension cords, which are really dangerous when the occupant of the house is tempted to use them as alleged "permanent" wiring. Hence, safety would be enhanced by the addition of such receptacles.

For example, the living room and bedrooms may simply need new receptacles added to existing lighting circuits; an existing 20-amp kitchen small appliance circuit may be all right if one or more additional circuits are installed, with new receptacle outlets for new appliances that were purchased after the original wiring was installed; or if a new 20-amp circuit is installed for laundry equipment that was originally connected to one of the kitchen small appliance circuits. Maybe a room air conditioner, a dishwasher, a waste disposal unit, or a deep-freeze unit has been connected to an existing circuit that was already fully loaded, instead of being provided with its own circuit when the appliance was purchased.

If the existing wiring is aluminum and is otherwise in good condition, you will want to replace all 15- and 20-amp receptacles and switches with devices marked "CO/ALR" and reconnect them properly, as explained in Chaps. 6 and 8. You will also want to replace any panelboard to which aluminum wiring is connected if its terminals are not marked "AL-CU," unless you are going to install a larger panelboard anyway, as is often necessary when modernizing is being done.

Plan the Job. Sections 220-2(d) and 220-31 of the Code cover methods of computing the loads for existing dwellings, which are substantially the same as the methods already explained for computing loads in new residential occupancies. So just plan and lay out your circuits as you would in a new building. Be sure to see that the cooking and dining areas have at least two 20-amp circuits, that each stationary or fixed appliance has its own circuit, and that the laundry area has its own 20-amp circuit. Preferably, each deep-freeze unit and refrigerator, especially a self-defrosting unit, should be on its own separate circuit. A homeowner can lose hundreds of dollars worth of food if a fuse blows or a breaker trips as a result of an overloaded appliance circuit that also supplies the deep-freeze and refrigerator. An individual branch circuit is a small price to pay for insurance against that kind of loss. Don't skimp on the number of branch circuits or on the number of receptacles. Remember, *most* modernizing could have been avoided by more liberal planning when the house was built. The older the house, the greater the likelihood that considerable modernizing may be needed to provide for many added appliances, some of which may not have been in existence when the house was built.

A 100-amp service will be large enough for most single-family dwellings that do not have all-electric heat. But you must make sure that the service you install is at least as large as the size required by the Code for the load involved. In any case, whether you must completely rewire the building, or merely add new circuits and new outlets, the chances are that you will have to change the service. If the building must be completely rewired, this can be done, and the old service and wiring can then be disconnected after the new service and wiring have been connected. But if some of the existing circuits (such as the lighting circuits) are to be retained, the new and the old circuits can be connected to the new service in one of two ways: (1) by retaining the old service equipment after altering it or (2) by replacing the old service equipment with a junction box.

Reconnecting Old Circuits. If the old service equipment panelboard is in good repair, if it is a 120/240-volt 3-wire panelboard, and if its grounded neutral busbar can be insulated from the panelboard enclosure by removing its main bonding jumper (by removing a screw or strap), then plan to alter and retain it. On the other hand, if the old panelboard is in poor repair, or if it is a 120-volt 2-wire panelboard, or if its grounded busbar is permanently bonded to the panelboard enclosure, then plan to replace it with a junction box or, if feasible, to convert it to a junction box.

Retaining the Old Panelboard. There are two possible ways of using the old panelboard without converting it to a junction box, depending on whether or not it has a main circuit breaker or fused switch. Let's call them "Option 1" and "Option 2."

Option 1: If the old panelboard has a main breaker or fused switch, whether the grounded busbar is bonded to the enclosure by a removable jumper (screw or strap) or is welded to the enclosure, this panelboard can be retained as a part of the total service equipment. The new panelboard, which will also have a main breaker or fuses, will also constitute a part of the total service equipment. The ampere rating of the *new* panelboard will then be based on the total load *other than* that part of the load connected to the old panelboard, but the size of the service-entrance conductors must be based on the total load connected to both panelboards, since the same set of service conductors will supply both panelboards.

Where both panelboards are used as service equipment, they must be grouped closely together. Up to six main service disconnecting means are permitted, so two disconnecting means—the main breaker or fuses in each panelboard—are acceptable. Where this is done and both panelboards are supplied by the same (new) set of service-entrance conductors, install an auxiliary gutter (a short section of wireway as discussed in Chap. 28) ahead of the two panelboards and connect the subset of service-entrance conductors from each panelboard to the main set of service-entrance conductors in the auxiliary gutter, as shown in Fig. 22-25.

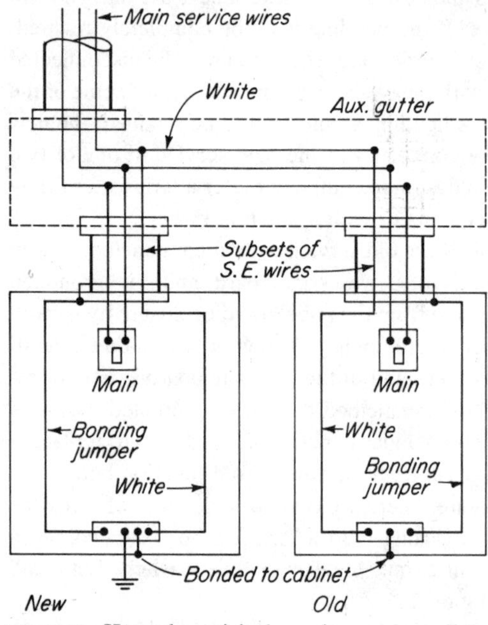

Fig. 22-25 Here the original service equipment becomes *part of* the total service equipment; the service equipment is now contained in two separate cabinets.

Since the old panelboard is retained as one of the two parts of the service equipment, its grounded neutral busbar must remain bonded to the panelboard enclosure, and the grounded neutral busbar in the new panelboard must also be bonded to the panelboard enclosure, since it, too, is part of the service equipment. The two grounded busbars will be bonded together (as required), since both will be connected to the grounded service-entrance conductor. The ground wire connected to the old panelboard will probably not be large enough for the entire enlarged service, so install one that is large enough, from the new panelboard grounded neutral busbar to the grounding electrode system.

Option 2: If the old panelboard does not have a main circuit breaker or main fused switch, it can still be retained as a load-center panelboard or fuse cabinet (downstream from the service equipment) if its grounded neutral busbar can be *insulated* from the panelboard enclosure by removing its main bonding jumper (screw or strap). In this case, the old panelboard will not be a part of the service equipment, so the grounded neutral busbar must be *insulated* from the panelboard enclosure. So remove the bonding screw or strap, and also remove the old ground wire after a new ground wire of the proper size has been connected to the new panelboard. The old metal enclosure must be grounded, but *not* through the neutral. Be sure that the metal raceway, metal armor, or grounding conductor in a nonmetallic cable properly grounds the old enclosure after insulating the neutral and removing the old grounding electrode conductor.

In this case, the *old* panelboard must be supplied by a *feeder* from a circuit breaker or set of fuses in the *new* panelboard. This feeder must have an ampacity rating at least as great as the ampere rating of the old panelboard, and the circuit breaker or fuses in the new panelboard protecting this feeder must have an ampere rating that is not greater than the ampere rating of the old panelboard. See Fig. 22-26.

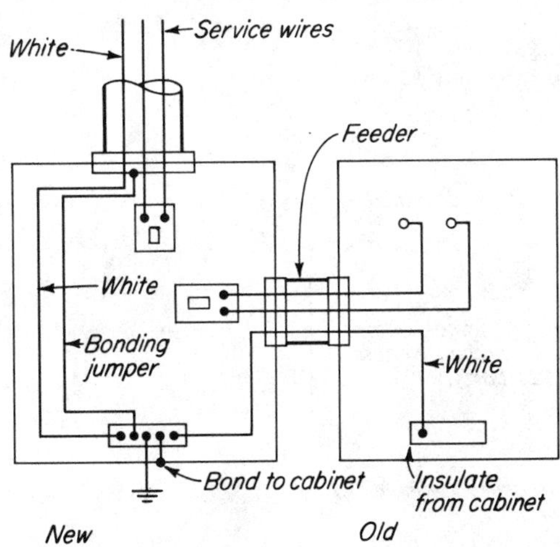

Fig. 22-26 Here the original equipment is *not* a part of the service equipment. It is only a "load center" for the original circuits.

In this case, the two panelboards may be any distance apart, such as in different rooms or areas of a basement. So if it is desirable to install the new service in a different location, this method may be preferable to the first method described, even if the old panelboard does have a main.

Removing or Converting Old Service Panelboard. Let us call this "Option 3." The new service equipment must have a main breaker or main fused switch, and must have branch-circuit breakers or fuses for each old circuit and for each new circuit, and preferably a few spares for future circuits. If the old panelboard cabinet or enclosure will be totally enclosed (and if every knockout that has been removed but will not be used is closed with a knockout closure as discussed in Chap. 10), the empty cabinet can be used as a junction box. However, if the cabinet is the very old type with the top open for connections to an old-style "A-base" meter, it must be discarded. Replace it with a junction box of adequate size, at least 6 × 8 in or preferably larger.

In either case, the old circuit wires are spliced to new circuit wires, installed from circuit breakers or fuses in the new panelboard. In other words, the old panelboard cabinet is either used as a junction box, or replaced by a suitable junction box. In either case, the new and old branch-circuit wires are spliced together, as shown in Fig. 22-27. Of course, if a new junction box is installed, the old cables must be disconnected from

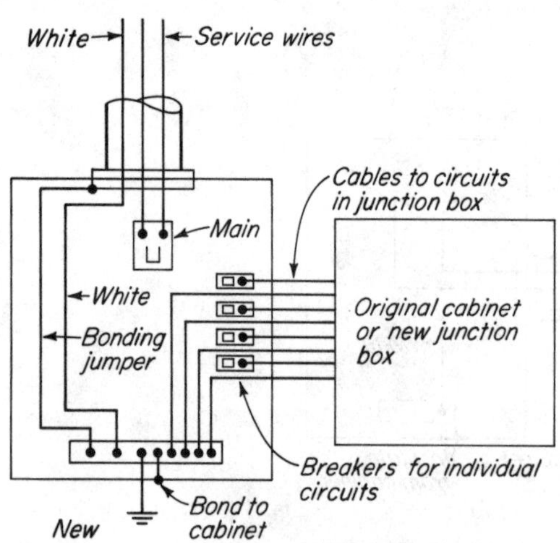

Fig. 22-27 Here the cabinet of the old equipment, if retained, serves only as a junction box.

the old panelboard cabinet and connected to the new junction box installed in its place. A new ground wire is installed in the new panelboard in the usual manner, and the old ground wire removed.

Changing Over. There are several ways in which you can make the changeover from the old to the new service, but an easy way is to have the new service equipment complete, with all *new* circuits connected, and the old system as nearly complete as possible without working on anything while it is hot. Then have your power supplier disconnect the old service and connect the new service. After that has been done, disconnect the old service-entrance conductors from the old panelboard and, *with the circuit breakers in the new panelboard turned off,* complete the job as follows:

Disconnect the existing circuits from the old panelboard, remove the interior of the old panelboard enclosure or replace the old panelboard with a junction box, connect the old and the new cables or conduit to the old cabinet or to the new junction box, and splice the old and new circuit conductors together. Remember, the white grounded wires of the old circuits must be insulated from the cabinet.

23
Farm Wiring

Whether we talk about a house in the city or one on the farm, appliances are used for the convenience they provide. But on farms, many other appliances besides those in the house are used in the business of farming —appliances bought as an investment on which the farmer expects dividends. These include water pumps, milking machines, milk coolers, hammer mills for chopping fodder, hay or crop dryers, water heaters to provide scalding hot water for the dairy, etc. These appliances of course constitute a very considerable load.

The wiring of farms involves problems already discussed, and many new ones. The maximum load in use at one time is likely to be much greater for the total farm than for a city home. There is a great deal of outdoor wiring, either overhead or underground, between the various buildings. Substantial distances are involved, which means that wires must be of sufficient size to avoid excessive voltage drop, and for mechanical strength in overhead runs. There may also be relatively poor grounding conditions. These and related factors will be covered in this chapter.

Preview. In a typical farm installation, wires from the power supplier's transformer end at a pole in the yard. (These are sometimes called ''May-poles'' to distinguish them from the power supplier's distribution poles, and because wires fan out from them in various directions.) From the top of the pole wires run down to the meter, then back to the top of the pole. Sometimes there is a circuit breaker, or a switch with or without fuses, just below the meter socket. The neutral wire, the meter socket, and any breaker or switch cabinet are always grounded at the pole.

From the top of the pole, a set of wires runs to the house, and other sets to barns and other buildings. Instead of overhead wiring, underground cables are being used more and more.

Adequacy. Adequate wiring as discussed in Chap. 12 applies equally to city or farm homes. Indeed, if anything, the farm home often deserves more attention than the city home, for, especially in the case of small farms, certain appliances are often installed in the house, instead of in other buildings as is the case on larger farms. Special attention must also be paid to adequacy of wiring in and between other buildings, and on the meter pole.

Few farmers whose farms are being wired for the first time can foresee all the different electrical appliances and motors they will be using in a few years' time. Too often the number of circuits originally installed turns out to be too small; wires between buildings turn out to be too small; wires on the pole should be larger. Always provide more capacity than is needed at the time of installation; doing so will increase the labor very little, but it will do away with later expensive alterations.

Overhead or Underground? A farm wired using overhead wires will have a large number of wires running all over the place. This gives the farm a very untidy appearance and invites problems of various kinds. Wires to low buildings can be damaged by moving vehicles. Many overhead wires on an isolated farm invite trouble from lightning. In northern climates where sleet storms are common, wires can break from the weight of accumulated ice; a broken wire on the ground is dangerous. Underground wires cost little more and do away with these problems.

If you are going to use overhead wiring, watch wire size carefully. Make sure the wire size selected is big enough to carry its load in amperes without excessive voltage drop. Make sure it is strong enough for the length of the span, which means that sometimes you must use wires larger than would otherwise be necessary for the load involved. Review Chap. 7. If you are going to use underground wires, use Type UF cable or raceway and follow the methods already discussed in Chap. 17 concerning the service entrance.

If several sets of underground wires are to enter the bottom of the meter socket at the pole, you will have a problem because the hub at the bottom of the socket is too small. In that case provide a weatherproof junction box just below the meter socket. Quite a number of runs can be terminated in the box, with only a single set of wires from the meter socket into the junction box. A junction box for this purpose is shown in Fig. 23-1. This particular junction box has terminals for the various runs

Fig. 23-1 A rainproof junction box of this type makes it easy to install several underground runs, starting at the meter socket. (*Hoffman Engineering Co.*)

of underground wires, doing away with the need for and nuisance of a lot of bulky splices in a box without terminals. A telescopic metal channel to protect a group of underground wires running up the side of a pole is shown in Fig. 23-2, installed with one of the junction boxes of Fig. 23-1.

Made Electrodes. The Code requirements for a "grounding electrode system" at each service, discussed in Chap. 17, apply to all occupancies, including farms. However, it is common on farms that none of the required "existing electrodes" are available, particularly at a yard pole, in which case a "made electrode" must be used.

Grounding on Farms. Chapter 9 covered the subject of grounding in general. You learned that for safety, the grounded circuit wire must be connected to the earth. A good ground connection offers protection not only against troubles in the wiring system but also against lightning. On a farm, there is considerably more danger from lightning than in a city. Yet a good ground connection is usually much harder to obtain on a farm than in a city. It is a most important subject, and often too little attention is paid to it. Study it well and fully grasp the principles outlined in the following paragraphs.

The Code in Sec. 250-84 specifies that the resistance between a driven pipe or ground rod, and the earth, should not exceed 25 ohms. A contin-

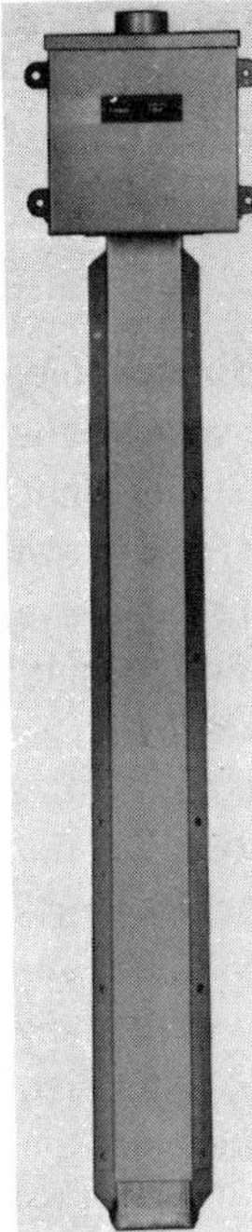

Fig. 23-2 The trough at left protects the above-ground portions of underground runs at the pole. The junction box is at the top. (*Hoffman Engineering Co.*)

uous underground metal water piping system usually provides a resistance to ground of less than 3 ohms. Local underground water or oil piping, or the frame of a metal-frame building with below-ground concrete footings, usually has resistance to ground of considerably under 25 ohms.

A study of over 200 farms in Minnesota was made in a rather dry year. Only 9 of 215 (1 in 23) had a resistance to ground of 25 ohms or less—and a 25-ohm ground is not a good ground. The other 22 out of every 23 had a resistance of over 25 ohms, some well over 100 ohms. In other words, perhaps one in a hundred farms had what would be considered a good ground connection in a city. The reason for the high resistance was reliance on a driven ground rod, which rarely provides as good a ground as an underground metal pipe system. The resistance to ground using a ground rod varies from time to time. The drier the earth, the higher the resistance, so if practicable, install the rod where rain from a roof will help to keep the earth wet. Usually a single rod is not sufficient; install two of them, at least 6 ft apart, and bond them together with No. 6 copper wire, which may be bare, using ground clamps at each end of the wire. Installing *more than two* rods will not significantly reduce the resistance to ground. Rods *longer* than the minimum required 8 ft probably *will* reduce the resistance to ground. If rock bottom is encountered, and a rod cannot be driven in 8 ft, the rod may be buried horizontally, as deep as possible.

The Meter Pole. There is a right and a wrong location for the meter pole. Why is there a pole in the first place? Why not run the service wires to the house and from there to the other buildings, as was sometimes done when farms were first being wired? That would lead to very large wires to carry the total load; very large service equipment in the house; expensive wiring to avoid excessive voltage drop, which is wasted power; a cluttered farmyard; and many other complications.

Locate the pole as near as practical to the buildings that use the greatest amount of power; on modern farms, the house rarely has the greatest load. Doing so will mean that the largest wires will be the shortest wires. In that way you will find it relatively simple to solve the problem of excessive voltage drop, without using wires larger than would otherwise be necessary for the current to be carried. The large, expensive wires to the buildings with the big loads will be relatively short, and the smaller, less expensive wires to buildings with the smaller loads will be relatively long. That keeps the total cost down.

Basic Construction at Pole. Of the three wires from the power supplier's transformer that end at the top of the pole, the neutral is usually but not always the top wire. Check with your power supplier. The neutral is always grounded at the pole. Wires from the top of the pole down to the meter socket, and back again to the top of the pole, can be installed in two ways. Use the method favored in your community.

In what is called "single-stack construction," the two hot wires and the neutral are run in one conduit down to the meter socket, and two more hot wires are run back to the top of the pole, all in the *same* conduit. The neutral ends at the meter socket, but at the top of the pole it is spliced to the neutrals of all the wires that run to various buildings, all as shown in Fig. 23-3.

In the second method, called "double-stack construction," all three wires are brought down from the top to the meter socket, in one conduit. Then all three wires *including the neutral* are run back to the top in a second conduit, as shown in the inset of Fig. 23-3. In this method the neutral from the power supplier's transformer is *not* spliced to the neutrals of the

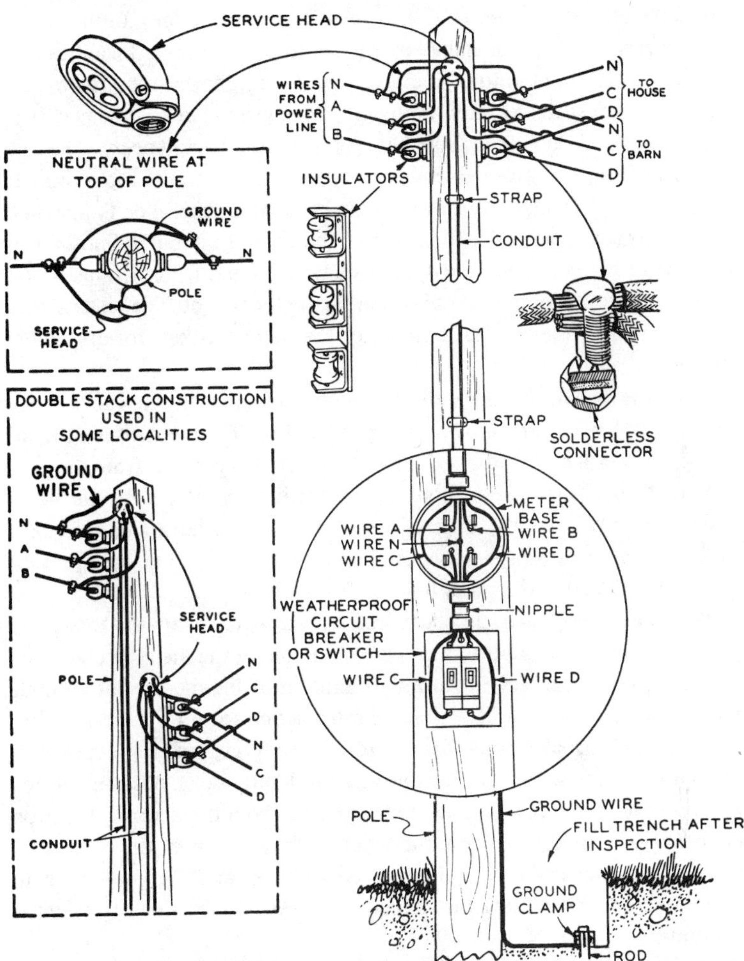

Fig. 23-3 Typical meter-pole installation on a farm.

wires running to various buildings; the neutral wire that runs back to the top of the pole from the meter socket of course is so spliced. (In this method, two lengths of 3-wire service-entrance cable may be used in place of two lengths of conduit.) Leave at least half the circumference of the pole clear so that line workers and repair workers can climb the pole without trouble.

Installing the Meter Socket. The meter socket is sometimes furnished by the power supplier but installed by the contractor. In other localities it is supplied by the contractor. Mount it securely at a height required by the local authorities; this is usually 5 to 6 ft above ground level. If a switch or circuit breaker is also used, install it just below the meter socket. The two hot wires from the power line always are connected to the top terminals of the meter socket. If the usual switch or circuit breaker is installed below the meter, the connections are as shown in the detail in Fig. 23-3. If a switch or breaker is *not* used, the two wires *C* and *D* in the illustration run directly to the bottom terminals of the meter socket. The neutral wire is connected to the middle terminal of the meter socket, for grounding purposes.

Insulators on Pole. Near the top of the pole, install insulator racks of the general type that were shown in Fig. 17-4. Provide one rack for the incoming power wires, and one for each set of wires running from the pole to various buildings. Remember that there is great strain on the wires under heavy winds, or in case of heavy icing in northern areas. Anchor the racks with heavy lag screws. Better yet, use at least one through-bolt, all the way through the pole, for each rack.

Ground at Meter Pole. The neutral wire always runs from the top of the pole, through the conduit to the center terminal of the meter socket. The neutral is not necessary for proper operation of the meter, but grounds the socket. At one time it was standard practice to run the ground wire from the meter socket (out of the bottom hub) to ground, but experience has shown that better protection against lightning is obtained if the ground wire is run outside the conduit. Run it from the neutral at the top of the pole, directly to the ground rod at the bottom of the pole. It is usually run tucked in alongside the conduit as far as it goes, then to ground. In some localities it is stapled to the pole on the side opposite the conduit.

Since at least 8 ft of a ground rod must be in contact with the soil, a 1-ft-deep trench about 2 ft long is dug from the pole or building, and the ground rod is driven into undisturbed soil. The top of the rod is a few inches above the bottom of the trench. The ground wire runs down the side of the pole or building to the bottom of the trench, then to the ground clamp on the rod. After inspection the trench is filled in, and the rod, the clamp, and the bottom end of the ground wire remain buried (see Fig. 23-4). Instead of a single rod, two ground rods, at least 6 ft apart, are recommended for a lower-resistance ground.

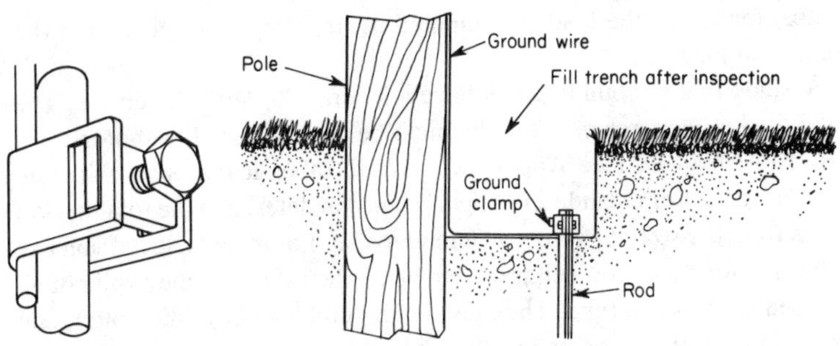

Fig. 23-4 In some localities, the top of the ground rod is below the surface of the ground. Fill the trench *after* inspection. If the ground rod is copper, use *only* a copper clamp.

How to determine the size of the ground wire has already been discussed in Chap. 17; do remember that if the ground wire runs to a made electrode, such as a ground rod, it never needs to be larger than No. 6.

Assembling Components on Pole. Whether you use the single- or the double-stack method, the general procedure is the same. At one time, the conduit often ended at a point below the insulators, which resulted in water following the wires into the service head. Now NEC Sec. 230-54 requires the service head to be *above* the topmost insulator, as also shown in Fig. 23-3. The service head itself will be of the general type that was shown in Fig. 17-8 with the right number of holes in the insulator. Run wires through the conduit, white or bare for the neutral, and usually black and red for the two hot wires, although these may be any color except white or green.

Everything on the pole (except wires from the power supplier's line and wires to buildings) is preassembled on the pole before it is erected. This includes the conduit with its service head and with wires already installed, the meter socket, the breaker or switch if used, the ground wire from the top of the pole, and all the insulators. Then when the pole goes up, it is ready for overhead wires to be installed on the insulators, and connected.

Wire Sizes: Pole to Buildings. Each building must have a service entrance quite similar to those discussed in earlier chapters, except without a meter. Wires from the pole to any building must be of the proper size. NEC Secs. 220-40 and 220-41 show how to compute the load for service conductors and service equipment for each farm building (except the house, for which the load is computed as for any other house) and the total farm load.

Assume that the building will have a 3-wire 120/240-volt service. The total load in amperes *at 240 volts* must be determined. For motors, see NEC Table 430-148 (see Appendix). For all other loads, start with the load in watts. For incandescent lights[1] you can determine the total load in watts from the size of lamps you intend to use. For receptacles, if you use 200 watts for each, you will probably be on the safe side; they will not all be used at the same time. Then divide the total watts by 240 (volts), and you will have the amperes at 240 volts.

Caution: Suppose that in a building you have six 120-volt 15-amp circuits for lights and receptacles. That theoretically makes a total of 90 amp at 120 volts, or 45 amp at 240 volts. But these circuits will not all be loaded to capacity at the same time, so determine the load in accordance with the following paragraphs.

For each building to which wires run from the pole (except the farm *house;* figure that as outlined in earlier chapters), first determine the amperes, at 240 volts, of all the loads that have any likelihood of operating *at the same time*. Enter the amperes under *a* of the tabulation below. Then proceed through steps *b* to *f* as outlined in the tabulation.

a. Amperes at 240 volts of all connected loads that in all likelihood will operate at the same time, including motors if any . . . . ______ amp

b. If *a* includes the *largest* motor in the building, add here 25% of the ampere rating of that motor (if two motors are the same size, consider one of them the largest) . . . . . ______ amp

c. If *a* does *not* include the largest motor, show here 125% of the ampere rating of that motor . . . . ______ amp

d. Total of $a + b + c$ . . . . ______ amp

e. Amperes at 240 volts of all other connected loads in the building . . . . ______ amp

f. Total of $d + e$ . . . . ______ amp

Now determine the minimum service for each building by one of the following steps:

[1] For fluorescent lights, add about 15% to the watts because the watts rating of a fluorescent lamp defines only the power consumed by the lamp itself. The ballast adds from 10 to 20%.

A. If f is 30 or less and if there are *not over two* circuits, use a 30-amp switch and No. 8 wire.[2] If there are *three or more* circuits, use a 60-amp switch and No. 6 wire.

B. If f is over 30 but under 60, use a 60-amp switch and No. 6 wire.

C. If d is less than 60 *and if f* is over 60, start with f. Add together 100% of the first 60 amp plus 50% of the next 60 amp plus 25% of the remainder. For example, if f is 140, add together 60, plus 30 (50% of the next 60 amp) plus 5 (25% of the remaining 20 amp) for a total of 60 + 30 + 5 or 95 amp. Use a 100-amp switch and wire with ampacity of 95 amp or more.

D. If d is *over 60 amp* start with 100% of d. Then add 50% of the first 60 amp of e, plus 25% of the remainder of e. For example, if d is 75 amp and e is 100 amp start with the 75 of d, add 30 (50% of the first 60 of e), and add 10 (25% of the remaining 40 of e) for a total of 75 + 30 + 10 or 115 amp. Use a switch or breaker of not less than 115-amp rating and wire with corresponding ampacity.

The wire sizes determined above will be the minimum permissible by the Code for each building. You would be wise to install larger sizes to allow for future expansion.

Overhead Spans. NEC Sec. 225-6(a) requires a minimum of No. 10 wire for overhead spans up to 50 ft, and No. 8 for longer spans. While the Code is silent on the subject, it is recommended that if the distance is over 100 ft, No. 6 should be used; if it is over 150 ft, it is best to install an extra pole. If the wires are installed in northern areas on a hot summer day, remember that a copper wire 100 ft long will be a couple of inches shorter the following winter when the temperature is below zero. Leave a little slack lest the insulators be pulled off the buildings during winter.

Triplex Cable. This material was shown in Fig. 17-5; it consists of two insulated wires wrapped spirally around a strong, bare neutral wire. It requires only one insulator for support. One triplex cable is usually considered preferable to three separate wires. Triplex cable is recognized by the Code as "Messenger Supported Wiring," Art. 321. It has been used by power suppliers as service drops for many years.

Total Load. The calculations just described determine the size of the wires from the pole to the separate buildings. To determine the size of the

[2] The Code in Sec. 230-41 requires a minimum of No. 8 if the building has not over two 2-wire branch circuits, and a minimum of No. 6 in all larger installations.

wires on the pole (from top, to meter socket, back to top of pole), proceed as follows using the load in amperes at 240 volts, as determined above, for each building (excluding the house until step 6):

1. Highest of all loads in amperes for an individual building: _______ amp at 100% . . . . _______ amp
2. Second highest ampere load: _______ amp at 75% . . . . _______ amp
3. Third highest ampere load: _______ amp at 65% . . . . . _______ amp
4. Total of all other buildings (except house): _______ amp at 50% . . . . _______ amp
5. Total of above . . . . _______ amp
6. House computed as discussed in other chapters, at 100% . _______ amp
7. Grand total including house . . . . _______ amp

Important: If two or more buildings have the same function, consider them as one building for the purpose above. For example, if there are two brooder houses requiring 45 and 60 amp, respectively, consider them as one building requiring 105 amp and enter 105 in line 1 above.

The total of line 7 above is the minimum rating of the breaker or switch (if used) at the pole, and the minimum ampacity of the wires that you must use on the pole. Do note that in listing the amperes for any building, the load in amperes used is the *calculated* load, not the rating of the switch or breaker used. For example, if for a building you calculated a minimum of 35 amp but you use a 60-amp switch (because there is no size between 30 and 60 amp), use 35, not 60 amp.

To connect the various wires at the top of the pole to each other, use solderless connectors of the general type that were shown in Figs. 8-17 and 8-18. Unless these connectors have snap-on insulating covers, the splices must be taped.

If the wires from the pole to the various buildings are to be underground, the wires from the bottom terminals of the meter socket will then, instead of running back to the top of the pole, be connected to the various underground wires.

Current Transformers. If the service at the pole is rated at 200 amp or more (line 7 of the preceding table), it means very large wires running from the top of the pole down to the meter, and back again to the top. That is both expensive and clumsy. It is not necessary to run such large wires down to the meter: Use a current transformer (abbreviated CT).

An ordinary transformer changes the *voltage* in its primary to a different voltage in its secondary. In a current transformer, the current flowing in its primary is reduced to a much lower current in its secondary. Most current transformers are so designed that when properly installed, the current in the secondary will never be more than 5 amp. A typical current transformer has the shape of a doughnut 4 to 6 in. in diameter. It has only a single winding: the secondary. The load wire (in which the current is to be measured) is run through the hole of the doughnut and becomes the primary. Assuming that the transformer has a 200:5 ratio, for use with a 200-amp service, the current in the secondary will be five two-hundredths of the current in the primary. If the current in the primary is 200 amp, 5 amp will flow in the secondary; if it is 100 amp in the primary, $2^1/_2$ amp will flow in the secondary. If the service were 400 amp, the transformer would have a 400:5 ratio.

One current transformer is installed at the top of the pole, with the two hot wires running through the hole. Four small No. 14 wires run from the top of the pole to the meter (which must be of the type suitable for use with a current transformer): two from the secondary of the current transformer and two from the hot wires, for the voltage. The meter operates on a total of not more than 5 amp but the dials of the meter will show the actual kilowatthours consumed. A wiring diagram is shown in Fig. 23-5.

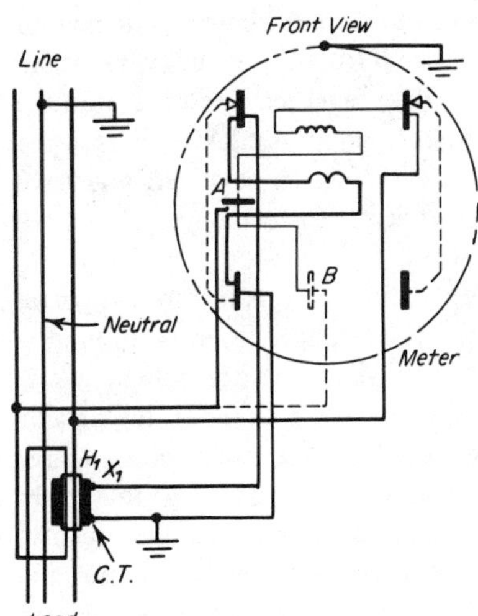

Fig. 23-5 Wiring diagram, using one current transformer, on a 120/240-volt service. (*General Electric Co.*)

Caution: The secondary terminals of the transformer must always be short-circuited while any current is flowing in its primary. If connected to a kilowatthour meter, the meter constitutes a short circuit. As purchased, the transformer will probably have a short-circuiting bar across its secondary terminals; this must *not* be removed until the transformer is connected to the meter. If you were to touch the terminals of such a transformer that is *not* short-circuited, you would find a dangerous voltage of many thousands of volts.

In an installation using a current transformer at the top of the pole, if you also want a switch at the pole, install it at the top. Such poletop switches are operated by a handle near the ground level, connected through a rod to the switch at the top as shown in Fig. 23-6. Note that the current transformer is installed in the same cabinet with the switch. Such switches are also available in the double-throw type, as required if a stand-by generating plant is installed for use during periods of power failure.

Your power supplier may require the current transformer to be accessible from the ground, and if so you may need a stub pole and wood backing for the service and metering equipment over 200 amp that is too large to fit on the pole.

Entrance at Individual Buildings. The entrance at any building served directly from the pole is made as already outlined in Chap. 17, except that the meter is omitted. Instead of running the ground wire from the neutral bar of the service equipment, you may run it from the neutral wire, at the point where it reaches the building, to the grounding electrode. Run the ground wire along the side of the service conduit or cable, just as at the pole, for maximum protection against lightning. If, however, you have used underground wires, ground the service in the usual way. Bond the neutral to the service equipment cabinet.

If a building has only one branch circuit and contains no equipment that must be grounded regardless of circumstances, then no ground is required. Necessarily the building must be wired with nonmetallic wiring and boxes, lighting fixtures must be of porcelain or other nonmetallic material, and there cannot be a receptacle, for receptacles must *always* be grounded. There are few buildings that would qualify; some that might qualify are an outhouse and a small shed. In a building that does qualify, you can use an ordinary toggle switch as the disconnecting means; the overcurrent protection may be in the building served or in the building from which it is served.

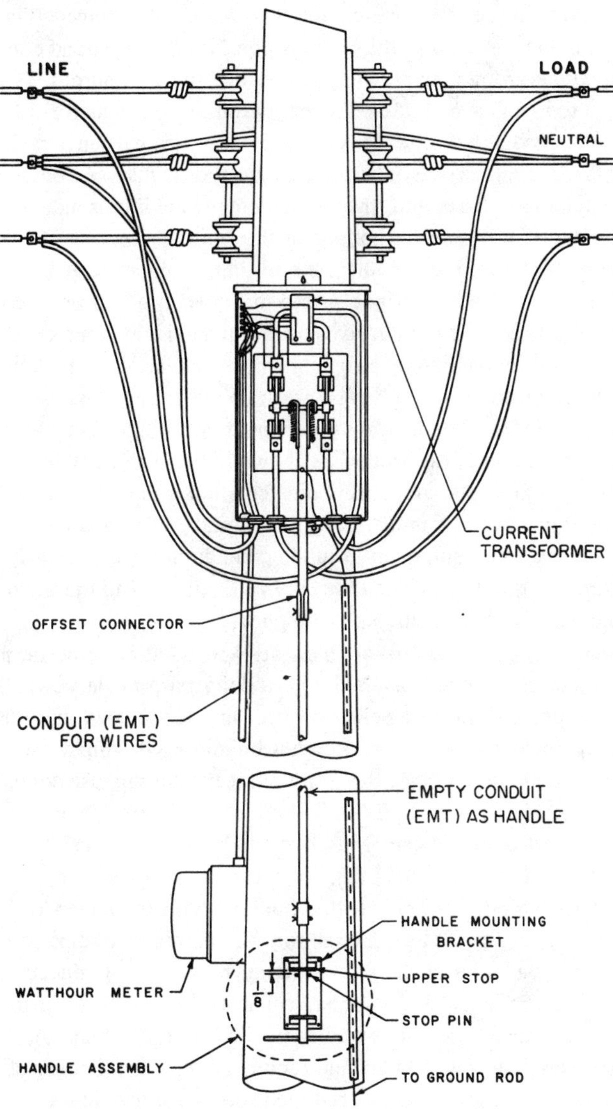

Fig. 23-6 If a switch is wanted in a high-capacity service, install it at the top of the pole. Note the current transformer in the cabinet of the switch. (*Hoffman Engineering Co.*)

In buildings with more than one circuit, you have two choices. The simpler is to provide a ground at the building consisting of a ground connection to any underground metal piping system and to a ground rod. The other choice is to install a separate equipment grounding conductor with the feeder to the second building. This might be an extra wire in a cable or single conductors either overhead or underground. It may be solid or stranded, insulated or bare, but the equipment grounding conductor *must* be insulated if livestock is housed in the building served. If the wire is insulated and No. 6 or smaller, the insulation must be green or green with one or more yellow stripes. Green insulated conductors larger than No. 6 may be difficult to obtain, so it is permitted to identify a grounding conductor larger than No. 6 by stripping off the insulation where connections are to be made or by applying a green color or green tape to the insulation. The size of this conductor is determined by the size of the overcurrent protection ahead of the feeder, as shown in NEC Table 250-95. At the building served this equipment grounding conductor must be bonded to any of the electrodes listed in NEC Secs. 250-81 and 250-83 that may be present. In this case the equipment grounding conductor must run with the circuit wires all the way back to the service equipment in the building from which the separate building is supplied.

Provide branch-circuit breakers or fuses at each building. One farm building might need only one branch circuit; a dairy barn might need 20 circuits. The equipment must be selected according to the load and the number of circuits. For a one- or two-circuit building, you might use a small fused switch as shown in Fig. 15-2; for a larger building, use equipment as shown in Fig. 13-4 or 13-8.

Tapping Service Wires at Building. Often two buildings, each with a nominal load, are located close to each other; they can then be served by one set of wires from the pole, running to the building with the larger load. Naturally the wires must be large enough for the combined load of both buildings. At the insulators of the first building, make a tap connection and run the wires to the second building, as shown in Fig. 23-7. At the second building, proceed as if the wires came directly from the pole.

If the second building is very small and requires only 120 volts, tap off only two wires (one hot wire and the grounded wire) as shown in the picture. If the second building has a considerable load so that 120/240 volts is desirable, make a tap connection to all three wires. Remember the requirement for a separate service equipment and a ground at the second building.

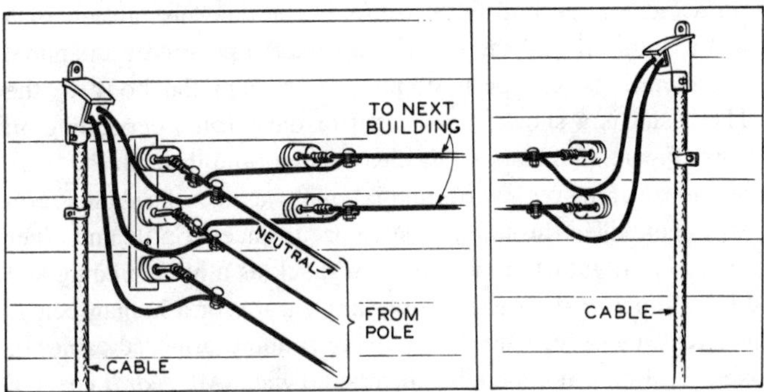

Fig. 23-7 If two buildings are near each other, and if the loads are small, tap the service wires where they are anchored on the first building and run them to the second.

Cable for Barn Wiring. In the early period of the 1930s when farms were first being wired in great numbers, the only type of nonmetallic-sheathed cable then made (now known as Type NM) was generally used in the wiring of farm buildings including barns. It gradually became clear that the usual high humidity and corrosive conditions in barns caused quick rotting of the cable (the outer fabric jacket and the fillers in the cable acted as wicks drawing moisture into the inside of the cable), leading to danger of short circuits, or shocks and fires. It became necessary to rewire many farm buildings after only a few years because of the short life of the cable.

This led to the development of what at first was called just "barn cable," which has now been standardized as Type NMC nonmetallic-sheathed cable. Later underground feeder cable Type UF was developed. This too has been found suitable for use wherever Type NMC (or Type NM) is otherwise used. Type UF may even be buried directly in the ground as has already been explained in Chap. 17.

In the wiring of barns and other farm buildings housing livestock, use only Type NMC or Type UF. There is no reason why the same types should not be used in all the buildings on a farm.

If you were to wire a barn using metal conduit or armored cable, the same corrosive conditions that rot away ordinary Type NM cable also attack the metal of the raceway or cable. As the metal rusts away, it destroys the equipment grounding conductor (since the raceway or armor serves as the equipment grounding wire). Suppose that in fact a metal system had been used and that the metal had been destroyed at some point. If an accidental ground fault occurs in the raceway, armor, or a metal box at a point beyond the break and if then a person or an animal touches the metal, the circuit is completed through the body to the ground. The result is a shock, unpleasant or dangerous, depending on many factors. Figure 23-8 makes this clear; the ground through the body is equivalent to touching both wires and is equally dangerous. Many farm animals have been killed through just such occurrences. Cattle and other farm animals cannot withstand as severe a shock as a human being and are killed by a shock that would be only unpleasant for a human being.

For these reasons never wire a barn using conduit, armored cable, or metal boxes. Use only nonmetallic-sheathed Type NMC,[3] or Type UF cable, with nonmetallic boxes.

Nonmetallic Boxes. Nonmetallic boxes are made of various plastic materials. Naturally face plates or covers used with them must also be nonmetallic. Nonmetallic boxes may be used only with Types NM, NMC, or UF cable.[4] Figure 10-4 shows a representative assortment of such boxes. Many types are available, including those with mounting brackets for installation on timbers of buildings. Nonmetallic boxes, being of one-piece construction, cannot be ganged, but multigang boxes are available.

Install such boxes as you would metal boxes, except that per NEC Sec. 370-7(c) connectors or clamps are not required at a single-gang box where the cable is secured within 8 in of the box, measured along the sheath. In other than single-gang boxes, cable clamps are required. In all cases the cable sheath must extend at least 1/4 in inside the box. Remember that in single-gang boxes without internal cable clamps, as explained in Chap. 10, one more wire may be installed in the box than in a similar steel box with clamps.

In the installation of nonmetallic boxes, one precaution is in order. Wood swells and shrinks in locations where dampness and humidity vary from time to time. Steel boxes, if solidly mounted on supporting timbers, do not present a mechanical problem as the timbers swell with moisture,

[3] In the house, Type NM cable or any other wiring system may be used. Usually, however, Type NMC is used throughout the entire farm. Also note that if Type NMC cable is hard to find, as is the case in many localities, Type UF cable (which costs only a trifle more than Type NMC) may always be used wherever Type NM or Type NMC may be used.

[4] While nonmetallic boxes were originally developed for farm use, in many areas they are now being extensively used in nonfarm installations.

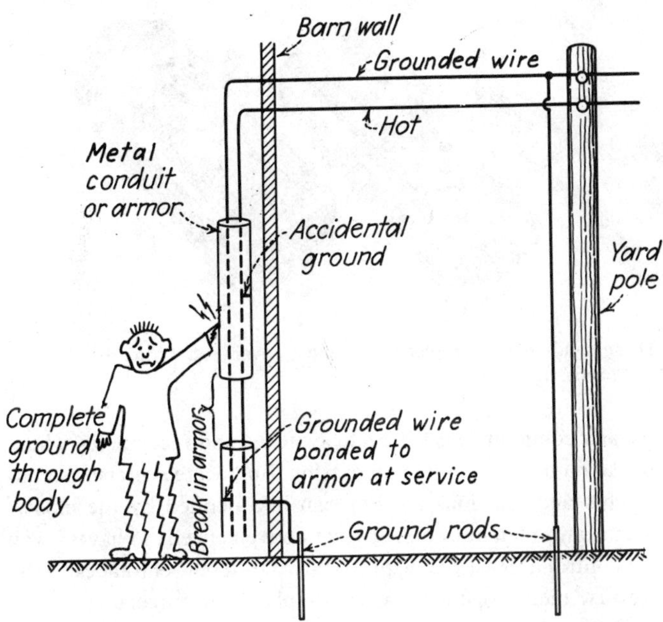

Fig. 23-8 If there were no break in the conduit or armor, ground-fault current would normally blow a fuse. Because the conduit or armor does have a break, a dangerous condition exists.

for the steel can "give" a little if required. Nonmetallic boxes, on the other hand, if screwed down tightly on dry timbers, have been known to break out their backs as the timbers swell with increasing moisture. So if you mount nonmetallic boxes on the surface of dry timbers, leave just a wee bit of slack; don't drive the mounting screws down completely tight. It is better to use boxes with external mounting brackets.

Another type of material for nonmetallic installations consists of devices such as those shown in Fig. 23-9. Each piece is a combination of outlet box and wiring device such as a switch, receptacle, etc. They can be used only with Types NM, NMC, and UF cable. Either one or two cables can be installed at each end of the combination. Screw terminals are provided for connections to the wires in the cables.

Actual Wiring of Barns. The physical makeup of the circuits, the combination of cable and boxes and wiring devices, is not different from that already discussed in other chapters. The chief points to be observed are

Fig. 23-9 These handy devices replace outlet box, cover, and wiring device, all in one piece. (*General Electric Co.*)

practicality and common sense. Locate switches and receptacles where they cannot be bumped by animals in passing. But locate switches at convenient heights and locations so they can be operated by the elbow; farmers' hands are often both full. A great convenience is to have 3-way switches to control at least one light from two different entrances to the barn. Never use metal sockets; always use plastic or porcelain.

Barns come in all sizes and descriptions. Provide lighting outlets and receptacles in proportion to the need. In Fig. 23-10 are shown some suggested wiring diagrams, from the "Agricultural Wiring Handbook." It is wise to locate outlets for lamps between joists, so that a lamp does not project too far into the aisle between stalls, where it might easily be damaged. Since most barn ceilings are dark and dusty, they reflect practically no light. Provide a reflector for every lamp; the area underneath will be lighted almost twice as well as without a reflector. Typical reflectors are shown in Fig. 23-11; they are reasonable in cost. Their use is a good investment. Keep them clean.

Provide a light to illuminate the steps to the haymow. In the haymow itself, inspectors often require so-called vaporproof fixtures of the type shown in Fig. 23-12. A "vaporproof" fixture is simply a socket for a lamp with a tight-fitting gasketed glass globe that covers the lamp and frequently a metal guard that protects the glass. Hay and the dust that arises in haymows are easily ignitable. If an unprotected lamp is accidentally broken, the lamp burns out instantly, but during that instant the filament melts at a temperature above 4000°F. This flash can start a fire, hence the requirement for enclosed and gasketed fixtures.

Cable in haymows should always be installed where it cannot possibly be damaged accidentally. In haymows, cover it with strips of board at

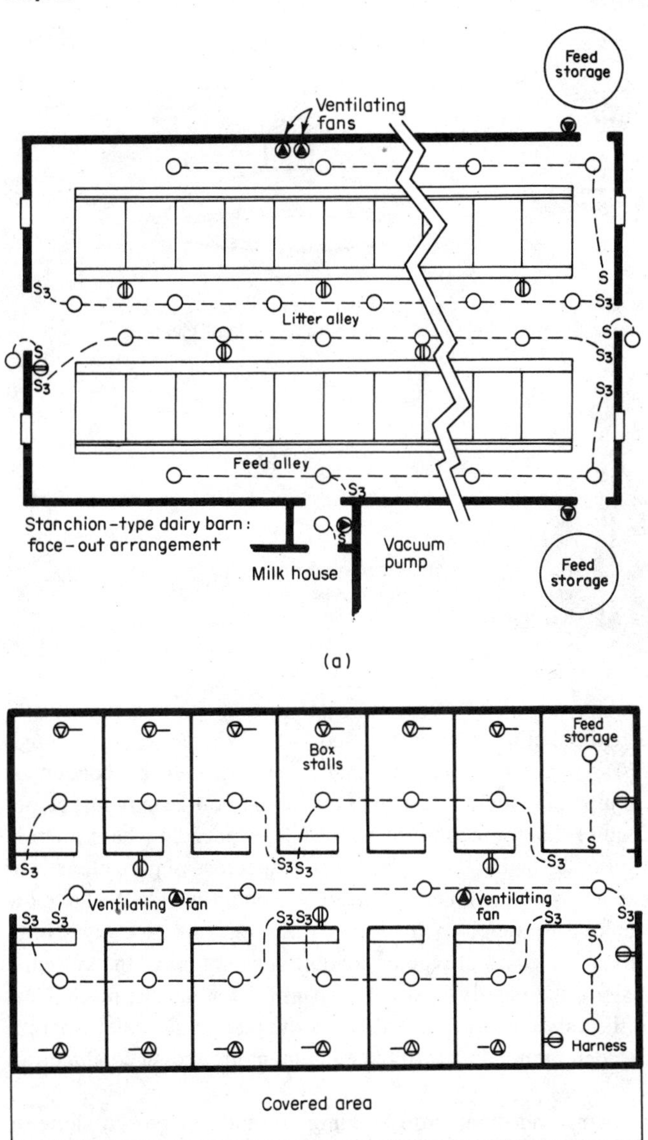

Fig. 23-10 Suggested wiring diagrams for farm buildings. (*National Food and Energy Council, Inc., 409 Vandiver West, Suite 202, Columbia, MO 65202.*)

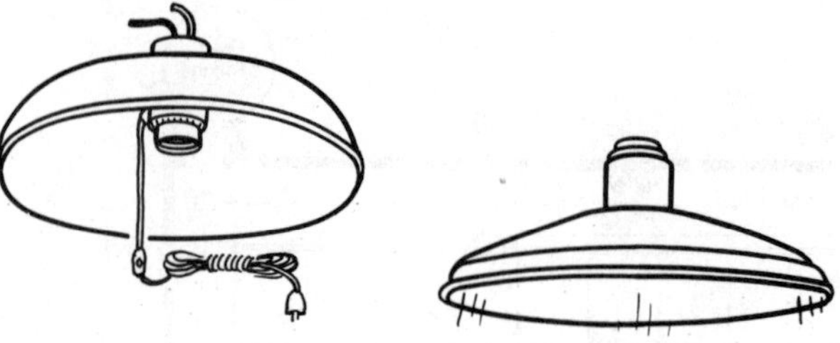

Fig. 23-11 A 60-watt lamp with a reflector is often as effective as a 100-watt lamp without a reflector. Reflectors must be kept clean.

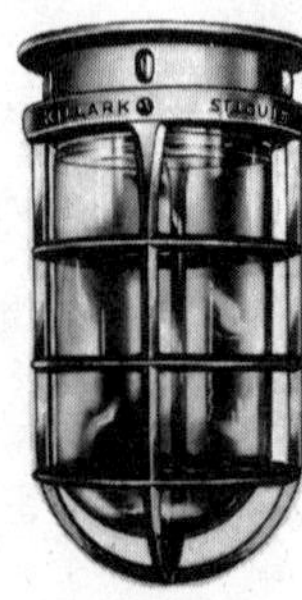

Fig. 23-12 A typical enclosed and gasketed fixture, often used in haymows. (*Killark Electric Mfg. Co.*)

points where it might be punctured by hayforks. Where it passes through a floor, NEC Sec. 336-6(b) requires that it be protected by conduit or other metal pipe to a point at least 6 in above the floor. Many inspectors sensibly require this protection for about 6 ft, especially where there is danger of forks damaging the cable. Some inspectors require conduit or EMT to extend from the haymow fixture to a point outside the haymow area. In such cases simply pull the cable through the raceway. Note, however, that this length of pipe or conduit does not make the system a conduit system; the pipe is used only for protection against mechanical damage to the cable. Be sure to terminate the pipe at an outlet where it can be grounded by bonding it to the equipment grounding conductor in the cable.

Cable in barns and other farm buildings should not be run along or *across* the bottoms of joists or similar timbers, because this exposes the cable to mechanical injury. The cable will receive good protection if it is run along the side of a joist or beam. More cable will be required to run the cable from the side of an aisle out to the middle for an outlet, then back to the side of the aisle, but consider the extra cost as insurance against damage. Figure 23-13 shows the details of recommended practice.

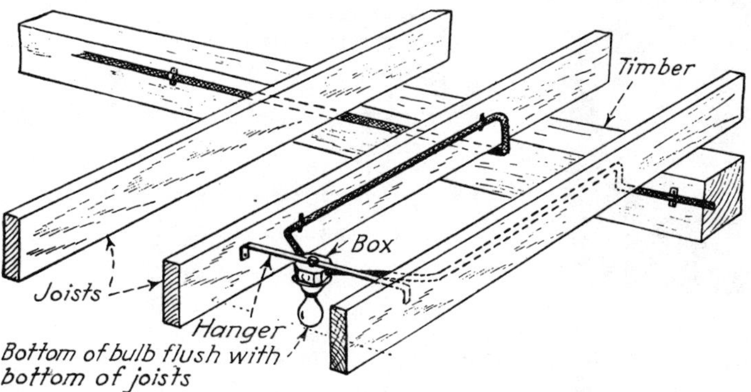

Fig. 23-13 Keep exposed cables away from centers of aisles. To avoid damage, it is best to keep lamps from projecting below the bottoms of joists.

Poultry Houses. One point that should be particularly noted is that special wiring is frequently required for lighting designed to promote egg production. It is well known that hens produce more eggs during winter months if light is provided during part of the time that would otherwise be dark. It is best to provide light at both ends of the day, morning and evening. Opinions vary as to the ideal length of the "day" but a 14- or 15-h period seems reasonably acceptable. Of course that means that the length of time the lights must be on varies from season to season; adjusting the period every 2 weeks seems a reasonable procedure.

In the evening, if all the lights are turned off suddenly, the hens can't or won't go to roost, but will stay where they are. Therefore it is necessary to change from bright lights to dim lights to dark. All this can be done by manually operated switches, but an automatic time switch of the type shown in Fig. 23-14 costs so little that manually operated switches shouldn't be considered. In the morning, the time switch turns on the bright lights at the time set, then when daylight appears turns them off.

In the evening, again at the time set, the switch turns on the bright lights. Later again at the time set, the bright lights go out and the dim lights are turned on; during this interval of dim lights the hens go to roost. Shortly thereafter all lights are automatically turned off. The wiring for

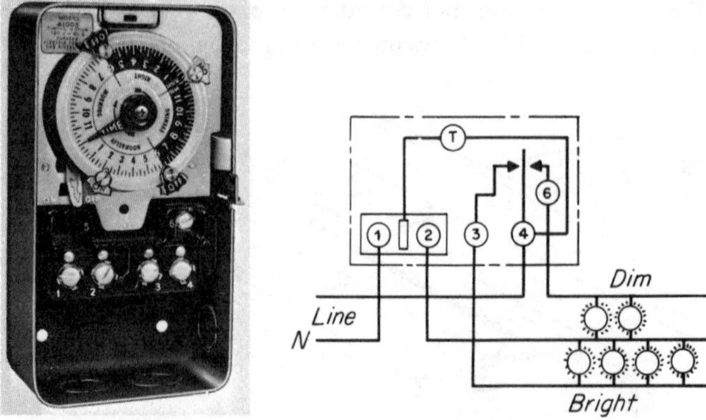

Fig. 23-14 A time switch automatically controls poultry-house lights, dim and bright, for forcing egg production. (*Paragon Electric Co., AMF Incorporated.*)

such switches is simple, and wiring diagrams are furnished with the switches.

Be sure that both the bright and the dim light falls on the roosts, for if the roosts remain in darkness when the lights come on, the hens may not leave their roosts. Neither will they be able to find the roosts in the evening if the roosts are in darkness, while the dim lights illuminate the rest of the pens. One 40- or 60-watt lamp for every 150 to 200 sq ft of floor area is usually considered sufficient for the ''bright'' period. Normal spacing is about 10 ft apart. For the ''dim'' period 15-watt lamps are suitable, but only about half as many as the number of ''bright'' lights are needed. It is a good idea to provide each light with a shallow reflector; otherwise a good share of the light falls on the ceiling and is lost.

Totally Enclosed, Environmentally Controlled Poultry and Livestock Confinement Systems. A number of fires in agricultural buildings where animals or poultry are permanently housed have been traced to a combination of dust, moisture and/or corrosive materials, and electrical wiring enclosures that are not dusttight and watertight. Dust enters the enclosure, and moisture or corrosive vapors make the dust conductive, resulting in unwanted current paths, arcing, and sometimes even in fire. To provide safety for these buildings, a new article, Art. 547, Agricultural Buildings, appeared in the 1978 Code. Your local inspector should be consulted where there is a question as to whether a particular building falls within these requirements.

Totally enclosed, environmentally controlled poultry and livestock confinement areas where litter or feed dust, corrosive vapors from animal excrement, moisture from cleaning operations or humidity, or any combination of these is present must be wired in Type UF or Type NMC cable. Open wiring on insulators is also permitted, as is conduit wiring for a Class II (dusty) Hazardous Location; open wiring, not discussed in this book, is difficult to protect adequately from physical damage, and all-metal systems are not recommended in corrosive locations. Use Type NMC or Type UF cable secured within 8 in of cast metal or molded plastic boxes with gasketed covers, similar to those shown at the top of Fig. 28-2, using watertight cable connectors of the type shown in Fig. 23-15. Be sure the oval opening in the neoprene grommet fits the cable being used. Liquidtight flexible metal conduit or Type S or SO cord must be used for flexible connections, as at motors. Lighting fixtures must be protected from physical damage or provided with lamp guards and must be

Fig. 23-15 Threaded end of watertight cable connector fits cast box hubs. For entry to nonmetallic boxes, use an adapter. (*T&B/ Thomas & Betts Corp.*)

designed or located so as to minimize the entrance of dust, moisture, or corrosive elements. Motor controllers, circuit breakers, fuses, time clocks, etc., must be enclosed in corrosion-resistant, weatherproof enclosures designed to minimize the entrance of dust, moisture, and corrosive elements. Motors must be totally enclosed or so designed as to minimize the entrance of dust, moisture, and corrosive particles.

Water Pump. Every farm will have a water pump. It not only will serve to provide water for all usual purposes but, in addition, will be a tremendous help in case of fire. But in case of fire, quite often power lines between buildings fail and fuses blow, so that the pump cannot run just when it is needed most. That can be avoided by simply considering the pump as a fire pump. The Code in Secs. 230-2 and 230-72(a) permits a fire pump to be connected through an independent service. In practice this simply means that wires are run to the pump house directly from the meter pole, ahead of the main fuses or circuit breaker. Then, even if the main fuses blow, the pump will still run. Simply run two wires from the pole, ahead of circuit breaker or main fuses, to the pump. There must be a fused disconnect switch (or a circuit breaker) for the pump only, and if mounted on the pole, it must be of the weatherproof type. Underground wires provide additional insurance against failure.

Isolated pump motors for irrigation will normally be served directly by the power supplier. Service equipment for these motors will include a fused switch or circuit breaker, meter mounting means, and a motor controller, and it should be marked both as "Service Equipment" and as "Intended for Remote, Isolated Pump Motors Only."

Motors. Stationary motors must be wired as discussed in Chaps. 15 and 30. For a portable motor, provide a receptacle appropriate for the voltage and amperage. If the motor is larger than 2 hp, the plug and receptacle must also be rated in horsepower. If the receptacle must be installed outdoors, use a cast Type FS box and one of the covers for it, of the proper size for the receptacle involved, as will be discussed in Chap. 28.

Yard Lights. Every farm will have at least one yard light. It could be installed on the meter pole, or on an outside wall of one or more buildings. The light should be controlled by 3-way switches at house and barn. Do note that the wires to the light must *not* be tapped off the service wires on the pole; the light may be mounted on the pole, but must be fed by separate wires from any branch circuit at house or barn. The yard light itself might be the old-time style of Fig. 23-16, or the modern type, shown

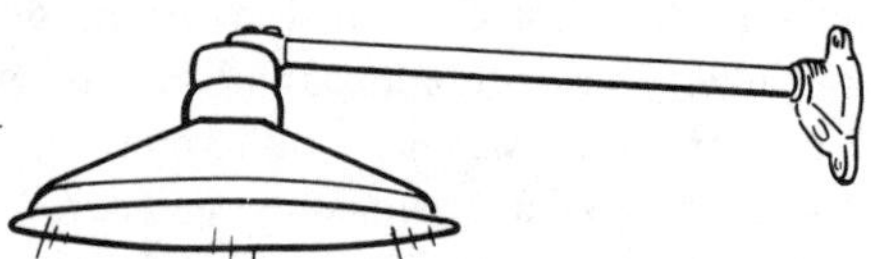

Fig. 23-16 Old-style yard light.

in Fig. 23-17. When the type with an exposed lamp is used, the lamp itself will be one of the Type R or PAR to be discussed in Chap. 29. Such lamps have internal reflectors and are available in either floodlight or spotlight type, depending on whether a large or a small area is to be lighted. Be sure the lamp you use has a "hard-glass" bulb, which will not be damaged when cold rain hits the hot bulb. Lamps with bulbs made of ordinary

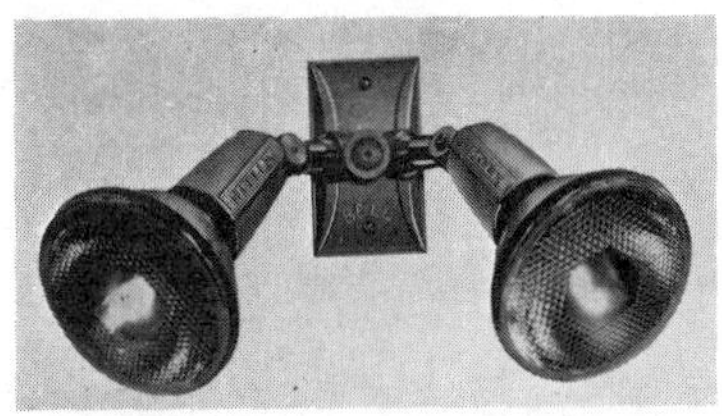

Fig. 23-17 Modern yard light using lamps with internal reflectors. (*Bell Electric Co.*)

glass shatter when hit by cold rain or snow and therefore should be used only indoors or where protected by an effective reflector.

A more modern yard light uses mercury-vapor lamps, to be described in Chap. 29. Such lamps produce about $3^1/_2$ times as much light *per watt* as ordinary lamps, cost about five times as much as ordinary reflector lamps of the hard-glass variety, but last at least twelve times longer. They require special fixtures. When used for yard lighting, they are often equipped with photoelectric controls that automatically turn the light on in the evening and off at daylight.

Lightning Arresters. While lightning-caused damage to electrical installations is quite rare in large cities, it is rather frequent in rural areas. The more isolated the location, the more the likelihood of damage. It occurs frequently on farms, and to a lesser extent in suburban areas and smaller towns. It is more frequent in southern areas, especially in Florida and other Gulf states.

Lightning does not have to strike the wires directly; a stroke *near* the wires can induce very high voltages in the wires, damaging appliances and other equipment as well as the wiring. Sometimes the damage is not apparent immediately, but shows up later as mysterious breakdowns.

While proper grounding greatly reduces the likelihood of damage, a lightning arrester, also called a secondary surge protector, properly installed reduces the probability of damage to a very low level indeed. Motor windings, in particular, are protected. Electronic equipment, such as TV and stereo, may not be completely protected, but the energy reaching them will be greatly reduced. Figure 23-18 shows an inexpensive arrester; it costs not very much over $10 and is smaller than a baseball. Three leads come out of it; connect the white wire to the grounded neutral busbar, the other two to the hot wires. At a meter socket, install the arrester as shown in Fig. 23-19. The meter socket has separate terminals for the three leads from the arrester, which is installed through a knockout and supported by a locknut in the bottom of the socket enclosure.

Fig. 23-18 A lightning arrester should be installed at the pole. Lightning arresters often are installed in the service equipment of individual buildings. (*General Electric Co.*)

Fig. 23-19 A lightning arrester installed at a self-contained meter socket. (*General Electric Co.*)

One should be installed at the meter pole. If the overhead feeders from the pole to the buildings are quite long, install another at each building. If separate terminals, as shown in Fig. 23-19, are not available in the equipment at the building, consult your local inspector for approval of the intended installation. The safety value of the protection offered by the arrester should far outweigh any technical objections to its method of installation where separate terminals are not available in the equipment.

24

Isolated Power Plants: Stand-By and Emergency

The Code does not define an isolated power plant. As used in this book, it will mean any local electric-power-producing system, usually of relatively small capacity, and with its output consumed by the owner. Such plants are used where electric power is not available from commercial power lines. Examples are isolated production or camp areas, repair or construction crews, boats, and mobile vehicles such as fire-fighting equipment. This chapter does not cover solar photovoltaic systems or wind-power generators which are sometimes interconnected with the serving utility. See page 508.

Engine-Driven Plants. This chapter will concern primarily engine-driven plants. In an isolated location, merely consider the terminals of the generator as the SOURCE, and wire the premises substantially as though they were served by a commercial power line; observe the safety requirements practiced in ordinary wiring. Some but not all engine-driven plants could be used not only as an isolated plant in locations mentioned in the preceding paragraph, but as a stand-by plant, or an emergency plant, as well.

Stand-By Plants. The Code in Arts. 701 and 702 covers stand-by plants, which generally mean plants installed primarily *not* for the safety of *people* but for economic reasons. An interruption of power for a few hours in a home may be a considerable nuisance but is not necessarily of grave consequence. In a chicken hatchery, a few hours' interruption may ruin a large load of eggs. In a greenhouse with electrically controlled heating equipment, loss of power may drop the temperature very low, with consequent great loss. Broadcasting stations need stand-by protection. On farms, power outages are often measured in days rather than hours; without power, milking machines are out of operation, pumps are idle, hot water is no longer available in sufficient amount, and heating equipment is usually dead. Continuous power is important in many hundreds of circumstances, and the only guarantee against an interruption is a stand-by generating plant (or an emergency generating plant, as will be discussed).

There is no Code requirement that a stand-by plant must be installed. But if by local, state, or federal law such a plant must be installed, then the requirements of NEC Art. 701 apply. The requirements are simple. The plant must be equipped with automatic means for transferring the load from the commercial power line to the generator; the plant must be capable of delivering full power within 60 s of the power failure; the fuel supply must be sufficient for at least 2-h operation at full load; and the plant must be tested at regular intervals. Wiring from the plant may be installed in the same raceways with the regular wiring.

Emergency Plants. Per Code Art. 700, an emergency power plant (as distinguished from a stand-by plant) is installed primarily for protection of *people,* although it may serve other purposes as well. The Code requirements apply only if the emergency plant is required by local, state, or federal laws. Emergency plants are installed in locations such as hospitals, where a loss of light in an operating room would be a catastrophe, as would loss of power in many other areas; in a windowless office, in a theater, or in a store, complete darkness due to power failure might very easily lead to panic with consequent injuries or even death. The same holds true of other places; in Code language, emergency systems are "generally installed in places of assembly where artificial illumination is required for safe exiting and for panic control in buildings subject to occupancy by large numbers of persons, such as hotels, theaters, sports arenas, health care facilities, and similar institutions."*

Where an emergency generating plant is required, the Code specifies that the plant must provide full power within 10 s of the power failure; the plant once started by a loss of power must continue to run for 15 min after power is reestablished (after power has failed, it is often restored for only a few seconds or a few minutes before it fails again, and this provision prevents the plant from starting and stopping repeatedly within those 15 min). Naturally every installation must include automatic means for starting the plant and transferring the load to it upon failure of commercial power.

The output of an emergency generating plant must be used to supply only special emergency circuits, to selected loads to be operated during the emergency, entirely independent of the usual wiring. Wires of the two systems may never enter the same raceway, cable, box, or cabinet of the usual wiring, except in exit or emergency lights supplied by both sources. Naturally wires from both systems must enter the automatic transfer equipment.

There are other requirements, and NEC Art. 700 should be thoroughly studied.

General Information. The most common generating plants are powered by a diesel or gasoline engine, the latter also being available for operation on natural or other gas. They produce 60-Hz power (or 50-Hz if required for foreign countries), in capacities from 500 watts to 500 kW or more. They are available in any required voltage, single- or 3-phase. Two of the smaller ones are shown in Fig. 24-1.

Since such plants are engine-driven, some means must be provided for starting the engine. The simplest is the hand-starting method, using a rope or crank; this method of course is practical only with the smallest plants. Usually a 12-volt heavy-duty automotive-type battery is used to

Fig. 24-1 Typical generating plants developing 5000 and 45 000 watts of power, single-phase or 3-phase, of any standard ac voltage. (*Onan Division of Onan Corp.*)

crank the engine. Smaller plants have a special starting winding in their ac generators, which also contain a dc winding for keeping the starting battery charged. On larger plants, the starting method is substantially the same as in an automobile.

For controlling plants, there are start-stop push buttons on the plant. Usually additional ones may be installed at a distance from the plant, up to several hundred feet away, connected using inexpensive control wire, similar to that used in wiring door chimes, thermostats, etc. The voltage in the wire is never more than the voltage of the cranking battery. There is also a "full automatic" plant (only in the smaller sizes up to 10 kW) that starts when a light or an appliance is turned on, and stops when everything is turned off. This type of control is especially popular on boats. The special type of control that starts a plant when commercial power fails will be discussed in the next paragraph.

Load Transfer Equipment. Under no circumstances may a plant be installed in such a way that its power will take care of the load to be operated, and at the same time also let it feed power back into the usual power line. That can be extremely dangerous to line workers who are working on what is supposedly a dead line. Several deaths of line workers have been traced to their working on such a supposedly dead line that was energized by a haphazardly installed stand-by plant. Remember, the same transformer that reduces 2400 volts to 120 or 240 volts also steps 120- or 240-volt power from a generating plant up to 2400 volts if the line is not disconnected while the plant is in operation.

In *every* case, install a double-throw switch. In one position, the switch connects the load to the power line; in the other position, it disconnects it from the power line and transfers it to the generating plant. If you have 120/240-volt service, you may usually use a 2-pole 3-wire *solid-neutral* switch, as shown in Fig. 24-2. In some localities, this is prohibited, and

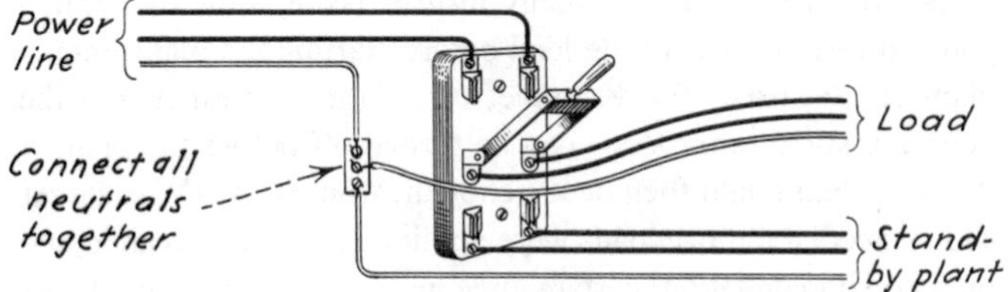

Fig. 24-2 In installing any stand-by or emergency generating plant, you *must* install a double-throw switch. If controls are automatic, the switch will be electrically operated. In most localities switches that do not switch the neutral are acceptable.

you must use a 3-pole 3-wire switch that does *not* have a solid neutral; the neutral conductor is then switched along with the hot wires, as shown in Fig. 24-3. Such switches can be either hand-operated, as shown in the illustrations, or electrically operated in the case of automatic equipment. Trace the wiring in Figs. 24-2 or 24-3, and you will see that there is no possible way for the generating plant to feed power back through the transformer into an otherwise dead line.

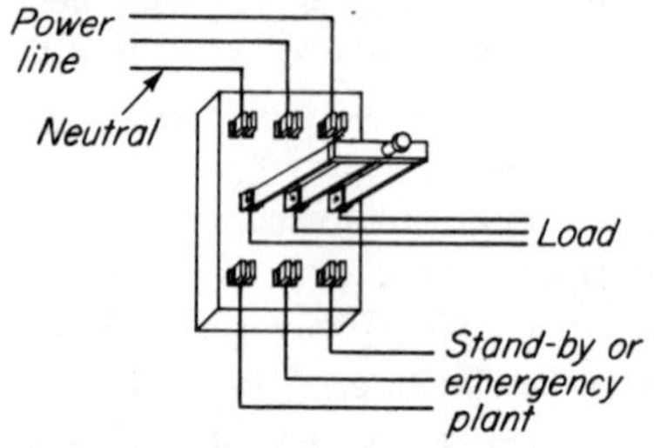

Fig. 24-3 In some localities, the neutral must be switched, as shown at left.

In selecting such a double-throw switch (whether hand-operated or electrically operated), do not make the mistake of selecting one that has an ampere rating large enough to handle only the current from the generating plant. For example, a 5-kW 120/240-volt plant will produce about 21 amp, so a 30-amp switch might appear to be big enough. However, the switch is also connected in the main line, so it must have an ampere rating large enough to carry the full current in that line, often 100 amp, during normal operation, that is, during all the time that the power has *not* failed. The ampere rating of the switch is determined by the maximum number of amperes it must carry during normal periods.

Advantages of Manual Control. A stand-by plant rarely has enough capacity to operate all the loads normally in use. Using a hand-operated switch, you can turn off part of the load before starting the plant; then a smaller plant can be used. For example, on a farm, the range and the water heater and some other loads can be turned off before the plant is started. A 5-kW plant might then be big enough, where a 10-kW or larger plant would be needed if those loads were not first turned off. In an industrial or commercial installation, certain large motors (air conditioning, for example) or other large loads could be turned off, thus making it possible to use a much smaller plant than would otherwise be needed. Thus a considerable saving in first cost can be achieved.

Advantages of Automatic Control. In many applications, including all those that the Code classifies as "emergency" installations, there is no choice; the transfer switch must be automatic. If the lights in a hospital operating room were to go out and stay out for even a few minutes during an operation, the delay could be serious indeed. If installed to provide power for heating systems, the automatic control provides power even at night or over the weekend, when no one is on the premises. Automatic installations are often made in isolated locations such as in microwave transmitting stations, where no caretaker may be seen for weeks at a time.

As already mentioned, a stand-by plant is rarely large enough to deliver all the power needed in the place where it is installed. In a chicken hatchery, it might be big enough to take care of specific loads (as with emergency plants). In other cases, automatic electric switches can be installed to disconnect selected loads from the line when the power fails, but not to reconnect them automatically when the voltage is restored by the generating plant. The stand-by plant then supplies only the remaining loads.

In addition to the basic advantage of being completely automatic, such transfer equipment also provides other conveniences. Among them are automatic charging of the cranking battery; test switches to test the equipment, either taking over the load or not, as desired; and "exercisers" to run the plant for, say, 15 or 30 min once a day, or on selected days during the week, without taking over the load, to keep the engine always ready to start instantly.

Installation Hints. Manufacturers' literature provides complete installation instructions, but some general hints will be in order. The plant must naturally be installed where the engine will cool properly. A small room is not considered a good location unless means are provided for adequate circulation of air to carry away the heat created by the engine. The fuel-supply system must be considered. If gasoline is used, an underground tank outside the building provides the best installation. Natural or manufactured gas may be preferred to gasoline, but the fuel supply must be on site. However, the inspector has the authority to approve the use of a public-utility gas system where the simultaneous failure of both the gas and the electric supply is judged to be unlikely. Figure 24-4 shows a properly installed plant.

Tractor-Driven Generators. For emergency use on farms, tractor-driven generators are often used. They are available in sizes up to 25 000 watts. Naturally they cost less than generating plants that have their own en-

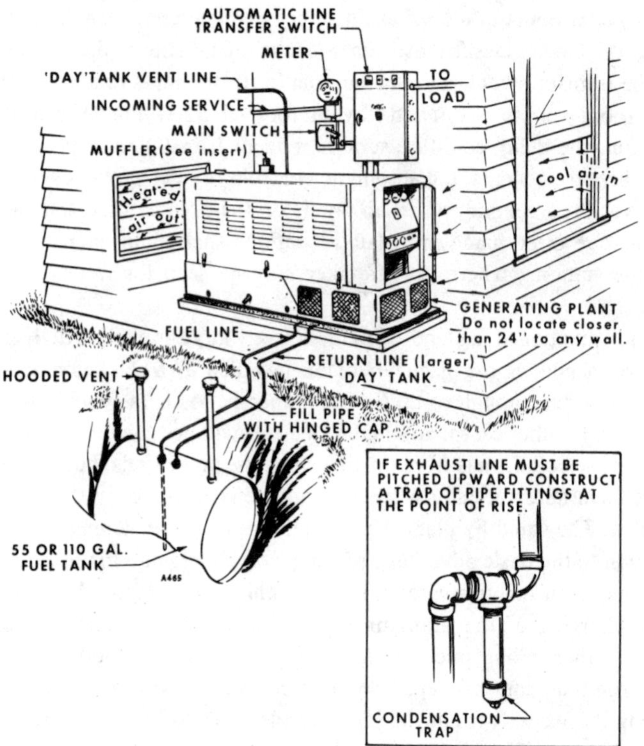

Fig. 24-4 Typical installation of engine-driven generating plant. (*Onan Division of Onan Corp.*)

gines. They do have the disadvantage of requiring time to line up the tractor with the generator, after spotting the generator in the right location; they also preclude the use of the tractor for other purposes during the period of power failure, which on farms sometimes lasts for rather extended periods.

Never install a generator, whether with its own engine or tractor-driven, without installing a double-throw switch. That is an absolute necessity for safety purposes.

In matching the generator with your tractor, be sure your tractor has sufficient horsepower at the power takeoff shaft, *at the speed required by the generator*. It should have at least 2 hp for each kilowatt of generator capacity; $2^1/_2$ hp for each kilowatt will enable you to better take advantage of the generator's overload capability. If you don't have enough horsepower in the engine, it will slow down, and so will the generator. The voltage, the frequency, and the capacity of the generator will drop off rapidly.

Battery-Powered Systems. Many years ago when few farms had commercial power available, many of them had their own engine-driven dc generators that charged a battery, usually 32 volts. Such systems necessarily had a very limited capacity, so that only a small number of lights and very few, if any, appliances could be operated. Today the use of such plants is extremely limited.

But in nonfarm applications, there are some uses where direct current is the normal power, as for telephone and other communications systems and for signaling purposes. Storage batteries of proper voltage are used, and they are kept charged from the usual ac supply. To protect against failure of the ac power, an ac stand-by generating plant is used to power the battery-charging apparatus.

But for emergency systems (as distinguished from stand-by systems) NEC Sec. 700-12(a) permits a storage battery instead of a generating plant to supply the emergency circuits that have been installed. The battery must be of a size that will supply the load for a minimum of $1^1/_2$ h.

The voltage will naturally drop slowly during any period of emergency but must not be lower than $87^1/_2\%$ of the normal voltage at the end of the $1^1/_2$-h period. If ordinary lead-acid storage batteries are used, they may not be of the ordinary automotive type, but must have transparent or translucent jars.

Naturally, automatic equipment must be installed to connect the batteries to the emergency circuits when normal power fails. The battery is kept charged from the usual ac power supply. To guard against prolonged outage periods, an ac generating plant can be installed to power the charging equipment.

25

Mobile Homes, Recreational Vehicles, and Parks

In any discussion of mobile homes and recreational vehicles, there is much confusion as to what is what. The Code classifies and defines the various kinds as either (1) mobile homes, or (2) recreational vehicles with four subdivisions: (a) travel trailers, (b) camping trailers, (c) truck campers, and (d) motor homes.

Definition: Mobile Home. In NEC Sec. 550-2 a mobile home is defined as "a factory-assembled structure or structures equipped with the necessary service connections and made so as to be readily movable as a unit or units on its own running gear and intended to be used as a dwelling unit(s) without a permanent foundation."* The mobile home is totally self-contained, and the phrase "without permanent foundation" implies that the mobile home can be moved from one location to another at the discretion of the owner. Most mobile homes once placed on a foundation are never moved again.

Definition: Recreational Vehicles. These are often called RVs, and that abbreviation will be used throughout this chapter. They are defined in NEC Sec. 551-2 as "a vehicular-type unit primarily designed as temporary living quarters for recreational, camping, or travel use, which either has its own motive power or is mounted on or drawn by another vehicle."*

A *travel trailer* as the name implies is towed by another vehicle. It must be of such size or weight so as not to require special highway-movement permits when drawn by a motorized vehicle. In general, that means not over 8 ft wide nor over 32 ft long. If larger, it becomes a mobile home. Its living area must be less than 220 sq ft, excluding built-in equipment such as wardrobes, closets, cabinets, kitchen units or fixtures, and bath and toilet rooms.

A *camping trailer* also is towed by another vehicle, and has *collapsible* partial side walls that fold for towing and unfold at the camp site.

A *truck camper* is a portable unit designed to be loaded onto and unloaded from the bed of a pickup truck.

A *motor home* is a self-propelled unit built on or permanently attached to a self-propelled motor vehicle chassis or on a chassis cab or van which is an integral part of the completed vehicle.

Interior Wiring: Mobile Home. This is discussed in NEC Art. 550, but it is always factory-installed, so it need not be discussed in detail here. Some general information, however, may be of value to you. The home does not contain a service entrance, but it does contain a distribution panel with a single disconnecting means and main overcurrent protection, which is usually a 50-amp circuit breaker, plus additional overcurrent protection for all the branch circuits. The neutrals of these panels must *not* under any conditions be grounded to the cabinets. They must be fully insulated just as the hot wires. The *service equipment* itself (consisting of main overcurrent equipment and disconnecting means, such as a circuit breaker or fuses on a pull-out block) is installed away from the home in a raintight housing. It may be installed on a pole and fed by overhead wires, or installed on a metal pedestal and fed by underground wires. The service equipment must be 120/240 volts and be rated at not less than 100 amp. It must contain provisions for connecting the feeder to the home by a permanent wiring method and may, and usually does, contain receptacles into which cords from the home are plugged. From the service equipment, power is supplied to the home by one of three methods.

*Reprinted with permission from NFPA 70-1984, National Electrical Code®, Copyright© 1983, National Fire Protection Association, Quincy, Massachusetts 02269. This reprinted material is not the complete and official position of the NFPA on the referenced subject, which is represented only by the standard in its entirety.

Smaller homes that do not require more than 50 amp are equipped with a permanently attached cord with a plug to fit the 50-amp receptacle in the service equipment. The cord must be between 21 and 36$^1/_2$ ft long.

If the home requires more than 50 amp, a permanent underground feeder is run from a junction box under the home, to the terminals for a permanent connection in the service equipment.

A third method used only in a few areas consists of a mast installed on the home, quite like masts on nonmobile homes discussed in Chap. 17. Wires run from the mast to the service equipment installed on a pole.

Whichever method is used, the feeder from the home to the service equipment must consist of four wires: two hot wires, an insulated white neutral wire, and an insulated green equipment grounding wire.

All non-current-carrying components in the home (including the cabinet of the distribution panel) plus the metal skin of the home must be interconnected with each other, and then connected to the green equipment grounding wire in the cord. The actual ground is in the service equipment at the supply end of the cord.

Service Equipment: Mobile Home. As already pointed out, the service equipment as such is not installed in or on the mobile home; it is installed at some distance from the home. An individual service equipment might contain only a 50-amp 3-pole 4-wire 125/250-volt receptacle into which the cord from the home is plugged, protected by a 2-pole 50-amp circuit breaker. Optionally, it may contain additional 20-amp 125-volt receptacles for use with equipment outside the home, such as an electric lawn mower or electric tools. Such receptacles must be protected by separate circuit breakers and *must* be protected by a GFCI.

A typical service equipment is shown in Fig. 25-1. It is rated at 100 amp and contains one 50-amp receptacle and one 50-amp breaker as already described, and also a 20-amp 2-pole 3-wire 125-volt receptacle of the grounding type (either 15- or 20-amp plugs will fit) protected by its own circuit breaker and GFCI. Note there is space to install a 100-amp 2-pole 125/250-volt breaker at a future date, should a home require more than 50 amp. In that case the 50-amp receptacle and 50-amp breaker would be disconnected, and the home would be served by a direct connection using no cords.

Note that while the service equipment must be rated at a minimum of 100 amp (having terminals suitable for wire with at least 100 ampacity), any receptacle and the breaker protecting it may not exceed a 50-amp rating. (Larger breakers may be installed if the home is not served by a cord

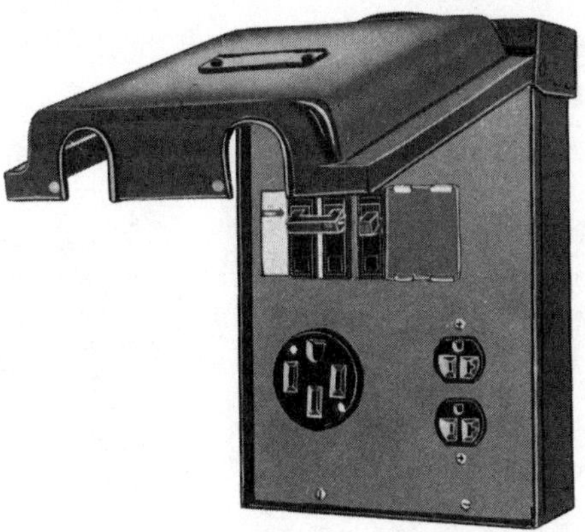

Fig. 25-1 Basic service-entrance equipment for a mobile home. Such equipment must have a minimum rating of 100 amp (the terminals must accept wires with an ampacity of at least 100). (*Midwest Electric Products, Inc.*)

but by direct permanent connection to terminals in the service equipment.) Each service equipment must have a grounding busbar for the neutral of the cord, the grounded wire of the smaller receptacle(s), and the green grounding wire of the cord. It must be bonded to its cabinet, and a ground wire run from it to an underground piping system if available, and to a ground rod.

In Fig. 25-2 is shown a similar equipment. Note that it contains a meter socket (which is optional on most service equipments for individual mobile homes). It is installed on top of a metal pedestal to be supplied by an underground feeder; the pedestal is optional on all similar units. The particular equipment shown is rated at 200 amp. It contains the usual 50-amp receptacle for one cord and a 50-amp breaker to protect it, and a 20-amp 2-pole 3-wire 125-volt grounding-type receptacle protected by a separate breaker and a GFCI. It contains space for a 200-amp 2-pole circuit breaker that can be installed at a future date, for a home with a permanent connection and no cord. Naturally they are not visible, but the unit contains terminals for "loop feed." Wires from the underground feeder to the service equipment end at a set of terminals in the equipment, and wires from another set of terminals are then run to the next service equipment, doing away with all need for splices underground.

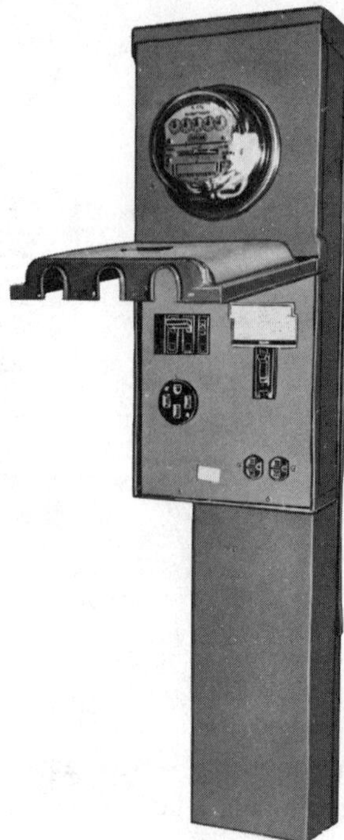

Fig. 25-2 A larger service entrance rated at 200 amp. It is equipped with a meter socket and supported by a metal pedestal for use with underground wiring. (*Midwest Electric Products, Inc.*)

The service equipments are best served by underground feeders, but this is not required. Overhead wires may be used, in which case the service equipments are installed on a pole supporting the overhead wires. Run service conduit from a hub in the top of the service equipment for about 8 ft up the pole, ending in a service head. Install the usual three wires. The neutral of the cord, as also the green equipment grounding conductor of the cord, must be connected to the grounding busbar in the cabinet, which in turn must be bonded to the cabinet, and grounded.

Mobile Home Parks. A mobile home *park* is a parcel of land used for the accommodation of occupied mobile homes. The park naturally contains a park office, and often facilities such as laundry equipment that might not be available in the individual homes, but which are made available to all the people living in the park. A mobile home *lot* is that space within a park, for one mobile home and accessory buildings or structures, for the exclusive use of the occupant.

Service Equipment: Entire Park. The service equipment for the entire mobile home park must be single-phase 120/240-volt. The details vary with the size of the park. In a *very* small park, the equipment might be located in the park office, with feeders to groups of individual lots. In most parks the supply to the park may be at a voltage higher than 240 volts, but with this higher voltage running only to a number of separate transformers each delivering 120/240 volts. Each transformer serves a feeder for a number of homes, just as in city wiring one transformer serves a number of houses.

Depending on the number of homes in the park, you might install a single feeder for all the homes if it is an exceedingly small park; or two to five in a medium-size park, each feeding 6 to 10 homes; or many more in a large park. To minimize voltage drop, it is best to install separate feeders for relatively small groups of homes. The feeders might be overhead, in which case service drops to individual lots would be much as in wiring a group of farm buildings as discussed in Chap. 23; underground feeders are more desirable.

How big must each feeder be? Start with an assumed load of 16 000 watts for each home, at 115/230 volts. Then apply the demand factors of NEC Sec. 550-22, reproduced here as Table 25-1. As you will note, the demand factor for a single home is 100%, but it drops off rapidly as the

TABLE 25-1 (NEC Table 550-22) Demand Factors and Volt-Amperes per Mobile Home Site (Minimum) for Feeders and Service-Entrance Conductors*

Number of mobile homes	Demand factor (%)	Volt-amperes per mobile home site (min.)
1	100	16,000
2	55	8,800
3	44	7,040
4	39	6,240
5	33	5,280
6	29	4,640
7-9	28	4,480
10–12	27	4,320
13–15	26	4,160
16-21	25	4,000
22–40	24	3,840
41–60	23	3,680
61 and over	22	3,520

*Reprinted with permission from NFPA 70-1984, National Electrical Code®, Copyright© 1983, National Fire Protection Association, Quincy, Massachusetts 02269. This reprinted material is not the complete and official position of the NFPA on the referenced subject, which is represented only by the standard in its entirety.

number of homes increases. For example, if there are five homes on one feeder, the demand factor is 33%, so that the minimum ampacity of the wires in the feeder for five homes would be

$$\frac{(16\,000)(5)(0.33)}{230} = \frac{26\,400}{230} = 115 \text{ amp}$$

Similar calculations show that the minimum ampacity for a feeder for 10 homes would be 188 amp.

Now suppose you have a park with 20 homes, and four feeders, one for each five homes. Will the service conductors for the entire park have to be four times the 110 amp of each feeder, or 440 amp? Not at all. Look at Table 25-1 again. The demand factor for 20 homes is 25%. So the minimum ampacity of the main service conductors must be

$$\frac{(16\,000)(20)(0.25)}{230} = \frac{80\,000}{230} = 348 \text{ amp}$$

Of course they should be somewhat larger than that to take care of the park office, and other requirements in the park for equipment that perhaps is not installed in the individual homes.

Interior Wiring: Recreational Vehicles. At this point you might do well to read again the definitions in the beginning of this chapter, to fully understand the difference between a mobile home and an RV. Some RVs have no electrical wiring at all. Those that are wired are factory-wired. The wiring might be of several types that will be discussed below.

Some RVs are equipped with wiring for use at 115 volts, thus having no electric power except while in an RV park. Such RVs are usually equipped with a permanently attached cord that must be at least 23 ft long for side entry and 28 ft for entry at the rear of the RV. The cord must be of a size dependent on the electrical load in the RV and be equipped with a plug of a type and rating again dependent on the load. An explanation of the acceptable plug and receptacle configurations will follow later in this chapter.

Other RVs are equipped with only a battery system, usually 12 volts and never over 24 volts, entirely apart from the battery of the vehicles on which they are mounted or towed. Necessarily the amount of power from such a system is very limited. The battery is kept charged from the usual 120-volt power. Other RVs are equipped with both a battery system that can be used when the vehicle is not in an RV park and also with a 120-volt system for use while it is in a park. The wiring of the two systems must be kept entirely separate; wires from the battery circuits must under no circumstances enter the same outlet boxes or similar fittings containing 120-volt wires.

Other RVs are equipped with an independent engine-driven generating plant delivering 115- or 115/230-volt 60-Hz power and are therefore completely independent of parks if the owners wish. However, there are often cases in which the owners may wish to take advantage of a park and the power there available instead of running their own plant for long hours. In that case, equipment must be installed in the vehicle so that commercial power from the park and power from the generating plant cannot possibly be connected to each other or to any piece of equipment at the same time.

The owners of an RV might spend just one night in an RV park, or they might spend several months. Some parks are equipped just for overnighters, but most are designed to accommodate RVs for extended periods of time. These latter parks are equipped with laundry, bathroom, and shower facilities, none of which are found on smaller RVs.

Recreational Vehicle Sites and Parks. A site is the space in a park occupied by one RV, and the park is an area intended to take care of a number of RVs.

The requirements for RV sites and parks are similar to those for mobile homes, but there are some very important differences. Each RV is equipped with a distribution panelboard with an insulated neutral busbar, and provided with a cord to separate equipment installed near the RV. Repeating, the size of the cord is dependent on the electrical load in the RV and may be either 3-wire 115-volt (one hot wire, the white grounded wire, and the green equipment grounding wire), if the load does not exceed 30 amp at 115 volts, or 4-wire 115/230-volt (two hot wires, the white neutral wire, and the green equipment grounding wire), and rated at 50 amp, in the case of larger or luxury RVs.

The equipment installed to supply power to an individual *mobile home* is per Code definition called service equipment. In the case of RVs, while the equipment at each site appears to be substantially the same as in the case of a mobile home, it is not called just a service equipment, but is in a special class called ''Recreational Vehicle Site Supply Equipment.'' It is a ''power outlet'' meeting special requirements, from which the RV is supplied. Groups of power outlets are served by feeders from the service equipment that supplies the park as a whole.

Figure 25-3 shows a typical power-outlet assembly for one RV. It contains one 20-amp 2-pole 3-wire 125-volt grounding receptacle and one 30-amp 2-pole 3-wire 125-volt grounding receptacle, each protected by a circuit breaker. The type of construction shown is by far the most popular type of equipment for RV parks using underground wiring.

Fig. 25-3 Typical power-outlet assembly for a recreational vehicle. When such assemblies are used in parks for recreational vehicles, there are important differences between them and the service-entrance equipment used for mobile homes. (*Midwest Electric Products, Inc.*)

Unlike service equipment in mobile home lots, the special power outlets for RVs must *not* be grounded at each site. The grounded wire (neutral in the case of 120/240-volt installations) must be insulated from the cabinet of the power outlet at each site, while the green grounding wire must be bonded to each cabinet. The white grounded wire must be continued from lot to lot until it reaches the service equipment for the entire park, where it must be grounded, as must the green equipment grounding wire also. But if the feeder to the individual power outlets at the various sites is in underground *metal conduit,* that conduit serves as the equipment grounding conductor.

Receptacles in Mobile Home Services and RV Power Outlets. Grounding receptacles of the 2-pole 3-wire 20-amp 125-volt type (either 15- or 20-amp plugs will fit) and those of the 50-amp 3-pole 4-wire 125/250-volt type are the general-purpose type shown in Figs. 21-1 and 21-2. But note that the 30-amp 2-pole 3-wire *125-volt* type is *not* a general-purpose type. It is shown in Fig. 25-4 and is used only in RVs, and must not be used for other purposes. The general-purpose type of the same rating may not be used in RV power outlets.

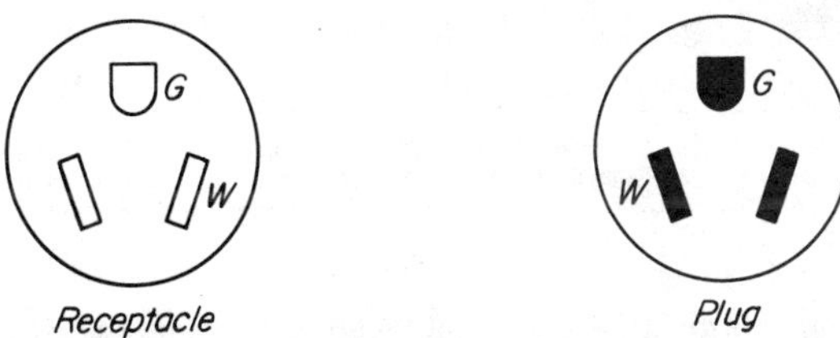

Fig. 25-4 The special-purpose 30-amp 125-volt 2-pole 3-wire receptacle that may be used *only* in recreational vehicle power outlets.

RV Park Services. As in the case of mobile home parks, the entire service to an RV park must be single-phase 120/240 volts, usually installed in the park office, with feeders to groups of power outlets for individual RVs.

The Code specifies that every power-outlet assembly for an individual site must contain at least one 20-amp 125-volt, 30-amp 125-volt, or 50-amp 125/250-volt receptacle, and that 75% of all the assemblies must contain 30-amp 125-volt receptacles. It is recommended that the modern RV park include the 50-amp 125/250-volt receptacle (in addition to the 20- and 30-amp types) in about 5% of the power outlets, to accommodate luxury motor homes that require a large amount of power. In addition to these *required* receptacles, each RV site supply equipment may contain additional receptacles of any of the three ratings, plus additional receptacles of *any* ratings for connection of equipment outside the RV. *All* 15- and 20-amp 125-volt single-phase receptacles must be GFCI-protected.

How big must each feeder be? Allow 3600 watts per site served by both 20- and 30-amp receptacles, 2400 watts for those supplied only by a 20-amp receptacle. Then apply the demand factor of NEC Table 551-44, reproduced here as Table 25-2. Let's assume a feeder supplies five sites

TABLE 25-2 (NEC Table 551-44) Demand Factors for Feeders and Service-Entrance Conductors for Park Sites*

Number of recreational vehicle sites	Demand factor, %	Number of recreational vehicle sites	Demand factor, %
1	100	10–12	47
2	100	13–15	45
3	100	16–18	44
4	89	19–21	42
5	71	22–40	40
6	63	41–100	39
7–9	53	101 and over	37

with equipment containing both 20- and 30-amp receptacles, and two with only 20-amp receptacles (remember these are 125-volt receptacles, not 125/250-volt receptacles as in the case of mobile homes). Apply the demand factor of Table 25-2. The minimum ampacity of each feeder then would be

$$\frac{[(3600)(5) + (2400)(2)](0.53)}{120} = \frac{(18\ 000 + 4800)(0.53)}{120}$$

$$= \frac{(22\ 800)(0.53)}{120} = \text{approx. } 101 \text{ amp}$$

What about the size of the service-entrance conductors serving the entire park? Assume the park has 35 sites, with five feeders as just calculated. That means 25 sites with two receptacles, and 10 with one receptacle. The calculation then is

$$\frac{[(3600)(25) + (2400)(10)](0.40)}{120} = \frac{(90\ 000 + 24\ 000)(0.40)}{120}$$

$$= \frac{(114\ 000)(0.40)}{120} = \frac{45\ 600}{120}$$

$$= 380 \text{ amp}$$

But the service will not be at 120 volts but rather at 120/240 volts, so the minimum size of the service equipment for the park would be 190 amp.

Naturally you would install somewhat larger service wires to take care of additional loads in the park, other than the feeders to the sites for the individual RVs.

Homes and RVs Used for Special Purposes. If a mobile home or an RV is used not for the purposes so far discussed but for a different purpose (such as portable offices, merchandise displays, dormitories on construction sites) *and* if it is intended to be supplied from outside 115- or 115/230-volt power, it must meet all the requirements discussed, except that the number and size of branch circuits need not be as normally required by the Code for factory installation.

26
Wiring Apartment Houses

Although a single apartment within an apartment building presents no new wiring problems of any consequence, the building as a whole does present such problems.

As a student, you will not be called upon to design the wiring layout of a large apartment building until you have mastered the problems in smaller projects. For this reason detailed discussions of *all* problems in the wiring of larger apartments are beyond the scope of this book. Rather, the information in this chapter should be considered only as a general guide and foundation, a sort of preview. You must also study NEC Art. 230.

Planning an Individual Apartment. To determine the minimum number of branch circuits for any single apartment, proceed as outlined for a single-family dwelling in Chap. 13. For lighting, allow 3 VA per sq ft. For example, a small apartment of 800 sq ft will require (800)(3) or 2400 VA for lighting and receptacle outlets, which means two circuits. To this must be added the two 20-amp small appliance circuits for the dining and cooking areas, just as in the case of the single-family house. (Laundry equipment is rarely installed in individual apartments, but if it is, a separate 20-amp circuit must be provided for it.) A separate circuit must be provided for each special appliance such as a range or dishwasher. The number of receptacles is determined as in separate houses. Don't overlook the requirement of GFCI protection for receptacles in bathrooms.

Service-Entrance Problems. In practically all cases there is but a single service for the entire building. In many cases each tenant pays for the power he or she consumes, so there is a separate meter for each apartment, plus another meter for "house loads": hall lights, water heaters, office loads, heating-system motors, etc. The service equipment must then supply a number of separate meters and disconnecting means. The Code in Sec. 230-72(d) requires that each occupant have access to his or her disconnecting means.

In just about all multiple-occupancy buildings, the service equipment must be grouped together in an accessible area, and the number of disconnecting means is limited to six for the entire building. If there are more than six disconnecting means, a separate main switch or circuit breaker must be installed ahead of them.

For specified large loads and *by special permission* for large-area buildings or for multiple-occupancy buildings having no available space for service equipment accessible to all the tenants, more than one service is permitted by NEC Sec. 230-2. Each service may have up to six disconnecting means (unless the service equipment is also a lighting and appliance branch-circuit panelboard, in which case not more than two service disconnects are permitted; see NEC Secs. 384-14 and 384-16).

One-Line Diagrams. Up to this point all diagrams have shown *all* the wires in an installation. If this scheme were used for larger buildings, the diagrams would become very cluttered and confusing. So in this chapter the diagrams will use only a single line to represent *all* the wires in the feeder or the circuit. While fuses are shown in some of the diagrams, circuit breakers may of course be used instead.

Grouped Service Equipment. The general rule is that service equipment for a multiple occupancy be grouped in a location accessible to all the tenants, as shown in Figs. 26-1 and 26-2. Branch-circuit disconnects and overcurrent protection are located in each individual occupancy. Figure 26-1 shows a single service disconnect. Figure 26-2 shows the maximum permitted, six, with taps made in an auxiliary gutter.[1]

Service Equipment in Each Occupancy. Where no space is available for

[1] For definition and discussion of auxiliary gutters, see Chap. 28.

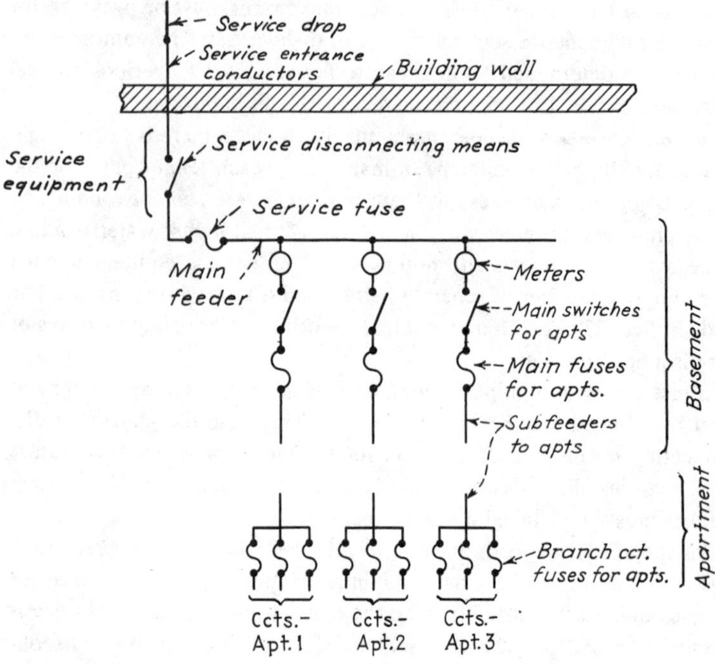

Fig. 26-1 Service equipment in a common location accessible to each occupant.

service equipment accessible to all the tenants, the service may be run to each occupancy, but only if *special permission* is granted by the inspector. One such arrangement is shown in Fig. 26-3, in which subsets of service-entrance conductors are run to each apartment. Usually the power supplier will require the meters to be grouped at one location.

Determining Wire Size to Individual Apartment. The wires to the overcurrent devices in individual apartments are called feeders or subfeeders if the service equipment is installed in a common location, as shown in Figs. 26-1 and 26-2; they are called subsets of service-entrance wires if installed as shown in Fig. 26-3. In either case, the ampere load (and therefore the wire size) is determined in the same way as was described in Chap. 13 for service-entrance wires. Allow 3 VA per sq ft of area for lighting, including the usual receptacles; add 3000 VA for the two required small appliance circuits; if laundry equipment is installed in the individual apartment, add 1500 VA. Total all the above; count the first 3000 VA at 100% demand factor and the remainder at 35% demand fac-

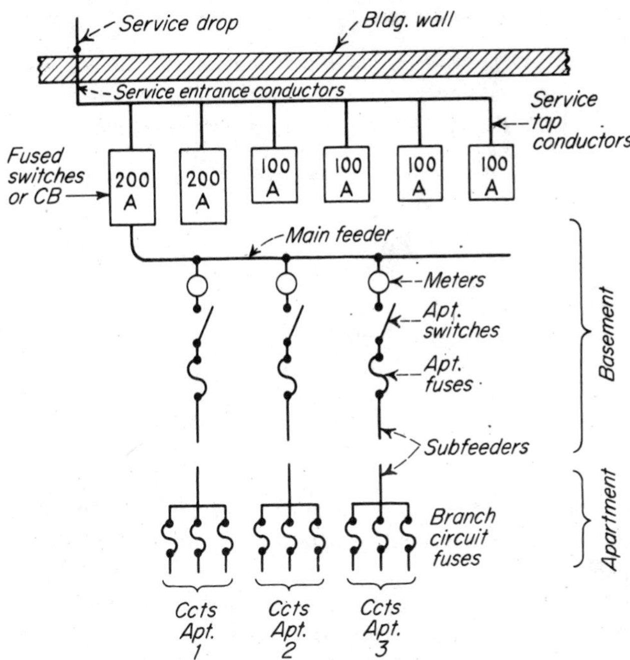

Fig. 26-2 Similar to Fig. 26-1, but with six separate service disconnecting means, the maximum number permitted.

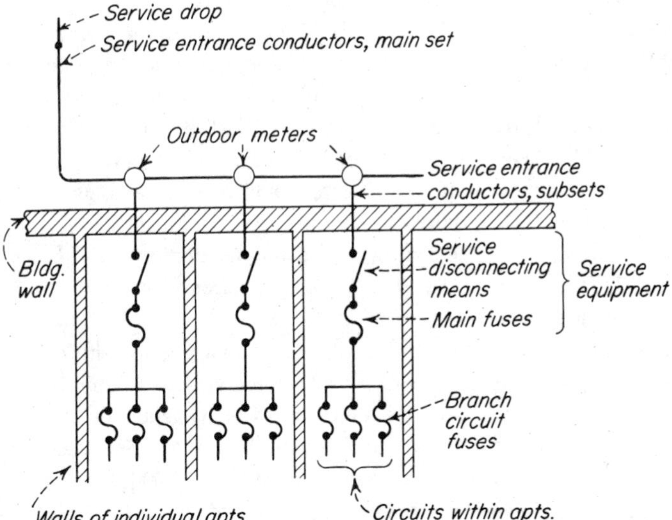

Fig. 26-3 If there is no common location accessible to all occupants, and by special permission only, service equipment may be installed in each apartment.

tor. Then add to that the watts rating of individual circuits to appliances such as range, air conditioner, and dishwasher, and you will have the total computed load. For an 800-sq-ft apartment, your calculation would look like this:

	Gross computed volt-amperes	Demand factor, %	Net computed volt-amperes
Lighting, 800 sq ft at 3 VA	2400		
Small appliances (minimum)	3000		
Total	5400		
First 3000 VA		100	3000
Remaining 2400 VA		35	840
Total			3840

At 120 volts, this is equivalent to $^{3840}/_{120}$, or 32 amp. This will require two 15-amp general-use lighting (and receptacle) circuits and two 20-amp small appliance circuits.

According to NEC Table 310-16, No. 8 wire is the smallest that may be used for 32 amp, but only if the wiring is as shown in Figs. 26-1 or 26-2, where the wires to the apartment are feeders. If installed as shown in Fig. 26-3, they are subsets of service-entrance wires, and NEC Sec. 230-42(b)(3) requires that they be a minimum of No. 6.

All the above applies to the 2-wire 120-volt supply to the apartment. If the main fuse protecting that apartment blows, the entire apartment will be without power and in darkness; the occupant will not be very happy. It would be much better to install a 3-wire 120/240-volt supply to the apartment. The 3840 VA at 240 volts would equal 16 amp, so at first glance it might appear that No. 12 wire would be large enough, but NEC Sec. 230-42 would still require a minimum of No. 6 for a subset of service-entrance wires per Fig. 26-3; however, if the installation is as shown in Figs. 26-1 or 26-2, then NEC Sec. 215-2(a) requires a minimum of only No. 10 for the feeder.

But as a practical matter, there will be few apartments with only the circuits discussed. Usually there are additional circuits to supply appliances, such as a range, an air conditioner, a dishwasher, or a garbage disposer. Assume that a range is to be added. Follow NEC Sec. 220-19 and add 8000 watts (if the range is rated at more than 12 000 watts, review Chap. 13) to the 3840 VA already calculated, making a total of 11 840 VA. Since a range is a combination 120/240-volt load, naturally a 3-wire supply must be installed. The 11 840 VA at 240 volts is equal to $^{11\ 840}/_{240}$, or 49.3 amp, for which No. 6 wire is suitable.

However, you have already learned that the neutral serving a range cannot be made to carry as many amperes as the hot wires carry. Consider only 70% of the allowance for the range. In the example above, the neutral would be figured this way: The range load is $^{8000}/_{240}$, or 33.3 amp, and 70% of 33.3 is 23.3 amp. Add this 23.3-amp load to the 16-amp lighting load originally figured, for a total of 39.3 amp, for which a No. 8 wire is suitable for the neutral.

Determining Service-Entrance Wires. The minimum size of the service-entrance wires for the entire building is determined by its probable total load at any one time. The method is similar to that used for a single-family dwelling: The lighting (including receptacles), the small appliance circuits, and the special loads such as ranges are simply computed for the entire building.

The greater the number of individual apartments in a building, the less the likelihood that all occupants will be using maximum power at the same time. Therefore the Code has established demand factors: for lighting and small appliances in Sec. 220-11, for clothes dryers in Sec. 220-18, for ranges in Sec. 220-19. The greater the number of apartments, the lower the demand factor.

Let us first consider the lighting and small appliance loads for the building as a whole. Allow 3 VA per sq ft of area in all the apartments; then add 3000 VA for the small appliance circuits in *each* apartment. Add the two figures together. Then apply the demand factors of NEC Sec. 220-11 as follows:

First 3000 VA at	100%
Next 117 000 VA at	35%
All above 120 000 VA at	25%

Next let us consider ranges. It is not at all likely that all occupants will be using their ranges at their maximum capacity at the same time. Decreasing demand factors are shown in NEC Table 220-19 (see Appendix). Note that the table specifies that its Col. A must be used at all times with the exception of those cases mentioned in the Notes following the table. Assuming that each range is rated at not over 12 000 watts, allow 8000 watts for one range, 11 000 watts for two ranges, 14 000 watts for three ranges. For four or more ranges, see Table 220-19.

So to determine the ampacity of the service-entrance wires, add together the volt-amperes required for the lighting and small appliance circuits (after application of the demand factor) and the watts required for the ranges. To this total must be added the "house load," which can be very considerable, to include hall and basement lights, laundry circuits at 1500 VA each, clothes dryers (the first four dryers must be included at full name-plate rating; if there are more than four, see NEC Table 220-18 in the Appendix), motors on heating or other equipment, water heaters if electric, and so on. For four or more fixed appliances, other than ranges, dryers, space heating, or air conditioning, on the same feeder or service, a demand factor of 75% may be applied to the name-plate-rating load.

Service Entrance for Three-Apartment Building. The maximum probable load can be computed in accordance with the following table:

	Gross computed volt-amperes	Demand factor, %	Net computed volt-amperes
Lighting, 2400 sq ft at 3 VA	7 200		
Small appliances, 3 apartments at 3000 VA	9 000		
Total gross computed volt-amperes	16 200		
First 3000 VA		100	3000
Remaining 13 200 VA		35	4620
Total net computed volt-amperes			7620

The 7620 VA covers only the apartments. The "house load" will probably come to 6000 VA, considering the need for the 20-amp laundry circuits. Adding the 6000 VA to the 7620 VA determined for the apartments makes a total of 13 620 VA, which at 240 volts is 56.7 amp. Accordingly a 60-amp service would be acceptable, and per NEC Table 310-16, No. 6 Type THW or RHW wire or No. 4 Type T or TW would be acceptable. However, remember that NEC Sec. 230-42(b) requires a 100-amp service for a single-family house, and while there is much more diversity if three families are involved than if only one is involved, most people would consider a 60-amp service for such an apartment building as quite skimpy, and prudence suggests that a 100-amp or larger service be installed.

If an electric range consuming not over 12 000 watts is added in each apartment, Col. A of NEC Table 220-19 (see Appendix) shows that 14 000 watts must be added for the three ranges in calculating the service. Adding 14 000 watts to the 13 620 VA already determined produces a total of 27 620 VA, which at 240 volts is equivalent to about 115 amp. The minimum would be a 125-amp service, using a circuit breaker as the disconnecting means; if fused equipment is installed, the minimum switch size would be 200 amp, since there are no switch sizes between 100 and 200 amp. The minimum ampacity of the service wires would in any case be 125 amp. A 150- or 200-amp service would be better. Review Chap. 13 concerning panelboards.

Optional Computing Methods. A simplified method of calculating feeder loads may be used for an individual apartment where the feeder has an ampacity of 100 or more. The method is the same as that for a single-family dwelling service, as discussed in Chap. 13.

Optional calculations for services or feeders to three or more apartments are permitted *only if all* the following conditions are met: (1) no apartment is supplied by more than one feeder; (2) each apartment has electric cooking; and (3) each apartment has electric space heating, or air conditioning, or both. For the calculations, assuming apartments of identical size and with the same appliances, proceed as follows (for a typical apartment):

a. For lighting and general-use receptacles, 3 VA per sq ft ______ VA
b. For each small appliance circuit (two required, minimum) and laundry circuit, 1500 VA .... ______ VA
c. Total of name-plate ratings of cooking appliances, dryers, water heaters, and all other fixed and stationary appliances not covered in *d*, below .... ______ VA
d. Name-plate rating of largest of air-conditioning or space-heating load .... ______ VA
e. Add *a*, *b*, *c*, and *d* to obtain total connected load (with smaller of air-conditioning or space-heating load omitted) per apartment .... ______ VA
f. Multiply *e* by the number of apartments on the feeder or service .... ______ × ______ apts.
g. Total .... ______ VA

Refer to Code Table 220-32 (see your copy of the Code) and apply the demand factor shown there for the number of apartments on the feeder or service. For example, if there are five apartments, use 45% of the total volt-amperes. If there are 12 apartments, use 41%. Divide the number of demand volt-amperes by 240, giving you the ampacity required for the hot wires of the service or feeder. The neutral may be smaller and may be calculated by NEC Sec. 220-22, which was discussed in Chap. 13.

See NEC Example 4(b) in the Appendix for a more detailed explanation, but do not be confused by the example, which applies the optional calculation to the two feeders for 20 apartments each and the main feeder for all 40 apartments but which calculates the individual apartment feeder and the neutrals by the conventional method.

The optional method may not be applied to house loads (community laundry, office, hallway lighting, pool, garden lighting, etc.), which must be calculated separately and added to the calculated load for the apartments themselves.

Larger Apartment Buildings. Based on the information already discussed, you should be able to determine the details of wiring larger buildings with more apartments. It would be well to study NEC Examples 4(a) and (b) (see Appendix).

The final installation of the service equipment in a larger building might well be shown in Fig. 26-4. This shows three sections, each containing four meters plus four circuit breakers, one for each of the four apartments. The three sections together take care of 12 apartments. Since this

Fig. 26-4 A typical group metering panel for use in larger apartment buildings. (*Square D Co.*)

is more than the maximum of six permitted disconnecting means, a main circuit breaker serving as the disconnecting means is provided ahead of all, in a separate section. If the wires from the utility transformer to the service equipment are very short, or the wires and the transformer are very large, refer to the discussion on interrupting capacity in Chap. 27 to determine whether the service equipment must be of special design in order to withstand the fault current that is available.

Low-Voltage Wiring. The usual doorbell and buzzer system (or chimes) will be installed in accordance with the principles already outlined in other chapters. However, in buildings of considerable size you must take into consideration the resistance of the wires in the longer lengths involved.

Sometimes the installation is made as shown in Fig. 26-5. Theoretically

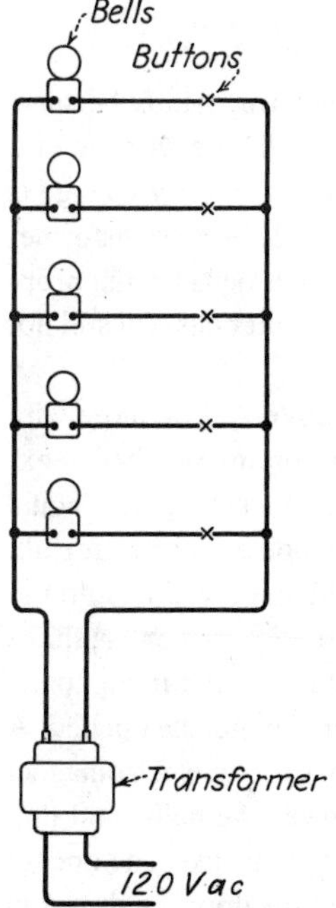

Fig. 26-5 If this wiring diagram is used, the distant bells ring too faintly, the nearby ones too loudly.

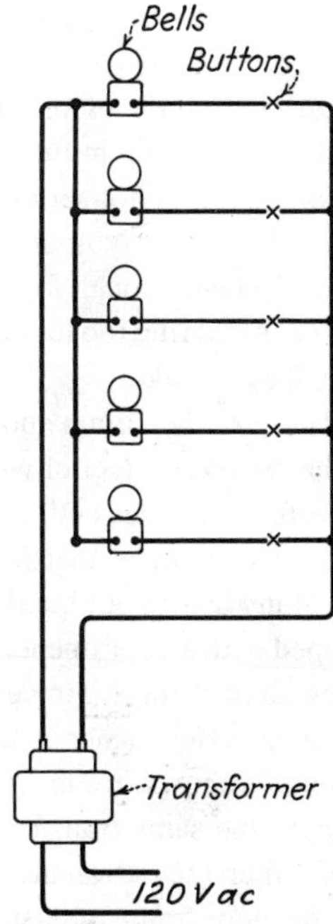

Fig. 26-6 If this diagram is used, all bells ring with equal volume. It requires only a little more wire than the diagram in Fig. 26-5.

Fig. 26-7 A typical door opener. (*Edwards Co. Inc., General Signal Corp.*)

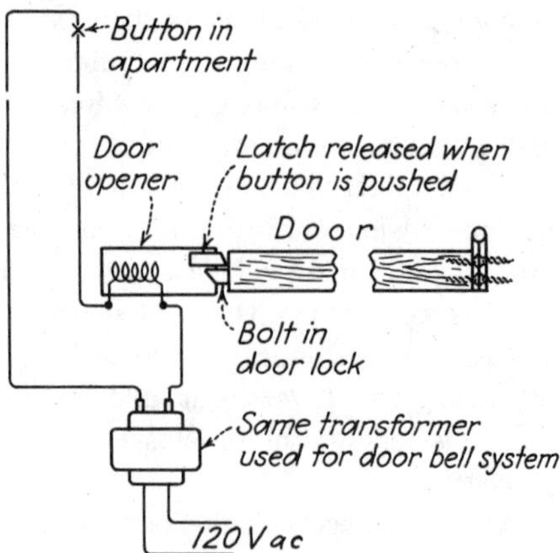

Fig. 26-8 Installation of door opener.

this diagram is correct, but the more distant bells will ring faintly because of the substantial voltage drop in the long run of small wire. A drop of 2 volts on a 120-volt circuit is not serious, but when the starting voltage is likely to be under 20 volts, a 2-volt drop is at least 10%. If the transformer voltage is stepped up high enough for the more distant bells to ring properly, the nearby bells will ring too loudly, but the distant bells will still not ring as loudly as they should.

The solution is to use the circuit shown in Fig. 26-6. A little more wire is needed, but the number of feet of wire involved for any one bell is exactly the same as for any other bell. Accordingly, all will ring with equal volume; a transformer voltage that is correct for one is correct for all.

Door Openers. Many apartment buildings have the door leading into the inside hall equipped with a door opener. Those with keys can enter; those who have no key cannot enter. However, pushing a button in an apartment releases a latch in the opener so that the door can then be opened. A typical door opener is shown in Fig. 26-7. It consists of an electromagnet that is energized by the same transformer that serves the bells, and that releases the latch when the button in an apartment is pushed. The opener is mortised into the door frame opposite the lock in the door, as shown in Fig. 26-8, which also shows the wiring diagram.

PART 3

Actual Wiring: Nonresidential Projects

Part 3 of this book covers the same subject matter as Part 2 but applies to nonresidential structures such as factories, offices, stores, schools, and institutions.

The basic principles covered in Parts 1 and 2 are equally applicable to commercial and industrial work. The major difference in the wiring of various occupancies is in the types of wiring methods and materials used, but we have space to cover only the most common of these. For the kinds of wiring discussed in Part 3, it is doubly important that you have a copy of the 1984 Code and study *it. Consider this Part 3 as a sort of introduction to the kinds of problems you will meet as you progress from working on residential and farm to commercial and industrial projects, for the coverage does not pretend to be complete. Wiring at voltages over 600, transformer vaults, complex control systems, and similar subjects are outside the scope of this book.*

Architects and engineers are more likely to be involved in the design of these projects than in residential work. Thus, the electrician's responsibility includes not only compliance with the Code, but compliance as well with plans and specifications prepared by others.

27
Planning Nonresidential Installations

Although the same basic principles that were covered in Parts 1 and 2 of this book also apply to nonresidential projects, there are some additional problems. Larger currents, higher voltages, 3-phase power in addition to single-phase power, and different kinds of materials and equipment may be involved.

This book cannot possibly cover all the details of larger projects. Some typical projects will be covered as a guide to help you in your career. *But the information in chapters from here to the end of the book must be considered more or less a preview of the kinds of problems that you, as a student, will encounter later in your career.*

Heavy-Duty Lampholders. A lampholder is any device by which current is carried to a lamp; as explained in Chap. 4, most people simply call them sockets. But lampholders are not necessarily the type used with lamps that have screw-shell bases; they include also those used with other kinds of lamp bases, some of which are described and illustrated in Chaps. 14 and 29. It might be well to note that ordinary medium screw-shell bases are not permitted on lamps larger than 300 watts; they must have mogul screw-shell (or other) bases. Heavy-duty lampholders are defined in NEC Sec. 210-21(a) as those rated at 750 watts[1] or more.

Branch Circuits: General Information. Branch circuits are covered by NEC Arts. 210 and 220. The ampere rating of a branch circuit is determined by the rating of the overcurrent protection (breaker or fuses) protecting the circuit. Naturally the ampacity of the wires in the circuit must be at least as great as the ampere rating of the overcurrent protection. Circuits supplying *two or more* outlets are rated at 15, 20, 30, 40, and 50 amp. If a *single load* consumes more than 50 amp, it must be supplied by a separate circuit of any required ampere rating; the circuit must supply only one outlet, except that in industrial establishments, where qualified people will service the equipment, multioutlet branch circuits larger than 50 amp are permitted.

In nondwelling occupancies, per NEC Sec. 210-6(c)(2), the voltage *between conductors* in a circuit may not exceed 150 volts if the circuit supplies one or more *medium-base* screw-shell lampholders, the kind used with ordinary incandescent lamps. The restriction therefore does not apply to electric-discharge lighting.

NEC Sec. 210-6(a) requires that branch circuits supplying lampholders, fixtures, or receptacles rated 15 amp or less must not exceed 150 volts *to ground* (review the definition of "voltage to ground" in Chap. 9), with exceptions. In industrial establishments the voltage may exceed 150 but may *not* exceed 300 volts to ground, if only competent persons will service lighting fixtures *and* if lighting fixtures do not have a readily accessible switch control on the fixture itself, have mogul screw-shell sockets or other types of lampholders approved for the purpose, and (if incandescent) are mounted at least 8 ft above the floor if possible. The second exception is similar: The voltage to ground may exceed 150 but may not exceed 300 volts on any premises other than within dwelling units if the circuits supply only permanently installed electric-discharge lamps that do not have switch controls on the fixtures *and* if electric-discharge lampholders of the screw-shell type are installed at least 8 ft above the floor.

These two exceptions permit fluorescent (or other electric-discharge) lighting fixtures operating at 277 volts to be used, under the conditions outlined. Lighting at 277 volts will be discussed elsewhere in this chapter.

[1] If the socket is the "admedium" size, it is a heavy-duty type if rated at 660 watts or more. But admedium bases and sockets are now practically museum pieces.

A demand factor may never be applied to a branch circuit. A demand factor may be applied only to feeders or service-entrance wires, where all the load is not likely to be in use at the same time. There are exceptions: A demand factor, as already explained in an earlier chapter, may be applied to a branch circuit serving a range, for it is not likely that all the heating elements will be used to the maximum at the same time; and a duty cycle, while not precisely called a demand factor, may be applied to some intermittent-duty motors (such as for an elevator) and to welders.

If the load on any circuit is continuous (3 h or more), remember that the maximum load on any circuit is limited to 80% of the normal rating of the circuit. This point will not be repeated in the discussion of various circuits in following paragraphs.

Electric-Discharge Lighting. This includes fluorescent lighting of all kinds as well as other types that will be defined in Chap. 29. When calculating the load on a circuit, remember that you must include not only the rated watts of the lamps but also the power consumed in the ballasts. For example, a fluorescent fixture with four 40-watt lamps consumes 160 watts in the lamps; the power consumed by the ballasts varies with the type and size of the lamps, but adding 15% would be a fair average. The four-lamp fixture mentioned would then consume 160 watts for the lamps, plus 15% of 160 or 24 watts in the ballast, for a total of 184 watts. A 2-amp allowance for each such fixture at 120 V would not be excessive, considering allowance for voltage drop, power factor, etc.

Permissible Loads. An individual branch circuit may serve any load for which it is rated. Branch circuits of 15 to 50 amp *serving two or more outlets* must comply with the following:

Fifteen-Ampere Branch Circuits. A 15-amp circuit may serve any kind of load in any kind of occupancy: incandescent or fluorescent lighting, receptacles, and so on. Either ordinary or heavy-duty lampholders may be installed on the circuit. Only 15-amp receptacles may be installed. The ampere rating of any cord- and plug-connected appliance may not exceed 80% of the rating of the circuit; for appliances fastened in place the limit is 50%. See NEC Sec. 210-23(a).

Compute all known loads, such as lighting, special appliances such as office copying machines, air conditioners, etc., and add a liberal allowance for unknown loads, such as maintenance equipment (floor polishers, for example, or vacuum cleaners). Receptacle outlets for which the load is unknown must be allowed for at 180 VA; a 2-amp allowance for each would not be excessive. Each heavy-duty lampholder, if its exact load is unknown, must be figured at 600 VA, about 5 amp. This is covered by NEC Sec. 220-2(c).

Twenty-Ampere Branch Circuits. Everything said about the 15-amp circuit is also correct for a 20-amp circuit, except that either 15- or 20-amp receptacles may be installed on it.

Thirty-Ampere Branch Circuits. These circuits may be used for appliances in any occupancy; the rating of any portable or stationary appliance may not exceed 80% of the rating of the circuit. They may be used for fixed lighting provided they serve only fixtures with *heavy-duty* lampholders. Since fluorescent lampholders are not of the heavy-duty type, they may not be used on 30-amp circuits but may be used only on 15- or 20-amp circuits. Only 30-amp receptacles may be installed.

Forty- and Fifty-Ampere Branch Circuits. These circuits may be used to supply cooking appliances as in dwelling occupancies and also infrared heating appliances and fixtures with heavy-duty lampholders.

Receptacles on 40-amp circuits may be of either the 40- or the 50-amp type (although there are no 40-amp receptacles presently available); on 50-amp circuits they may be only of the 50-amp type.

All Fifteen- to Fifty-Ampere Branch Circuits. Fixed outdoor electric deicing or snow-removal units may be supplied by any of these circuits, provided the circuit supplies no other load. This is covered by NEC Sec. 426-4, which also specifies that such loads must be considered *continuous* loads, which, as already discussed, requires derating of the nominal capacity of the circuit.

Taps on Branch Circuits. Many times it is not practical or necessary to run wires of the same size as the circuit wires to individual loads on the circuit. This subject is covered in NEC Sec. 210-19(c). Examples are wires to individual lighting fixtures or to individual heavy-duty lampholders of two or more floodlights supplied by the same branch circuit. Such taps must not be more than 18 in long (in the case of lighting fixtures, measured from the point where the wires emerge from the fixture), and must always have an ampacity sufficient for the load. In any event, taps must have an ampacity of not less than 15 if on 15-, 20-, 25-, or 30-amp circuits, and not less than 20 if on 40- or 50-amp circuits.

In wiring recessed lighting fixtures, as already discussed, ordinary wire may be used in wiring the circuit up to a junction box, from which a tap 4 to 6 ft long, ending at the fixture, is permitted and often required, as covered in NEC Sec. 410-67(c).

Flexible cords if approved for use with specific appliances may be No.

18 on 15- or 20-amp circuits, No. 16 on 30-amp circuits; on 40- and 50-amp circuits they must have an ampacity not less than 20. See NEC Sec. 240-4.

277-Volt Lighting. Over a period of years, the lighting in all areas has tended to higher and higher levels, consuming more and more watts per square foot of floor area. That in turn requires more and more circuits, or larger wires. So, instead of using 120-volt circuits, why not use a higher voltage, thus reducing the amperage of any given load? Then any size of wire will carry more watts than at a lower voltage. For example, No. 14 wire with a load current rating of 15, installed on a 120-volt circuit, can carry 15 × 120 or 1800 watts. If installed on a 277-volt circuit, it can carry 15 × 277 or 4155 watts. (If the loads are continuous, requiring derating as already discussed, the figures would be 1440 and 3324 watts.) Obviously fewer circuits would be required for any given load in watts, with consequent great savings of cost in the number of conduits needed, in the amount of wire needed, and especially in installation labor. The conditions under which 277-volt lighting may be installed have been discussed earlier in this chapter.

Today 277-volt fluorescent lighting is very common. If 277 volts seems a peculiar voltage, remember that if a building is served by a 3-phase wye-connected 480Y/277-volt service, the voltage between the neutral and any phase wire is 277 volts. The 3-phase 480-volt power is still available for 3-phase loads. Install the 277-volt lighting system as you would a 120-volt system, but be sure switches and branch-circuit breakers are UL Listed for the higher voltage. Since the voltage *to ground* is over 150 volts, plug fuses may not be used.

In all 277-volt wiring, a word of caution is in order. Each 2-wire circuit for 277-volt fixtures of course consists of a grounded wire connected to the neutral of the 480Y/277-volt system, and one of the three hot or phase wires of the system. If the circuit is 3-wire, it will include two hot wires. In an installation of any size, some of the circuits will contain the hot wire from phase *A*, others the hot wire from phase *B*, and still others the hot wire from phase *C*. Now the voltage between the grounded wire and any hot wire is 277 volts, but between any two hot wires is 480 volts. If switches to control the lighting are grouped in a multigang box, NEC Sec. 380-8(b) requires that the voltage between any two adjacent switches must not exceed 300 volts. If the switches cannot be arranged to accomplish this, the box must be equipped with permanent barriers between any two switches where the voltage between them would be not 277 but 480 volts.

Temporary Wiring. On large projects temporary power and lighting is often necessary during the period of construction before the permanent wiring system is completed. All the usual rules regarding services, feeders, overcurrent protection, grounding, etc., apply, but the wiring itself may be multiconductor flexible cords or cables adequately supported up off the floor and protected from accidental damage. Lamps must be provided with lamp guards or otherwise located to protect them from accidental contact or breakage. Receptacles and lighting must be on separate circuits, and the receptacles must be GFCI-protected, unless the grounding-conductor continuity of the circuits, flexible cords, and cord- and plug-connected equipment is checked by a designated individual before use, if damaged, after repairs, and at intervals not exceeding 3 months. All the rules for temporary wiring are found in NEC Art. 305.

Service to a Building. It would be well at this time to review the Code requirements for services. NEC Sec. 230-2 states that a building must be supplied by only one set of service-drop or service-lateral conductors. However, there are seven exceptions to this rule.

As defined in NEC Art. 100, a building is a structure that stands alone or that is separated from any adjoining structure by a fire wall. Any opening(s) in such a fire wall must be protected by approved fire doors. In other words, if a portion of a building structure is cut off from all other portions of the structure by walls that are classified by the authority having jurisdiction as fire walls, you must consider the cut-off portion as a separate building. Each such building is entitled (but not required) to be supplied by its own service drop or lateral just as though it were a separate structure. The authority having jurisdiction for classifying buildings, fire walls, fire doors, etc., is usually a building official or a fire department official. Where there is any question, the electrical inspector will usually consult with such an official.

The exceptions to NEC Sec. 230-2 that allow additional drops or laterals to a building are as follows:

Exceptions Nos. 1 and 2 allow a separate drop or lateral for a fire pump as also for emergency or stand-by systems, as required for some types of occupancies.

Exception No. 3 allows a separate service for a multiple-occupancy building (such as a shopping center or an apartment house) where there is no available space for service equipment that would be accessible to all occupants. This requires special permission. In other words, if there is no other reason for having more than one service, the inspector will usually want to judge for himself or herself whether or not there is a common accessible service equipment location.

Exceptions Nos. 4 and 5 pertain to installations of large capacity and buildings of large area, conditions you are not likely to encounter, but it is well to know of them for future reference.

Exception No. 6 allows a separate service for two or more services of differing characteristics, such as for different voltages, phases, frequencies, or rate structures. In other words, a single building could have a 480Y/277-volt 3-phase 4-wire service, a single-phase 120/240-volt service, and a 240-volt single-phase service for a water heater on a different rate schedule; it could conceivably also have a dc service or a 400-Hz service. Such an extreme case is not likely to be encountered, but you should be aware of what is permitted.

Exception No. 7 allows two to six sets of underground conductors, size 1/0 and larger, connected together at their supply end but not at the load end, where they feed two to six separate service disconnects at the same location, to be considered as one service lateral.

Ground-Fault Protection of 480-Volt Services. Experience shows that arcing faults to ground are more likely to be sustained at 480 volts than at 240 volts. The current in the arc is generally below the level that would be seen as overcurrent by a fuse or circuit breaker above 800 amp (especially if the normal load is small in relation to the overcurrent device). Such arcing faults have resulted in much damage, particularly in 480-volt service switchboards. The Code, therefore, in Sec. 230-95, requires that ground-fault protection be provided for each 480/277-volt wye service disconnecting means rated 1000 amp or more. The equipment is similar in principle to the GFCI discussed in Chap. 9, but is set to respond to current to ground in amperes (generally 200 amp or more, with a Code maximum of 1200 amp) rather than milliamperes, and also incorporates a time-delay setting to allow for coordination with overcurrent devices or ground-fault protection on feeders downstream. A ground fault exceeding the trip and time settings of the sensor will cause the service disconnect to open. Ground-fault protection may be used on other than 480/277-volt systems, and at other locations than the service, such as feeders or individual circuits. Hospitals having ground-fault protection on the service must also have it for every feeder with at least a 6-cycle time delay to reduce the likelihood of a remote fault shutting down the entire service.

A balance between protection and continuity of service requires engineering judgment in making the trip-out and time-delay settings. Each installation must be tested when installed, and should be tested periodically thereafter.

Service Equipment Grounding and Bonding Connections. The Code requirements for the relative location of the grounding and bonding connections at service equipment become critical when ground-fault protection is used, for fault current taking an unintended path with relation to the ground-fault protection equipment can affect the sensitivity of the system, or render it inoperative. Figure 27-1 is a drawing of a service switchboard, simplified by showing only one "hot" conductor. The neutral disconnect shown is a removable section of bus, but in smaller equipment a neutral wire that can be lifted from its termination will satisfy NEC Sec. 230-75. A common type of ground-fault protection called "zero sequence" is shown, with a transformer sensor *A* encircling *all* of the circuit conductors, including the neutral. When the current through the sen-

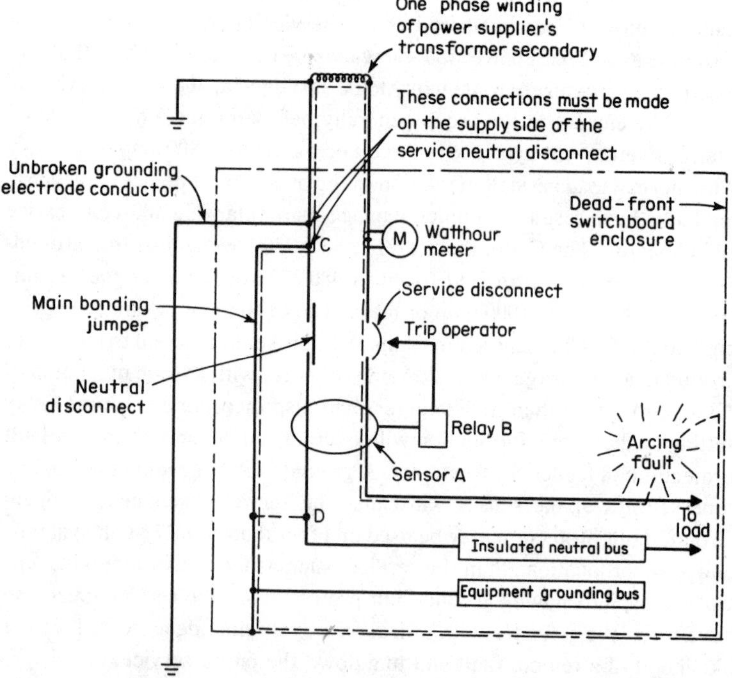

Fig. 27-1 The relative location of service grounding and bonding connections, disconnects, and ground-fault sensors is critical.

sor is not balanced—that is, when some of the current is diverted through a ground fault—any sensor current above the setting operates relay *B*, which in turn trips the main disconnect. NEC Secs. 250-23(a) and 250-61(a) require that the connections of the grounding electrode conductor and the main bonding jumper to the neutral be made *on the supply side* of the neutral disconnect. Both of these conductors are sized in accordance with NEC Table 250-94, but where the service-entrance conductors are larger than 1100 kcmil (mcm, MCM) the main bonding jumper must be increased in size in accordance with NEC Sec. 250-79(c).

An arcing fault is shown at the right side of the switchboard, with the path of fault current shown in a dashed line. It can be seen that the current through the sensor is not balanced, and when the time-delay and current settings of the relay are exceeded, the main breaker will be tripped. It can also be seen that if the main-bonding-jumper connection to the neutral conductor were moved from point *C* to point *D*, the total fault current would then flow through the sensor, and the ground-fault protection would not function.

The relative location of these connections at service equipment is important even if ground-fault protection is not used; but it is *critical* when using ground-fault protection, whether of the zero-sequence or any other type.

Interrupting Capacity and Withstand Ratings. A fault can bypass part of the conductor-and-load impedance (see Chap. 2) of a circuit, and allow the power source to deliver a large number of amperes during the time it takes for ordinary fuses or circuit breakers to open. The fault current is limited only by the voltage, the size and impedance of the transformer feeding the service, and the impedance of the conductors and equipment between the transformer and the fault, and could be upwards of 10 000 amp and sometimes as much as 200 000 amp. Ordinary fuses, breakers, switches, etc. do not have an interrupting capacity above 10 000 amp, and at higher currents could literally explode due to the intense magnetic fields set up. Such magnetic fields are directly proportional to the *square* of the current, so the magnetic forces when 100 000 amp are flowing are 100 times as great as when 10 000 amp are flowing.

In Sec. 110-9 the Code requires all devices that interrupt current to have sufficient interrupting capacity, and in Sec. 230-98 specifically requires service *equipment* to be suitable for the short-circuit current available. Equipment assigned a short-circuit rating (sometimes called the withstand rating) has been tested to determine that clearances between live parts are not reduced below a safe distance (short-circuit magnetic forces tend to bend busbars); that insulating supports, even if broken, continue to function safely; and that cables are not pulled from their connectors (the magnetic forces tend to whip cables about).

Calculation of available short-circuit current is beyond the scope of this book, but you should be aware of the conditions that may make it a factor in selection of equipment and overcurrent devices. As an example, for a single-phase 120/240-volt 3-wire service, if the transformer is 37.5 kVA with 1.7% or less impedance, and your wires from the transformer to the service equipment are No. 1/0 aluminum not over 10 ft long, ordinary 10 000-amp service equipment will not be suitable. This is a borderline example, and if the transformer or wire size were smaller or the transformer impedance or the distance were greater, then 10 000-amp service equipment would be suitable. Underground wye-connected network utility distribution systems often are capable of delivering large amounts of short-circuit current. Whenever the conditions suggest the need, your power supplier should be consulted for a determination of the current available at your location.

Overcurrent devices are available with interrupting capacities up to 200 000 amp. One means of using equipment with a withstand rating less than the available current is to use current-limiting overcurrent devices that interrupt the current before it reaches the maximum available. Fuseholders for current-limiting fuses will not accept non-current-limiting fuses, thus assuring continued protection where required. (See Chap. 5 for a discussion of fuse classes.) In very large installations, available fault current may be a consideration at locations in the system beyond the service, such as feeders.

Switchboards and Panelboards. Service-entrance wires where they enter a building may end at one or more switchboards, or one or more panelboards. All necessary disconnecting equipment and overcurrent protection, covering the building as a whole, is located in this equipment. Look at such a system as you would a big tree. The trunk is the service-entrance wire, the point where the trunk breaks up into half a dozen or more branches is the service equipment protecting the building as a whole, and the points where the larger branches in turn split into smaller branches are feeder or branch-circuit distribution points. But what is the difference between a switchboard and a panelboard? And between a service-distribution panelboard and a branch-circuit panelboard? The Code in Art. 100 defines panelboards and switchboards as follows:

Panelboard. A single panel or group of panel units designed for assembly in the form of a single panel; including buses, automatic overcurrent devices, and with or without switches for the control of light, heat, or power circuits; designed to be placed in a cabinet or cutout box placed in or against a wall or partition and accessible only from the front.

Switchboard. A large single panel, frame, or assembly of panels on which are mounted, on the face or back or both, switches, overcurrent and other protective devices, buses, and usually instruments. Switchboards are generally accessible from the rear as well as from the front, and are not intended to be installed in cabinets.*

Generally speaking, then, if the equipment is installed in or on a wall, it is a panelboard, which can range in size from the small fuse cabinet of Fig. 27-2 to the larger cabinet in Fig. 27-3, with many circuits, to still larger ones. If the equipment is installed away from a wall and is accessible from either the front or the back, it is a switchboard. From this statement and the Code definition, you could easily infer that switchboards are never enclosed, but most switchboards now being installed are of the metal-enclosed type. Switchboards are usually installed only in very large buildings.

The distribution riser diagrams of Fig. 27-4 show a one-story building at *A* and three-story buildings at *B* and *C*. These might be factories, office buildings, stores, or any other large building. These diagrams are typical of installations where a separate switchboard is installed.

Interior Distribution. In most houses and many other buildings of moderate size, the branch-circuit wires run directly from the service equipment. In such cases there are no feeders between the service equipment and the branch circuits, although the short copper busbars within the service equipment are technically feeders. In larger installations, the distances involved, the size of the total load, and the number of branch circuits make it impractical to start all branch circuits from the service equipment. So, in large buildings install various panelboards, which might be (1) service equipment panelboards, (2) distribution panelboards, or (3) branch-circuit panelboards. A single panelboard could combine two of these functions, or all three, in a single cabinet. Branch circuits

Fig. 27-2 This simple fuse cabinet is a small panelboard. (*Square D Co.*)

Fig. 27-3 A much larger panelboard, containing many circuit breakers. (*Square D Co.*)

may start from any kind of panelboard, and if all three kinds are installed in a building, could start from any of them.

If the panelboard you install is a service equipment panelboard (whether or not its cabinet also contains feeder or branch-circuit overcurrent protection), it must be of a type suitable for service equipment, and be so marked.

A building may have up to six separate disconnecting means and main overcurrent protection equipments; each of these can be a service equipment panelboard which may also contain overcurrent protection for feeders or branch circuits. From this equipment, run feeders to branch-circuit panelboards located in various parts of the building; the branch circuits start from there. If the building is quite large, one or more distribution panelboards can be installed between the service equipment and the branch-circuit panelboards. In other words, from the service equipment, run feeders to the distribution panelboards located where you

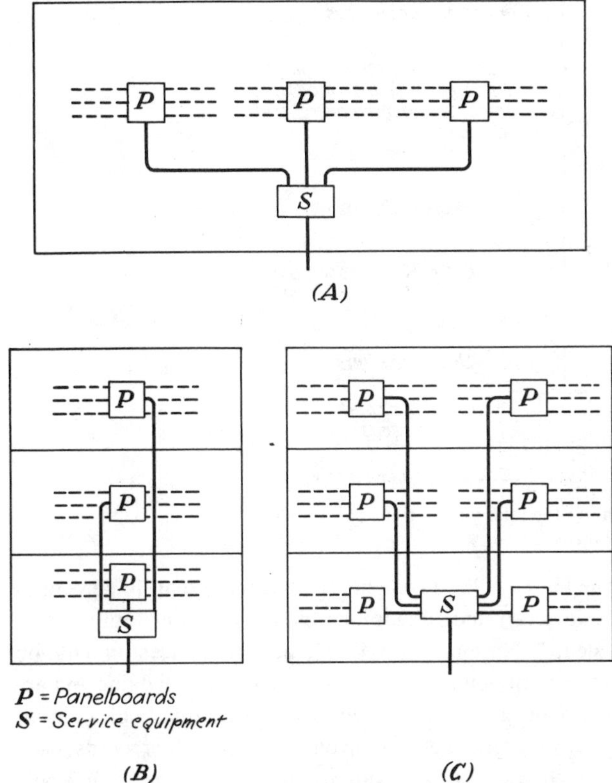

Fig. 27-4 Typical distribution riser diagrams for large buildings.

wish. From each distribution panelboard, run subfeeders to the branch-circuit panelboards.

If all that sounds complicated, see Fig. 27-5. Each single line in that diagram represents whatever number of wires might be involved in the run. Equipment shown in dashed lines represents equipment that may be installed, if you wish.

Both switchboards and panelboards are covered in NEC Art. 384; NEC Secs. 384-14 to 384-16 cover important points regarding types of panelboards and the overcurrent protection required for them.

Lighting and Appliance Branch-Circuit Panelboards. This is the kind of panelboard that you will encounter most often, and it has special restric-

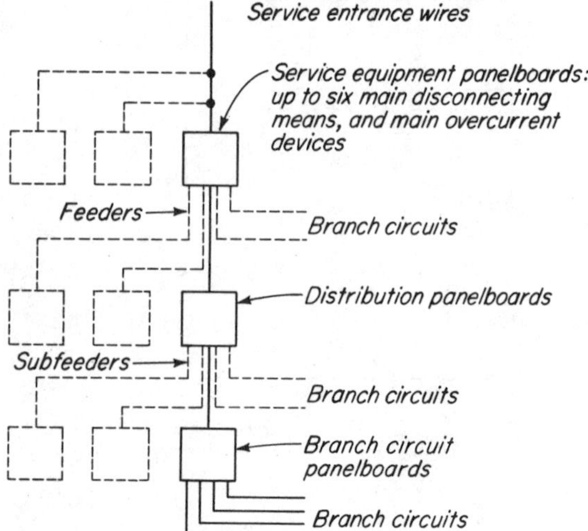

Fig. 27-5 This shows the relative positions of three different kinds of panelboards.

tions. It is defined in Sec. 384-14 as one "having more than 10 percent of its overcurrent devices rated 30 amp or less, for which neutral connections are provided."* No more than 42 overcurrent devices (in addition to the *main* overcurrent devices) may be installed in any lighting and appliance branch-circuit panelboard. Each fuse and each *pole* of a circuit breaker is counted as a separate overcurrent device. In other words, each 2-pole breaker is counted as two overcurrent devices, and each 3-pole breaker as three overcurrent devices.

A physical means must be provided to prevent the installation of more overcurrent devices than the panelboard is designed, rated, and approved for. In other words if "standard" width breakers *can* be replaced by the half-size "thin" breakers, this is taken into account in the total number of spaces allowed in the panelboard and in its ampere rating.

Each lighting and appliance branch-circuit panelboard must be protected on its supply side by not more than two sets of *main* overcurrent devices (two breakers or two sets of main fuses). There are two exceptions to this rule.

The first exception provides that if the overcurrent device (breaker or fuses) protecting the feeder to the panelboard has an ampere rating not greater than the ampere rating of the panelboard, no further main overcurrent device is required for the panelboard. In some cases, a feeder supplies two or more panelboards, in which case it is necessary to have main overcurrent protection at each panelboard. The overcurrent device may be inside as an integral part of the panelboard, or a separate circuit breaker or fused switch may be installed ahead of the panelboard. If a feeder supplies only one panelboard, it is usually the practice that the overcurrent device protecting the feeder have an ampere rating not greater than that of the panelboard; in that case the overcurrent device also provides the required overcurrent protection for the panelboard.

Under the second exception, if the panelboard is used as service equipment for an existing individual residential occupancy (a house or an individual apartment), up to six mains are permitted, but note that this provision applies only to *existing* installations. For *new* installations, lighting and appliance branch-circuit panelboards used as service equipment must have *not more than two* mains.

In Chap. 13, Figs. 13-4 and 13-8 show lighting and appliance branch-circuit panelboards with a single main; Fig. 13-5 shows one with two mains. Instead of supplying circuits within the panelboard as shown, one of the two mains in Fig. 13-5 could supply a large branch circuit, such as a range, or a feeder to another branch-circuit panelboard as shown in Figs. 27-2 and 27-3.

The panelboard that was shown in Fig. 13-5 is called a split-bus panelboard because it has two sets of buses, an upper and a lower set.

Other Panelboards. If the panelboard is *not* a "lighting and appliance branch-circuit panelboard," there is no restriction on the number of overcurrent devices it may contain, and they may be of any size (so long as not over 10% of them are 30-amp or smaller with a neutral connection, which would make the panelboard a lighting and appliance branch-circuit type). No overcurrent protection is required at the panelboard, provided the feeder that supplies the panelboard has overcurrent protection not exceeding the rating of the panelboard.

Such panelboards may also be used as service equipment, provided there are not over six main disconnecting means, and the panelboard is suitable for service equipment.

Derating of Panelboards. If most of the load served by a panelboard is continuous (3 h or more), it must not be loaded to more than 80% of its rated capacity. This is not required if the panelboard and all its overcurrent devices are approved for continuous duty at 100% of rating; such combinations are rarely available.

Neutral Busbar: Grounded or Insulated? Every piece of service equipment must have what is usually called a neutral busbar. This must be bonded to the cabinet as already explained in other chapters. This busbar contains connectors for the neutral of the service wires and the ground wire and usually smaller connectors for the grounded wires of all the feeders or branch circuits. All the equipment grounding wires must also be connected to a grounded busbar, which must be bonded to the neutral busbar; in service equipment the connectors for these equipment grounding wires are often installed along with the other connectors on the neutral busbar.

But a panelboard that is *not* used as service equipment but is connected on the load side of (''downstream from'') the service equipment cabinet, whether an inch or a hundred feet away, must have the busbar for all the ground*ed* (white) wires *insulated* from the cabinet. But if the wiring is by means of nonmetallic-sheathed cable or nonmetallic raceway containing a bare equipment ground*ing* wire, the cabinet must contain a *separate* busbar for all the ground*ing* wires, and that must *not* be insulated from the cabinet but rather must be bonded to the cabinet. Such a separate busbar is shown in Fig. 27-6.

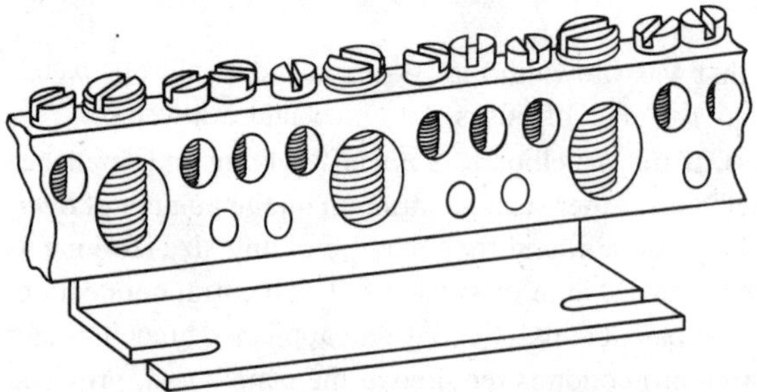

Fig. 27-6 Sometimes ground*ed* circuit wires must be insulated from the cabinet, but if equipment ground*ing* wires are required, they must be bonded to the cabinet. In such cases use a grounding bar of the general type shown, for all the equipment grounding wires, and bond it to the cabinet.

Three-Phase Lighting. In larger buildings the service is usually 3-phase, but the two wires that run to any one lighting fixture are necessarily single-phase. How then is the lighting to be handled? There are four ordinary ways, as follows:

1. A separate 120/240-volt single-phase service may be installed in addition to the 3-phase service, to supply all the usual 120- or 240-volt loads such as lighting, receptacles, and small single-phase motors.

2. If the service is wye-connected 3-phase 4-wire 208Y/120-volt, then 120-volt single-phase 2-wire circuits are directly available by connecting to the neutral of the service and one of the phase wires, as shown in Fig. 27-7. At *A* of Fig. 27-8 is shown a diagram such as is usually sup-

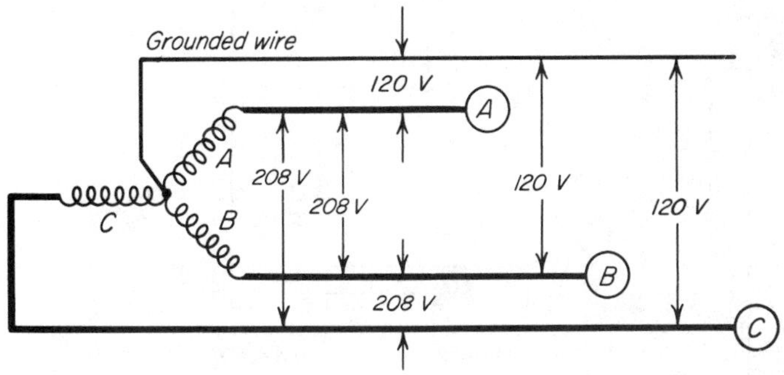

Fig. 27-7 A 208Y/120-volt system can supply both 120-volt single-phase circuits, and 208-volt 3-phase circuits, at the same time.

plied with such a 3-phase panelboard, and at *B* a diagram that may be easier for beginners to understand. These two illustrations illustrate a 3-phase service used *only* for 120-volt lighting, which is rarely the case. More usually the 3-phase service supplies both single- and 3-phase circuits; then you would install one or more 3-phase breakers in the service equipment to one or more 3-phase loads. This will be explained in more detail later in this chapter.

In Fig. 27-7, one of the wires is marked "grounded wire." That wire while grounded remains just a grounded wire; it becomes a neutral only if it becomes part of a 4-wire circuit.

3. When 120-volt current isn't directly available from the service equipment, transformers must be installed to step the higher voltage down to 120 or 120/240 volts. All the wiring beginning with the *secondary* of such a transformer constitutes a "derived system," the proper installation of which will be discussed later in this chapter. The primaries of such transformers are connected to two of the phase wires, whether their volt-

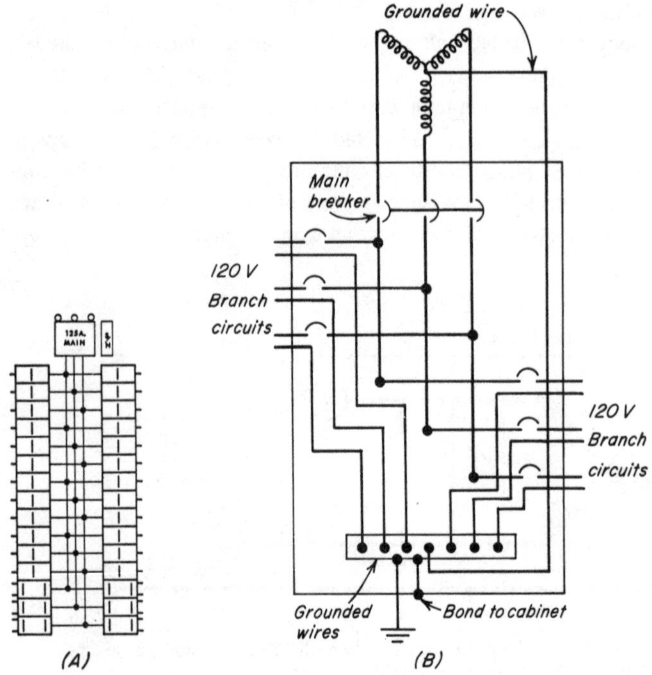

Fig. 27-8 At *A*, diagram for a 3-phase lighting panelboard; at *B*, a more pictorial diagram.

age is 277, 480, or 600 volts. Of course, a 3-phase transformer, or transformer bank, could be used, each of the secondaries then supplying 208Y/120-volt current for lighting, receptacles, or other single-phase loads such as small single-phase motors.

4. If the service is wye-connected 3-phase 4-wire 480Y/277-volt, fluorescent lighting operating at 277 volts is usually installed. The wiring would be as shown in Figs. 27-7 and 27-8, except that each single-phase circuit would operate at 277 volts instead of 120 volts. Of course, step-down transformers would still have to be installed to provide 120/240-volt power for incidental 120-volt lighting (incandescent lighting is prohibited by NEC Sec. 210-6 on 277-volt circuits except for mogul-base lampholders in industrial establishments), for receptacles for office machines and similar 120-volt loads, and to operate small single-phase motors.

Types of Distribution Systems. You have already studied the manner of computing the size of services, feeders, etc., including the size of the neutral conductor, and how to install them. You have learned that a grounded wire is not always a neutral wire. Where it is not a neutral wire, it carries the same current and must be the same size as the ungrounded wires. However, you need to know more about arranging feeders and branch circuits of various systems.

Circuit Arrangements: 208Y/120-Volt Systems. Such systems can provide 120-volt single-phase and 208-volt 3-phase circuits, either voltage alone or both at the same time, as shown in Fig. 27-7. Figure 27-8 shows the system providing only 120-volt circuits for lighting. Now see Fig. 27-9, which shows how the system can supply five different kinds of circuits at the same time, as follows:

1. Two-wire 120-volt single-phase circuits can be supplied by simply connecting the grounded wire of each circuit to the neutral *N*, and the ungrounded or hot wire of each circuit to either *A*, *B*, or *C*. So far as is possible, the number of ungrounded wires should be divided equally between *A*, *B*, and *C*. Note that the grounded wire of any such 120-volt circuit is *not* a neutral even if it is connected to a neutral wire. (If all four wires were connected in a circuit, the grounded wire would then be a neutral.) Each such 2-wire circuit is of course a single-phase circuit.

2. Two-wire 208-volt single-phase circuits can be supplied by simply connecting the two ungrounded wires of the circuit (whether feeder or branch circuit) to any two of wires *A*, *B*, or *C*. So far as it is possible, connect the same number of circuits to wires *A* and *B* as you do to *B* and *C*, or to *A* and *C*. Such 208-volt circuits are used to supply, for example, water heaters rated at 208 volts.

3. Three-wire 120/208-volt single-phase circuits (feeders or branch circuits) can be supplied by connecting the grounded wire of each such circuit to the neutral *N*, and two ungrounded wires to either *A* and *B*, or *B* and *C*, or *A* and *C*. If there are to be a number of such circuits, they should so far as practical be connected so that wires *A*, *B*, and *C* carry equal loads. Such 3-wire 120/208-volt circuits are quite common, for they make available the usual 120-volt current for lighting, receptacles, and so on, and also 120/208-volt current for ranges designed for that voltage. Do note that the grounded wires of such 120-volt circuits are *not* neutrals, since the grounded wire in this case carries the same current as the ungrounded wires.

4. Three-wire 208-volt 3-phase circuits (feeders or branch circuits)

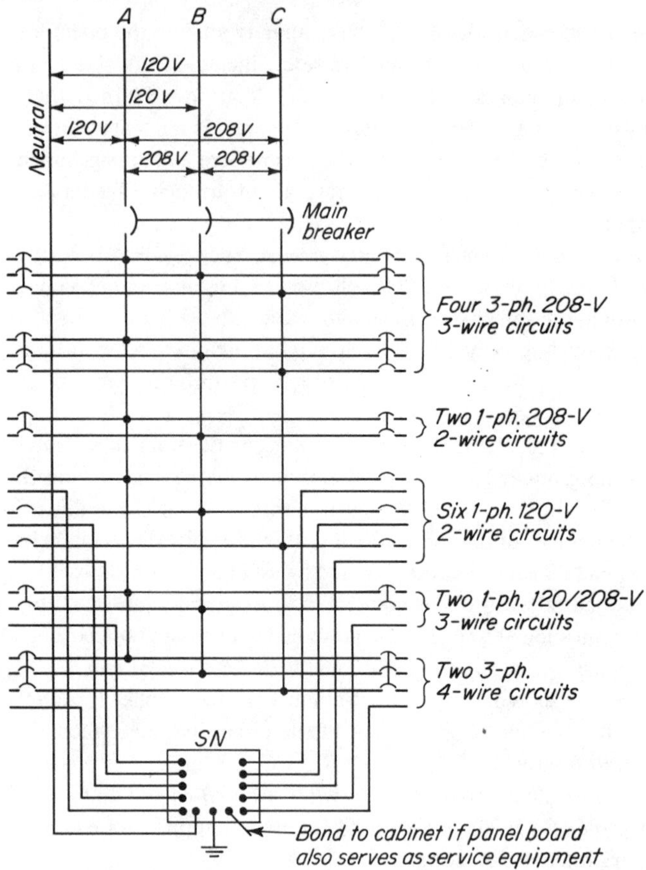

Fig. 27-9 This shows how a 208Y/120-volt system can supply five different kinds of circuits.

can be supplied by simply connecting the three ungrounded wires to the three wires A, B, and C. Such circuits are for operating 3-phase motors, 3-phase heating loads, and so on.

5. Four-wire 208Y/120-volt 3-phase circuits (feeders or branch circuits) can be supplied by connecting the grounded wire (which is now a neutral) to the neutral wire N, and three ungrounded wires to wires A, B, and C. Such feeder circuits are often used to supply one or more 3-phase panelboards in commercial and industrial buildings (as well as in larger

apartment houses); each such panelboard can supply branch circuits or other feeders operating at any of the voltages discussed in circuits 1 to 4 above.

Often, such 4-wire 3-phase feeders are installed as "risers," each extending from the service equipment (which is usually in the basement) to several upper floors. At each floor, a 3-wire single-phase subfeeder, as described in circuit 3 above, is installed to supply 120- or 208-volt loads. In installing such subfeeders, be sure to observe the 10- and 25-ft-tap-wire rules discussed in Chap. 5. Remember too that the grounded wire in such a circuit derived from a 4-wire source is *not* a neutral and that the grounded conductor busbar must be insulated from its cabinet.

Circuit Arrangements: 480Y/277-Volt Systems. Everything that has been said about 208Y/120-volt systems is also correct for 480Y/277-volt systems, except that where "120 volts" appears you must change it to "277 volts," where "208 volts" appears you must change it to "480 volts," and where "208Y/120 volts" appears you must change it to "480Y/277 volts." Do note, however, that water heaters and ranges are not available for voltages higher than 240 volts. As already discussed elsewhere, 277-volt circuits are used primarily for fluorescent lighting.

Circuit Arrangements: Delta-Connected 3-Wire 3-Phase Systems. If the service is at 240 volts, you will have 240-volt 3-phase power available by connecting to all three of the wires, or 240-volt *single*-phase by connecting to any two of the three wires. If the service is at 480 volts, you will have 480-volt 3-phase power available by connecting to all three of the wires, or 480-volt *single*-phase power by connecting to any two of the three wires. Single-phase 480-volt power is not often used, but it could be used to serve the primary of a transformer of a derived system, as will be discussed later in this chapter.

Such a system sometimes has one of the wires grounded, and it is then known as a "*corner-grounded*" system because the grounding point is at one "corner" of the delta. The grounded wire is not a neutral; it is just a grounded circuit wire. Such a grounded wire must never have a fuse or breaker installed in it. Like other grounded circuit wires it must be white, or for wires larger than No. 6 (which are not available in white), it must be painted white or marked with white tape at all connection points. This does not change the voltage relationship between wires. It simply means that the grounded wire will be at zero voltage to ground, and that the other two will have a voltage of 240 (or 480) volts *to ground*.

An ungrounded delta system should have a ground indicator (light, horn, etc.) to alert responsible maintenance personnel to take immediate steps to locate and clear the ground. The first ground will not affect operations, but a second ground on a different hot leg, and possibly at a different location in the plant, is the equivalent of a short circuit with fault current flowing by unpredictable paths between the two faults. The current in piping and equipment could cause an injury or fire. Production will be interrupted until one of the grounds is cleared.

Circuit Arrangements: Delta-Connected 4-Wire 3-Phase 240-Volt Systems. Such a system, as you have already learned, is basically a 3-wire 240-volt 3-phase system, with one of its transformer secondaries tapped at its midpoint. That makes 120/240-volt single-phase power available as was shown in Fig. 3-12. The wire from the midpoint of that secondary must per NEC Sec. 250-5 be grounded. It is a neutral so long as it continues throughout the system in *3-wire* circuits only. At any point where a 120-volt circuit is led off from the three wires, it ceases to be a neutral and becomes merely a grounded wire.

Power at 240 volts *single*-phase is obtained by running wires from any two of the three wires, *not* including the grounded wire. Power at 240 volts 3-phase is available by connecting to all three of the wires, again not including the grounded wire.

In such installations, the transformer with the tap at its midpoint is

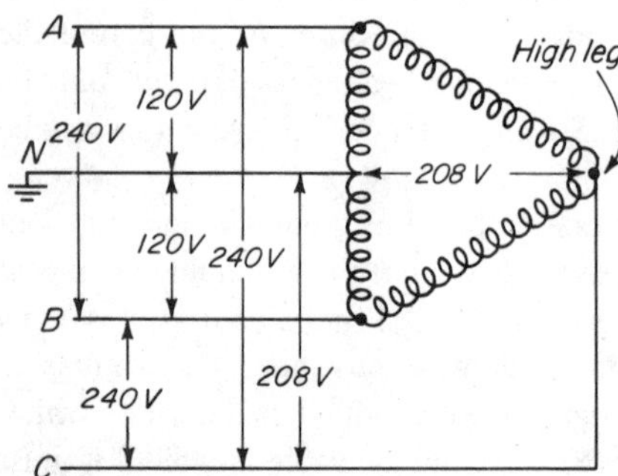

Fig. 27-10 In a 240-volt delta-connected 3-phase system, the voltage from the neutral to the "high leg" is 208 volts. Use extreme care never to connect what you want to be a 120-volt circuit between the neutral and the high leg.

usually larger (of a higher kilovolt-ampere rating) than the other two, because it must provide all the extra power for the single-phase circuits. The wiring diagram is shown in Fig. 27-10, in which the grounded wire is marked *N* and the three phase wires *A*, *B*, and *C*.

High Leg. But one caution is very important. In Fig. 27-10, the voltage between *N* and either *A* or *B* is 120 volts. But the voltage between *N* and *C* is 208 volts ($120 \times \sqrt{3} = 208$, or $120 \times 1.732 = 208$). The wire *C* is then called the *"high leg"* (high-voltage leg, or the wire with the higher voltage to ground; it is also called by other names such as "wild leg" or "stinger"). The high leg does not necessarily have to be wire *C*; it is whichever wire is *not* connected to the transformer secondary that has a tap at its midpoint.

In wiring 120-volt circuits on such a 4-wire delta system, it has often happened that instead of connecting the two wires of the circuit to wires *N* and *A* or *B*, they are accidentally connected between *N* and wire *C*; then what was intended to be a 120-volt circuit becomes a 208-volt circuit, which would of course ruin 120-volt equipment connected to it. Therefore NEC Sec. 215-8 requires that the wire in the high leg must be orange-colored. Instead of using an orange color throughout, you may instead identify the wire at any accessible point *if* the neutral is also available at that point, which you can do by tagging it or using other effective means. That would, for example, include painting it orange at such points or wrapping orange-colored tape around it.

The Code further requires, in Sec. 384-3(f), that in a switchboard or panelboard the high-leg bus be the one in the center unless the switchboard or panelboard section includes a self-contained meter socket in which the high leg is on the right. However, a great many existing installations do not comply with this rule (which first appeared in the 1975 NEC), so whenever you are working on a 3-phase 4-wire delta system, a voltage tester should be used to positively identify the high leg.

One disadvantage of the 4-wire delta-connected 3-phase system (besides the danger of accidentally connecting 208 volts to what is intended as a 120-volt circuit) is that one of the three transformers must be larger than the other two and that the wires from the transformer secondary with the midpoint tapped must be larger than the third or high-leg wire. That makes it difficult to properly protect all three wires. The simplest solution is to use the same size of wire throughout, in other words making wire *C* larger than needed. The other solution when using a smaller wire from the high leg *C* than from *A* and *B* requires a very special circuit breaker with a smaller trip coil in one leg than in the other two or, if fused equipment is used, with a smaller fuse (sometimes requiring an adapter that permits only a fuse of a smaller rating than the other two to be installed). In some localities, this type of system is no longer installed.

Open-Delta 3-Wire 3-Phase System. Suppose one transformer of three in a 3-wire 3-phase system fails and must be replaced. While one transformer is missing, the two remaining ones constitute what is called an open-delta system (sometimes called a "V" system), as shown in Fig. 27-11. The open-delta system with two transformers will still deliver 3-

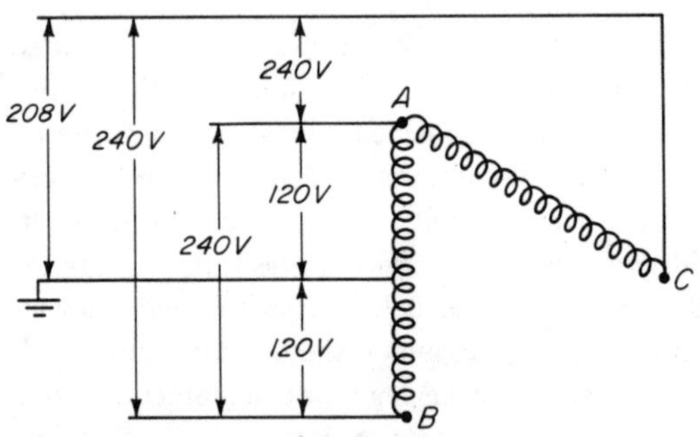

Fig. 27-11 A 120/240-volt "open-delta" 3-phase system. The transformer supplying single-phase power should be larger than the other one.

phase current, but only 57% of the normal power available from all three transformers. That reduced power may be an inconvenience to the user of the power, but it usually does not totally disable the user's business.

But this characteristic can be advantageous in some installations. If the power need is likely to be on the small side for some time, the power supplier may install only two transformers, connected open-delta. If each of the two transformers is rated at 50 kVA, they can deliver 85.5 kVA of power (85.5 kVA is 57% of the power that three 50-kVA transformers could deliver). Then, when the power needs of the installation increase, by adding one 50-kVA transformer, the power supplier boosts the total available power from 85.5 to 150 kVA.

One of the two transformers of an open-delta system may have its secondary midpoint tapped, as shown in Fig. 27-11, to supply 120/240-volt single-phase power. That transformer should of course have a higher capacity than the other one in the system.

Neutrals of Feeders, Single-Phase. In a 2-wire feeder, both wires must be the same size. If a 3-wire feeder serves only 120-volt loads, all wires must be the same size. But if a 3-wire feeder serves both 120-and 240-volt loads, the Code in Sec. 220-22 permits the neutral in some circumstances to be smaller than the hot wires. The "feeder" under discussion might be the service-entrance conductors serving the entire installation, or a feeder serving part of the load.

See Fig. 27-12, which shows a feeder serving both 120- and 240-volt loads. Assume now that the 120-volt loads are disconnected; in your

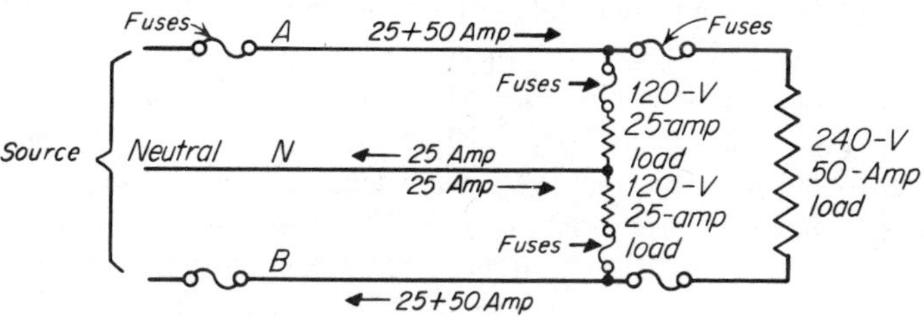

Fig. 27-12 A feeder serving both 120-volt loads and 240-volt loads.

mind erase them from the circuit. The neutral does not run to the 240-volt load; it is not necessary to the operation of the 240-volt load. Therefore the 240-volt load does not need to be considered so far as the size of the neutral is concerned. The neutral needs to be large enough only to carry the "maximum unbalance" which in Sec. 220-22 is defined as the maximum connected load between the neutral and any ungrounded wire.

Reverting now to Fig. 27-12, if the 120-volt loads are 25 amp each, and the 240-volt load is 50 amp, each hot wire must carry 25 + 50 or 75 amp. If the two 120-volt loads are identical, for example 25 amp as in the illustration, the neutral carries no current at all, but if one of the 120-volt loads is turned off, the neutral must carry the current for the other 120-volt load, in this case 25 amp, and must be sized accordingly.

In ignoring the 240-volt load, an electric range or dryer will give you a problem, for they are not 240-volt appliances but 120/240-volt appliances. At low heat the range operates at 120 volts; at high heat, at 240 volts. Only the blower motor and timer operate at 120 volts in the usual dryer. Calculate the feeder neutral as follows:

A. Total load in watts (including the range and/or dryer at the value normally required by the Code) . . . . ________watts

B. Deduct all loads operating at 240 volts, but not including the range and/or dryer . . . . ________watts

C. Deduct 30% of the watts that you have included in *A* for the range and /or dryer . . . . ________watts

D. Remainder (*A* minus *B* minus *C*) . . . . ________watts

The total wattage of *A*, divided by 240, gives you the ampacity of the minimum size of wire acceptable for the hot wires. The remaining watts of *D*, divided by 240, gives you the required ampacity for the neutral, assuming that the 120-volt loads are evenly balanced between the two hot wires. If for some unusual reason it is impossible to divide the 120-volt loads into two substantially equal portions, determine the wattage of the larger of the two groups and divide that by 120, which gives you the maximum unbalance and determines the minimum ampacity of the neutral.

If the maximum unbalance is over 200 amp, then per Sec. 220-22, count only 70% of the amperage above 200. For example, if the total unbalance is 300 amp, the total to use is 200 amp plus 70% of the remaining 100 amp, or 70 amp, for a total of 270 amp. If, however, part of the load consists of fluorescent lighting, the 70% factor may *not* be applied to that portion of the load consisting of electric-discharge lamps (fluorescent or HID lamps, discussed in Chap. 29). Follow the example of later paragraphs concerning 3-phase feeders.

It must be noted that all the above is correct only if the service wires come (a) from a *single-phase* transformer as in *B* of Fig. 3-9 in Chap. 3, or (b) from a 3-phase *delta-connected* transformer as in Fig. 3-12 in Chap. 3, and then only if the wires come from that leg of the transformer which has the center tap, so that the voltage between either of the hot wires and the neutral is 120 volts, or (c) from a 3-phase 4-wire *wye-connected* transformer and all four wires including the neutral are used.

Occasionally the premises are served by the neutral and only two of the hot wires of a 3-phase 4-wire wye-connected system. In that case the neutral must always be of the same size as the hot wires. If the voltage of the system is assumed to be 208Y/120, the voltage between the hot wires will be 208 volts, and between either hot wire and the neutral it will be 120 volts.

Neutrals of Feeders, Three-Phase. Now assume a 3-phase 4-wire feeder of 120/208 or 277/480 volts, serving both 3-phase loads connected to the three hot wires, and also one or more single-phase lighting loads. Any one single-phase load will of course be connected to the neutral and one of the hot or phase wires. For the purpose of determining the neutral, ignore the 3-phase loads such as motors.

Now consider each of the three single-phase loads (lighting, appliances, single-phase motors, etc.) separately. Determine the amperage of each, but exclude the amperage for any electric-discharge lighting load. Select the amperage of the largest of these three loads; assume it is 350 amp, *excluding* the electric-discharge lighting.

If the unbalance is over 200 amp, the Code in Sec. 220-22 permits a demand factor of 70% for the unbalance over 200 amp, whether the feeder is single-phase or 3-phase, but *not* for any portion of the load that consists of electric-discharge lighting. (The reason for excluding electric-discharge lighting is rather involved technically, and its explanation is beyond the scope of this book.) Assume that the unbalance for the electric-discharge lighting load determined separately is 90 amp. To determine minimum neutral size, proceed as follows:

First 200 amp of load other than electric-discharge lighting, demand factor 100%	200 amp
Remaining 150 amp of load other than electric-discharge lighting, demand factor 70%	105 amp
Maximum unbalance, excluding electric-discharge lighting	305 amp
Unbalance for electric-discharge lighting only, demand factor 100% in every case	90 amp
Final maximum unbalance	395 amp

The neutral must have an ampacity of at least 395 amp.

Calculating Different Occupancies. With the above general discussions it should be possible to calculate almost any type of building, taking into consideration the requirements of Table 220-11 of the Code (see Appendix). However, some types of buildings will be separately covered in some detail in Chaps. 31, 32, and 33.

Grounding. The theory and importance of grounding were discussed in Chap. 9, and the actual method of grounding small projects in Chap. 17. It will be well for you to review the subject there before proceeding. You must consider separately (a) system grounding and (b) equipment grounding. *System* grounding refers to grounding one of the current-carrying wires of an installation. *Equipment* grounding refers to grounding metal non-current-carrying components of the system.

System Grounding. This subject is covered by Code Sec. 250-5. Alternating-current systems *must* be grounded if this can be done so that the voltage *to ground*[2] is not over 150 volts. In practice this includes single-phase installations at 120 volts and at 120/240 volts (also 240-volt installations if the service is derived from a transformer that is grounded, as is usually the case) and also 3-phase installations at 208Y/120 volts.

Grounding is required on 480Y/277-volt systems; it is not required for a 480-volt delta ungrounded system. It is also required on all 240-volt 3-phase delta-connected installations if one of the phases is center-tapped as in Fig. 27-10. Grounding is also required, regardless of voltage, if a bare wire is used for the neutral in the service.

Occasionally there is a building that has only 240-volt single-phase loads, or only 3-phase loads. Such loads do not require a neutral for proper operation. Nevertheless the Code in Sec. 250-23(b) requires that if there is a ground at the transformer or transformers serving the building (as there usually is), the grounded wire must be brought into the building. It ends at the service switch, where it is grounded in the usual way. This grounded wire need not be larger than the size of ground wire required for the system. The purpose of this requirement is to provide a metallic path (the grounded wire) between raceways, cable armor, or equipment enclosures on the premises and the grounded terminal on the power supplier's transformer for current from accidental grounds. Otherwise, part of the path of fault current on the premises would be through the earth (between the service grounding electrode and the grounding electrode at the transformer) and might have an impedance so high that not enough current would pass through it to open an overcurrent device.

Separately Derived Systems. If there is no way to provide 120/240-volt single-phase circuits for lighting, receptacles, and other small 120- or 240-volt loads, from the service installed in a building, as is the case with 480- or 480Y/277-volt services, some other way must be found to provide those circuits. The usual way is to install what is called a "separately derived system." This consists of a transformer (or several as required) whose primary is connected to two 277- or 480-volt wires of the installation. The secondary delivers the 120/240-volt power that is wanted. See Fig. 27-13.

The secondary only of the transformer, and all wiring connected to it,

[2] "Voltage to ground" in the case of grounded systems means the maximum voltage between the grounded neutral wire and any other wire in the system; in the case of ungrounded systems, it is the maximum voltage between *any* two wires.

constitute the separately derived system. The terminals of the secondary of the transformer become the SOURCE for all wiring of the separately derived system. All the wiring in the separately derived system must be installed under the same conditions, as if the power for the system came from a separate 120/240-volt service entrance.

Note in Fig. 27-13 that the primary of the transformer is supplied by two 480-volt wires. If those wires are in metal conduit, or in armored cable, the separate equipment grounding wire shown in the diagram would not be installed, because the conduit or the armor serves as the grounding wire. If it were supplied by the grounded neutral and one wire

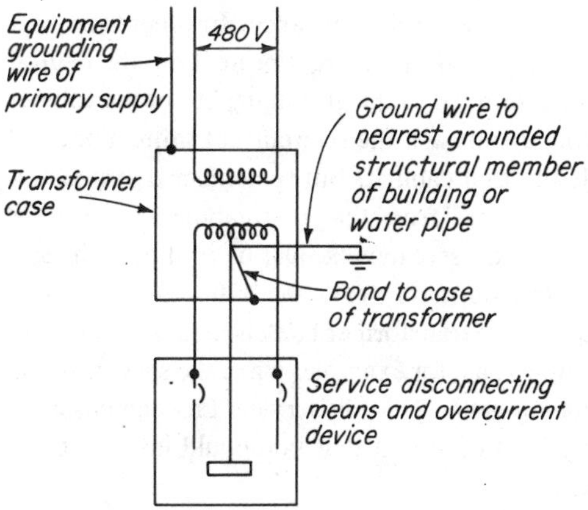

Fig. 27-13 Proper method of installing and grounding a "derived system."

of a 277-volt circuit, the same is true. But the grounded neutral in the circuit serving the transformer would of course have to be fully insulated all the way, and insulated from the non-current-carrying parts such as the transformer case. The equipment grounding wire shown would be installed only if the primary were supplied by nonmetallic-sheathed cable or wires in a nonmetallic raceway. In any of these cases, if a ground fault developed between the wires serving the transformer (or the primary itself) and the metal raceway of the wires supplying the transformer, the fault current would travel all the way back to the service equipment and disconnect the circuit.

Note, as shown in Fig. 27-13, that any grounded wire in the basic higher-voltage system does not extend to the derived system; the connection is lost, for there is no electrical connection between the primary and the secondary of the transformer. Therefore the separately derived system must be grounded as provided in NEC Secs. 250-5(d) and 250-26. The size of the ground wire can be determined by NEC Table 250-94 (see Appendix).

Since the secondary of the transformer delivers 120/240-volt power, its midpoint is a neutral. The neutral of the secondary of the transformer must be bonded to the transformer case and also to the nearest metal structural member of the building or to the nearest water pipe. The secondary of the transformer must be treated as has already been mentioned, just as if it were a separate service entrance. In other words, it must have a disconnecting means and overcurrent protection; the neutral from the secondary must be grounded to the grounding bar of the cabinet containing the disconnecting means and the overcurrent protection, and that in turn must be bonded to that cabinet. But remember that if panelboards are installed "downstream" from that first cabinet, the neutral busbar must be *insulated from* those cabinets, just as in other wiring.

Instead of one or more transformers providing 120/240-volt single-phase power, sometimes a 3-phase transformer bank is installed to supply a separate 208Y/120-volt supply, as, for example, when the service to the building is supplied by a 480- or 480Y/277-volt service. Proceed basically as outlined, and install any kind of circuit that you could install from a 208Y/120-volt service entrance.

The separately derived system is not necessarily fed from one or more transformers. It could be supplied by a motor-generator—a motor operated from the available voltage, driving a generator producing the desired voltage. In that case the generator and all wires leading from it constitute the derived system. It could be fed from a converter, a device that looks like a motor but has in a single case the motor itself, fed from the available voltage, and other windings producing the desired voltage. In that case the windings producing the desired voltage, and all wiring from their terminals, constitute the derived system.

An emergency or stand-by generator, *if feeding entirely separate* circuits that are in no way connected to the ordinary wiring of the building, constitutes a separately derived system.

On-Site Generation. The increasing concern for energy conservation has resulted in electric energy developed on the premises, such as wind-power generators and solar photovoltaic systems, and in the need to interconnect these systems with the normal utility service. When the output of the on-site source exceeds consumption on the premises, the excess power is purchased by the utility. The interconnection and the need to protect the equipment of both the on-site and the utility systems are complex matters. Should you contemplate installing such a system, consult both your local inspector and your power supplier before proceeding.

28
Nonresidential Wiring Methods and Materials

In nonresidential occupancies (and also in high-rise apartment buildings) the wiring is most often installed in raceways such as rigid metal conduit, intermediate metal conduit (IMC), or EMT rather than in the form of cables. These raceways when properly installed provide a good continuous equipment grounding conductor and also provide considerable flexibility, in that circuits may be changed, wires added, and breakdowns repaired by pulling new wires into existing raceways. Where exposed, these raceways are neat and afford ample protection for wires.

Conduit Fittings. For exposed runs of conduit, ordinary outlet boxes may be used, but cast conduit fittings of the type shown in Fig. 28-1 are sturdier and neater and are often used. They are known by many trade names such as Electrolets, Condulets, Unilets, etc. These devices are really specialized forms of outlet boxes, but instead of having removable knockouts, they have one or more threaded openings or hubs. Accordingly, with a very few basic body shapes, dozens of different combinations of openings are available. Each opening is threaded for the size of conduit for which the fitting is designed. Such fittings are also available with built-in connectors for threadless conduit or for EMT. However, regular connectors can be used with EMT, for the thread of an EMT connector will fit the threaded opening of the fitting.

A few of the more common types are shown in Fig. 28-1. While the illustration shows only one cover, naturally you must install a cover on

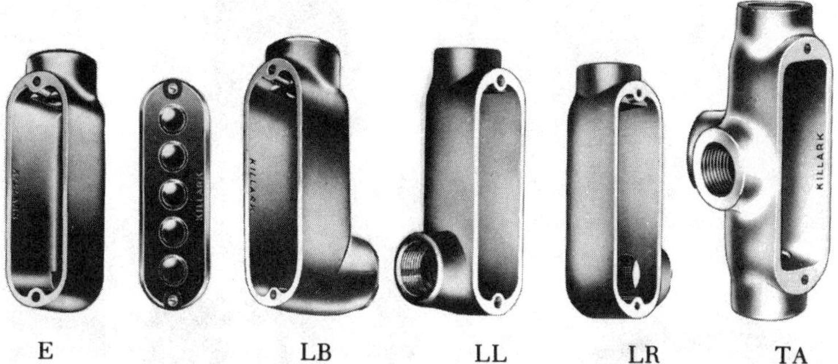

Fig. 28-1 For exposed runs of conduit, fittings of the type shown here are commonly used. There are dozens of different shapes or types. (*Killark Electric Mfg. Co.*)

each fitting after the installation is completed. The Type E with the cover at its right is used at the end of a run to some types of motors, transformers, or similar equipment not having provisions for a conduit connection. The other fittings shown avoid awkward bends in conduit. The Type LB is commonly used where a run of conduit comes along and then must run through a wall or ceiling; it is equally useful in going around a beam or similar obstructions. Types LL and LR are handy for 90-deg turns where the mounting structure is not suitable for an LB. However, the wires are more difficult to feed into an LL or LR than into an LB. There are very many kinds of combinations, some as complicated as the Type TA with four openings, which obviously is not used very often; Type T is similar but has only three openings, without the opening in the back of the fitting.

In a different style of body there are available many types similar to the Type FS shown in Fig. 28-2, which is used mostly for the mounting of switches, receptacles, and similar devices, using, for example, the weatherproof cover with a spring-loaded flap cover for a receptacle. The same Fig. 28-2 also shows a Type E with a special gasketed cover that permits installation of a toggle switch outdoors. On exposed runs of conduit, lighting fixtures can be mounted on round Type P fittings. In general, cast boxes and fittings are subject to the same internal space requirements

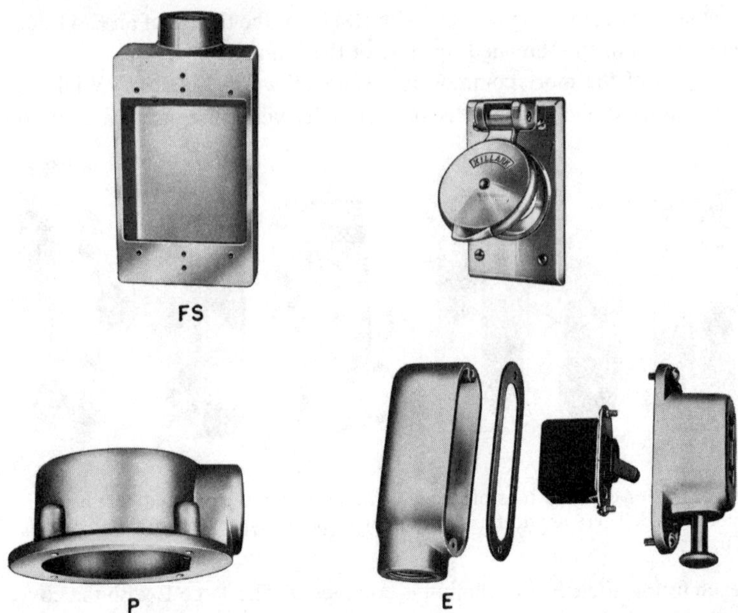

Fig. 28-2 Larger fittings are used to support switches, receptacles, fixtures, etc. (*Killark Electric Mfg. Co.*)

(cubic-inch content for number of conductors and devices, and distance between conduit entries when used as pull boxes) as are outlet, pull, or junction boxes of sheet steel. See Chap. 10.

Other Wiring Systems. The Code in Arts. 318 through 366 covers all the recognized wiring methods or systems. In addition to those discussed in this book, there are many others, but you will not need information about all of them until you have mastered the more common systems described in this book.

Junction Boxes. Wires of the smaller sizes are sufficiently flexible to be readily pulled into conduit. If the runs are long, boxes of the type described in Chap. 10 as well as the conduit fittings shown in Figs. 28-1 and 28-2 may (within their capacity limits) be installed as pull boxes. All boxes must be installed where they will be permanently accessible without damaging the structure of the building in order to gain access to them.

The larger the wire, the more difficult the pulling process becomes. For wires No. 6 and larger, and particularly in the large kilocircular-mil (kcmil, mcm, MCM) sizes, it becomes increasingly difficult to pull them into conduit, especially if there are long runs or several bends. Then it becomes necessary to install pull boxes in strategic locations; many times they are used in place of conduit elbows. A pull box, as the name implies, is a box so located that the wires can be pulled more easily into the conduit from one outlet to the next. A single box is often used for a number of runs of conduit, as shown in Fig. 28-3. Boxes are also used where wires have to be spliced. In many cases a box is used as both a splice box and a pull box. All such boxes are called junction boxes.

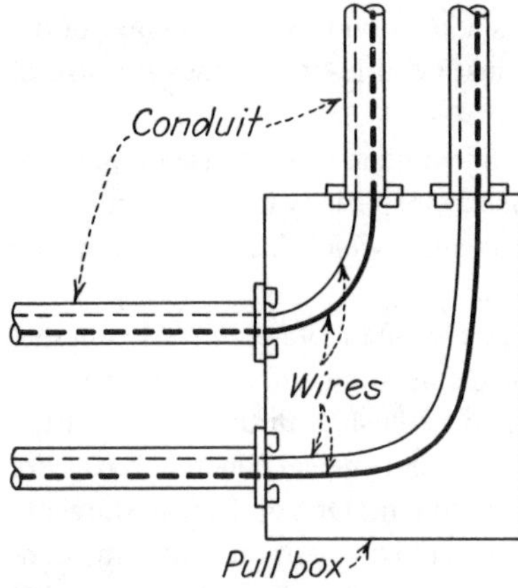

Fig. 28-3 Junction boxes are used with conduit to make it easier to pull wires through long lengths of conduit, or for splicing wires, or for both purposes in the same box.

For wires No. 6 and smaller, boxes already discussed in Chap. 10 may be used. If the conduit is $^3/_4$-in trade size or larger, *and if* it contains No. 4 or larger wires, *or if* cables[1] are used containing No. 4 or larger wires, then NEC Sec. 370-18 specifies the minimum dimensions of junction boxes. For *straight pulls* (conduit entering one wall of a box and leaving by the opposite wall) the length of the box must be at least eight times the trade size (not the outside diameter) of the largest conduit entering the box; several lengths of conduit may run into and out of the box. If the largest size of conduit is the 3-in size, the box would have to have a minimum length of 24 in. Often a length greater than the minimum will be found to be more practical than the Code minimum, especially if several conduits enter the box.

For *angle* or *''U'' pulls,* if only one conduit enters any one wall of the box, the distance to the opposite wall must be at least six times the con-

[1] If cables are used, determine the size of the conduit that would be required to contain the wires in the cable, if they were separate wires, and pretend that that size of conduit is being used instead of the cable, to determine the required size of the junction box.

duit size. But if more than one conduit enters the box, the minimum distance to the opposite wall must be six times the size of the largest conduit, plus the sizes of each of the other conduits entering the box. So, if, for example, only one 2-in conduit enters the box, its minimum length must be 12 in. But suppose two 2-in conduits, plus two $1^1/_2$-in conduits, enter the box. Then the minimum length of the box is $12 + 2 + 1^1/_2 + 1^1/_2$ or 17 in.

There is one other requirement. The distance from the center of the conduit by which a set of wires enters a junction box to the center of the conduit by which it leaves the box must be at least six times the size of the conduit.

All dimensions discussed are Code minimums. It is often good practice to use boxes larger than the minimum size (especially if a large number of conduits enter the box) to avoid crowding, which increases labor and leads to possible damage to insulation.

Concrete Boxes. Walls and ceilings are often of reinforced-concrete construction. The conduit and the boxes must be embedded in the concrete, if the devices in the boxes are to be flush with the surface of the walls. Ordinary outlet boxes may be used, but special concrete boxes are preferred and are available in depths up to 6 in. One of these is shown in Fig. 28-4. The conduit and the boxes must be in position before the concrete is poured. These boxes have special ears by which they are nailed to the wooden forms for the concrete. When installing them, stuff the boxes tightly with paper to prevent concrete from seeping in. When the forms are removed, the conduit and the boxes are solidly embedded; the interiors of the boxes are clean and ready to use. Figure 28-5 shows an installed box.

Deflection of Wires. If wires are bent sharply and press against a protruding surface (as that of a conduit bushing) where they emerge from a conduit, the insulation might be damaged and ground faults might develop. This is especially true of larger wires because of their weight. For that reason, NEC Sec. 373-6(c) requires an insulating bushing (or the equivalent) where No. 4 or larger wires enter or leave the conduit, such as at a box, cabinet, or similar enclosure.[2] A metal conduit bushing with an insulated throat (molded insulating material on the inside) is shown in Fig. 28-6. Such bushings are also available with a connector for a bonding jumper, as shown in the same illustration. A bonding jumper is a com-

[2] A conduit fitting (such as those shown in Figs. 28-1 and 28-2) or a threaded hub in a box is so designed that this extra protection is not needed.

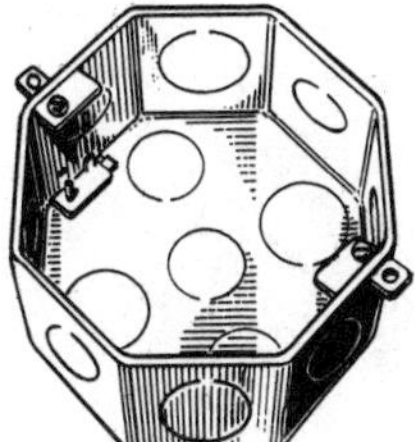

Fig. 28-4 A concrete box, designed to be embedded in concrete.

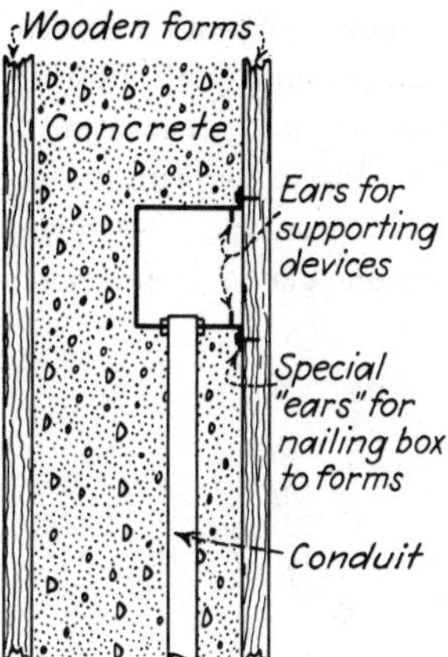

Fig. 28-5 Concrete boxes are nailed to wooden forms before the concrete is poured.

monly used means of assuring the required grounding continuity between a service conduit and the service equipment cabinet.

Figure 28-7 shows a bushing made entirely of insulating material. When using such bushings, you must install two locknuts, one on the outside of

Fig. 28-6 Metal grounding bushing with insulated throat and terminal for bonding jumper. (*Union Insulating Co.*)

Fig. 28-7 Conduit bushing made entirely of insulating material. (*Union Insulating Co.*)

the box or cabinet and another on the inside, before the bushing is installed. This ensures grounding continuity between the conduit and the box or cabinet.

An acceptable substitute for a bushing with an insulated throat is an insulating liner similar to the fiber bushings ("redheads") used with armored cable but of much larger size. They must be constructed so that they snap into place and are not easily displaced. Usually this type is used only on existing installations where an insulating bushing as described in preceding paragraphs was overlooked in the initial installation.

Locate cabinets so that the incoming runs of conduit will be so placed as to require minimum deflection of wires where they emerge from the conduit, and so that in vertical runs the weight of the wire will not be supported by a bend in the wire at the end of the conduit run. Figure 28-8 shows the wrong and Fig. 28-9 the right method. Let the bends in the wire be sweeping and gentle rather than abrupt.

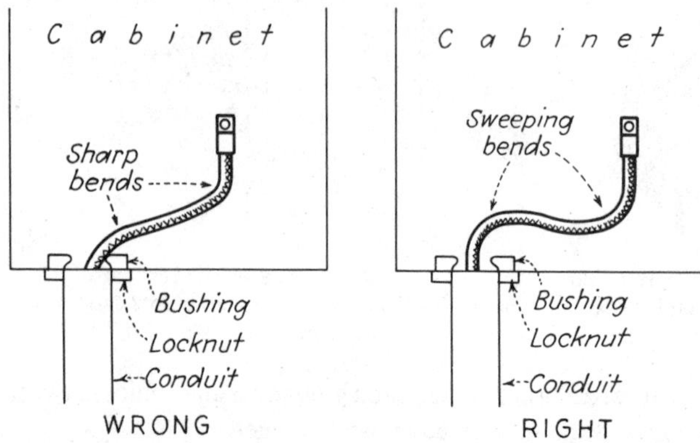

Fig. 28-8 The bend where the wire emerges from the conduit is too abrupt. The wire rests on the conduit bushing, which could lead to damage to the insulation.

Fig. 28-9 The bend in the wire should be gentle and sweeping. This tends to prevent ground faults where wires emerge from conduit.

Supporting Vertical Runs of Wire. Terminals on panelboards and similar equipment are not designed to support any substantial weight. Where there is a vertical run of wire, the weight of the wire is quite considerable, especially in the larger sizes. If such vertical runs are connected directly to terminals, damage may result. The Code in Sec. 300-19(a) requires that wires be supported at the top of a vertical raceway, or as close as is practical to the top, and also at additional intervals, as shown in Table 28-1. If the vertical portion of a raceway is less than 25% of the spacing shown in Table 28-1, no supports are required.

One way of accomplishing the required support is to use the fitting shown in Fig. 28-10. Boxes or cabinets are installed at the required intervals, and the fitting is installed in the conduit as it enters the bottom of the cabinet. Be sure one is installed at the topmost cabinet.

Grounded Circuit Wires. The Code in Sec. 200-6 requires the grounded circuit wire (whether it is a neutral or not) to be white. But No. 4 and larger wires are not available in white. So in the case of No. 4 and larger wires the Code permits you to use any color (except green) if at every

TABLE 28-1 [NEC Table 300-19(a)]* Spacings for Conductor Supports

		Conductors	
		Aluminum or copper-clad aluminum	Copper
No. 18 thru No. 8	Not greater than	100 feet	100 feet
No. 6 thru No. 0	Not greater than	200 feet	100 feet
No. 00 thru No. 0000	Not greater than	180 feet	80 feet
211 601 thru 350 000 cmil	Not greater than	135 feet	60 feet
350 001 thru 500 000 cmil	Not greater than	120 feet	50 feet
500 001 thru 750 000 cmil	Not greater than	95 feet	40 feet
Above 750 000 cmil	Not greater than	85 feet	35 feet

Fig. 28-10 This fitting is very effective in supporting vertical runs of wire. (*O-Z/Gedney Co.*)

terminal the grounded wire is painted white, or wrapped with white tape. If the grounded wire is No. 6 or smaller, it must be white throughout its entire length (unless it is a bare wire, which is acceptable under certain circumstances as explained elsewhere).

Continuity of Ground. Preceding chapters have shown how the various runs of conduit or armored cable tie outlet boxes or other equipment into one continuously grounded system. However, if one or more of the wires in a feeder or branch circuit have a voltage above 250 volts *to ground,* NEC Sec. 250-76 requires special bonding means. Any of the methods required for service equipment and raceway bonding, such as grounding bushings, grounding locknuts, grounding wedges, or threaded hubs, as described in Chap. 17, may be used. If the knockout is of the same size as the raceway or fitting (not oversized and with no concentric or eccentric knockout parts remaining), the double-locknut arrangement shown in Fig. 28-11 is acceptable. If EMT or a cable connector is used, the shoulder on the connector serves in place of the locknut on the outside of the cabinet. While not required, a grounding locknut as shown in Fig. 28-12 may be substituted for the connector locknut to achieve a better job.

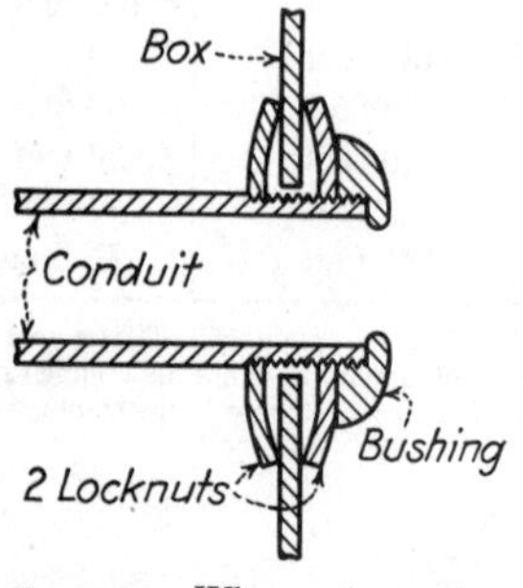

Fig. 28-11 Where the voltage *to ground* is over 250, use the double-locknut construction shown here instead of the ordinary construction shown in Fig. 10-12, but only if the knockout is not oversized or does not include unremoved concentric or eccentric parts.

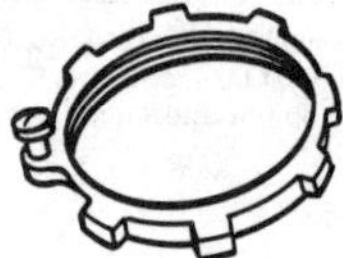

Fig. 28-12 A grounding locknut. It is not required, but is recommended in place of the locknut on cable and EMT connectors.

Surface Metal Raceways. In nonresidential buildings, the wiring is usually in conduit embedded in the concrete. To add new outlets, or move existing outlets, is almost impossible if the conduit is to be concealed, and conduit that is installed on the surface is very unsightly and rarely acceptable. Yet changes are often necessary, especially in offices, stores, and other buildings where layout changes are frequent.

Instead of conduit, use metal[3] surface raceways; they are attractive and blend well with the background surfaces. Such raceways may be used not only for power wires but for telephone and similar circuits as well, but both must *never* be installed in the same channel.

Two styles are available. One is called the one-piece type (although it actually consists of two pieces, preassembled at the factory), and the other is called the two-piece type. The one-piece type is installed empty;

[3] Nonmetallic surface raceways are also available.

the wires are then pulled into it just as in conduit. Using the two-piece type, the base member is installed first, the wires are then laid in place, and then the cover is installed. The Code does not list the number of wires that may be installed in any given size and type of surface raceway, but NEC Sec. 352-3 states that the *size* of any wire may not be larger than that for which the raceway was designed, and Sec. 352-4 states that the *number* of wires installed may not exceed the number for which the raceway was designed. All that sounds rather vague but the answer is simple. UL tests the raceway for the "designed" size and number of wires and Lists the product only if suitable. So you merely have to consult the tables furnished by manufacturers for such information.

Splices and taps normally are made only in junction boxes, but NEC Sec. 352-7 permits them at any point of a metal surface raceway of the removable-cover type. The wires including the splices and taps must not occupy more than 75% of the cross-sectional area of the raceway at the point where the splice is made.

The Code in Art. 352 limits both one- and two-piece types to *exposed* runs in dry locations, although the material may be run through (but not inside of) dry walls, partitions, and floors if the raceway is in a continuous length where it passes through. The voltage between any two wires must not exceed 300 volts unless the material in the raceway is at least 0.040 in thick, in which case the voltage may be up to 600 volts.

The one-piece type, into which the wires are pulled after the raceway is installed, is shown in Fig. 28-13. This figure also shows the dimensions of the smallest size; larger sizes are available. It is held in place by several methods, using clips, couplings, or straps as shown in the same illustration. There are available many appropriate kinds of elbows, adapters, switches, and receptacles to fit each size of raceway. Some of these are shown in Fig. 28-14.

The two-piece type is shown in Fig. 28-15, which also shows the dimensions of the smallest size; larger sizes up to $4^3/_4$ in wide by $3^9/_{16}$ in deep are available. Many fittings are available similar to those shown in Fig. 28-14, but designed for this type of raceway. One of the larger sizes of raceway is available with a metal divider in the base, thus providing two separate channels. Use one channel for power wires, the other for telephone or other nonpower purposes. This makes a neat installation and meets the Code requirement that power wires and other wires must occupy separate channels.

The Code states that metal surface raceway must not be installed

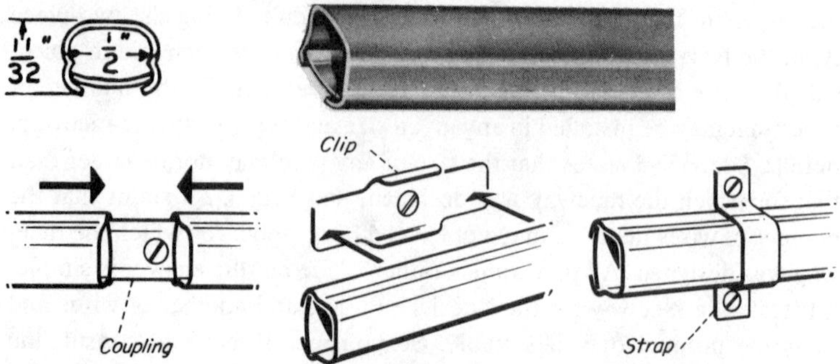

Fig. 28-13 Typical "one-piece" metal surface raceway. Wires are pulled into place after the raceway is installed. (*The Wiremold Co.*)

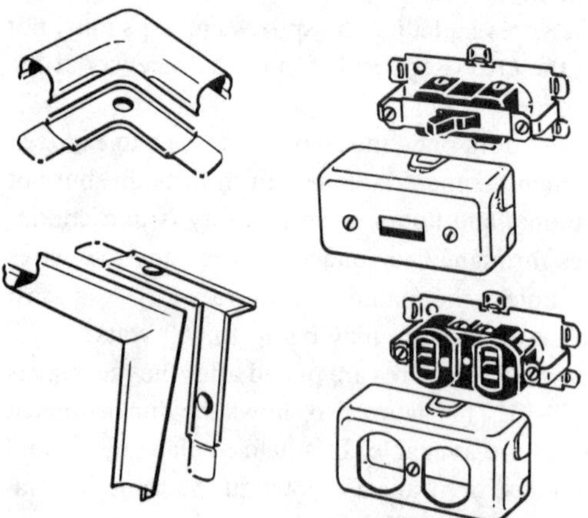

Fig. 28-14 Fittings used with the raceway shown in Fig. 28-13. (*The Wiremold Co.*)

Fig. 28-15 Typical "two-piece" metal surface raceway. Install the channel, lay the wires in place, then snap the cover into place. (*The Wiremold Co.*)

where subject to severe physical damage unless approved for the purpose. There are many cases where the material must be run on the floor, which certainly is a location where physical damage might be expected.

For this purpose there is available a "pancake" raceway, shown in Fig. 28-16, which also shows the dimensions of the smallest size available. As

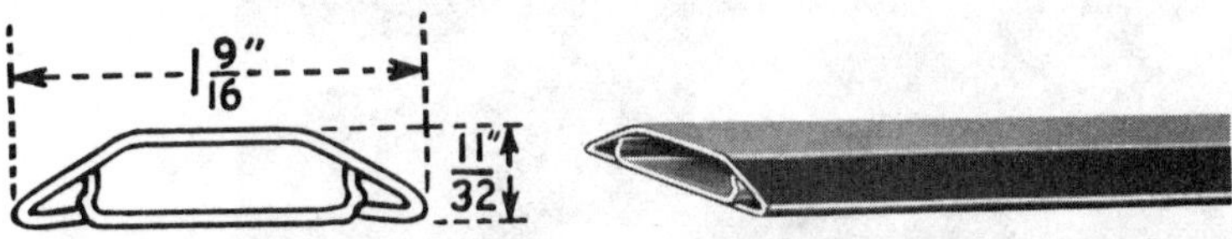

Fig. 28-16 "Pancake" type of metal surface raceway for installation on floors. (*The Wiremold Co.*)

in the other two-piece type, the base is installed first, the wires are laid into place, and the cover is then installed. As with other surface raceways, channels may contain power wires, or telephone and similar wires, but never in the same channel. As with other types, fittings such as elbows and so on are available.

Multioutlet Assemblies. Especially in stores, schools, laboratories, and similar locations, there is often need for very many receptacles spaced quite close together. Such receptacles are not necessarily all in use at the same time, but having them available at reasonably close intervals is most convenient, does away with the need for extension cords, and permits, for example, many floor lamps in a store to be plugged in at the same time, so that any one of them can be turned on instantly to show to a customer. Much installation labor is involved if a number of such receptacles are installed using ordinary wiring methods.

To reduce the cost very considerably, use the special raceway shown in Fig. 28-17. The covers are prepunched for receptacles so that special fittings for the receptacles are not needed. This type of raceway (called a multioutlet assembly in the Code) is available with prewired receptacles in place, or may be obtained empty, with the receptacles prewired on separate long lengths of wire. Several spacings are available, with receptacles from 6 to 60 in apart. A typical installation is shown in Fig. 28-18.

The Code covers multioutlet assemblies in Art. 353. The material may be installed like other types of metal surface raceways, but if it runs

Fig. 28-17 Raceway with many receptacle outlets, at regular intervals. If many receptacles are needed, this saves much time over other installation methods. (*The Wiremold Co.*)

Fig. 28-18 A typical installation of the multioutlet assembly shown in Fig. 28-17. (*The Wiremold Co.*)

through a wall or partition, no receptacle may be within the wall or partition. Moreover, it must be installed so that the cover can be removed from any portion outside the wall or partition, without disturbing the sections inside the wall or partition.

This kind of material is used not only in commercial establishments, but is equally useful in homes, especially in the small appliance circuits discussed in Chap. 13. It is equally suitable for home workshops or other locations where numerous receptacles are desirable.

Except for homes or guest rooms of hotels and motels, each 5 ft of a multioutlet assembly must be considered as one outlet of not less than 180-VA capacity. If, however, it is likely that a number of appliances may be used at the same time, each foot must be considered an outlet of not less than 180-VA capacity. See NEC Sec. 220-2(c) Exc. 1.

Remote-Control and Signaling Circuits

In Art. 725, the Code covers such circuits, divided into Classes 1, 2, and 3. These include circuits such as the wiring between a motor controller and its push-button stations, low-voltage circuits for switching as discussed in Chap. 20, thermostat circuits, and doorbell and chime circuits.

Class 1 Circuits. This has two subdivisions: (a) power-limited circuits at not over 30 volts and not over 1000 VA with the power source, such as a transformer, protected by an overcurrent device rated at not over 167% of the ampere rating of the source; and (b) remote-control and signaling circuits of up to 600 volts such as the wires between a motor controller and its push-button stations.

Class 2 and 3 Circuits. These are circuits limited to not over 100 VA, usually at not over 30 volts, such as circuits for doorbells, chimes, thermostats, and low-voltage switching. Such circuits have already been discussed; the power in the circuit is either inherently limited by the design of the transformer to a very low total (usually far below the 100-VA permitted maximum), or limited by a combination of a power source and overcurrent protection, as specified in NEC Table 725-31(a). However, a Class 2 circuit may operate at up to 150 volts if its current is limited to not over 0.005 amp (5 mA).

Wires for such circuits must not be run with the usual power supply wires at 120 volts or higher in the same conduit or cable.

Wiring: Wet Locations

When boxes or cabinets are installed outdoors, NEC Sec. 300-6(a) requires that they be suitable for the purpose. In locations that are more or less permanently wet, such as laundries and dairies (creameries), *or* in locations where walls are frequently washed or have wet or damp absorbent surfaces such as damp wood, NEC Sec. 300-6(c) requires that all boxes, raceways, cabinets, fittings, and so on, be mounted so that there will be an air space of at least 1/4 in between the equipment and the supporting surface.

Wiring: Corrosive Conditions

The Code in Sec.300-6 specifies that boxes, cabinets, raceways, cable armors, and so on, must be suitable for the environment in which they are installed. Ordinary materials used elsewhere are not suitable or acceptable in locations where corrosive conditions exist. Such locations include (but are not limited to) areas where acids and alkali chemicals are handled or stored, meat-packing plants, some stables (as already discussed in Chap. 23), and areas immediately adjacent to a seashore.

They must have special finishes suitable for the purpose, or be made of materials that require no further protection such as brass or stainless steel, or be nonmetallic.

Outdoor Wiring

This subject is covered by NEC Art. 225. Underground wiring and overhead wiring have been discussed in Chap. 17 in connection with services and in Chap. 24 concerning farm wiring. Likewise, the minimum sizes of wires between supports and their clearances above the ground, above roofs, and from windows, porches, and the like have also been discussed there. Review the subject in those chapters. Other aspects of outdoor wiring will be discussed here.

The ampacity of overhead spans is as shown in NEC Tables 310-17 and 310-19 (see Appendix). But a word of caution: When you use the higher-ampacity "free-air" tables, watch for excessive voltage drop if the wires are loaded to their full ampacity, especially if the distance is substantial. Review voltage-drop discussions in Chap. 7.

Festoon Lighting. This is defined in NEC Sec. 225-6(b) as "a string of outdoor lights suspended between two points more than 15 ft apart."* Such a string of lights consists of two or three wires (depending on whether a 2- or 3-wire circuit is used, but usually it is a 2-wire circuit) with weatherproof sockets or pin-type sockets attached to the wires at close intervals. Figure 28-19 shows a weatherproof socket, and Fig. 28-20 shows a pin-type socket.

The minimum-size wire permitted for festoon lighting is No. 12, unless the string is supported by a messenger wire, in which case No. 14 is the minimum size permitted. For spans of over 40 ft, a messenger wire *must* be used. (A messenger wire is a nonelectric supporting wire from which the wires are suspended.) The wires are individually supported by single insulators or by a rack with an insulator for each wire. The means of support must *not* be a fire escape, downspout, or plumbing equipment.

For a *permanent* festoon-lighting installation, use weatherproof sockets as pendants suspended from the festoon wires. The socket leads are connected in staggered fashion to the wires by means of small mechanical splicing devices of the type shown in Fig. 8-18. The joint is then insulated with tape. For a temporary festoon installation, the *pin*-type sockets may be used *only* if the festoon wires are *stranded*. The socket is opened up, the wires are laid into the grooves in the lower section, and the top section is then screwed up tight, causing the pins in the socket to puncture the insulation and penetrate into the strands of the stranded conductors. For installations with a messenger wire, the upper portion of the pin-type socket has a wire rack or hook to hang over the messenger wire. These hooks support the sockets and the festoon wires. Such installations are frequently used for temporary outdoor display areas. The per-

Fig. 28-19 A weatherproof socket for outdoor use. (*Leviton Mfg. Co., Inc.*)

Fig. 28-20 Using sockets of this type, just lay the wires into the grooves, and screw on the cover. Use only with *stranded* wires. (*Pass & Seymour, Inc.*)

manent type is often used above outdoor used-car sales lots. A word of caution: if the lighting is for temporary use on a construction job, the U.S. Occupational Safety and Health Administration (OSHA) requires that heavy-duty cord be used. Pigtail sockets are designed to be installed on individual conductors, so where construction employees are under the protection of OSHA, factory-assembled cord sets with molded-on lighting sockets, lamp guards (or suitable reflectors), and supporting means must be used.

Wireways

Often it is necessary to run large wires for a considerable distance or to run a large number of wires to a central location or to several locations. Instead of using large conduit, it is quite common practice (and more economical) to use a wireway. The Code in Art. 362 defines wireways as "sheet-metal troughs with hinged or removable covers for housing and protecting electric wires and cable and in which the conductors are laid in place after the wireway has been installed as a complete system."* Figure 28-21 shows a short section of wireway and some

*Reprinted with permission from NFPA 70-1984, National Electrical Code®, Copyright© 1983, National Fire Protection Association, Quincy, Massachusetts 02269. This reprinted material is not the complete and official position of the NFPA on the referenced subject, which is represented only by the standard in its entirety.

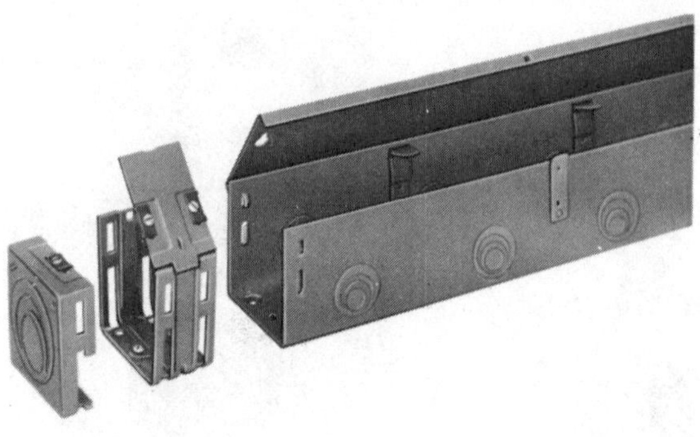

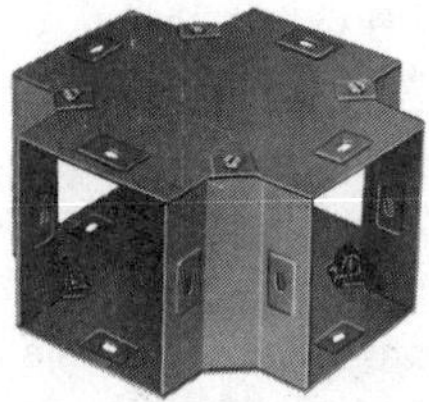

Fig. 28-21 Wireways of this kind are very useful, especially when one is installing large sizes of wire. (*Square D Co.*)

fittings used with it: elbows, crossovers, end sections, and so on. Wireways are available in lengths of 1, 2, 3, 4, 5, and 10 ft, and in cross sections from $2^1/_2 \times 2^1/_2$ in up to 8×8 in. Raintight types are available for outdoor installations. The various sections and fittings are assembled in place as a complete raceway system, then the wires are laid in place.

A wireway may be run any distance, and must be supported every 5 ft, unless approved for support at 10-ft intervals. An exception allows vertical runs to be supported at up to 15-ft intervals where there is only one joint between supports. Of course all joints must be made up tight. A wireway may be run through a wall if in an unbroken section. The total cross-sectional area[4] of all the wires in a wireway must not exceed 20% of the internal area of the wireway, except at points where there are splices

[4] The cross-sectional area of the more ordinary wires may be found in Table 5 of NEC Chap. 9 (see Appendix).

or taps, where the limit is 75%. Splices and taps must be made where they will be permanently accessible.

A wireway may not contain more than 30 current-carrying wires at any one point unless the correction factors of Note 8 to NEC Tables 310-16 through 310-19 have been applied; in that case there is no limit to the number of wires provided they do not occupy more than the 20% limit mentioned in the preceding paragraph.

Busways

A busway is a factory-made assembly of copper or aluminum busbars that are insulated from each other and from a sheet-steel enclosure, which may be ventilated or unventilated; busways are available in standard lengths of 10 ft or less. Long used for services and heavy feeders in industrial applications, busways are now being more widely used in commercial and high-rise residential occupancies. Busways may be used as branch circuits (including "trolley" types, with rolling power-takeoff devices) for the supply and support of lighting fixtures or for the supply of portable hand tools; as service-entrance conductors; and as feeders. Ampere ratings range from 50 to 6000 amp.

Plug-in types are available, with regularly spaced access openings, usually 12 in apart, where a fused switch or circuit breaker may be plugged in. Plug-in units should be placed where safe access, by means of a rolling platform, may be had for maintenance purposes. The plug-in switch or breaker must be operable from the floor by means of ropes, chains, or hook sticks. Branch circuits extending from these plug-in disconnects may be in metal raceways, or where feeding portable or movable equipment, in a special bus-drop flexible cable.

Busways must be exposed, except that totally enclosed nonventilated types, without plug-in units for other than individual lighting fixtures, and with joints accessible for maintenance, may be installed behind removable panels (as above a lift-out ceiling) provided the space is *not* used for air handling purposes.

Busways installed vertically, as for a feeder in a high-rise building, must be designed for vertical support. For other requirements, see NEC Art. 364.

Auxiliary Gutters

Sometimes it is necessary to run large wires for a relatively short distance but to make numerous taps in the wires. For an example, consider

Fig. 26-2, where the service-entrance wires must run to six switches. Unless very special switches are used (which is rarely done), it is not permissible to run wires to the first switch, make splices within the switch and run them on to the second, make more splices there and run on to the next, and so on. Instead, install an auxiliary gutter, which the Code covers in Art. 374. An auxiliary gutter is similar in appearance to a wireway, and a short section of a wireway is often used as an auxiliary gutter. But the function of an auxiliary gutter is more nearly that of a long junction box.

Auxiliary gutters may contain only wires or busbars but must *not* contain switches, overcurrent devices, or similar equipment. They must not run more than 30 ft beyond the equipment they supplement; if necessary to go beyond 30 ft, use a wireway. Gutters must be supported every 5 ft. The gutter is first installed; then the wires are laid into place. Gutters having hinged covers make the installation of wires quite simple. The restrictions as to the number of wires and the space they may occupy are the same as for wireways.

Because of the endless range in size and shape required, gutters are often made to fit a particular space. But if they are available in a size that will serve the purpose, a section of wireway is entirely suitable for use as an auxiliary gutter.

Underfloor Raceways

For large floor areas without partitions where it is expected that locations for power may be frequently or periodically changed, such as offices, research facilities, and light manufacturing, one of the solutions is an underfloor raceway system. Future access to branch circuits and to communication circuits at practically any location on a floor is possible through the use of underfloor raceways (Art. 354), as shown in Fig. 28-22, cellular metal floor raceways (Art. 356), or cellular concrete floor raceways (Art. 358). Cellular metal floor raceways provide a series of parallel cells formed of sheet metal and used as the structural support forms for a poured-concrete floor, usually in steel-frame buildings. Cellular concrete floor raceways consist of a series of parallel voids precast in structural-concrete floor slabs. Underfloor raceways are individual metal troughs, usually rectangular in cross section, which are installed either flush with or beneath a finished floor of wood or, more commonly, concrete. All three types require that means be provided initially to route conductors to the underfloor raceways from cabinets and panelboards, usually wall-

Fig. 28-22 Underfloor raceway before concrete floor is poured, showing junction boxes and preset inserts for power and communications. (*Walker Division of Butler Manufacturing Co.*)

mounted at the perimeter of the area served. Service pedestals surface-mounted on the floor or service access wells mounted flush with the floor provide access for the mounting of receptacles and/or communications outlets. These service pedestals may be preset at the time of the original installation or can be set at any future time by cutting through the floor into the raceway or into preset access points at regular intervals along the cell. Markers, usually flat-head brass screws, are installed at the ends of runs to facilitate the future location of the cells. Header ducts crossing at right angles are often provided at intervals to facilitate future interconnection between cells. Power and communications conductors must not occupy the same cell. All materials, supports, fittings, junction boxes, header duct, etc., must be listed by UL as compatible with the brand of raceway used, and the manufacturer's instructions must be followed carefully.

Flat Cable Assemblies

For the lighting of warehouses and factories a wiring method which provides a neat appearance, ease of installation, and convenience for future changes is flat cable assemblies, covered in Code Art. 363. A surface metal raceway in the form of a $1^5/_8$-in × $1^5/_8$-in U-shaped channel is the supporting means for Type FC cable, an assembly of four No. 10 stranded conductors in a web of PVC insulation. Devices which tap the cable through sets of pins that puncture the insulation and also provide support for lighting fixtures can be placed at any point along the channel. Installed on a single 4-wire 480Y/277-volt 20-amp branch circuit, one run of Type FC cable can supply 13 296 watts of lighting (continuous load). When the assembly is installed at 8 ft from the floor or higher, it is not necessary to cover the open bottom of the channel. See Fig. 28-23.

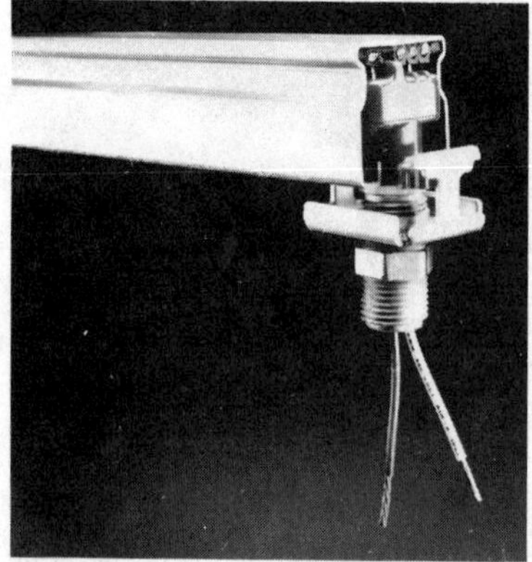

Fig. 28-23 Cutaway view of Type FC cable installed in metal surface raceway, showing tap device to both supply and support a lighting fixture. (*The Wiremold Co.*)

Cable Trays

In the service areas of commercial buildings, in factories, in electric-generating stations, and in other locations where there are large numbers of cables in one location, cable trays, covered in Art. 318 of the Code, are often used. The most common form is a ventilated steel bottom with two sides and an open top (see Fig. 28-24), but these trays may also be made of aluminum or other noncombustible materials and in the form of

Fig. 28-24 Metal-clad cable being installed in a cable tray. (*Husky Products, Inc.*)

ladders or solid bottoms and may be covered as well. Cable trays are primarily a support system for recognized cable wiring methods, but metal cable trays of sufficient cross section may be classified by UL to serve as an equipment grounding means. The greatest advantage of cable trays is the ease with which cables may be removed, added, or replaced, when compared with runs of cable each individually supported. Consult Art. 318 of the Code for the spacing of cables, because cable ampacity must be reduced unless space is maintained between them to aid in the dissipation of heat. Except where passing through floors or walls, cable trays must be exposed, and the contained cables must be accessible.

29
Nonresidential Lighting

Chapter 14 discussed the fundamentals of lighting, and the kinds of incandescent and fluorescent lamps used mostly for residential and some nonresidential lighting. The same types of lamps are also used in large nonresidential occupancies. But other types of lighting are often used in nonresidential occupancies, and these will be covered in this chapter.

Extended-Service Incandescent Lamps. In a home or in small offices it is a simple matter to replace a burned-out lamp. In larger establishments, a maintenance person may have to stop some other task, carry a ladder perhaps a long distance or up stairways, and then replace the lamp, a very expensive procedure. In such locations it may be wise to use lamps designed for a voltage higher than the actual voltage—for example, 130-volt lamps on a 120-volt circuit. This will greatly extend their life but will reduce the lumens per watt, thus increasing the cost of light. But this higher cost may be fully justified by the savings in maintenance costs.

The use of 130-volt lamps on 120-volt circuits entails a good deal of guesswork as to the actual watts consumed, the lumens of light produced, and the lamp life. It is therefore much more logical to use what are called "extended-service" lamps. These are designed for a 2500-h life when used at their rated voltage and will consume their rated watts. They produce from 5 to 20% fewer lumens per watt than ordinary lamps. Their cost is higher than that of ordinary lamps.

Group Relamping. In buildings other than homes, it often becomes an expensive procedure to replace lamps one at a time, as already explained. If a hundred were replaced at a time, the cost per lamp would be much lower. For that reason, many large establishments make a practice of replacing all lamps in a building (or one area of a building or plant) at one time. This is often done when 20% of the lamps have burned out. The percentage will vary with costs and efficiency requirements. The extra cost of the new lamps is less than the cost of replacing lamps one at a time.

"R" and "PAR" Lamps. These lamps have an aluminum or silver reflector deposited directly on the *inside* of the glass bulb, as a permanent part of the lamp. Aluminum and silver are excellent reflectors; being part of the lamp itself, the reflectors are permanently bright and untarnished. They cannot get dusty or dirty, thus eliminating maintenance costs, and preventing the loss in efficiency that comes with dirty separate reflectors. They reduce the initial cost because there is no need for separate reflectors or elaborate fixtures, and they are designed for much longer life than ordinary lamps of corresponding size.

Hence they are widely used in commercial work such as for merchandise displays. They may be in recessed ceiling fixtures, or in fixtures mounted on an electrified track. They are also suitable for supplementary lighting in industrial installations, especially buildings with very high ceilings. There are two types: R and PAR, shown in Fig. 29-1. Type R lamps are available in sizes from 15 to 1000 watts and have a substantially standard bulb of the required shape. In the spotlight type the light is concentrated into a more or less circular beam from 35 to 60 deg in diameter. In the floodlight type the beam is from 80 to 120 deg in diameter. Most Type R lamps have bulbs made of ordinary or "soft" glass and can

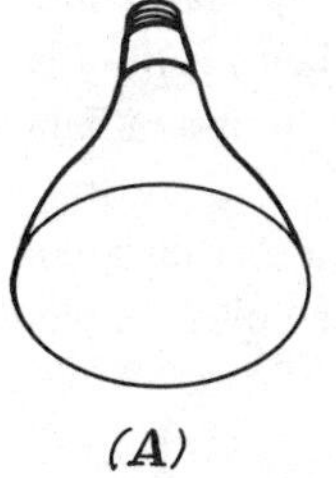

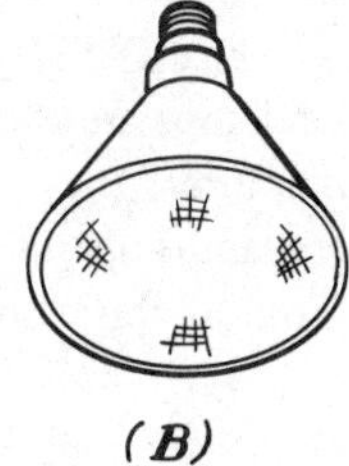

Fig. 29-1 Type R and PAR lamps have internal reflectors that stay permanently clean.

be used only indoors, because cold rain or snow falling on the hot glass would crack it. A few have "hard" glass, which withstands rain or snow. Type R lamps are used widely in merchandise displays, lighting special areas such as walls, pictures, or other points deserving special lighting.

Type PAR lamps are available in sizes from 75 to 500 watts (or up to 1000 watts in the Quartzline construction which will be described later). They are constructed of molded two-part glass bulbs. The two-piece construction permits more accurate positioning of parts, with the result that beams can be controlled to a greater degree than in Type R lamps, in turn permitting better concentration of light, and narrower beams. In the floodlight type most of the light is concentrated into a beam about 60 deg in diameter, but some sizes are available with wider or narrower beams. In the spotlight type the light falls in a beam about 30 deg in diameter, while in the larger sizes the beam is rather oval in shape, ranging from 13 × 20 deg to 30 × 60 deg. Some sizes are available in a choice of beams. All are designed for a 2000-h life.

All PAR lamps are made of a special hard glass that is not affected by cold rain or snow and consequently may be used indoors or outdoors. They are widely used in lighting buildings or landscapes, construction projects, signs, sports lighting, and similar undertakings. While R and PAR lamps cost several times as much as ordinary lamps, that higher cost is more than offset by the elimination of expensive reflectors and their maintenance and by their longer life.

"Tungsten-Halogen" Lamps. As you have observed, an ordinary lamp blackens during its life; near the end of its life the inside of the bulb is quite black. As the lamp is used, the tungsten of which the filament is made gradually evaporates; the evaporated tungsten deposits on the inside of the bulb, blackening it. That in turn reduces the transparency of the bulb, and is responsible in great part for the fact that the light output of an ordinary lamp is considerably reduced toward the end of its life.

It has been found that a bit of iodine introduced into a bulb will, through a rather complicated chemical action, prevent the blackening. The tungsten that evaporates during its use is redeposited on the filament, and the bulb remains clear. Such lamps are called "tungsten-halogen" type. Depending on the manufacturer, they are called by trade names such as Quartzline®,[1] or IQ. Such lamps operate at a very high temperature, high enough to soften any kind of glass; so the bulbs are gen-

[1] Registered trademark, General Electric Co.

erally made of quartz, which withstands exceedingly high temperatures without softening.

Such lamps are usually designed to produce about the same number of lumens per watt as ordinary lamps, but with a life two or three times that of ordinary lamps. Their lumen output is maintained at almost the initial value; ordinary lamps toward the end of their life usually produce only about 85% of their initial output.

Tungsten-halogen lamps are not used for general lighting in homes or offices, but rather for special purposes as, for example, floodlighting, sports lighting, spotlighting displays, and similar applications.

In the general-purpose type, tungsten-halogen lamps are available in sizes from 150 to 2000 watts. The bulb is tubular, less than $^1/_2$ in. in diameter and, for general lighting types, from $3^1/_8$ to $10^1/_{16}$ in. in length. Bases vary a great deal with the size and type; three common types are shown in Fig. 29-2. One has a contact at each end, another has both contacts at

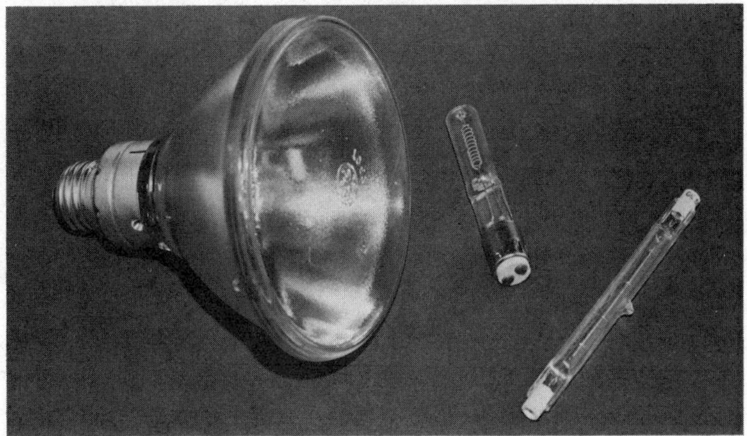

Fig. 29-2 Tungsten-halogen lamps are long and slim but sometimes are enclosed in larger glass bulbs, sometimes of the PAR shape. (*General Electric Co., Nela Park.*)

one end, and one is enclosed in a PAR-shape spot with a medium screw-shell base. Special-purpose lamps are available in other sizes, in different-sized bulbs, and with different kinds of bases, including some types that have merely a wire lead at each end.

The actual area occupied by the light-producing filament is very small, resulting in an exceedingly bright and small source of light. This in turn makes it possible to concentrate the light into very narrow, intense beams. With the proper reflectors the beam can be as narrow as 6 deg in one direction, but quite wide in the other, making possible a much higher degree of illumination in a given area than is possible using ordinary lamps of equal rating in watts. This makes the light especially suitable for lighting athletic fields. The field is well lighted, yet there is no glare for spectators sitting on the side of the field opposite the lamps. But because of the intense concentration of light, such lamps must be used in fixtures or floodlights and in locations where it is not likely that one will look directly at the lamps from a short distance. Their high operating temperature also makes it necessary to use fixtures or floodlights especially designed for them.

Size of Light Source. For general lighting in homes, offices, and similar locations, we need adequate light, but from widely diffused areas so that the light does not come from a small area, for we would then get the effect of reading in direct sunlight; reading in the shade of a tree is much more comfortable than reading in direct sunlight, even if the degree of illumination is lower there. In other words, we need light sources of low surface brightness.

But there are times when we need concentrated light in one area, not for general seeing but to emphasize, for example, a piece of sculpture, a necklace in a store, or a painting. Then we use a reflector to concentrate the light, to direct all the light into a small area. Using the best of reflectors, it is impossible to concentrate the light from an ordinary lamp into a really small area. The reason for this is that the source of light is not small enough. The smaller the source of the light (not the glass bulb but the area of the filament or arc tube that produces the light in the bulb), the greater the degree of concentration that the reflector can accomplish, and the narrower the beam.

Therefore to concentrate the maximum of the available light into a very narrow beam, we must use a lamp in which the filament is concentrated in as narrow or small an area as is possible. One way of accomplishing this is discussed in the following paragraphs.

6- and 12-Volt Lighting. If you need a light source concentrated in a very small area, use special 6- or 12-volt lamps, operated by a transformer that steps the usual 120-volt circuit down to 6 or 12 volts. Such lamps are available in the R and PAR types with internal reflector, as already discussed, in sizes from 15 to 240 watts. The available beam shapes vary from flood to very narrow spot. The narrowest available type concentrates 50% of the available light into a beam as narrow as 4 × 6 deg.

This concentration is made possible because the filament in a low-voltage lamp is very short. In an ordinary 60-watt 120-volt lamp the filament is about 21 in long. It is first wound into coil form, which greatly reduces the length. Then the coil itself is coiled so that the over-all dimensions are small enough to fit into the bulb of the 60-watt lamp. A low-voltage lamp of the same size has a much *shorter* but much *thicker* filament. The *area* occupied by the filament therefore is much smaller in the low-voltage lamp, and makes possible the very narrow beam already described.

Other Incandescent Lamps. Since a large manufacturer may make well over 10 000 kinds of lamps, only the most commonly used types can be discussed here. To show the range, you can buy a "grain-of-wheat" lamp (used in surgical instruments) that consumes a fraction of a watt, or a large lamp that consumes 10 000 watts. You can buy a lamp for deep-sea diving that will not be damaged by 300 lb of water pressure per square inch on the bulb. You can buy "black-light" lamps that produce no visible light, but when their "black light" falls on properly painted surfaces, it makes them glow in spectacular colors.

More Information on Fluorescent Lamps. Chapter 14 discussed only the bare fundamentals of ordinary fluorescent lamps. This chapter will discuss many other details concerning such lamps: kinds of bases, various starting methods, efficiencies, and other characteristics.

Bases on Fluorescent Lamps. Figure 29-3 shows the various kinds of

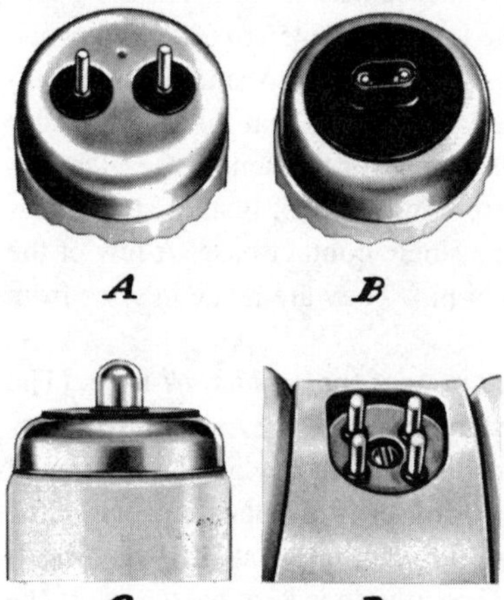

Fig. 29-3 Common types of bases on fluorescent lamps. (*General Electric Co.*)

bases used on fluorescent lamps, depending on their size and starting method. Later paragraphs will define the particular kind used on each kind of lamp. The bi-pin base is shown in *A* of Fig. 29-3; there are three diameters (miniature, medium, and mogul) depending on the diameter of the lamp. At *B* is shown the recessed double-contact type (made in two diameters), at *C* the single-contact type. The 4-pin type shown at *D* is used only on circular lamps.

Types of Fluorescent Starting. There are three types of starting in common use: preheat, instant-start, and rapid-start. Lamps designed for one type of starting will not (with a few exceptions) fit sockets designed for a different kind. In every case a ballast must be used, and the ballast must be carefully matched to the kind of starting and to the particular lamp under discussion, as well as to the circuit voltage.

The *preheat* type of starting was the kind used when fluorescent lighting was first introduced, and is still used. It was described in Chap. 14. Lamps for preheat starting have a filament or cathode at each end of the tube. When the lamp is first turned on, the current heats these filaments or cathodes. Then the starter opens, and current flows as an arc through the tube, limited to proper value by the ballast; this flow of current keeps the cathodes hot as long as the current flows, as long as the lamp is lighted. Fluorescent lamps designed for preheat starting have bi-pin bases, and are available in sizes from 4 to 100 watts.

Lamps of the *instant-start* type also have a filament or cathode at each end, but each one is short-circuited inside the lamp. Obviously then a current can't be made to flow through the filament or cathode before the lamp lights. No starter is needed, but the ballast is such that the open-circuit voltage before the lamp is turned on is from 450 to 600 volts; when the lamp is turned on, that voltage is high enough to start the lamp instantly. The current consumed by the lamp is such that the ballast under load drops the voltage to that which is proper for the lamp. After lighting, the cathode is kept hot by the arc current in the tube. Instant-start lamps are mostly of the Slimline type using a single-contact base. A few of the earlier type used a medium or mogul bi-pin. They are made in sizes from 40 to 75 watts.

Most lamps in new installations are now of the *rapid-start* type. (The difference in the time between the instant-start and the rapid-start is very small; the former is truly instant; in the latter there is a lag of a fraction of a second.) Rapid-start lamps require no starter; the ballast must be of the proper type. The lamp has a cathode at each end, but it is not short-circuited as in the instant-start type; the cathode is kept heated while the lamp is in operation by special $3^1/_2$-volt windings in the ballast, plus the heat of the arc current in the tube. This reduces the voltage necessary to start the lamp. Starting is based on much the same principle as in the instant-start type. The ballast delivers an open-circuit voltage of about 250 to 400 volts at the instant of starting and then controls the voltage to the proper value for normal operation. Rapid-start lamps do not have a uniform kind of base. Larger sizes have the recessed double-contact type; small sizes, the bi-pin. Rapid-start lamps are available in sizes from 30 to 215 watts and include the so-called high-output and Power Groove®,[2] types.

It should be noted that there is available a 40-watt, 48-in lamp of the *preheat/rapid-start* type. It may be used in either a fixture with a ballast (and starter) designed for preheat lamps, or in a fixture with a ballast (no starter) designed for rapid-start lamps.

The first fluorescent lamps made were of the preheat variety; then the trend was to the instant-start as that type was developed. Now the trend is to the rapid-start in new installations.

Efficiency of Fluorescent Lamps. The lumens per watt delivered by a fluorescent lamp depend on many factors and particularly on the color of the light produced by the lamp. From Chap. 14 you will remember that there are many kinds of "white." Tables showing the lumens produced are generally based on the cool-white color. The nominal watt ratings of fluorescent lamps are based on the power consumed by the lamp itself, not including the power consumed by the ballast. The power consumed by the ballast, as compared with that consumed by the lamp, varies a great deal. If the fixture and ballast are designed for a single lamp, the power consumed by the ballast is a much higher percentage of the total than it is in fixtures designed for two (or four) lamps. The proportion for the smaller lamps is much greater than for the larger lamps. In the smallest sizes and in single-lamp fixtures, the ballast may consume a third of the total watts of the circuit. In the largest sizes and with two lamps per ballast, it may be less than 10%. For the most commonly used lamps it is probably safe to use a figure of 15%. Actually, then, the total watts in a circuit serving fluorescent lamps will be approximately 15% higher than the watts indicated by the ratings of the lamps. However, with a power factor of other than 100%, the reactive current must also be added. This can be automatically taken care of by multiplying the volts by the power factor, before dividing the watts by the volts to obtain the amperes.

[2] Registered trademark, General Electric Co.

The light output of a brand-new fluorescent lamp is somewhat unstable. Efficiency ratings are based on the output after 100 h of operation, at which time the lamp has stabilized. After 100 h the output drops slowly but continuously so that toward the end of the lamp's life its output may be as much as 30% below the 100-h figure. During its life the lamp delivers an average of perhaps 85% of its 100-h rating.

Energy-Saving Fluorescent Lamps. Starting with the 1972 oil crunch, lower levels of illumination have been recommended. To make this easily possible, special fluorescent lamps have been developed that consume 86 to 88% of the watts consumed by the lamps they replace but which deliver 88 to 92% of the light. They are available in sizes from 25 to 185 watts and *fit into the fixtures using the lamps they replace*. Their average rated life is about the same as that of standard lamps. There may be a tendency to shorten ballast life, in which case matching high-efficiency ballasts can be used.

The efficiency of fluorescent lamps also varies with the type and size; in general the larger sizes produce more lumens per watt than the smaller ones. The actual efficiency varies from 45 to 100 lumens per watt; note these ratings are based on the power consumed by the lamp itself, not including the power consumed by the ballast. As pointed out in Chap. 14, the deluxe cool white and deluxe warm white produce about 30% fewer lumens per watt than the other types.

Lumens per Foot of Length. Entirely aside from the lumens *per watt*, the lumens *per foot of length* of a fluorescent lamp are another important consideration. Most commonly used types produce from 600 to 750 lumens per ft. Thus a fixture with four of the 40-watt 48-in lamps will have altogether 16 ft of total lamp length, and will produce about 12 000 lumens. To produce the footcandle levels of lighting demanded today for efficient work in offices and other areas, many lighting fixtures must be installed.

Why not produce lamps that will produce more lumens per foot of length? Such lamps are available. Lamps called "high-output" produce from 1000 to 1050 lumens per ft; others called "extra-high-output" or "Power Groove" produce from 1700 to 2000 lumens per ft. (Note that "high output" does not refer to lumens *per watt* but rather to lumens *per foot of length*.) Such lamps are available in 48-, 60-, 72-, and 96-in lengths. Using such lamps will not result in an installation consuming fewer watts for any given footcandle level of lighting, but will permit a smaller number of fixtures and lamps to be used. This leads to a definite saving in installation cost, less maintenance, and higher footcandle levels of lighting.

The output per foot of length is controlled by the current flowing through the lamp. The current in the ordinary 40-watt fluorescent lamp is about 0.430 amp. In the high-output type it is about 0.800 amp, while in the very-high-output type it is about 1.5 amp. The current is controlled by the proper selection of ballasts.

U-Shaped Fluorescent Lamps. These lamps, shown in Fig. 29-4, can be used in place of the usual type to provide attractive ceiling patterns and

Fig. 29-4 The U-shaped fluorescent lamp has advantages in some types of lighting. (*General Electric Co.*)

lighting effects. The length of the tube is about 48 in, but it is bent into the U shape so that the space occupied by the lamp is only about 24 in. Two of these can be installed in a single 24 × 24-in ceiling module. The output in lumens per watt is somewhat less than that of the 40-watt straight lamp, but higher than would be obtained by using straight 2-ft 20-watt lamps that would fit into the same space.

Fluorescent Type Designations. The numbering scheme used for fluorescent lamps can be quite confusing. Two basic schemes are used, one for lamps with bi-pin bases, and another for lamps with single-pin or *recessed* double-contact bases.

A typical designation for a bi-pin lamp is "F30T8." The "F" means "fluorescent"; "30" means "30 watts"; "T" means "tubular"; "8" means the tube is $^{8}/_{8}$ or 1 in. in diameter. (In the combination preheat/rapid-start type, which is $^{12}/_{8}$ or $1^{1}/_{2}$ in. in diameter, you would expect the designation "T12" but it does not appear, the lamp designation being merely "F40.") The type designation often has additional letters indicating color or other special construction.

In lamps with single-pin or recessed double-contact bases, the number in the designation, instead of standing for the watts rating of the lamp, indicates the nominal length in inches; the watt designation does not appear at all. Thus an "F48T12" lamp is a fluorescent 48 in long, in a tubular bulb $^{12}/_{8}$ or $1^{1}/_{2}$ in. in diameter.

Effect of Voltage on Fluorescent Lamps. Ordinary incandescent lamps must be selected for the circuit voltage on which they are to be operated. For maximum efficiency a 120-volt lamp should not be operated on a 115- or 125-volt circuit. Chapter 14 outlined the very considerable effect that off-voltage has on their life and efficiency.

Fluorescent lamps, on the other hand, are not rated in volts; the same lamp is used on a 120-volt circuit as on a 240-volt circuit. However, the ballast used must be carefully selected to match the voltage of the circuit on which the ballast and the lamp are to be used. Suppose the proper ballast has been selected and is used with a lamp on a 120-volt circuit. What is the effect on the lamp if the circuit voltage changes? No precise data can be given because the effect is dependent on the size of the lamp, the kind of lamp, and the particular ballast used.

On a 40-watt fluorescent lamp, however, a 10% overvoltage will increase the light output about 10% and reduce the efficiency about 5% but affect the life of the lamp very little. However, the ballast will overheat, reducing its life. A 10% undervoltage will have the opposite effect, but the lamp may not start properly. In other words, fluorescent lamps are not so radically affected by incorrect voltages as incandescent lamps. But for the reasons cited, it is very important to match the rated voltage of the ballast to the line voltage.

Color of Light from Fluorescent Lamps. As already mentioned in Chap. 14, there are many kinds of white fluorescent lamps. The deluxe varieties of white are about 30% less efficient than the other kinds of white. The cool-white lamp represents about 75% of all the fluorescent lamps sold and is quite satisfactory where color discrimination is not overly important.

In general the cool white and deluxe cool white produce light that simulates natural light; the warm white and deluxe warm white produce light that is more nearly like the light produced by incandescent lamps and emphasizes the red, orange, and brown colors.

For lighting merchandise displays in stores, the deluxe cool white probably gives the best over-all effect, more or less simulating natural daylight. But if the merchandise on display tends toward the red, orange, and brown colors, the deluxe warm white is preferred by many.

There are situations where it is important that the object observed will have the same appearance as when observed under natural light. Examples are clothing stores, color printing establishments, and parts of museums. This can be accomplished by using Chromaline®,[3] fluorescent lamps, which are available in 20-, 30-, 40-, 75-, and 110-watt sizes and which produce light that substantially duplicates natural light. However, the light output in lumens per watt is considerably lower than that of ordinary fluorescent lamps.

For decorative purposes and specialized lighting, some fluorescent lamps are available in colors such as blue, green, gold, pink, and red. Ordinary incandescent lamps in color are very inefficient, for they are ordinary lamps with color on the glass that absorbs most of the light, allowing only the desired color to pass through. Fluorescent lamps on the other hand have the proper phosphor on the inside of the tube, which creates primarily the color desired. For that reason they are many times as efficient as colored incandescent lamps.

Tabulation of Characteristics of Fluorescent Lamps. Table 29-1 shows the characteristics of the more common types of lamps. Necessarily this must be an abbreviated table, and those needing more information about a particular lamp or about lamps not shown can obtain it from lamp manufacturers.

Electric-Discharge Lamps. In any ordinary incandescent lamp, the current flows through a tungsten filament, heating it to a high temperature, often above 4000°F. At that temperature light is produced.

In an electric-discharge lamp the current flows in an arc inside a glass or quartz tube from which the air has been removed and a gas or mixture of gases has been introduced. The most ordinary example is the fluorescent lamp already discussed. Here the current is relatively low, from a small fraction of an ampere to a maximum of $1^1/_2$ amp. It flows through a relatively long length, up to 96 in. The gas pressure in the tube is relatively low, almost a vacuum.

HID Lamps. There are other kinds of electric-discharge lamps in which the current flow is as high as 10 amp, but it flows through a very short arc tube, just a few inches long. The gas pressure in the tube is very much higher than in a fluorescent lamp. Such lamps are called high-intensity discharge (HID) lamps. There are three different kinds in common use, which will be described later.

In all HID lamps, the arc tube operates at a very high temperature; indeed the high temperature is necessary for the proper operation of the

[3] Registered trademark, General Electric Co.

TABLE 29-1 Characteristics of Fluorescent Lamps

Variety	Watts	Bulb type	Diameter, inches	Length, inches	Type of starting	Lumens* Total	Lumens* Per watt	Type of base
Ordinary	15	T-8	1	18	Preheat	870	59	Medium bi-pin
	30	T-8	1	36	Preheat	2 200	73	Medium bi-pin
	15	T-12	1½	18	Preheat	800	53	Medium bi-pin
	20	T-12	1½	24	Preheat	1 250	63	Medium bi-pin
	40	T-12	1½	48	Preheat/rapid	3 150	79	Medium bi-pin
Slimline	40	T-12	1½	48	Instant	3 000	75	Single-pin
	55	T-12	1½	72	Instant	4 600	84	Single-pin
	75	T-12	1½	96	Instant	6 300	84	Single-pin
High output	60	T-12	1½	48	Rapid	4 300	72	Recessed double contact
	85	T-12	1½	72	Rapid	6 650	78	
	110	T-12	1½	96	Rapid	9 200	84	
Extra-high output .	110	PG-17	2⅛	48	Rapid	7 450	68	Recessed double contact
	165	PG-17	2⅛	72	Rapid	11 500	70	
	215	PG-17	2⅛	96	Rapid	16 000	74	

* Total lumens and lumens per watt are for cool-white lamps after 100 h of use. The average during the useful life of the lamps will be about 15% less. The figures are based on watts consumed by the lamps, not including power consumed by the ballasts. *Important:* See also paragraph regarding energy-saving fluorescent lamps.

lamp. Air currents must not be allowed to affect the temperature; so the small arc tube is enclosed in a much larger glass bulb, which determines the over-all dimensions of the lamp. None of these lamps can be connected directly to an electric circuit; they must be provided with a ballast to match the type and size of lamp involved, and the voltage of the circuit. When first turned on, the lamp starts at low brightness, gradually increasing; it may require 2 to 10 min for the lamp to reach full brilliancy. If turned off, it must cool for 1 to 15 min before it will relight.

This fact must be considered if HID lamps are planned for locations where power failures are frequent, or where violent voltage fluctuations occur, for considerable time is lost in relighting the lamps, once they have gone out. For this reason, often a few incandescent lamps are installed as "insurance" in areas that are otherwise lighted only by HID lamps. Even if a power interruption is very short, the HID lamps go out. When the power is restored a few seconds or few minutes later, the incandescent lamps provide some degree of illumination instead of total darkness, during the time it takes for the HID lamps to restart. In this way the incandescent lamps tend to prevent accidents, or panic, during the time that total darkness would otherwise prevail.

Although all HID lamps have long life, that life depends on the number of times the lamp is turned on. Hours of life shown are based on burning the lamp at least 10 h every time it is turned on. If burned continuously, the life is longer. Note too that the number of lumens produced diminishes rather quickly at first before leveling off. The number of lumens shown for each lamp is the figure after burning 100 h, as in the case of fluorescent lamps; the lumens per watt are based on the watts consumed by the lamps, not including the ballasts. In the larger sizes the ballast watts are very roughly 10% of the lamp watts, but the percentage is higher for the smaller sizes.

The principal advantages of HID lamps are high efficiency (lumens per watt), very long life, and high watt output (and high number of lumens) from single fixtures, thus reducing installation and maintenance costs. They are widely used in factories, service stations, gymnasiums, parking lots, street lighting, and, in general, locations where large areas must be lighted. The deluxe colors are also being used in stores and offices.

HID lamps are made in many types, three of which are in common use: (1) mercury-vapor, (2) metal-halide, and (3) high-pressure sodium. All operate on the principles already outlined, but the details of each will be discussed separately.

Mercury-Vapor Lamps. These lamps, usually called just mercury lamps, were introduced in 1934. Since that time they have been vastly improved. Their life now is probably ten times that of the original lamps; output in lumens per watt has been greatly increased; and their present cost is a fraction of their original price. Mercury lamps have a short arc tube with some argon gas in it, and some mercury, which is a liquid. When the lamp is turned on, an arc develops through the argon gas, in a short *part* of the tube. This slowly heats the mercury so that it gradually vaporizes, and then the arc develops through the entire length of the arc tube. When all the mercury is vaporized, the lamp operates at full brilliancy. This starting procedure takes about 10 min. Every lamp requires a ballast, which is carefully matched to the lamp being used and the voltage of the circuit.

Mercury lamps are available in sizes from 50 to 3000 watts; the 175- and 400-watt sizes are the most common. Figure 29-5 at *C* shows the 400-watt

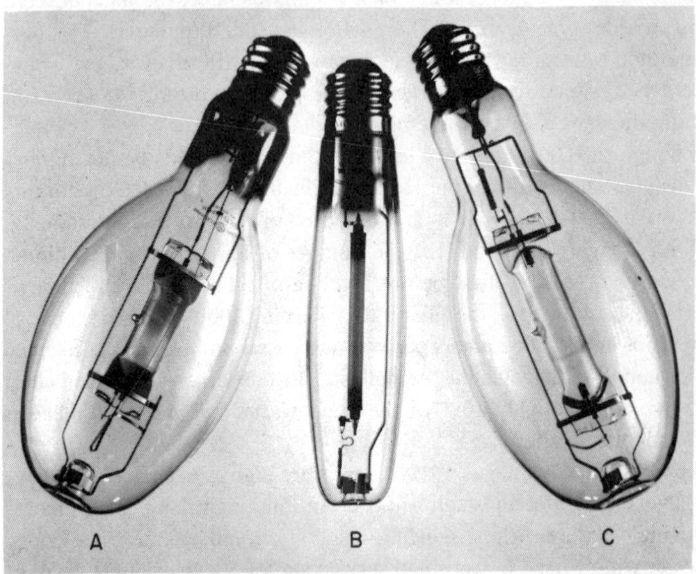

Fig. 29-5 Three types of HID lamps: metal-halide, high-pressure sodium, and mercury. (*General Electric Co.*)

size, which has an over-all length of about 11¼ in. The size of the bulb varies a great deal with the wattage of the lamp. Figure 29-6 shows several shapes, although *not* in proportion to the actual sizes. Many of the sizes are available in the "R" (reflector) type.

The life of mercury lamps is extremely long. In sizes above 100 watts,

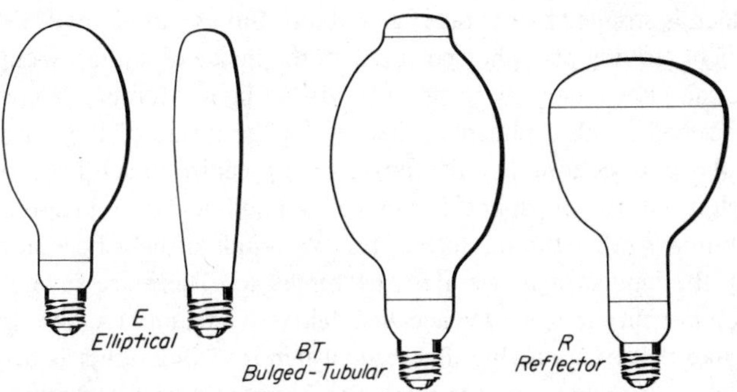

Fig. 29-6 Mercury lamps are made with many shapes of bulbs. (*General Electric Co.*)

it is at least 24 000 h (almost three years of continuous burning); for the smaller sizes it is about 16 000 to 18 000 h. Their efficiency in lumens per watt is much greater than that of incandescent lamps, very roughly about the same as that of larger fluorescent lamps. However, their efficiency drops off faster than in other lamps. In the 400-watt size, at the end of 4000 h the output is from 90 to 95% of the original value; after 12 000 h it is 85 to 90%; after 16 000 h it is 80 to 90%. Therefore in planning an installation using mercury lamps, the number of lamps planned should be based not on their initial output, but rather on the average during the time they are retained in the system, or on a minimum level of illumination.

Their comparatively high light output coupled with their relatively small size, and especially their long life, make these lamps especially suitable where it is difficult and expensive to install fluorescent lighting providing an equivalent amount of light. Since it is usually very expensive to replace a lamp, the very long life of mercury lamps suggests their use in such locations.

Originally, the only type of mercury lamp available produced light that had very little red and orange in it, the result being a greenish-bluish light that was acceptable for many purposes, but totally unsuitable where color was a factor. People had a corpselike appearance under this light; blue and green colors were emphasized and strengthened, orange and red were distorted to appear brown or black. The only way to correct this shortcoming was to use a combination of mercury lamps (deficient in red) and incandescent lamps (strong in red). This original type of lamp is still in limited use.

In a mercury lamp with a clear glass bulb, you see the peculiar color just described, but the arc also produces a good deal of *invisible* ultraviolet light, which is stopped by the outer glass bulb. But as stated, invisible ultraviolet light striking phosphor powders on the inside of a fluorescent tube makes the phosphor glow to produce visible light. Mercury lamps are now also made with a phosphor coating on the inside of the glass bulb. This phosphor coating permits most of the greenish-bluish light to pass through it, but the phosphor also creates a good deal of orange-red light in addition, so that the mixture of the two kinds of light becomes more nearly the kind of light we are accustomed to. There are several kinds of such mercury lamps. One is called deluxe white, and its color is very much like that of cool-white fluorescent lamps. Still another is deluxe warm white; its color output is much like that of warm-white fluorescents. The deluxe white produces about 10% more lumens per watt than the deluxe warm white.

Metal-Halide Lamps. This is a type first introduced in 1964. The lamps are known by various trade names (depending on the manufacturer), such as Multi-vapor®,[4] or Metalarc. In appearance they are quite similar to ordinary mercury lamps; Fig. 29-5 at *A* shows the 400-watt size. The principle of operation is substantially that of the mercury lamp, except that the arc tube, in addition to argon gas and mercury, contains other ingredients such as sodium iodide, thallium iodide, indium iodide, or scandium iodide. This leads to a very high efficiency of 80 to 100 lumens per watt. Suitable ballasts are required, and the restarting time is about 10 min.

Metal-halide lamps are available in two types. One has a clear glass bulb, producing light of acceptable color, approaching that of cool-white fluorescent lamps. The other type has a phosphor coating on the inside of the bulb, producing light that is substantially identical with that of cool-white fluorescent lamps.

These lamps are available in five sizes: 175, 250, 400, and 1000 watts, plus a special 1500-watt size for sports lighting. Their life is much less than that of mercury lamps, about 20 000 h for the 400-watt and 10 000 h for the 1000-watt size. At the end of about two-thirds of their life, the lumen output drops to about 80% of their initial output. Their cost is higher than that of mercury lamps.

In spite of these apparent disadvantages, metal-halide lamps have two important advantages. Their output in lumens per watt is about 60% greater than that of mercury lamps, thus reducing the cost of the light and the number of fixtures that must be installed. The other advantage is that

[4] Registered trademark, General Electric Co.

the light can be directed into a relatively small area with reflectors. The mercury lamp in the deluxe white type presents a light source of considerable area to the reflector, so that narrow beams are not possible. The metal-halide lamp of the clear-glass type (but not the type with a phosphor coating on the inside of the bulb) presents a light source of an area that is a small fraction of the outer glass bulb of the mercury type, thus making a narrow beam practical, where that is desirable.

Formerly in changing from mercury to metal-halide lamps, it was necessary to change the ballasts also, making for a somewhat expensive changeover cost. Now there are available *special* 400- and 1000-watt metal-halide lamps that can be used with *some* ballasts designed for mercury lamps, but not with others. If the ballasts originally installed for mercury lamps are the right kind, the new metal-halide lamps can be substituted for the mercury lamps, with an increase in light output of about 50%, but no increase in watts consumed. It must be noted that the life of these special lamps for use with mercury-lamp ballasts is lower than that of ordinary metal-halide lamps used with metal-halide ballasts. But be sure that before you undertake to switch from mercury to metal-halide, a really knowledgeable person checks to make sure that the ballasts are suitable for these special metal-halide lamps. In one brand, these special lamps are known as I-line Multi-vapor.

High-Pressure Sodium Lamps. This HID lamp was introduced in 1965. Again there are several trade names, depending on the manufacturer, such as Lucalox®,[5] and Ceramalux. The arc tube operates at an exceedingly high temperature, around 1300°C; so it is made not of glass or quartz but of a very special translucent ceramic material that can withstand the temperature involved. The arc tube contains sodium in addition to xenon gas and mercury. While lamps containing sodium normally produce a very yellowish light, the high-pressure sodium lamps produce light that is quite rich in orange and red, much like that of warm-white fluorescents or the smaller sizes of ordinary incandescent lamps.

The principle of starting is quite different from that of other HID lamps; special ballasts are required. The lamp lights to full brilliancy in less time (3 min) than other HID lamps. It is at present available in sizes from 50 to 1000 watts. The 400-watt size is shown at *B* in Fig. 29-5; it is $9^3/_4$ in long. Its light output is the highest of all known electric light sources: 80 to 140 lumens per watt, almost double that of fluorescent or mercury lamps and about five times that of 500-watt incandescent lamps.

[5] Registered trademark, General Electric Co.

The life of a high-pressure sodium lamp is 24 000 h if 10 h of use per start is assumed. Its output in lumens is maintained at a very high level, about 90% after 8000 h.

While this is an expensive lamp (about four times the cost of a mercury lamp of the same size), its very high output in lumens per watt makes the cost of the electric power for the light very low. The total lumens from one lamp are so high that a smaller number of fixtures or reflectors is often possible than with less efficient lamps. The small size of the arc tube permits narrow beams to be projected, making the lamp suitable for either floodlighting or spotlighting. It is often used in buildings such as factories with high ceilings, and for outdoor locations such as street lighting, parking lots, and athletic fields, as well as ''washing'' (floodlighting) the outer contours of a building. As is usual with any new type of lamp, undoubtedly many improvements will come about over a period of years, making the lamp suitable for many more applications. It is already being used in some offices and schools, with very careful design of the lighting and color scheme, which must be done by a competent lighting engineer.

See Figs. 29-7 and 29-8 for an example of relighting a manufacturing plant by replacing fluorescent fixtures with high-pressure sodium. At 5 cents per kilowatthour annual energy savings amount to $3360. Production rose by 7% and rejects dropped by 40% following the relighting.

Fig. 29-7 Manufacturing area lighted to about 50 fc by using high-output fluorescents, with a connected load of 59.5 kW. (*General Electric Co., Nela Park.*)

Fig. 29-8 With only minor wiring alterations and a change to high-pressure sodium lamps and fixtures, the fc level was more than doubled, to 100 fc, while the connected load was reduced to 42.7 kW. (*General Electric Co., Nela Park.*)

Note that the walls and ceiling were painted in light colors to take advantage of the increased light and to reduce contrast.

Comparison: Three HID Types. The basic characteristics of the three types of HID lamps are shown in Table 29-2; all data are based on the 400-watt size of each type.

TABLE 29-2 Comparison of Three HID Lamp Types

	Mercury	Metal-halide	High-pressure sodium
Life, h	24 000	20 000	24 000
Lumens per watt	55	85	125
Percent of initial lumens after hours indicated	16 000 h 70–90%	10 000 h 75%	12 000 h 90%
Color of light	Good	Good, green and yellow emphasized	Yellow, orange
Cost of lamp	Lowest	Medium	Highest
Cost of light	Highest	Medium	Lowest

Low-Pressure Sodium Lamps. These lamps have limited application owing to their monochromatic characteristic, producing only yellow light. Their efficiency is about 50% higher than that of high-pressure sodium lamps, so they are used where color discrimination is less important than efficiency, as, for example, in highway lighting.

Luminaires. Everybody knows what a lighting fixture is. In engineering jargon and in many technical publications, fixtures are called *luminaires*. In this book they will simply be called *fixtures*.

Good Lighting. Chapter 14 outlined the fundamental fact that one lumen of light falling on one square foot of area always produces one footcandle of illumination. From this statement it is easy to jump to some very wrong conclusions. For example, if a room with 100 sq ft of floor area is lighted by a lamp producing 5000 lumens, there are obviously 50 lumens for every square foot of floor area, but the illumination will be very much less than 50 fc. That is because only a part of the light that is generated reaches the place where it is to be used.

Based on the preceding paragraph, you can see that theoretically all you need to do to determine the number of watts required per square foot, for any number of footcandles, is to use the simple formula

$$\frac{\text{Number of footcandles wanted}}{\text{Lumens per watt of lamps used}} = \text{watts (VA) per square foot needed}$$

But that is a theoretical figure, and in practice you will have to provide from two to four times as many watts as the answer in the formula above indicates. If you multiply the answer by 3, you will probably be "in the ball park" and not hopelessly wrong.

The Code in Sec. 220-2(b) requires service and feeder capacity for a minimum of anywhere from 1/4 VA to 5 VA per sq ft for lighting in various occupancies. It would be very useful to be able to translate "volt-amperes per square foot" directly into a predictable level of illumination, but unfortunately this can't be done. Of the total light produced by a lamp, some is absorbed by the fixture and some by the room surfaces. The part that reaches the surfaces to be lighted depends on many factors. There is no dependable, simple method that will give you the level of illumination in a room, knowing only the light sources. Volt-amperes per square foot has been used as the basis for rule-of-thumb estimates, but accurate predictions using this method depend on years of experience.

Computing Requirements of Various Areas. To compute the lighting requirements of industrial, commercial, and similar spaces involves many factors such as coefficient of utilization, room ratio, reflectance of ceilings and walls, maintenance factors, types of lamps, and types of fixtures. This kind of determination is usually undertaken by a lighting engineer, not by the contractor who does the wiring, and the procedure therefore will not be covered in this book.

Comparative Lumen Output of Various Kinds of Lamps. The output of each kind of lamp in lumens per watt has already been outlined in other parts of this chapter in considerable detail. Remember that for fluorescent and HID lamps the figures given were for lamps after 100 h of use and were based on the watts of the lamps only, not including the power consumed by ballasts or other auxiliary equipment. A brief comparison of the various types follows:

Incandescent (filament type)	10.5 to 22 lumens per watt
Fluorescent	58 to 100 lumens per watt
Mercury	46 to 63 lumens per watt
Metal-halide	80 to 115 lumens per watt
High-pressure sodium	80 to 140 lumens per watt
Low-pressure sodium	100 to 183 lumens per watt

Location of Reflectors. Especially in direct-lighting installations, the spacing and location of reflectors become very important. Usually it is sound practice to mount them as far apart as their height above the floor. This results in a reasonably even level of illumination at the work level of tables or benches. If reflectors are mounted too far apart, dark areas will result.

Point-of-Work Lighting. It is quite customary to provide additional localized lighting, beyond the general lighting, where work of a very exacting nature is performed. Such locations are common in manufacturing establishments—assembly of very small parts, inspection stations, toolmakers' stations, etc. In work such as is shown in Fig. 29-9, 500 fc is by no means too high; investment in such lighting pays dividends. Special reflectors or reflector-type lamps described earlier in this chapter are often used.

The illumination in the concentrated area, however, should not be too great as compared to the immediately surrounding area. Otherwise the contrast will be extreme, glare will become a factor, and much of the advantage of the additional lighting will be lost.

Lighting of Athletic Fields. The lighting of an athletic field is a much bigger project than is usually realized. For example, consider what is required for a baseball field for major-league games. American League stan-

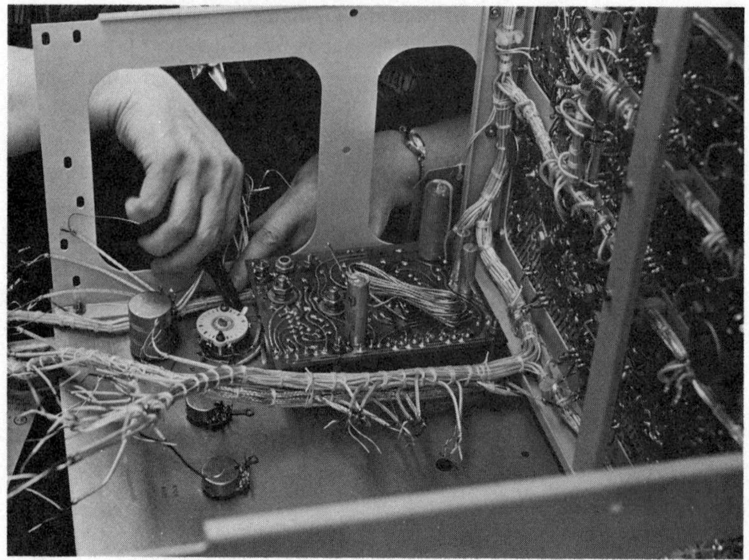

Fig. 29-9 In areas where exacting detailed work is performed, provide 300 to 500 fc of light. (*General Electric Co., Nela Park.*)

dards are 150 fc for the infield, and 100 fc for the outfield. But some major-league stadiums have twice as much as the minimum, in the belief that it will contribute to the spectacle and make it easier for spectators to see at long distances. For good TV broadcasting, authorities recommend 150 vertical fc, a good deal less than the lighting recommended some years ago, thanks to the greatly improved sensitivity of TV cameras.

Some fields, formerly lighted with incandescent lamps, consumed over 2000 kW. Most have now been reworked, using HID lamps, with both significant reductions in load and higher levels of light. For example, at Cleveland Municipal Stadium the original installation consisted of 1318 incandescent lamps, each consuming 1500 watts, for a total of roughly 2000 kW. The installation was revamped, retaining 398 of the 1500-watt lamps but replacing 920 of them with 1000-watt metal-halide lamps. The total power consumption dropped from about 2000 to about 1500 kW. In spite of the reduction in power, the lighting on the infield increased from about 180 to almost 300 fc and in the outfield from about 140 to almost 200 fc.

From this example you can see that by using HID lamps it is possible to greatly increase the footcandles of lighting while reducing the total amount of power used and thus to avoid the extremely expensive procedure of installing larger wires, conduits, transformers, and so on, that would be required if additional incandescent lighting were added.

Even fields for lesser events than major-league baseball require more power for lighting than most people would think necessary. A high school football field should have at least 75 kW, and college football requires from 150 to 500 kW. A tennis court should have 10 kW for ordinary recreational playing and 20 kW or more for more serious games. Really meaningful figures cannot be given, for so much depends on the kind of lamps used, the fixture design, and the location of the fixtures.

From this it should be apparent that the lighting of athletic fields in general involves problems much greater than might be expected. The design of such installations is not for the beginner.

30
Wiring for Motors

Chapter 15 covered the wiring of ordinary types of motors as used in residential and farm applications. Review that chapter now, for some of the points covered there will not be repeated here. This chapter will cover the wiring of motors in commercial and industrial applications (at not over 600 volts) so that when installed, the motor will have proper operating characteristics, proper control, and proper protection.

This chapter is divided into four parts:

Part 1: General information.

Part 2: One motor and no other load on one circuit.

Part 3: Two or more motors (with or without other load) on one circuit.

Part 4: Hermetic refrigerant motor-compressors (motor and compressor enclosed in common enclosure, as used in air-conditioning equipment).

PART 1: GENERAL INFORMATION

In reading this chapter, you will make it easier for yourself if you will think in terms of one specific motor; then reread the chapter thinking about a much smaller or much larger motor, or a different kind of motor.

The Code requirements for wiring motors are rather complicated and involved, and the wiring of motors ranging from a fraction of a horsepower to thousands of horsepower can never be reduced to a few simple rules. This chapter does not pretend to cover every last detail covered by Arts. 430 and 440 of the 1984 Code. Rather, it is intended to pertain to the more ordinary types of motors and the more ordinary types of installations, which should cover a very large percentage of all motors being installed. For the more difficult and intricate types of installations, consider this chapter a sort of preview of the kinds of problems that will be met; you will not be called upon to design such installations until you have had considerable experience in the more ordinary jobs. But it can help you to understand the basic principles of the rules so that you can use the Code itself to better advantage.

Voltage Ratings of Motors. While circuits used in ordinary wiring are rated at 120, 240, and 480 volts, motors are still rated at 115, 230, and 460 volts, but are designed to operate at voltages 6 to 10% above or below their rated voltage. Although the power supplier may deliver, say, 240 volts to the premises, it is more than likely that a motor out on the end of a branch circuit will see closer to 230 volts. It is for this reason that motors are realistically rated at a voltage *lower* than the voltage of the system to which they are connected. Motors rated at 200 volts are also available, for use on 208-volt systems.

Switches. Various kinds of switches will be mentioned in this chapter, so a good understanding of the various types is essential. Study the following definitions, which are quoted from Art. 100 of the Code.

> **General-Use Switch.** A switch intended for use in general distribution and branch circuits. It is rated in amperes, and is capable of interrupting its rated current at its rated voltage.
>
> **General-Use Snap Switch.** A form of general-use switch so constructed that it can be installed in flush device boxes or on outlet box covers, or otherwise used in conjunction with wiring systems recognized by this Code.[1]
>
> **Isolating Switch.** A switch intended for isolating an electric circuit from the source of power. It has no interrupting rating, and it is intended to be operated only after the circuit has been opened by some other means.
>
> **Motor-Circuit Switch.** A switch, rated in horsepower, capable of in-

[1] In other words, these are switches used in controlling lights and so on in ordinary house wiring.

terrupting the maximum operating overload current of a motor of the same horsepower rating as the switch at the rated voltage.

Circuit Breaker. A device designed to open and close a circuit by nonautomatic means and to open the circuit automatically on a predetermined overcurrent without injury to itself when properly applied within its rating.[2]*

Other Definitions. The following definitions, while not directly quoted from the Code, are essential for a thorough understanding of the Code rules for motors.

"In Sight from." Both visible *and* not more than 50 ft apart.

Time Rating. The Code in Sec. 430-7(a) requires a considerable assortment of information on the name-plate of a motor, and one item is the "time rating." Each motor must be identified as continuous rated or 5-, 15-, 30-, or 60-min rated. A motor with a "continuous" rating, of course, may be used even if it is not expected to run continuously, but a motor not rated continuous must *never* be used if it could under some circumstances run continuously as explained in the next paragraph. Such use would frequently lead to a burned-out motor.

"Continuous Duty." The words "continuous duty" as applied to a motor must not be confused with "continuous load" as applied to loads other than motors, and as discussed in Chap. 7. As applied to motors, Code Sec. 430-33 reads, "Any motor application shall be considered to be for continuous duty unless the nature of the apparatus which it drives is such that the motor cannot operate continuously with load under any condition of use."* That means that if it is possible for the motor to run continuously it must have a continuous-duty rating. In other words, the *time rating* on the name-plate must be *continuous*. This is true even if the motor normally would not run continuously. For example, a manual controller could be left in the on position, which would result in the motor running continuously.

The practical aspect is that just about every motor must be considered "continuous duty" except special types discussed in the next paragraph.

Noncontinuous-Duty Motors. Examples of such motors are those on hoists or elevators; they can't operate continuously under load. Such

[2] A circuit breaker is a combination switch and an overcurrent device. It can be manually closed or opened, but trips automatically on overload.

motors are considered protected by the branch-circuit short-circuit and ground-fault protection, provided its rating does not exceed that shown in NEC Table 430-152. But it is wise to protect such motors with motor overload protection as required for the continuous-duty type if this can be done without nuisance tripping.

Motor Branch Circuit. The elements that make up a motor branch circuit are as follows:

A. Motor-branch-circuit wires. The wires from the circuit's source of supply, such as a panelboard, to the motor.

B. Disconnecting means. The switch or circuit breaker to disconnect all the ungrounded wires to the motor and its control.

C. Motor-branch-circuit short-circuit and ground-fault protection. The circuit breaker or set of fuses that protects the wires, the controls, and the motor itself against overcurrent *due to short circuits or ground faults only*.

D. Motor controller. The device itself (magnetic contactor or manual switch) that is used to start and stop the motor, reverse it, control its speed, and so on.

E. Control circuit. Wires leading from a controller to a device such as a push-button station at a distance. It does not carry the main power current, but only small currents to signal the controller how to operate.

F. The push-button station. As mentioned in preceding paragraph.

G. Motor over*load* protection. A device to protect the motor, its controller, and its circuit only against possible damage from higher-than-normal currents caused by overloading the motor or in case the motor does not start when turned on. Such devices are *not* capable of opening short-circuit or ground-fault currents.

H. Secondary control circuit. The wires to a separate controller by which the speed of a wound-rotor motor can be regulated. This type of motor is not used often enough to warrant discussion in this book.

I. Secondary controller. The controller mentioned in the preceding paragraph.

Figure 30-1 shows all these elements, lettered as above, if the motor is on a separate circuit. In practice, often one element can be made to serve two or even three different purposes. At other times, two or more elements may be installed in a single enclosure.

Motor Current. The current consumed by a motor varies with the circumstances. The starting current is very high, but it gradually drops as the motor speeds up. When the motor reaches full speed, delivering its rated horsepower, at rated voltage, the motor carries what is called "full-

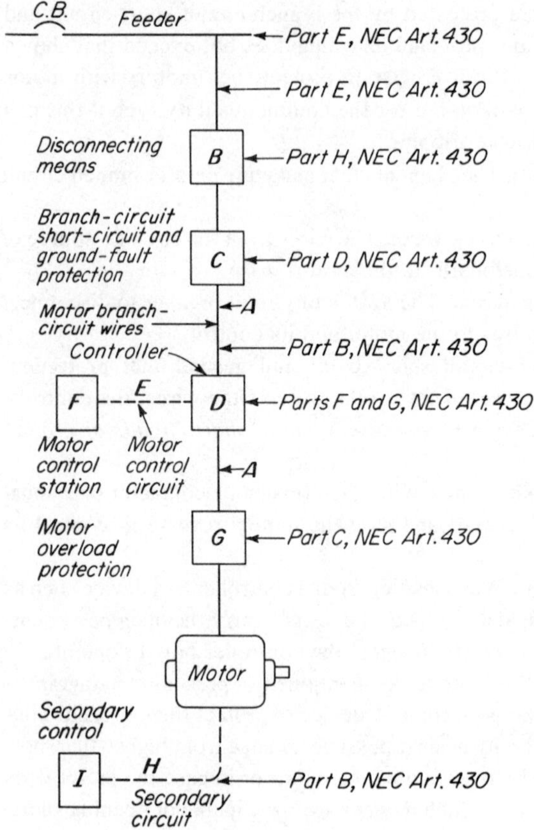

Fig. 30-1 Elements of a motor branch circuit.

load current." The current increases, however, if the motor is overloaded.

Full-Load Current. This is also called "running current." This is the current consumed by the motor while delivering its rated horsepower, at its rated voltage. It is the current in amperes stamped on the name-plate of the motor. However, when an installation is being planned, the motor has usually not yet been obtained, so the name-plate information is not available. Therefore consult NEC Table 430-148 for single-phase motors and Table 430-150 for 3-phase motors; both are in the Appendix of this book. (NEC Table 430-147 covers dc motors, and Table 430-149 covers *2-phase* motors; neither type is discussed in this book, so the tables are not shown in the Appendix.)

If you are using motors rated at 208 volts, on 208-volt circuits, increase the values in Tables 430-148 and 430-150 by 10% as required by the notes to the tables.

Overload Current. As already discussed in Chap. 15, no motor should be installed with the expectation that it will continuously deliver more than its rated horsepower. The motor can deliver more than its rated horsepower, but it will consume more current and will overheat. If the overload is large enough, the motor will burn out. So the Code requires a motor over*load* device that will disconnect the motor if the overload is large enough to damage the motor if continued indefinitely. It will also disconnect the motor if it fails to start when turned on. Overload devices will be described later in this chapter.

Starting Current. When a motor is first connected to the line, the current consumed is very much higher than after the motor comes up to speed. The ratio of starting current (also called "locked-rotor" or "stalled-rotor" current) to running current varies greatly with the size and type of motor and may be 600% or more. An ordinary single-phase 1-hp 230-volt motor consuming only 8 amp while running at full speed and full load will require as much as 48 amp for a second or two when first started. For larger motors the ratio is usually lower, but starting current is nevertheless always higher than running current.

The ratio of starting current to running current for any given motor also depends on the kind of machinery the motor is driving. If the load is hard to start, the starting amperes will be higher than if the load is easy to start. Moreover, the higher starting amperage will persist for a longer time, for the motor will not come up to full speed as quickly as with an easy-starting machine.

Any ac motor of $^1/_2$ hp or more manufactured since 1940 carries on its name-plate a "Code letter" that indicates the approximate kilovolt-amperes consumed *per horsepower* with the motor in a locked-rotor condition. Locked-rotor means that the motor is locked so that the rotor cannot turn. Of course, a motor is never operated that way, but at the moment when a motor is first connected to the power line, it is not turning, so that locked-rotor condition does exist until the motor starts to turn. The current consumed by the motor until it starts to turn is very high. Then the current drops off until the motor reaches full speed, at which time normal current is established. From the Code letter you can determine the maximum current required by a motor while starting, which is useful information in establishing the size of the various components in a motor circuit. The table in Code Sec. 430-7(b) covering the Code letters follows.*

Code letter	Kilovolt-amperes per horsepower with locked rotor	Code letter	Kilovolt-amperes per horsepower with locked rotor
A	0 to 3.14	L	9.00 to 9.99
B	3.15 to 3.54	M	10.00 to 11.19
C	3.55 to 3.99	N	11.20 to 12.49
D	4.00 to 4.49	P	12.50 to 13.99
E	4.50 to 4.99	R	14.00 to 15.99
F	5.00 to 5.59	S	16.00 to 17.99
G	5.60 to 6.29	T	18.00 to 19.99
H	6.30 to 7.09	U	20.00 to 22.39
J	7.10 to 7.99	V	22.40 and up
K	8.00 to 8.99		

Generally speaking, the larger the motor, the lower the locked-rotor current per horsepower.

To establish the approximate maximum amperes required by a specific motor while starting, first determine the actual Code letter on the nameplate of the motor; assume that it is J and that the motor is 3 hp. Refer to the table and you will find that J shows from 7.10 to 7.99 kVA per hp. Call the average 7.55 kVA; or 7550 VA. For 3 hp the total is 7550 × 3 or 22 650 VA. If the motor is single-phase, divide by the voltage. This $^{22\,650}/_{230}$ produces a result of about 98 amp. In other words, as the motor is first thrown on the line there is a momentary inrush of about 98 amp, diminishing gradually to about 17 amp (Table 430-148) when the motor is running at full speed and delivering its rated 3 hp.

If the motor is a 3-phase motor, first multiply the volts by 1.73 (see Chap. 3); in the case of a 230-volt motor, the result is 397.9. Use the number 400, which is easy to remember. Then divide 22 650 by 400, giving about 57 amp, which is the inrush when the motor is first turned on, diminishing to about 8 amp when the motor reaches full speed.

PART 2: ONE MOTOR AND NO OTHER LOAD ON ONE CIRCUIT

Motor-Feeder Wires

If the motor is the only load on the circuit, there is no feeder, only a branch circuit, whether it starts from the service equipment or from a panelboard that is fed by a feeder from the service equipment. Sometimes a feeder supplies more than one motor, and information concerning such cases will be furnished later in this chapter.

Motor-Branch-Circuit Wires

Branch-Circuit Wires for One Motor. This subject is covered by Part B of NEC Art. 430. Branch-circuit wires for one motor must have a minimum ampacity of 125% of the full-load current of the motor—as determined from NEC Tables 430-148 or 430-150 (see Appendix), or, for multispeed motors, torque motors, and adjustable-voltage motors, as determined from the current rating marked on the motor name-plate, and for shaded-pole or permanent-split-capacitor fan or blower motors, as determined from the full-load current marked on the equipment name-plate. Why must the wires have a larger ampacity than the load they are expected to carry? The over*load* device (which is also installed in the circuit and will be explained later in this chapter) usually has a rating 25% higher than full-load current, to allow for nominal overloads. So the circuit wires must be large enough to carry at least 25% more than the normal full-load current.

Assume you are going to install a 2-hp 230-volt single-phase motor; its full-load current per Table 430-148 is 12 amp. Multiply that by 1.25 (125%) to reach 15 amp. So use wire with an ampacity of at least 15. If you are going to install a 10-hp 230-volt 3-phase motor, its full-load current per Table 430-150 is 28 amp. Multiply that by 1.25 to reach 35 amp. Use wire with an ampacity of at least 35.

If the application is other than continuous duty, the minimum size of the motor-circuit wires ranges from 85 to 200% of the full-load current, as shown in NEC Table 430-22(a) Exception. Consult this table in your copy of the Code. Article 100 of the Code defines the various types of duty.

Voltage Drop. The Code does not take voltage drop into consideration in specifying the minimum size of wires to be used. It is not considered good practice to permit voltage drop over $2^1/_2$% in a motor circuit (with the motor running at full speed at normal power), for voltage drop is merely wasted power, which may cost a considerable sum during the life of the motor. Too great a voltage drop also means the motor will not deliver full power,[3] start heavy loads, or accelerate so rapidly as under full rated voltage. There is additional voltage drop in the feeder, if one is involved.

Don't overlook the starting current of the motor. If you have sized your circuit wire for a 2¹/₂% drop during normal running, but the motor consumes four times normal current while it is starting, the drop during that period will be 10%, not 2¹/₂%. This may be of little importance if the motor starts without load, the load being applied after the motor comes up to speed. But if the motor must start against a heavy load, voltage drop can become quite important. However, there is one compensating factor. Since the wires must have an ampacity of at least 125% of the normal full-load current, the voltage drop will be *less* than when figured on the basis of full-load current. So in quite short runs, the minimum size of wire will probably not result in excessive drop; in longer runs, you will be wise to calculate the drop. How to do so was discussed in Chap. 7.

Disconnecting Means

This subject is discussed in Part H of Art. 430 of the Code. A motor disconnecting means (circuit breaker or switch) is required to disconnect the motor and its controller from the circuit; this is in the interest of safety so that the motor and its controller can be totally disconnected from the electric power while working on either the motor or the machinery that it drives. It must plainly indicate whether it is in the closed (on) or the open (off) position. It must be readily accessible (if there are more than one, at least one must be readily accessible). It must open all ungrounded wires to the motor *and its controller*. It must always be "in sight from" the controller.

Type of Disconnecting Means. A motor disconnecting means must be either a circuit breaker or a motor-circuit switch. There are, however, six exceptions to this rule, as follows:

1. The disconnecting means for a cord- and plug-connected motor of any size may be an attachment plug and receptacle.
2. For a stationary motor of not over ¹/₈ hp, the branch-circuit overcurrent device (including a plug fuse) may be the disconnecting means.
3. For any *ac* motor not over 2 hp and not over 300 volts, the disconnecting means may be an ac (but not ac-dc) general-use *snap* switch, if

[3] A motor develops power in proportion to the *square* of the voltage. At 90% of rated voltage, the motor will deliver 0.90×0.90 or 0.81 (81%) of the power it would deliver at rated voltage.

the ampere rating of the switch is at least 125% of the full-load current of the motor.

4. For a stationary motor of not over 2 hp and not over 300 volts, the disconnecting means may be a general-use switch with an ampere rating *not less than twice* the full-load current of the motor. Figure 30-2 shows a switch suitable for a small motor.

Fig. 30-2 This 3-pole SN 100-amp switch is rated as the disconnecting means for a 15-hp motor at 230 volts, 3-phase. (*Square D Co.*)

5. For a stationary motor rated at over 100 hp, the disconnecting means may be a general-use switch, or it may be an isolating switch, which must be marked "Do not open under load."

6. For a motor rated at more than 2 hp up to and including 100 hp, if it has an autotransformer-type controller, the disconnecting means may be a general-use switch if the motor drives a generator and other specified conditions are met, as outlined in NEC Sec. 430-109 Exc. 3.

Except for motors rated 2 hp or less, most motors do not come within these exceptions and *must* have either a circuit breaker or a motor-circuit switch for a disconnecting means, as *any* motor *may* have.

Ampere Rating. The disconnecting means for *any* motor must have an ampere rating of not less than 115% of the full-load current of the motor. In most cases, it has to have a higher rating than that; and, as you have learned, it must be much higher (200%) where a *general-use* switch is used. Possible locked-rotor conditions require a disconnecting means to have a high interrupting capacity.

Motor-Branch-Circuit Short-Circuit and Ground-Fault Protection

This subject is covered by Part D of NEC Art. 430. This motor-branch-circuit overcurrent protection is necessary to protect the wires of the circuit against overloads greater than the starting current, in other words, against short circuits and grounds. At the same time, it protects the motor controller, which is designed to handle only the current consumed by the motor and which would be damaged by a short circuit or ground if branch-circuit protection were not separately provided. Grounds are often practically equivalent to short circuits.

As in other branch circuits, the protection in a motor branch circuit may take the form of either fuses or circuit breakers. The maximum ampere rating permitted by the Code for the overcurrent device in a *motor* branch circuit is sometimes higher when fuses are used than when circuit breakers are used.

Maximum Rating of Motor-Branch-Circuit Short-Circuit and Ground-Fault Protection. This subject is covered by Code Sec. 430-52. The maximum rating depends on the type of motor, the kind of overcurrent protection provided, and the kind of starter used. For both single- and 3-phase motors most often used, if they are provided with starters of the kind discussed in this book, the limits expressed as a percentage of the full-load motor current are, per NEC Table 430-152, as follows:*

Type of motor	Time-delay fuses	Non-time-delay fuses	Circuit breakers
With no Code letter (see page 562), or with Code letters F to V	175%	300%	250%
With Code letters B to E	175%	250%	200%
With Code letter A	150%	150%	150%

To determine the maximum rating for less common types of motors, or starting methods other than those discussed in this book, first refer to Tables 430-148 or 430-150, then refer to Table 430-152 to determine the maximum rating of the overcurrent device. These tables are in the Appendix of this book.

If applying these various percentages leads to a nonstandard rating of fuse or breaker, the next larger standard rating may be used.

Where the ratings shown in NEC Table 430-152 will not allow the motor to start, such as where a motor may have to start under extremely heavy load conditions, time-delay fuses may be increased up to 225% of full-load current. Circuit breakers may be increased up to 400% of full-load current, if that current is not over 100 amp, and up to 300% if it is more than 100 amp.

Since Sec. 430-22 requires that the motor-branch-circuit wires must have an ampacity of *at least* 125% of the motor *full-load current,* and since Sec. 430-52 permits overcurrent protection in some cases to be up to 400% of the motor full-load current, it follows that the branch-circuit overload protection may be as much as $^{400}/_{125}$, or 320% of the ampacity of the wire. This is totally contrary to general practice; hence it should be well understood that this is permitted only in the case of wires serving motors that have motor over*load* protection rated at not over 125% of the full-load motor current, which protects both the motor and the wires from damage caused by overload.

All the foregoing has reference to the maximum setting permitted for the overcurrent device. In practice it should be set as low as possible while still being able to carry the maximum current required by the motor while starting or running. However, Sec. 430-57 requires that when fuses are used, the fuseholder must be capable of holding the *largest* fuse permitted. Occasionally this will necessitate using an adapter to permit, for example, 60-amp fuses to be used in fuseholders designed for the larger 70- to 100-amp fuses. However, if time-delay fuses are used, this requirement is waived. It makes sense to use only time-delay fuses.

Manufacturers' Recommendations. Where the overload relay selection table in a controller or specifications on the name-plate of a piece of motor-driven equipment include a maximum branch-circuit rating, that rating should not be exceeded even when it is lower than that permitted by the Code. Similarly, the *type* of branch-circuit overcurrent device is sometimes specified, such as "use 15-amp time-delay fuse," and *must* be complied with. See NEC Sec. 110-3(b).

Motor Controllers

This subject is covered by Part G of NEC Art. 430. A motor controller according to Code definition is any switch or device normally used to start and stop the motor. The controller may be a manually operable device, or it may be an automatic device as found on refrigerators, air conditioners, furnace motors, and similar appliances.

Requirements for Controllers. In general the controller must be capable of starting and stopping the motor it controls, and of interrupting the stalled-rotor current of the motor. These basic requirements are of chief interest to manufacturers. Users and contractors will find that approved controllers supplied by manufacturers will meet these requirements, *if the proper selection is made* by the user. Nevertheless it will be well to be familiar with the requirements.

The usual form of controller is what most people call just a motor starter; it can be the manual type or the magnetic type. The starter also contains the motor over*load* protection devices (which will be discussed separately and which must not be confused with branch-circuit short-circuit and ground-fault devices). Figure 30-3 shows a manual starter that can be controlled only at the starter; Fig. 30-4 shows a larger one for bigger motors. A magnetic starter is shown in Fig. 30-5, while Fig. 30-6 shows a separate start-stop station that may be located at a distance from the controller or motor.

Controllers have sturdy contactors that will withstand many closings and openings of the contacts as the motor is started and stopped; they

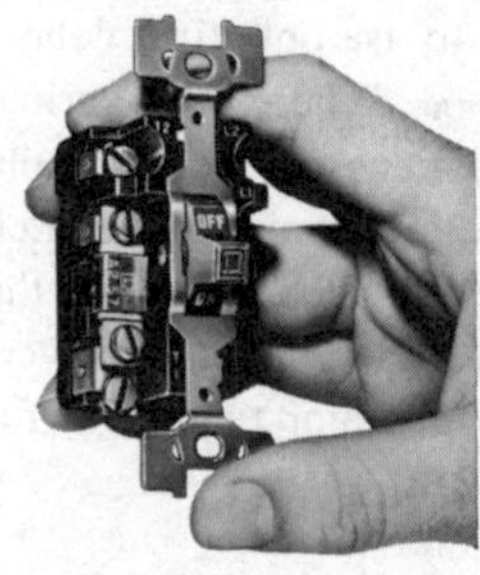

Fig. 30-3 Small starters of this kind are used with fractional-horsepower motors. (*Square D Co.*)

Fig. 30-4 Another manual starter for large motors. (*Square D Co.*)

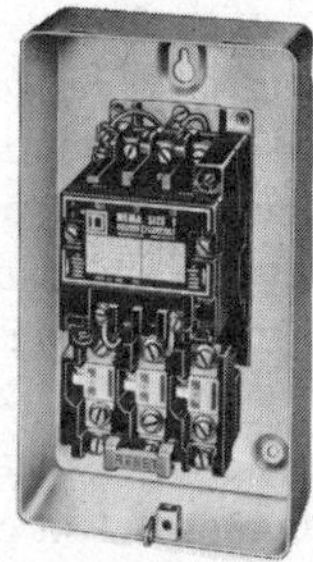

Fig. 30-5 A magnetic starter, including overload devices. (*Square D Co.*)

Fig. 30-6 A start-stop push-button station, used with magnetic starters. (*Square D Co.*)

last very much longer than the contacts of ordinary switches would last under the arcing that occurs when a motor is stopped.

The types described connect the motor directly to the full voltage of the power supply. There are other kinds that start the motor at a reduced voltage, the full voltage being applied only as the motor approaches full speed. These types are not very widely used and will not be discussed in this book.

Which Wire to Open. Unless the same circuit breaker is used as both controller and disconnecting means (which will be discussed later), the controller need not open all ungrounded wires,[4] only enough to start and stop the motor. On a single-phase motor therefore it need open only one wire, on a 3-phase motor only two wires. Actually, however, more often the controller disconnects all ungrounded wires; in the case of 3-phase motors supplied by a corner-grounded delta-connected supply, it also opens the grounded wire.

Controller "in Sight from." The controller must be in sight from, and within 50 ft of, the disconnecting means already discussed, and there are no exceptions. It must also be in sight from the motor and its driven machinery unless (1) the motor disconnecting means can be locked in the open (off) position or (2) an *additional* manually operable disconnecting switch is installed in sight from the motor location. This is covered by NEC Sec. 430-86. The purpose of this, of course, is safety, to make sure a motor is not started accidentally while somebody is working on the motor or its driven machinery.

Type of Controller Required. In general a controller or starter must have

[4] The controller may also open the grounded wire if it is so constructed that it simultaneously opens all the ungrounded wires.

a *horsepower* rating not lower than that of the motor. There are some exceptions, the most important ones being:

1. Stationary motor of 1/8 hp or less. For a stationary motor rated at 1/8 hp or less that is normally left running and is so constructed that it cannot be damaged by overload or failure to start, such as clock motors and the like, the branch-circuit overcurrent device may serve as the controller.

2. Portable motor of 1/3 hp or less. For a portable motor rated at 1/3 hp or less, the controller may be an attachment plug and receptacle.

3. Stationary motor of 2 hp or less. For a stationary motor rated at 2 hp or less, *and 300 volts or less,* the controller *may* be a general-use switch having an ampere rating at least twice the full-load current rating of the motor. A "general-use ac-only" switch may also be used, provided the full-load current of the motor does not exceed 80% of the rating of the switch; the switch must have a voltage rating at least as high as that of the motor.

4. Circuit breaker as controller. A branch-circuit circuit breaker, rated in amperes only, may be used as a controller. When this circuit breaker is also used for running overcurrent protection, it must also meet the requirements for running overcurrent protection.

How Motor Starters Operate. Some manually operated starters for small motors look like a large toggle switch, as shown in Fig. 30-3, and are turned on and off in the same way. Others, as shown in Fig. 30-4, have push buttons in the cover. A mechanical linkage from the buttons closes the contacts when you push the start button and opens them when you push the stop button. Each type contains one or more over*load* devices that open the contacts if the motor is considerably overloaded for some time, or if it fails to start when turned on. If these devices stop the motor, find out why, correct the problem, and reset by turning the switch back on or by pushing a stop-reset button.

Magnetic starters, as shown in Fig. 30-5, can be stopped by start-stop buttons located at a distance, sometimes also by buttons in the cover. Inside the starter is a magnetic coil or solenoid that closes the starter contacts when the start button is pushed. When the stop button is pushed, the circuit to the coil is interrupted, the coil loses its energy, the starter contacts open, and the motor stops. The motor over*load* device(s) in the starter stop the motor on continued overload, or failure to start. After a trip-out, reset by pushing the reset button.

The principle of operation of a magnetic starter can be better under-

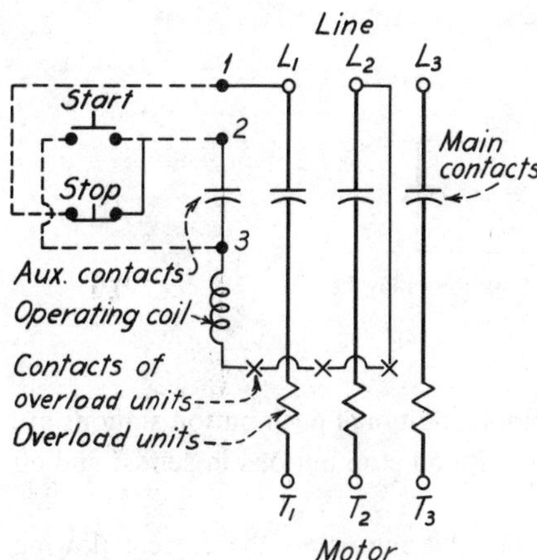

Fig. 30-7 Circuit diagram for the starter and push button shown in Figs. 30-5 and 30-6.

stood by studying the diagram of Fig. 30-7, which is for a 3-phase motor. The wires shown as heavy lines carry the power current; those shown as light lines carry only a very small signal current flowing through the start-stop buttons. The dashed lines show the wiring, either inside or outside the controller enclosure, to the push-button station. Note the difference between the two buttons: The start button is the normally open type, and pushing the button *closes* the circuit; the stop button is the normally closed type, and pushing the button *opens* the circuit. Both are of the momentary-contact type.

The magnetic starter is an electrically operated switch. When the operating coil is energized by pushing the start button, it closes the three main contacts and starts the motor. It also closes a small auxiliary contact at the same time. (In the diagram, the operating coil is not energized and the motor is not running.) Study the diagram and you will see that pushing the start button lets current flow through the operating coil, which energizes it, thus closing all four contacts. That starts the motor. When you remove your finger from the start button, that appears to open the circuit, but the coil remains energized because the circuit through the coil, at first energized by the current flowing through the start button, now remains energized by the current flowing through the auxiliary contact. At all times while the motor is running, the current that energizes the coil *flows through the stop button,* which remains closed until the motor is stopped by pushing the stop button.

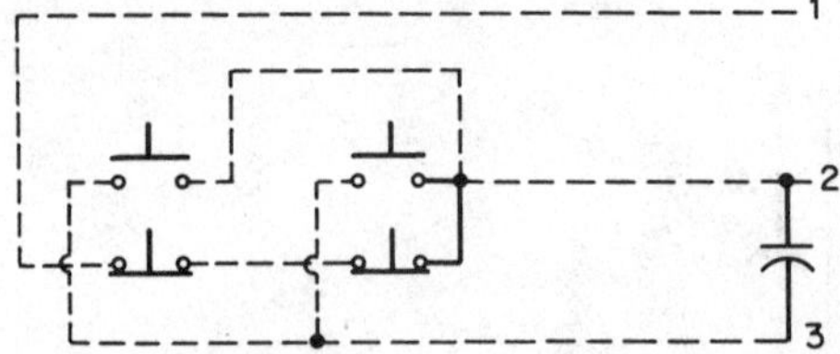

Fig. 30-8 Any number of push-button stations may be added, as shown.

For control at two or more points, additional push-button stations are connected as shown in Fig. 30-8, with all stop buttons in series, and all start buttons in parallel.

When you push the stop button, this interrupts the current flowing through the coil, deenergizing it, and all four contacts open, stopping the motor. Should the voltage fail or drop to a very low value while the motor is running, the coil will not have enough power to keep the contacts closed. The contactor will "drop out," stopping the motor, just as if you had pushed the stop button. It can be started again only by pushing the start button. This is a safety measure, for a motor running at a greatly reduced voltage would probably burn out. Moreover, if the contactor did not drop out, the motor would automatically restart when the power is restored—most dangerous if motor-driven machinery starts unexpectedly.

Motor-Control Circuits

This subject is covered by Part F of NEC Art. 430. A motor-control circuit is defined as "The circuit of a control apparatus or system that carries the electric signals directing the performance of the controller, but does not carry the main power current."* In other words, it consists of the wires between a controller and one or more start-stop stations (in Fig. 30-7 the start-stop station is shown next to the controller, but it could be in the same case as the controller; the station usually is located at a distance from the controller).

Such wires should be protected against physical damage, such as by running them through a raceway, for example. The wires must have over-

*Reprinted with permission from NFPA 70-1984, National Electrical Code®, Copyright© 1983, National Fire Protection Association, Quincy, Massachusetts 02269. This reprinted material is not the complete and official position of the NFPA on the referenced subject, which is represented only by the standard in its entirety.

current protection by breaker or fuses rated at not more than the Table 310-16 through 310-19 ampacities, without derating factors, for No. 14 and larger and not more than 7 amp for No. 18 and 10 amp for No. 16. However, there are exceptions: If the control circuit is tapped from the motor circuit, as shown in Fig. 30-7, and does not extend beyond the controller enclosure, it is only necessary to provide short-circuit and ground-fault protection, and it is permitted to be protected by motor branch-circuit overcurrent devices not exceeding 400% of the ampacity of the control-circuit conductors. When No. 14 and larger control-circuit conductors extend beyond the controller enclosure, this is reduced to 300% of the Table 310-16 ampacities for 60°C conductors, including the limitations given in the dagger (†) footnote. When a control circuit is supplied from a transformer with a 2-wire secondary, the overcurrent protection in the transformer primary is acceptable for protecting the secondary conductors when the primary overcurrent protection is based on the secondary-conductor ampacity multiplied by the transformer secondary-to-primary ratio.

In any case, the motor-branch-circuit short-circuit and ground-fault protective device is considered to also protect the control-circuit conductors if opening the control circuit would create a hazard, as for example with the control circuit of a fire-pump motor controller.

Motor- and Branch-Circuit Overload Protection

This subject is covered by Part C of NEC Art. 430. As discussed earlier in this chapter, the overcurrent protection in the motor branch circuit protects the circuit, the motor, and its control apparatus against short circuits and ground-fault current *only*. But that protection is usually no protection at all for the motor itself against current higher than normal *running* current, caused by overloading the motor, or by failure of the motor to start when turned on.

Need for Motor Overload Protection. A motor that requires, for example, 10 amp while delivering its rated horsepower may require 40 amp while starting. Once the motor has come up to speed, if it is not overloaded, the current will drop to 10 amp. But there may be times when the machine that the motor drives will need more than the rated horsepower of the motor; the motor will in almost all cases be capable of delivering more than its rated horsepower, for a short period of time, but consuming a much higher current while doing so. Under overload, the motor normally consuming 10 amp may draw 15 amp or more. If allowed to draw 15 amp continuously or for a considerable period of time, the motor will in all likelihood be damaged, reducing its life, or it may even burn out.

So far, only motor-branch-circuit short-circuit and ground-fault protection has been discussed, and that may be as high as 400% of the normal running current of the motor, or 40 amp in this case. Obviously a 40-amp overcurrent device will in no way protect the motor against damage if an overload makes the motor consume 15 amp instead of 10 amp.

Therefore the Code requires (in addition to the branch-circuit overcurrent protection) another device to protect the motor and its controller against current smaller than short-circuit or ground-fault current, but higher than the normal full-load current of the motor, and caused by overloading the motor or failure of the motor to start when turned on. The Code calls this "motor and branch-circuit overload protection," which is much too long a phrase to use repeatedly, so to distinguish it from breakers and fuses in general (and specifically those used as *branch-circuit* overcurrent protection) in this book it will be called just an over*load* device. Just remember that when an over*load* device is mentioned, it refers to the device that protects the motor and its controller *only* against too high a current caused by overloading the motor or by failure of the motor to start. The device can be integral with (built into) the motor, or it can be a separate unit. It is usually installed in the motor starter.

When Required. Overload devices are always required except for (1) noncontinuous-duty motors as defined earlier in this chapter; (2) a manually started motor of 1 hp or less if it is *not* permanently installed and is in sight from its controller (any such motor may be used on any 120-volt circuit that has branch-circuit overcurrent protection rated at not over 20 amp); (3) a motor that is part of an approved assembly that does not normally subject the motor to an overload, if the motor is protected against failure to start either by a protective device integral with the motor, or by the safety controls of the assembly itself (such as the safety combustion controls of a domestic oil burner); or (4) a motor that has sufficient impedance[5] in its windings to prevent overheating due to failure to start, when it is part of an approved assembly.

In Which Wire. If fuses are used as over*load* devices, one must be inserted in each ungrounded wire. In the case of 3-phase motors supplied

[5] There are many impedance-protected ac motors, usually $^{1}/_{20}$ hp or less, such as clock motors and some fan motors. Motors in UL Listed assemblies of this type are marked on their name-plates as "Impedance Protected" or by the letters "ZP" ("Z" is the symbol for impedance).

by a corner-grounded delta system, one *must* also be inserted in the grounded wire.

In the case of 3-phase motors, one consideration in favor of *not* using fuses as overload devices (or in the motor branch circuit) lies in a characteristic of all 3-phase motors. Three wires run to every 3-phase motor; if only two were connected, the motor would not start. But if the motor is once started, and one of the three wires is disconnected (as in the case of a blown fuse), the motor will continue to run; it is said to "single-phase." It will deliver less than its rated horsepower and draw more than normal current. If not quickly stopped by overload devices, it could be very seriously damaged in a very short time.

If overload devices *other than fuses* are used, you must install one in the ungrounded wire to any 115-volt single-phase motor; one, in either wire, to any 230-volt single-phase motor; three, one in each wire, to any 3-phase motor.

Types of Overload Devices. Fuses, while permitted, are not generally used except for smaller motors, and usually only if the branch-circuit fuse is rated as required for a motor branch circuit, and is *not* rated higher than permitted for over*load* protection. Ordinary fuses are totally unsuitable, but time-delay fuses will sometimes serve the purpose.

Circuit breakers may be used, but standard ratings seldom match the required settings for motor overload protection.

Another type is the integral, built-in device in a motor, and that will be discussed under a separate heading.

The most usual overload device is shown in Fig. 30-9. It is called by various names, such as thermal units or heater units. In almost all cases they are installed on motor controllers or starters; they could be in separate cabinets. While starters are rated in *horsepower,* the overload devices are rated in *amperes,* and must be selected to correspond to the *running* current of the motor. Suppliers of controllers have information that will permit selection of proper ratings of the overload devices. In general, motors of 1 hp or less if automatically started, and all motors of

Fig. 30-9 An overload unit (heater coil) of the melting-alloy type, used in motor starters to protect against overloads. (*Square D Co.*)

more than 1 hp, may have overload devices rated at not more than 125% of the motor running current if the motor has a service factor of 1.15 or more, *or* is marked for a temperature rise of not over 40°C; if not marked with one of these two factors, the limit is 115%.

If, however, overload devices so rated do not permit the motor to start or carry the load, the next higher rating may be used, provided it does not exceed 140% where 125% is the normal maximum, and 130% where 115% is the normal maximum.

The overload devices are basically heating elements, the degree of heat reflecting the motor current. When the heat exceeds the rating of the device, a metal alloy or a bimetallic strip is activated which converts the electrical-thermal energy to mechanical energy and releases (opens) the contacts in the holding-coil circuit, which drops out, stopping the motor. (There are also solid state overload devices with no moving parts.) In the case of a 3-phase motor, the three separate contacts are all in series, as shown in Fig. 30-7. No matter which of the three overload units operates to open the contact, the control circuit is opened and the motor stops, just as if the stop button had been pushed. The motor can be started again only by first pushing a reset button to close the overload-device contact; this can be done only after the overload device has cooled off. Some overload devices reset automatically, but such a type should not be used if the controller can be reclosed (as by a limit or pressure switch) automatically *unless* the unexpected automatic restarting of the motor could not result in injury to persons. For example, a table-saw motor overload device should *not* reset automatically.

Integral Overload Devices. Some motors have built-in overload protection, making the need for separate overload protection unnecessary. This built-in device is a component of the motor, and operates not only on the heat created by the current flowing through it, but also on heat conducted to it from the windings of the motor. If the motor is already hot from a long period of running at full load, then the device will disconnect the motor more quickly if an overload develops than if the motor started cold and immediately became overloaded to the same degree. When such devices shut off a motor, they usually must be manually reset after the motor cools off. Automatically resetting built-in motor protectors are also available, but should *not* be used unless the unexpected restarting of the motor could not result in injury to persons. The Code requirements for such integral protectors are of interest primarily to motor manufacturers.

Combining Several Components of Motor Branch Circuits

Branch-Circuit Short-Circuit and Ground-Fault Protection and Motor Overload Device Combined. As you have already learned, the motor-branch-circuit short-circuit and ground-fault protection may in most cases have an ampere rating several times as great as the maximum permitted for the overload device. However, some motors start so readily that their starting current isn't much higher than their running current. In that case, per NEC Sec. 430-55, the overload device may be omitted if the rating of the branch-circuit overcurrent protection *is not higher* than the maximum permitted for the overload device. When this situation exists, the motor-branch-circuit overcurrent protection can be a circuit breaker or time-delay fuses. Ordinary fuses will not serve the purpose.

Controller and Overload Device Combined. If ordinary motor starters of the type shown in Figs. 30-3 to 30-5 are installed, they usually include overload devices of the type shown in Fig. 30-9, or similar ones. In that case the controller serves both purposes; it would be more proper to say that the overload devices are combined with the controller. This is covered in NEC Sec. 430-39.

If the controller is a circuit breaker and its ampere rating does not exceed the maximum permitted for the overload device, no separate overload device is required; see NEC Sec. 430-83 Exc. 2.

If the controller is a fused switch, and its fuses are of the time-delay type of an ampere rating not greater than permitted for the overload device, no separate overload device is required. This condition is discussed in NEC Sec. 430-90.

Disconnecting Means, Branch-Circuit Overload Protection, and Controller Combined. Per NEC Sec. 430-111, if a circuit breaker is installed meeting the conditions required for branch-circuit overcurrent protection, it may also be used as the disconnecting means and the controller (not overlooking the "in sight from" and similar requirements already discussed). The circuit breaker must be operable by using your hand on a lever or handle.

If, however, a fused switch is used (to replace the overcurrent protection ordinarily provided by the breaker), the fuses must of course be of an ampere rating not exceeding that required for short-circuit and ground-fault protection of the motor branch circuit.

Disconnecting Means, Branch-Circuit Overcurrent Protection, Controller, and Overload Devices Combined. Often, a single enclosure will contain a circuit breaker serving as the disconnecting means and branch-circuit short-circuit and ground-fault protection (or a fused switch serving the same purpose) *and* the motor controller, which usually contains the overload devices. This is not really a case of one device serving several purposes, but rather a case of several components, each meeting the requirements that would be required for each component if installed separately, being installed in a single enclosure.

PART 3: TWO OR MORE MOTORS ON ONE CIRCUIT

The Code permits two or more motors on one circuit, but the requirements for the components of the circuit can be quite complicated, so unless it is *desirable*[6] to have several motors on the same circuit, it is generally best to provide a separate circuit for each motor.

"Approved for Group Installation." In any discussion concerning two or more motors on a single circuit, you will frequently see the expression "Approved for group installation." As you have already learned, controllers and over*load* devices ordinarily used in a circuit supplying one motor are not capable of opening short-circuit or ground-fault currents; they are protected by the motor-branch-circuit short-circuit and ground-fault device. But when several motors are connected on a single circuit, that overcurrent device must have a rating high enough to protect the entire group of motors. It then might exceed 400% of the running current of the *smallest* motor in the group, and a short circuit or ground-fault current would damage or destroy the controller or overload device for that motor. To offset this possibility, special controllers and overload devices (including thermal cutouts and overload relays) have been developed that are capable of opening a much higher current than would an ordinary controller or overload device. When separate devices are assembled together in the field, they must be UL Listed for such use and the manufacturer must provide instructions for their use with each other. The maximum rating of the fuse or inverse-time circuit breaker permitted to protect the circuit must be marked on each controller listed for group installation.

Several Motors on GENERAL-PURPOSE Branch Circuit. This subject is covered by NEC Sec. 430-42, which, however, makes reference to several other sections that must be considered in interpreting it.

[6] An example is a single piece of machinery powered by two or more motors. If for any reason only one of the motors stopped, it might damage the machine. Worse yet, if a person were caught in the machine, it is most important that the *entire* machine be shut down at the same time.

If the motors are cord- and plug-connected, and are 1 hp or less, the attachment plug may not be rated at more than 15 amp at 125 volts, or 10 amp at higher voltages; separate overload protection is not required. But if it is larger than 1 hp, overload protection is required but it must be integral with the motor, or the motor-driven appliance.

If the motors are not cord- and plug-connected, and are *not over 1 hp,* they may be connected to a circuit with branch-circuit overcurrent protection not over 20 amp at not over 125 volts, or 15 amp at more than 125 but not over 600 volts, but only if (1) the full-load rating of each motor is not over 6 amp; (2) each motor has individual over*load* protection *if* it would be required for the motor if installed on a separate circuit, as discussed earlier in this chapter; and (3) the branch-circuit overcurrent protection is not larger than the maximum shown on any controller.

If the motors are not cord- and plug-connected but are larger than 1 hp or have a full-load rating exceeding 6 amp, each must have (1) individual overload protection that would be required if installed on a separate circuit, but approved for group installation, and (2) a controller listed for group installation; and (3) branch-circuit overcurrent protection must not be larger than the maximum stated on the controller and the overload device.

Several Motors on One MOTOR Branch Circuit. This subject is covered by NEC Sec. 430-53. If each motor is not over 1 hp, the conditions are as already outlined for several motors on a *general-purpose* branch circuit.

Two or more motors of any size (or one or more motors plus other load) may be connected to a *motor* branch circuit, if *all* the following conditions are complied with: (1) Each motor must have an overload device and controller listed for group installation and for use with each other and showing the maximum permissible rating of the branch-circuit overcurrent device; (2) any circuit breaker in the circuit must be listed for group installation; and (3) the branch-circuit fuses or circuit breakers must not be larger than would be required for the largest motor if the motor were connected to a separate circuit, plus the full-load current of each of the other motors, plus the ampere ratings of any other loads on the circuit.

In this connection, note that while ordinary overload devices are *not* suitable for interrupting short-circuit and ground-fault currents, the types listed for group installation have a much higher interrupting capacity than those not so listed. Hence, the branch-circuit overcurrent protection may be larger than would otherwise be permitted if the motors are equipped with controllers and overload devices listed for group installation. When several motors are connected on the same circuit, you will see references to the largest motor in the group. The largest motor is not necessarily the one with the highest horsepower; it is the one having the highest full-load current per NEC Tables 430-148 and 430-150. If two motors are identical, consider one of them as the "largest."

We shall now discuss the various components of a motor circuit supplying two or more motors, in the same order as discussed in the case of a single motor on its own circuit.

Feeders for Two or More Motors. The ampacity of the feeder is computed this way: Start with 125% of the full-load current, per NEC Tables 430-148 and 430-150, of the largest motor supplied; add the full-load current of all the other motors supplied; add the number of amperes required for any other load supplied. The total is the minimum ampacity of the feeder wires. But if the circuitry is so interlocked that the motors cannot all start and operate at the same time, consider only the largest combination that can operate at one time. This subject is covered by NEC Secs. 430-24 and 430-25.

The overcurrent protection required for such a feeder is discussed in NEC Secs. 430-61 to 430-63. The maximum rating of the overcurrent device is the sum of (1) the maximum rating permited *by* NEC *Table 430-152* (see Appendix) for the largest motor in the group and (2) the total of full-load currents of all the other motors in the group. If in addition to motors the feeder also supplies other loads, the amperes needed to supply those loads must be added to the sum of (1) and (2) above.

Motor-Feeder Circuit Taps. Figure 30-10 could be interpreted as three motors on one branch circuit or as three motors on three separate circuits supplied by feeder *A-B-C*. If overcurrent protection located at points *B* and *C* would be difficult to reach (overcurrent protection must be located where readily accessible), NEC Sec. 430-28 permits an alternative. Omit the overcurrent protection at the points where the circuits

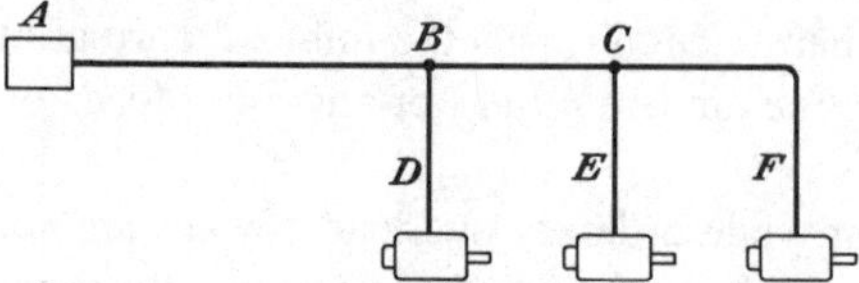

Fig. 30-10 A feeder supplying three motors.

begin, and install it at the ends of the taps (*D*, *E*, *F*), provided the tap for each motor has the ampacity required for a motor on its own circuit and, additionally, meets one of the following four conditions: (1) not over 10 ft long and enclosed in a raceway or motor-controller enclosure; (2) not over 25 ft long, with an ampacity at least one-third of the ampacity of the wire in the feeder, and protected from physical damage; (3) not over 25 ft long horizontally, 100 ft total, in high-bay manufacturing buildings, where the conductors, No. 6 copper or No. 4 aluminum or larger, have an ampacity at least one-third of that of the feeder conductors and are enclosed in a raceway; and (4) the tap wire has the same ampacity as that of the wire in the feeder. Following this procedure avoids the complications inherent when several motors are supplied by one branch circuit.

Branch Circuits for Two or More Motors. The ampacity of such a branch circuit is determined exactly as just outlined for a feeder. The overcurrent protection for the circuit is also computed as for a feeder.

Combination equipment that has two or more motors (or a combination of motor and other loads) of the type that operates as a unit must be marked to show the maximum rating of the branch-circuit breaker or fuses that may be used. Such equipment must also be marked to show the minimum branch-circuit ampacity.

Disconnecting Means. Each motor may have its own disconnecting means, but the Code permits a group to be handled by a single disconnecting means under one of three conditions:

1. If several motors drive different parts of a single machine, as, for example, metalworking or woodworking machines, cranes, hoists.
2. If several motors are in a single room within sight of the disconnecting means.
3. If several motors are protected by a single set of branch-circuit overcurrent devices.

The rating of such a common disconnecting means must not be smaller than would be required for a single motor of a horsepower equal to the sum of the horsepower of all the individual motors (or a full-load current equal to the sum of the full-load currents of all the individual motors).

Motor Overload Protection. Select an overload protection device for each motor as you would if that motor were the only motor on the circuit, except that it must be of the type listed for group installation. This is required by Code Sec. 430-53(c)(1).

Controllers. Each controller must be of the type listed for group installation. Several motors may be handled by a single controller under the same conditions as outlined for disconnecting means.

PART 4: HERMETIC REFRIGERANT MOTOR-COMPRESSORS

In Art. 440, the Code deals with electric motor-driven air-conditioning and refrigerating equipment,[7] the major component of which is the hermetic refrigerant motor-compressor, which is defined in Sec. 440-1 as "A combination consisting of a compressor and motor, both of which are enclosed in the same housing, with no external shaft or shaft seals, the motor operating in the refrigerant."* In this book it will be called just a motor-compressor.

Ordinary motors are cooled by radiating the heat that develops during operation into the surrounding air. But, in a motor-compressor, the motor is closely coupled to the compressor and both are sealed in a common case. See Fig. 30-11. While the motor-compressor is running, the refrigerant is continuously entering the case in a gaseous state at a tempera-

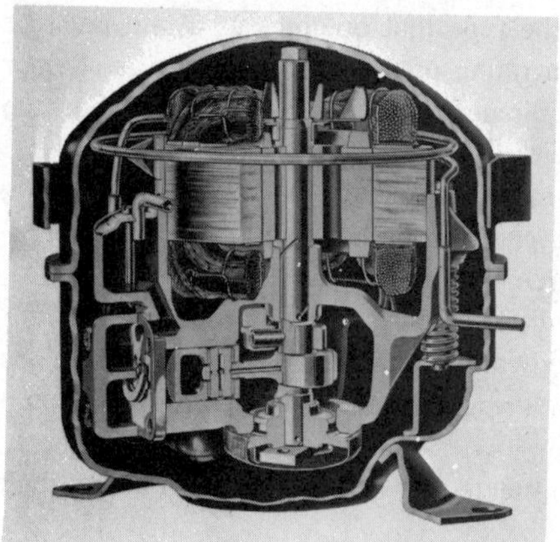

Fig. 30-11 Cross-sectional view of a typical motor-compressor. (*Tecumseh Products Co.*)

*Reprinted with permission from NFPA 70-1984, National Electrical Code®, Copyright© 1983, National Fire Protection Association, Quincy, Massachusetts 02269. This reprinted material is not the complete and official position of the NFPA on the referenced subject, which is represented only by the standard in its entirety.

[7] If such equipment is driven by an ordinary motor, not a motor-compressor, install the motor as explained in Parts 2 and 3 of this chapter. Room air conditioners of the type discussed in Chap. 21, as well as equipment such as household refrigerators, freezers, drinking water coolers, and beverage dispensers, are subject to the rules there discussed, not to the rules of this chapter.

ture *lower* than the ambient, and so it cools the motor. Because of that unusually efficient cooling, a motor of any particular physical size can safely consume more amperes (and deliver more horsepower) than an ordinary motor of the same physical size and cooled in the usual way. But when such a motor-compressor is started after a period of idleness, the refrigerant is at the ambient temperature and provides relatively poor cooling; the motor then heats up faster than an ordinary motor of the same size. For that reason, the requirements for motor-compressors differ in some respects from those covering other motors.

Information on Name-Plate. A horsepower rating appears on the name-plate of an ordinary motor, but never on a motor-compressor. Instead, the name-plate shows:

1. Rated-load current. This is defined as "the current resulting when the motor-compressor is operated at the rated load, rated voltage, and rated frequency of the equipment it serves."*

2. Branch-circuit selection current. Sometimes this is shown, sometimes not. If required to, the manufacturer of the motor-compressor will show it on the name-plate. If it does appear, you must always use it in place of rated-load current when sizing some of the components of the circuit supplying the motor-compressor.

3. Locked-rotor currents. If the motor-compressor is 3-phase, this must always be shown. If it is single-phase, it is required only if the rated-load current is over 9 amp at 115 volts, or over 4 1/2 amp at 230 volts.

Using Name-Plate Data. In installing an ordinary motor, the ratings of various components in the circuit are dependent on the horsepower rating of the motor. But the motor-compressor does not have a horsepower rating. Where the requirements for a component are different from those of an ordinary motor, they will be pointed out in following paragraphs.

As an example, assume you are going to install a 60-Hz 230-volt 3-phase motor-compressor with a rated-load current of 25 amp (and in some cases a branch-circuit selection current of, for example, 30 amp). Remember, if it has both, always use the higher of the two. Assume it has a locked-rotor current of 125 amp.

Motor-Branch-Circuit Wires. The minimum ampacity of these wires must be 125% of the rated-load current of 25 amp, or 31.25 amp. But since the

*Reprinted with permission from NFPA 70-1984, National Electrical Code®, Copyright© 1983, National Fire Protection Association, Quincy, Massachusetts 02269. This reprinted material is not the complete and official position of the NFPA on the referenced subject, which is represented only by the standard in its entirety.

motor-compressor we are discussing also has a branch-circuit selection current of 30 amp, the minimum ampacity becomes 125% of 30 or 37.5 amp.

Motor-Branch-Circuit Short-Circuit and Ground-Fault Protection. The branch-circuit fuses or circuit breaker must be able to carry the starting current of the motor-compressor. They must normally have a rating not over 175% of the rated-load current or the branch-circuit selection current. In the case of the motor-compressor under discussion, that would be 175% of 30 amp or 52 1/2 amp. If, however, that will not carry the starting current, it may be increased to 225% or 67 1/2 amp.

Disconnecting Means. Ordinarily the disconnecting means depends on the horsepower of the motor involved, but the motor-compressor under discussion does not have a horsepower rating, so we must work backward. For the moment, let's pretend it does *not* have a branch-circuit selection current, but only a rated-load current of 25 amp. Turn to NEC Table 430-150 in the Appendix and look in the "230-volt" column. It does not have a horsepower rating corresponding to 25 amp; the next highest figure in the column is 28 amp, which corresponds to 10 hp, so temporarily call it a 10-hp motor. But the motor-compressor under discussion does have a branch-circuit selection current of 30 amp, and Table 430-150 does not show a motor corresponding to 30 amp, so use the next higher figure of 42 amp, which corresponds to a 15-hp motor. So temporarily call it a 15-hp motor.

But the motor-compressor under discussion has a locked-rotor current of 125 amp. Go to NEC Table 430-151 (see Appendix); look in the 3-phase 230-volt column for 125 amp. You won't find it, but the next higher number is 132 amp, which corresponds to a 7 1/2-hp motor.

If Tables 430-150 and 430-151 lead to two different answers, you must use the larger of the two. If the motor-compressor under discussion did *not* have a branch-circuit selection current, you would provide a disconnecting means suitable for a 10-hp motor; since it does have a branch-circuit selection current, you must provide disconnecting means suitable for a 15-hp motor.

Controller. When you have determined the answer for the disconnecting means, you also have the answer for the controller. For this motor-compressor, provide the same controller that you would provide for a 10-hp (or 15-hp) motor. In most cases, as, for example, in a household air-conditioning system, the controller will be part of the air-conditioning unit.

Motor Overload Protection. Proceed as in the case of an ordinary motor, but base your conclusions on the branch-circuit selection current of the motor-compressor, if it has one; if it does not have one, base it on the rated-load current. The maximum rating is 140% if a thermal protector is used or 125% if a circuit breaker is used. Usually the overload device is part of the equipment, frequently integral with (built into) the motor. Equipment that is thermally protected must be marked "Thermally Protected" or "Thermally Protected System."

31
Wiring Commercial Occupancies

Stores, offices, and warehouses are generally called commercial occupancies. There are combinations such as a factory with a retail store in front, which combines both industrial and commercial occupancies under one roof; and of course most factory buildings include an office. In the older parts of many cities are found retail stores with living quarters above or to the rear. Local building or fire codes may require that different occupancies in the same building be separated by fire-rated walls or floors and that the wiring methods differ from one occupancy to the other within the same building; so consult the proper authority before proceeding to plan the wiring for a mixed occupancy. In this chapter some of the things you will encounter in working in commercial occupancies will be discussed.

Use the methods already explained in other chapters to determine the requirements for services, feeders, panelboards, and circuits.

See NEC Tables 220-2(b) and 220-11 (see Appendix) for the *minimum* volt-amperes per square foot of area that must be allowed for lighting. Using the minimum figures will rarely provide the degree of illumination that is desirable. The levels of illumination mentioned in this chapter for various occupancies are those recommended by authorities in the field; however, read Chap. 14, section "Footcandles Required for Various Jobs," and Chap. 29, section "Good Lighting."

Offices

We will discuss only individual offices, and smaller buildings with a relatively small number of individual offices. A small modern, well-appointed office building will of course have adequate lighting, and usually an assortment of electric equipment such as water coolers, copying machines, possibly room-type air conditioners, and so on.

Lighting. The development of more efficient light sources, coupled with the need to conserve energy, has resulted in a reduction in the minimum Code requirements for office lighting from 5 to 3.5 VA per sq ft. This capacity must be calculated in service, feeder, and panelboard sizes, but the actual lighting installed governs the outlets and branch circuits needed.[1] Depending on the many factors outlined in Chap. 29, 3.5 VA per sq ft, using fluorescent lighting, may provide from 50 to 80 fc. The recommended minimum is 70 fc; 100 fc is better. For accounting and drafting work, the minimum recommendations are 150 and 200 fc respectively.

In planning an office installation of some significant size, do consider the advantages of 277-volt fluorescent lighting discussed in Chap. 27.

Lighting is such a visible target for energy conservation efforts that worker efficiency may suffer, and any reduction may be more costly in the long run. One of the approaches to this problem is "task lighting," placing the lighting fixture as close as is practical to the job being done. Remembering the law of inverse squares discussed in Chap. 14, you will see that the amount of light required to maintain the same footcandle level at the work plane is greatly reduced as the source is moved closer to the work. Problems with contrast between the work area and the surrounding space must be worked out, but considerable energy savings have been demonstrated by using task lighting. Means for switching off unused lighting and periodic cleaning of reflectors and diffusers also save energy.

One means of providing task lighting at the workplace, plus convenience receptacles, consists of using relocatable wired partitions. When

[1] An additional load of 1 VA per sq ft must be included for general-use receptacles. To that portion of the receptacle load exceeding 10 kVA, calculated on the basis of 180 VA per outlet, a demand factor of 50% may be applied.

office rearrangement takes place, the partitions, which do not extend from floor to ceiling, can be relocated. Up to six individual contiguous units can be interconnected by means of hard-usage flexible cords and plugs. These partitions are UL Listed under the category "Office Furnishings" and are covered in the Code in Art. 605.

In individual offices, provide adequate light at the work station, and less in other parts of the office. The light source should generally be more or less above the worker's head, but not so far back that it will cast a shadow of the head upon the work being done. Nor should it be so far forward that the light can shine directly into the worker's eyes.

It is always best to consult a competent lighting engineer in designing an office lighting layout.

Suspended Ceilings. The space between a suspended ceiling and the structural floor above is often used as part of the heating, cooling, and ventilating system. For that reason there is concern regarding the materials placed in this location, out of view, which can either provide fuel for a fire or can give off smoke and gases that quickly spread to occupied spaces in the building. Code Sec. 300-22(c) limits the wiring methods permitted to metal raceways and metal-clad cables. Signaling and communications cables must either be installed in metal raceways or be UL Listed as having fire-resistant and low smoke-producing characteristics.

Fluorescent fixtures for installation in a suspended ceiling are marked "Fluorescent Suspended Ceiling Fixture." Fixtures marked "Recessed Fluorescent Fixture" may also be used, but these fixtures must meet more stringent UL requirements, since it must be assumed that thermal insulation may be installed above them. The use of thermal insulation on a suspended ceiling is considered to be unlikely.

Manufactured Wiring Systems. A system designed to simplify relocation of lighting fixtures in a suspended ceiling consists of lengths of $^3/_8$-in flexible metal conduit or Type AC armored cable with No. 12 copper conductors that have plugs and receptacles of a special configuration factory-assembled to the ends. These are used to supply and interconnect fluorescent fixtures which are flush-mounted in the ceiling. Cables may also extend into hollow walls and partitions for switches and to supply receptacles. The raceway or cable must be supported every $4^1/_2$ ft and within 12 in from each fitting, as specified in Code Art. 604. Some parts of such a system are shown in Fig. 31-1.

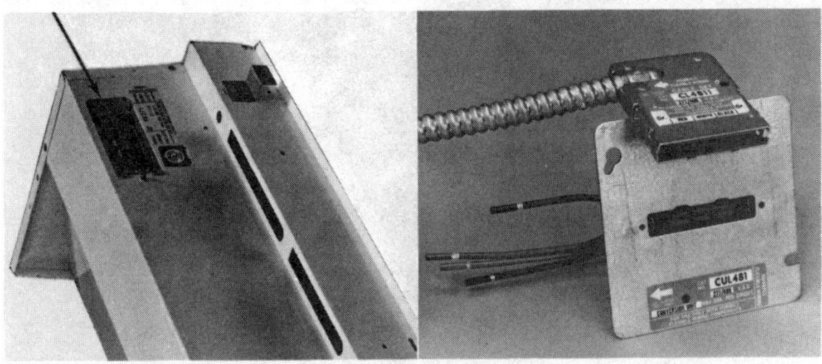

Fig. 31-1 At left, a lighting fixture with a factory-installed receptacle. At right, a plug and connector for supplying the manufactured wiring system from a permanently installed branch circuit. (*Harvey Hubbell, Inc.*)

Receptacles. The Code contains no specific requirements for the *number* of receptacles in offices, but most of them have entirely too few. Every individual office should have an absolute minimum of two; one for every 10 lineal feet of wall space is not too many considering the wide use of dictating machines, electric typewriters, calculators, and similar equipment. If the building is not totally air-conditioned, each individual office should have a circuit for a room or window air conditioner, which will probably be installed sooner or later. Outside the individual offices, don't overlook receptacles for equipment such as copying machines, water coolers, and maintenance equipment.

Flat Conductor Cable. Where an underfloor raceway as discussed in Chap. 28 has not been installed, another method of providing power at a desk in an open office area employs Type FCC cable, covered in Code Art. 328. Flat conductor cable consists of three or more flat copper conductors, usually rated at 20 amp, placed edge to edge and sealed in a flat jacket of insulation. Because of the low profile it can be installed under carpet squares (not over 30 × 30 in) with a bottom shield and a grounded metal top shield, as shown in Figs. 31-2 and 31-3. Special splicing means are used at branches, at feed points, and at surface-mounted floor outlets. All the materials used for the installation must be provided by the same manufacturer.

Poke-Through Fittings. Where a receptacle is needed away from a wall and there is a suspended ceiling for the story next below, the wiring for a receptacle surface-mounted on the floor can be run under the floor.

Fig. 31-2 By using Type FCC cable under carpet squares, outlets can be relocated or added with a minimum of disruption. (*Thomas & Betts Corp.*)

Fig. 31-3 Three-conductor flat cable, showing metal top shield and hold-down tape, feeding a duplex receptacle in a power pedestal. (*The Wiremold Co.*)

Where the floor is concrete and it is necessary to maintain its fire-resistance rating, there are available poke-through fittings such as that shown in Fig. 31-4 which will accomplish this. A smooth round hole must be cut through the concrete. After the fitting has been installed, it seals the opening against the passage of smoke and gases and, in addition, surrounds the conductors with an intumescent material which expands when subjected to heat and will then fill the space left by insulation burned off the conductors. These fittings are Listed by UL with fire-resistive ratings of up to 4 h.

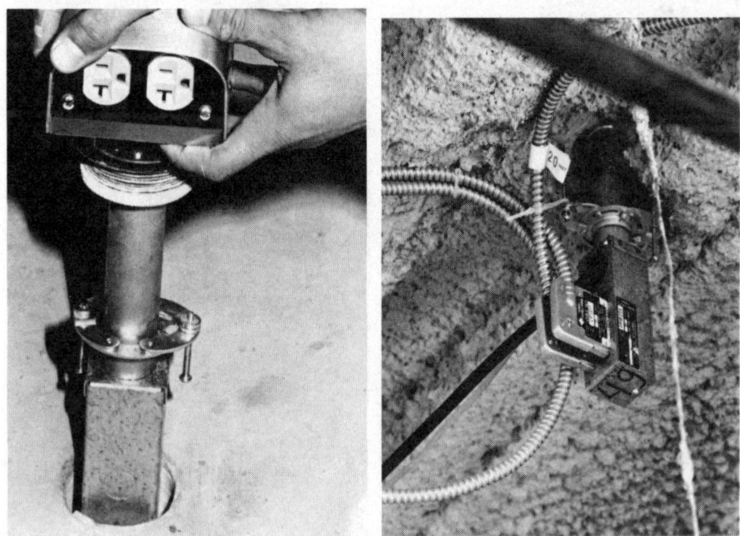

Fig. 31-4 At left, a fire-rated poke-through assembly is lowered through a 3-in hole in the concrete floor. At right, in the space between the floor and the ceiling suspended below it, supply connections are quickly made by using a manufactured wiring system as just discussed. (*Harvey Hubbell, Inc.*)

Power Poles. Another means of getting power and communications outlets to open locations in offices is by means of what one manufacturer calls Tele-Power poles. The material is essentially a rectangular raceway in cross sections from $^3/_4$- × $1^1/_2$-in to $2^3/_4$- × $2^7/_8$-in, in 10- to 15-ft lengths, steel or aluminum, with a variety of finishes to go well with most office decors. The pole is installed vertically between the floor and a suspended ceiling, wherever desired. It is supported at floor and ceiling, and a variety of appropriate fittings is available. The pole has an internal divider providing two separate channels, one for power wires and the other for telephone or similar low-voltage wires. All wires are fed into the channels from the top. Usually the power wires are factory-installed in the power channel, with two or more duplex receptacles near the bottom end. Figures 31-5 and 31-6 show the material as purchased and in an installation.

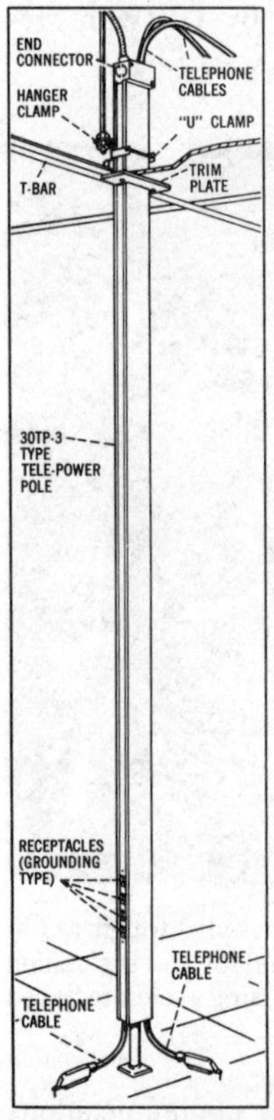

Fig. 31-5 A Tele-Power pole. It contains both power wires and telephone or similar wires, in two separate channels. It can easily be moved as office layouts change. (*The Wiremold Co.*)

Fig. 31-6 Typical installation of Tele-Power pole. (*The Wiremold Co.*)

Signaling Equipment. In larger offices it will be necessary to provide raceways, usually concealed, for low-voltage wires for telephones, buzzer systems, call systems, etc. Such conduit or surface raceway must never contain wires that are part of the regular electrical system operating at 120 volts or higher.

Emergency Lighting. The Code does not require an emergency source of power, but it may be required by local laws. Whether required or not, emergency lighting of at least exit signs and exitways is most desirable. Review Chap. 24.

Stores

Stores come in all sizes, from small specialty shops or corner groceries to multistory department stores or shopping centers. The principles explained here can be more readily understood if applied to the smaller stores. Don't overlook the Code restriction on circuit ratings if the loads are *continuous,* as is normally the case in any store.

Lighting. The Code in Table 220-2(b) requires 3 watts per sq ft in areas open to shoppers. But if you plan your installation based on the Code minimum, you will probably not achieve the level of illumination needed from the standpoint of good merchandising. The minimum lighting in the average store must be at least 80 fc in areas open to the public in order to

Fig. 31-7 A well-lighted store. Note the combination of general lighting from fluorescent fixtures and down lighting from recessed ceiling fixtures. (*General Electric Co., Nela Park.*)

provide attractive displays. In larger stores and in better locations, the minimum is 100 fc. For self-service counters 150 to 200 fc may be needed.

If these levels of illumination at first seem high, remember that mediocre lighting may be sufficient to sell the customer the specific things that he or she has decided to buy before coming into the store. No merchant, however, will be a great success by depending only on that type of business. Good lighting draws attention to the merchandise on display, encouraging customers to buy what they want rather than only the things they need. So, good lighting becomes an investment rather than an expense. See Fig. 31-7, which shows a well-lighted store. General lighting is provided by recessed fluorescent fixtures. Show cases and wall cases are down-lighted from recessed ceiling fixtures by using incandescent lamps; in some instances HID lamps (described in Chap. 29) are being used.

The figures above are those that were recommended before the energy crisis, and somewhat lower levels of light may be tolerated. Consideration should be given to less light in nonshopping areas. Regular and frequent cleaning of the reflective surfaces of fixtures will also help a great deal. In any event, the wise course is to consult a competent lighting engineer.

In many cases, one or more branch circuits may be needed to supply receptacles for plugging in show cases lighted by fixtures inside the case.

Show-Window Lighting. The Code in Secs. 220-2(c) and 220-12 requires an allowance of at least 200 VA for every linear foot of show window, measured horizontally at the base. For better stores, this is not enough. Usually an array of Type R or PAR lamps is very effective. The installation must be made so that the light source can't be seen from the street, if an effective display of merchandise is to be obtained. Figure 31-8 shows a well-lighted window.

Show-window lights can be controlled manually, but it is better to install a time switch of the general type shown in Fig. 31-9, which will automatically turn the lights on and off at a preset time. The particular switch shown has a Sunday and holiday cutoff device so that any particular day of the week can be automatically skipped. It also has an astronomical dial, which means that the switch can be set to turn the lights on and off at a predetermined interval of time before and after sundown. Once set, it automatically compensates for longer or shorter days as the seasons change.

Type of Lighting Equipment. For a modern store located in perhaps the most competitive area of a large city, there is little choice except to design the lighting system specifically to suit the size, structure, and layout

Fig. 31-8 A well-lighted show window.

Fig. 31-9 Automatic time switches are most convenient in controlling show-window lighting. (*Paragon Electric Co., AMF Incorporated.*)

of the building, the type of merchandise sold, the effects desired, and many other factors. This can be accomplished only by a competent lighting specialist, who would select the type and number of fixtures, and their arrangement, after analyzing all the factors involved.

In general, lighting fixtures will be of the fluorescent type, with some supplementary incandescent lighting. The fixtures might be suspended from the ceiling, mounted on the ceiling, or recessed in the ceiling. Where the fixtures are installed directly on a low ceiling, select fixtures that direct most of the light downward. In the case of higher ceilings with the fixtures suspended below them, it is better to select fixtures that direct part of their light upward, to avoid a somewhat cavernous appearance that results from dark, unlighted ceilings.

Often it is desirable to direct extra light on some particular item on display. Recessed ceiling fixtures with incandescent lamps are often used for this purpose. Types R and PAR lamps are particularly effective.

When installing fluorescent fixtures, be sure to select the particular kind of ''white'' that goes best with the type of merchandise on display. Review Chap. 29 concerning this.

Emergency Lighting. Follow the suggestions made for offices earlier in this chapter.

Signs and Outline Lighting

This topic is covered by NEC Art. 600. Such equipment is made and installed by competent sign shops. The circuit to a sign may be rated at not over 20 amp, unless it supplies only transformers for electric-discharge lighting (neon signs, for example), in which case the maximum is 30 amp.

Each commercial building or occupancy with outdoor ground-floor footage accessible to pedestrians must be provided with at least one 20-amp branch circuit, with an accessible outdoor outlet for use with the sign.

Each sign must be provided with a disconnecting means to disconnect all ungrounded wires from the sign. It can be either a switch or a circuit breaker and must be in sight from the sign or sign controller. In addition to the disconnecting means, the sign may have other control devices such as flashers; any such device must be of a type approved for the purpose *or* have an ampere rating at least double the ampere rating of the transformer it controls. However, if it controls no motors, the control device may be an ac general-use snap switch rated at no less than the ampere load it controls.

32 Wiring Industrial Occupancies

Industrial wiring differs from that in other occupancies mainly in that the wiring is mostly exposed and that wires, raceways, apparatus, etc., are generally larger. You will be more apt to use power tools such as a power hack saw, a power pipe threader, a hydraulic conduit bender, and a winch for wire pulling. Use these machines carefully and safely; use your head and save your back.

The principles you have learned about branch circuits, feeders, services, calculating loads, and overcurrent protection are the same. In this chapter you will find a few things which apply particularly to industrial wiring.

Machine Tools. A machine tool used to shape or form metal or plastic and complying with NFPA Standard No. 79-1981 is to be considered as an individual load even though it may employ several motors and/or heaters. All that is necessary, according to Code Art. 670, is to provide a branch circuit and a disconnecting means of adequate size.

Cranes and Hoists. Most industrial plants have one or more cranes or hoists for the moving of heavy materials. An overhead crane generally employs at least three motors: a hoist motor to raise and lower the hook, a carriage or truck motor to move the hook and load from side to side on the bridge, and a bridge motor to move the whole thing from end to end on a set of rails. The requirements for the wiring of the crane itself are found in Code Art. 610.

Running along the crane runway is a set of conductors from which power is supplied to the crane from collectors that slide or roll along the conductors. This set of runway contact conductors may be individual wires of various shapes supported by insulators and must be located so that they will not inadvertently be touched by persons, or they may be a special form of U-shaped busway which provides an outer steel enclosure, in which case the collector is a trolley which rolls along inside the busway. The runway conductors must be provided with a motor-circuit switch or a circuit breaker arranged to be locked in the open position and located to be operable from ground level and within sight of the crane and the runway contact conductors.

Electric Welders. Depending on the type of welder (ac transformer and dc rectifier, motor-generator, resistance), the ampacity of the conductors of a branch circuit to a welder and of the feeder to two or more welders need not be based on the full rated primary current of the welder, depending on the duty cycle. The overcurrent protection for an individual welder may be rated or set at 200% of the rated primary current of arc welders and at 300% of the rated primary current of resistance welders. This is contrary to what you have learned about overcurrent protection of conductors, but it is safe in this case because the conductors, even though carrying more than their rated ampacity for short periods, have time to cool while the work is being set up for the next weld. Details are found in NEC Art. 630.

Integrated Electrical Systems. Where the unexpected shutdown of an industrial process would present a hazard to personnel and equipment, where the system is serviced by qualified persons, and where effective alternate safeguards are provided, there are several places in the Code which permit the relaxing of some rules, such as those on overcurrent protection, grounding, and disconnect location. These rules are referenced in Code Art. 685. Be sure that the authority having jurisdiction for enforcement of the Code is consulted before using any of these means of assuring an orderly shutdown.

Type MI Cable. The letters MI stand for mineral insulation. The cable consists of solid copper conductors insulated from each other and from a seamless copper sheath by tightly packed magnesium oxide. Owing to the completely inorganic nature of the cable, it will outlast any system using the usual types of insulation, and it is highly resistant to heat, cold, liquids, and corrosion. Because there are no voids in the cable, it is suitable for use in hazardous locations (see Chap. 33) without the need to seal against the passage of gases, vapors, or flames.

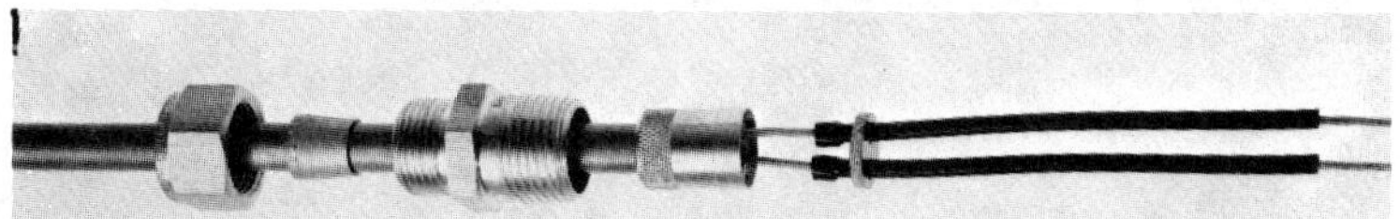

Fig. 32-1 Termination of Type MI cable, showing, from left to right, a gland nut, a compression ring, a gland body, a sealing pot (which self-threads onto cable sheath and which must be filled with a plastic or epoxy sealing compound), a washer (which must be crimped to the end of the pot), insulated sleeves, and bare copper conductors. (*Pyrotenax USA Inc.*)

It is installed like other cable, and being of relatively small diameter it can be readily installed in crowded quarters. Code Art. 330 recognizes Type MI cable for practically any use. The copper sheath is corrosive-resistant, but where the corrosive condition would be harmful, the cable can be had with a factory-installed outer jacket of plastic.

Because the magnesium oxide is highly absorbent, it is necessary to seal the ends whenever the cable is cut in order to prevent moisture in the air from being absorbed. See Fig. 32-1, which shows the cable and the special connector used with it. A sealing compound must be placed at the end of the mineral insulation, and the bare conductors must be provided with insulating sleeving. The cable must be supported every 6 ft, and bends must have a radius of at least five times the cable diameter.

Pin-and-Sleeve Devices. Plugs, receptacles, and cord connectors of the pin-and-sleeve type are of heavier construction than the usual straight-blade or locking type of wiring devices discussed in Chap. 21. They are made in ratings of up to 400 amp and 600 volts.

Round pins fit into corresponding sleeves, as seen in Fig. 32-2. Either the pins or the sleeves are slotted to provide a spring action for good contact. Housings are of cast metal or high-impact plastic. In some makes one configuration of pins and sleeves may serve on any size of circuit within its amperage and voltage rating; so it is up to the installer and the user to see that on any one premises receptacles will not be used

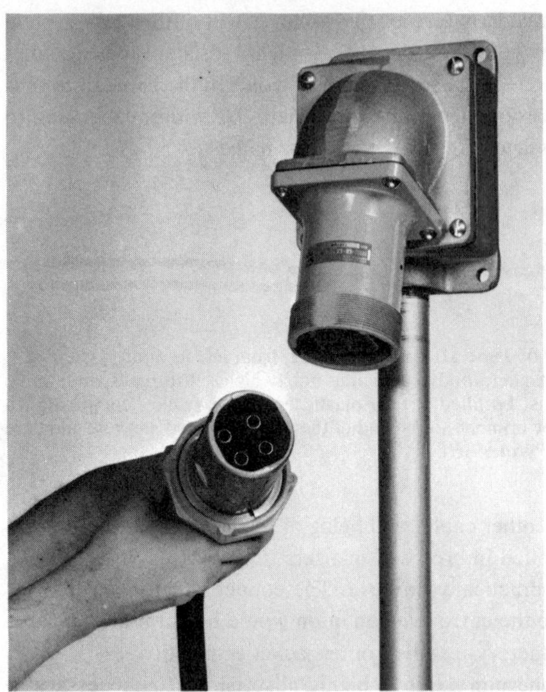

Fig. 32-2 A 100-amp, 600-volt pin-and-sleeve plug and receptacle. (*Crouse Hinds Co.*)

with plugs of a different amperage or voltage rating, as required by Code Secs. 210-7(f) and 410-56(f). There is an international standard which establishes noninterchangeable configurations for pin-and-sleeve devices, which some United States manufacturers are following.

33 Wiring Special Occupancies

There are occupancies neither commercial nor industrial which we are calling "special" occupancies. Some are specifically covered in the Code, such as hazardous locations, Arts. 500 through 516, health-care facilities, Art. 517, and places of assembly, Art. 518, while others, such as institutional occupancies (prisons, schools, etc.) have no separate Code article but often have special requirements under local building codes and fire codes.

Places of Assembly

This subject is covered in NEC Art. 518. A "place of assembly" is defined as

> buildings or portions of buildings or structures designed or intended for the assembly of 100 or more persons. Places of Assembly shall include, but are not limited to: Assembly halls; auditoriums, including auditoriums in schools, mercantile, business and other occupancies; exhibition halls; armories; dining facilities, including restaurants; church chapels; dance halls; mortuary chapels; museums;

skating rinks; gymnasiums and multipurpose rooms; bowling lanes; pool rooms; club rooms; places of awaiting transportation; courtrooms; and conference rooms.*

The chief restriction is that in such places wiring must be in metal raceways, MI cable, Type MC cable (not covered in this book), or nonmetallic raceways encased in not less than 2 in of concrete. But if the local building or fire codes do not require the building, or portions thereof, to be of fire-rated construction, then rigid nonmetallic conduit, armored cable, or (if the building does not exceed three floors above grade) nonmetallic-sheathed cable may be used.

If the building or part thereof is used as an exhibition hall (as for trade shows), the wiring to the display booths may be approved cables or cords used as temporary wiring. Emergency lighting, at least for exit signs and exitways, may be locally required. Even if not required, it is very desirable, using any method described in Chap. 24.

Schools

It is well known that many students have impaired vision, the percentage increasing progressively from grade school to high school to college. Experience indicates that good lighting contributes materially toward improving the academic performance of students. Often those who have lagged behind respond favorably when proper illumination is provided. Good lighting is therefore a tremendous asset for the student, and has been found to reduce the cost of providing education on a "per student per year" basis.

Footcandles Required. The number of footcandles required varies greatly with the nature of the work done in a particular area. Classrooms need 50 fc for reading reasonably large print, 100 fc for smaller print or handwritten or duplicated material; blackboards need 100 fc. Domestic science rooms need 50 fc; sewing rooms need 150 fc. Shops and drafting rooms need 100 fc; for halls and corridors 20 fc is usually considered sufficient. Figure 33-1 shows a well-lighted classroom.

Again, the levels suggested are those recommended before the energy crisis. Somewhat lower levels may now be tolerated. Special attention should be paid to areas where reading or writing is not being done most of the time, such as corridors, halls, cafeteria, etc.

*Reprinted with permission from NFPA 70-1984, National Electrical Code®, Copyright© 1983, National Fire Protection Association, Quincy, Massachusetts 02269. This reprinted material is not the complete and official position of the NFPA on the referenced subject, which is represented only by the standard in its entirety.

Fig. 33-1 A well-lighted schoolroom using recessed fluorescent lighting. Note the absence of shadow and glare and the supplementary lighting for the chalkboard. (*General Electric Co., Nela Park.*)

For cafeterias, 30 fc is adequate; auditoriums (except for the stage) require 15 fc, and gymnasiums 30 to 50 fc. But it would be wise indeed to provide those areas with circuits permitting much higher levels of illumination, even if the footcandle levels suggested above are not normally exceeded. Such areas are often pressed into temporary use as study halls, lecture rooms, for giving examinations, and similar purposes; often such areas must be permanently converted if the number of students increases beyond the original estimates.

In auditoriums, it is good practice to install dimmers so that the light can be controlled as the area is used for various purposes. Then a high level of illumination can be used when the area is used for nonauditorium uses, or can be dimmed when not needed; such dimmers are available for fluorescent lighting, the type most widely used.

Code Requirements for Schools. The Code requires a lighting load allowance of 3 VA per sq ft of area, except that only 1 VA is required for assembly halls and auditoriums, 1/2 VA in corridors, and 1/4 VA in storage areas. The demand factor is always 100%, and the lighting should be considered continuous. But if you provide circuits for only what the Code requires as a minimum, you will probably not achieve the levels of illumination discussed. To provide adequate lighting, be sure to install enough circuits and feeders.

In addition to the lighting circuits, you must provide circuits for other loads, and in modern schools that constitutes a very significant total. Take into consideration not only usual loads such as heating, ventilation, possibly air conditioning, but other loads such as audiovisual equipment, appliances in domestic science departments, and special loads in shops and laboratories. Don't overlook plenty of receptacle outlets, including those in halls and corridors, where they are most essential for maintenance crews.

Emergency Lighting. Auditoriums and the like are "places of assembly," and most local authorities will require, at a minimum, emergency lighting for exit signs and exitways so that the occupants can leave a building safely in case of emergency. See Chap. 24.

Churches

The Code requires an allowance of 1 VA per sq ft for lighting, with 100% demand factor. That will not provide the footcandles of illumination now considered minimum.

Illumination below 10 to 15 fc in the main worship area would be considered inadequate. To provide this level may require 2 to 4 VA per sq ft, depending on many factors such as ceiling height, reflectance of ceilings and walls, and especially the type of lighting fixtures selected.

In older churches, one frequently sees lighting fixtures that seem to have been designed primarily to conform to the architectural scheme of the church, with less thought given to the matter of good lighting. Sometimes such fixtures are designed to use a number of small lamps, providing much less light than a single lamp consuming the same total number of watts.

Today in many newer churches, the major part of the lighting is usually provided by simple, unobtrusive fixtures at the ceiling. They are installed pointing forward so that they are not visible to the people in the congregation, unless they happen to look toward the rear of the church. By using lamps of the PAR type, good light distribution is obtained without using large, clumsy reflectors. But because the lamps are not easy to replace, it may be well to burn them at less than their rated voltage, thus prolonging their life, and certainly lamps should be used with longer than usual rated life (for example the metal-halide Quartzline type).

Additional general lighting is often provided by suspended fixtures or lanterns that fit into the architectural scheme of the particular church.

The sanctuary and the altar constitute the focal point and must receive attention. This area should be lighted to 50 to 100 fc or even higher; lighting is usually provided by floodlights, or Types R and PAR lamps, concealed from the congregation. This may require a minimum of 1500 watts depending on the area, the arrangement, and the desired effect. Figure 33-2 shows the First Presbyterian Church in Oklahoma City; it is a splendid example of a well-designed church lighting project. But each church represents an individual problem, and the design of the installation should be left to one well versed in the art of church lighting. In new construction, lamps of the HID type (discussed in Chap. 29) are now often being used. Again, the services of a competent lighting engineer are recommended; such design work is not for an ordinary contractor to handle.

If the church has a choir loft, at least 50 fc or more should be provided

Fig. 33-2 An unusually well-lighted church. Note the combination of down lighting, and fixtures for general illumination. (*General Electric Co.*)

there. An organ motor will probably require its own circuit. Conveniently located receptacles must be installed to operate pulpit lights, public address systems, and similar equipment, not overlooking those required for maintenance work. Outdoor receptacles are desirable for lighted displays during festive seasons.

Emergency Lighting. Churches are "places of assembly" and deserve emergency lighting of at least exit signs and exitways. See Chap. 24.

Hazardous Locations

Locations are classified as hazardous because of the presence or handling of explosive gases or liquids or easily ignitable fibers. The basic rules for wiring in such locations are covered by NEC Arts. 500 through 503, and more specific information is covered in Arts. 510 through 517. It is obvious that this book can discuss only the barest details of this intricate subject.

Before you attempt to install any wiring or equipment in a hazardous location, you will need to have considerable experience under the supervision of experts, and to become completely familiar with the Code requirements involved. The intent of the information about to be set forth is to give you *some* of the fundamental facts of the various classifications, and the Code rules applying to them.

How Classified. The Code uses three basic *classes* to indicate the *type* of hazard involved, and Classes I and II are further subdivided into *groups* to further indicate the *exact* type of hazard involved. Each of the three classes is divided into two *divisions* to indicate the *degree* of hazard involved.

Class I Locations. A Class I location is one in which flammable gases or vapors are or may be present in the air in quantities sufficient to produce explosive or ignitable mixtures.

A Class I, Division 1 location is one in which hazardous concentrations of gas (1) exist under normal operations, (2) exist frequently because of maintenance or leakage, or (3) might exist because of breakdown or faulty operation that might also result in simultaneous failure of electric equipment. Examples are paint-spraying areas, systems that process or transfer hazardous gases or liquids, portions of some cleaning and dyeing plants, and hospital operating rooms where flammable anesthetics are used.

A Class I, Division 2 location is a location (1) in which flammable gases or volatile flammable liquids are normally confined within containers or closed systems from which they can escape only in case of breakdown, rupture, or abnormal operation; (2) in which hazardous concentrations are prevented by mechanical ventilation from entering but which might become hazardous upon failure of the ventilation; or (3) that is adjacent to a Class I, Division 1 location and might occasionally become hazardous unless prevented by positive-pressure ventilation from a source of clean air by a ventilation system that has effective safeguards against failure. Examples are storage in sealed containers and piping without valves.

Class II Locations. A Class II location is one having combustible dust.

A Class II, Division 1 location is one in which combustible dust is in suspension in the air in sufficient quantity to produce an explosive or ignitable mixture (1) during normal operation, (2) as a result of failure or abnormal operation that might also provide a source of ignition by simultaneous failure of electric equipment, or (3) in which combustible dust that is electrically conductive may be present. Examples are grain elevators, grain-processing plants, powdered-milk plants, coal-pulverizing plants, and magnesium-dust-producing areas.

A Class II, Division 2 location is one in which combustible dust is not normally in suspension in the air and is not likely to be under normal operations in sufficient quantity to produce an explosive or ignitable mixture but in which accumulations of dust (1) may prevent safe dissipation of heat from electric equipment or (2) may be ignited by arcs, sparks, or burning material escaping from electric equipment. Examples are closed bins and systems using closed conveyors and spouts.

Class III Locations. A Class III location is one in which there are easily ignitable fibers or flyings, such as lint, but in which the fibers or flyings are not in suspension in the air in sufficient quantity to produce an ignitable mixture.

A Class III, Division 1 location is one in which easily ignitable fibers or materials that produce combustible flyings are manufactured, used, or otherwise handled, such as textile mills, cotton gins, woodworking plants, and sawmills.

A Class III, Division 2 location is one in which such fibers are stored, such as a warehouse for baled cotton, yarn, and so forth.

Groups. Equipment for use in Class I locations is divided into groups A, B, C, and D; for Class II, into groups E and G. Each group is for a specific hazardous material, which you can determine by consulting your copy of the Code. All equipment for Class I, Division 1 and Class II, Division 1 hazardous locations must be marked to show the class and the group for which it is approved; some equipment is suitable for more than one group and is so marked.

Methods of Protection. Where electrical equipment cannot be located outside hazardous areas, there are several methods of protection. Intrinsically safe equipment and circuits, which cannot release enough energy even under fault conditions to ignite an explosive vapor, can be used, but generally only for control or signaling functions because of the low power involved. Another method is the use of ordinary arc- and spark-producing equipment within a purged and pressurized enclosure, where clean air or an inert-gas atmosphere is maintained inside the enclosure to prevent any flammable vapors from entering. By far the most common methods are those which follow.

Basic Principles Applied to Equipment. Enclosures for equipment in Class I locations must be *explosionproof*. This does not mean that they are made so that explosive gases or vapors cannot enter into them. They are made so that such gases or vapors can enter the enclosure, where they do ignite and explode, but the enclosure is made so that it can withstand and contain the force of the explosion. Moreover, the hot exploded gas does escape, but not until after it has passed through a tight joint that is either threaded or has a wide ground-finish flange. In either case, before it finally escapes to the outside of the enclosure, it has cooled to a temperature below the ignition temperature of the gas in the surrounding atmosphere. The cooling takes place while the gas passes through the long circuitous path of a threaded joint, or across the wide, tight-fitting ground-finish flange. Motors shall be approved for Class I locations, or shall be: totally enclosed and pipe-ventilated, and interlocked with the ventilation system; totally enclosed and pressurized with an inert gas, and interlocked with the gas supply; or of a type submerged in a liquid that is flammable only when mixed with air. Ordinary motors with no brushes, switches, or similar arc-producing devices are permitted in Class I, Division 2 locations.

For Class II locations, enclosures must be dust-ignitionproof. Dust can be prevented from entering enclosures by means of gaskets, and enclosures can be made with large exposed surfaces for more rapid heat dissipation. Motors shall be dust-ignitionproof or totally enclosed pipe-ventilated, and in Division 1 must be approved for Class II locations.

If a Class II dust-ignitionproof enclosure is used in a Class I location, gas can get in, explode, and blow the enclosure to pieces; this might set off a larger explosion in the general area, leading to fire or injury to people. Likewise if a Class I explosionproof enclosure is installed in a Class II location, it can overheat when blanketed with dust and start a fire. Therefore it is important that equipment be labeled for the *specific* location and division where it is installed.

For a Class III location, equipment merely has to be totally enclosed to prevent the entry of fibers and flyings, and to prevent the escape of arcs, sparks, or hot particles. Motors shall be totally enclosed, with some exceptions where the inspector judges the conditions to be suitable for open motors without arcing contacts.

Proper Maintenance. In all cases, enclosures such as lighting fixtures and panelboard enclosures must be kept reasonably clean from accumulations of residue, fibers, dust, or whatever is contained in the atmosphere. And they must be properly installed. For example, if one of four screws in the cover of an explosionproof switch enclosure is left untightened, or if the flanged joint of a cover or box is scratched, exploded gas will escape before it has sufficiently cooled, and an explosion or fire could result.

A threaded joint must have at least five full threads fully engaged. Figure 33-3 shows an explosionproof box with threaded hubs for threaded rigid metal conduit and a threaded cover. Figure 33-4 shows an explosionproof receptacle and plug. The box containing the receptacle has a ground-finish flange joint with four screws, and a threaded hub for conduit. The plug and receptacle are designed so that the plug can be with-

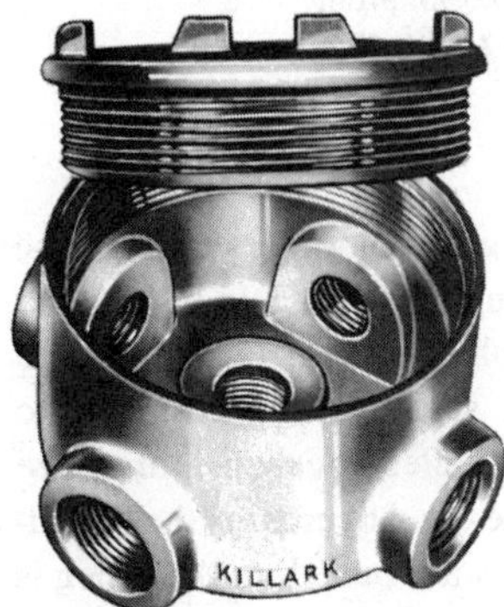

Fig. 33-3 A typical explosionproof fitting. (*Killark Electric Mfg. Co.*)

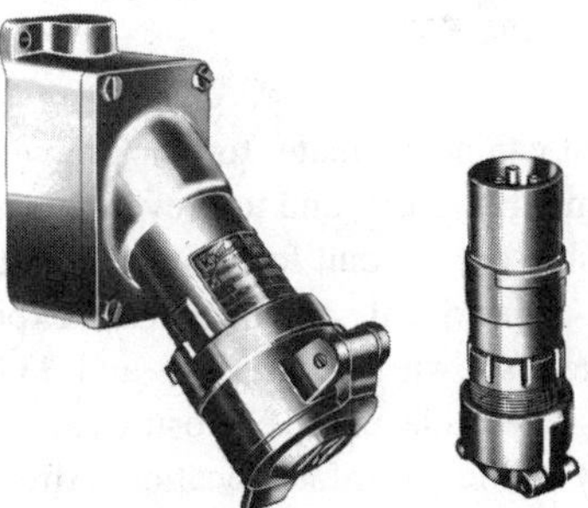

Fig. 33-4 An explosionproof receptacle and plug to fit. (*Killark Electric Mfg. Co.*)

drawn only part of the way, which breaks the circuit; an arc, if it forms, will explode the small amount of vapor in the interior. This takes place during the brief period of time it takes to twist the plug before it can be withdrawn.

Some plugs and receptacles are interlocked so that the plug cannot be inserted or removed unless a switch is in the off position and the switch cannot be turned on unless the plug is fully inserted.

An explosionproof incandescent light fixture is shown in Fig. 33-5. Explosionproof fluorescent lighting fixtures are also available, as are explosionproof panelboards, disconnecting means, circuit breakers, motors, and various other equipment—even telephones.

Sealing Fittings. Explosive gases or vapors can pass from one enclosure to another through conduit. The conduit itself can contain a substantial amount of such material. To minimize the quantity of explosive vapors

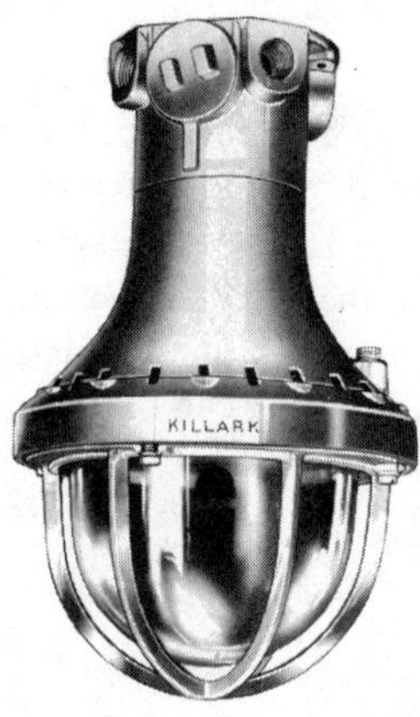

Fig. 33-5 An explosionproof lighting fixture. (*Killark Electric Mfg. Co.*)

that can accumulate, to block movement of gases or vapors through the conduit system, and to prevent pressure-piling (a building up of pressure inside the conduit following a contained explosion, sometimes followed by subsequent, more violent, explosions), sealing fittings, similar to those shown in Figs. 33-6 and 33-7, are installed. Sealing fittings are installed adjacent to enclosures and where the conduit crosses the boundary of the hazardous location. After the wires are pulled in, the fitting is dammed with a fibrous material and then filled with a sealing compound (both supplied by the fitting manufacturer), which effectively prevents explosive or exploded gases from passing from one part of the electrical

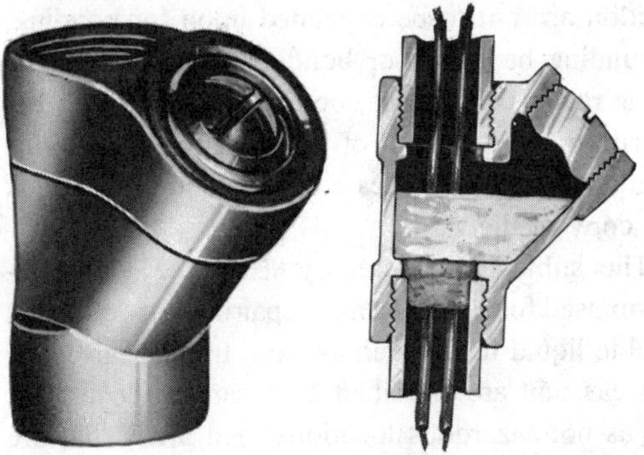

Fig. 33-6 By pouring sealing compound into a fitting of this kind, one run of conduit is sealed off from another, minimizing danger of explosion. (*Killark Electric Mfg. Co.*)

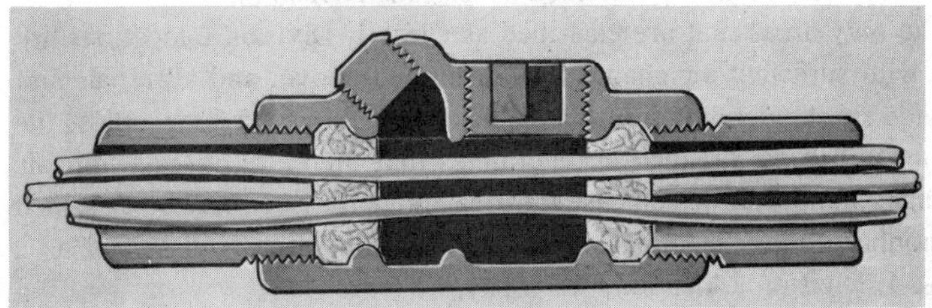

Fig. 33-7 Cutaway view of sealing fitting suitable for either vertical or horizontal conduit run. (*Crouse-Hinds Co.*)

installation to another, or from passing from a hazardous location to a nonhazardous location, where unexploded gases might reach a source of ignition.

The fitting shown in Fig. 33-6 is for a vertical run of conduit, while that in Fig. 33-7 may be used for either a vertical or a horizontal run. The fiber dam and sealing compound can be seen in the cutaway views. Individual wires should be separated so that the sealing compound will flow all the way around each wire.

Wiring Methods. Threaded rigid metal conduit, intermediate metal conduit, and Type MI cable are acceptable in all locations. All exposed non-current-carrying metal parts, enclosures, and raceways must be grounded. At least five full threads must be engaged where conduit enters a threaded hub in a cast enclosure. The usual locknut-bushing or double-locknut types of connection are not to be depended upon for bonding; grounding locknuts, grounding bushings, or bonding jumpers must be installed at all intervening raceways, fittings, boxes, cabinets, etc., between the hazardous location and the point of grounding at the service equipment. In some locations, cables of types not discussed in this book are permitted. See your copy of the Code.

Commercial Garages. This subject is covered by NEC Art. 511. A commercial garage is a location used for servicing and repairing self-propelled vehicles that use flammable liquid fuels, such as cars, trucks, buses, or tractors. A garage has areas that are classified as hazardous locations, areas that are classified as nonhazardous locations, and areas that are classified as above the hazardous locations.

The only areas that are classified as Class I, Division 1 are gasoline-dispensing areas (which will be discussed under "Service Stations") and any pit below floor level unless the pit has at least six air changes an hour, in which case it may be classified as Class I, Division 2.

The only areas that are classified as Class I, Division 2 locations are pits with sufficient air changes, as described above, and all repair and service areas extending from the floor to 18 in above the floor, unless the authority having jurisdiction determines that there are at least four air changes an hour, in which case the area above the floor may be classified as nonhazardous. In most cases, however, this area is classified as a Class I, Division 2 location.

Office areas, stock rooms, toilets, etc., are classified as nonhazardous locations if they have at least four air changes an hour, or if they are effectively cut off from the repair area by partitions. Areas that are adjacent to the repair area but not cut off by partitions and do not have four air changes an hour may be classified as nonhazardous locations if they have either sufficient ventilation for ordinary needs, an air pressure differential, or spacing such that, in the judgment of the authority having jurisdiction, no hazard exists. [A pressure differential means that the air pressure is slightly higher in the adjacent area than in the hazardous (repair) area so that no vapors will flow from the hazardous area to the adjacent area.]

Wiring Methods. The wiring in nonhazardous areas of commercial garages may be any type discussed in earlier chapters. Exceptions: In an area *above* a hazardous location that is not cut off by a ceiling, the wiring must be in a metal raceway, Type MI cable, or several other types of cable not discussed in this book. Armored cable (Type AC) is *not* permitted. EMT is very commonly used in such areas.

Lighting fixtures in the area above vehicle lanes must be not less than 12 ft above the floor, or be so constructed as to prevent sparks or hot particles from falling to the floor. Where fluorescent lighting is used and the fixtures are neither 12 ft above the floor nor equipped with a glass or plastic bottom, there are plastic sleeves and similar devices available to hold a broken fluorescent tube (lamp) and prevent the hot cathode ends of the lamp from falling to the floor. Other equipment above the hazardous location, such as motors or switches, must be *totally enclosed* if less than 12 ft above the floor.

Wiring in the hazardous location must be of a type described in this chapter for Class I, Division 1 or Division 2 (whichever it is). A sealing fitting is required to be so installed where the wiring enters or leaves the hazardous location that there is no coupling or other fitting between the sealing fitting and the boundary between the hazardous and the nonhazardous location. A sealing fitting is also required at an equipment enclosure, such as a receptacle or switch box, that is within the hazardous location, but it is common practice to keep all such wiring and equipment above the 18-in level where it is practical to do so. Equipment can almost invariably be kept above the 18-in level, but sometimes the wiring is brought in from underground or installed in the floor slab. In all such cases, install threaded metal conduit, use *gasoline- and oil-resistant wire,* and seal the conduit so that there is no coupling or other fitting between the sealing fitting and the 18-in level. Also install a sealing fitting at the panelboard end as the last fitting above grade.

Battery chargers and their control equipment must *not* be installed in the hazardous location. If batteries are charged in a separate compressor room, tire-storage room, or similar area, the room must be well ventilated to allow dissipation of the hydrogen gas. (There is not enough hydrogen gas involved for a battery room to be classified as a hazardous location, but the small amount that is developed must be allowed to dissipate.)

Service Stations. NEC Art. 514 covers gasoline-dispensing and service stations. The service (such as "lubritorium") area, office, storage room, toilets, etc., are treated the same as similar areas in a commercial garage. The gasoline-dispensing areas are classified as follows:

The area within the gasoline dispenser (pump) up to 4 ft above its base and all space beneath the dispenser are Class I, Division 1 locations. The space surrounding and within 18 in of the dispenser enclosure and extending 4 ft above the dispenser base and the area beyond the dispenser enclosure and extending 20 ft horizontally in all directions from the dispenser enclosure and extending up to 18 in above grade are Class I, Division 2 locations. This includes any indoor area not suitably cut off by partitions. Below grade, the area beneath any Class I, Division 1 or Division 2 location is a Class I, Division 1 location.

The area within 10 ft horizontally of a storage-tank fill pipe and extending to 18 in above grade is a Class I, Division 2 location; the below-grade area beneath this is a Class I, Division 1 location.

The spherical volume within a 3-ft radius of the discharge point of a storage-tank *vent pipe* is a Class I, Division 1 location; the spherical volume between the 3-ft and a 5-ft radius is a Class I, Division 2 location.

Any wiring to pole lights, signs, etc., must be kept out of the hazardous locations or be made to comply with the rules for the location, which includes a sealing fitting at each end of a conduit that passes through a hazardous area. It also means that any below-grade wiring within any hazardous location must be in rigid or intermediate metal conduit and be nylon-covered gasoline- and oil-resistant wire, since gasoline can and often does get into below-grade conduit in hazardous locations. (Under some conditions, rigid nonmetallic conduit and Type MI cable may also be installed underground.) For wiring to a gasoline dispenser, a sealing fitting must be the first fitting in the conduit where the conduit emerges from below grade at the dispenser and also where the conduit emerges from below grade at the panelboard. An explosionproof flexible fitting is frequently required between the sealing fitting and the equipment inside the dispenser, since the conduit may not quite line up with the internal fittings, and there is not much room for bending rigid metal conduit underneath the dispenser.

Any branch circuit supplying a dispenser or passing through a dispenser (such as to supply a lighting standard near the dispenser) must have a circuit breaker or switch that disconnects *all* conductors of the circuit, including the grounded (white) conductor. There are special circuit breakers made for this specific purpose; they have a switching pole for the grounded conductor as well as for the ungrounded conductor, but there is no overcurrent device in the grounded pole of the breaker. If such a circuit breaker is not used, a switch without a fuse in the grounded conductor, but one that opens all conductors (including the grounded conductor) simultaneously, must be used.

Paint-Spraying Booths. NEC Art. 516 covers finishing processes, which include paint-spraying booths and areas. Some repair garages have a body shop with a paint-spraying booth. No electric equipment and no wiring except threaded rigid or intermediate metal conduit or Type MI cable that contains no splices or terminal connections are permitted in a paint-spraying booth unless the equipment and the splice and terminal enclosures are approved for both Class I, Division 1, Group D locations and for accumulations of readily ignitable residues. The only equipment meeting both these requirements that is available (as of this writing) is lighting fixtures. So how do you wire such a booth? Proceed as follows:

Lighting fixtures (in conjunction with transparent baffles, exhaust-air movement, and careful placement) can often be installed where they will not be covered with paint residue; but it is sometimes difficult to do and not always acceptable to some inspectors. So use lighting fixtures that are approved both for Group D and for paint residue. Then wire them with threaded rigid or intermediate metal conduit (or Type MI cable) without splice boxes or switches inside the booth. Explosionproof boxes, as shown in Fig. 33-3 may be used as long as there are no splices in the box; and switches must be installed *outside* the booth.

If fixtures other than the type that is approved for residue are used, install them outside the booth so that the light will enter the booth interior through extra-strength glass panels in the roof or sides of the booth (or both). Wire these fixtures as required for the garage location in which they are located.

The exhaust fan (which must have nonferrous blades, that is, not iron or steel, but rather aluminum or brass) must have the fan motor installed outside the exhaust stack and connected to the fan by a metal shaft or by a staticproof belt. Wire the motor as required for the location. That is, if it is inside the garage repair area, wire it accordingly; if it is outdoors, wire it for an outdoor location. (In some cases one wall of a booth may be the exterior building wall; so in some cases, the motor could be outdoors.)

Caution: Remember, all wiring in hazardous areas is subject to many rules and restrictions. Thoroughly familiarize yourself with all the applicable Code requirements, and carefully follow the manufacturer's instructions for material and equipment installation.

Bibliography

"Agricultural Wiring Handbook," National Food and Energy Council, Inc., Columbia, Mo.

"ANSI Y32.9-1972," American National Standards Institute, New York.

"Getting the Most from Your Lighting Dollar," National Lighting Bureau, Washington, D.C.

"Handbook of Efficient Lighting Practices," General Electric Co., Cleveland, Ohio.

"IES Lighting Handbook," Illuminating Engineering Society, New York.

"Light and Color," General Electric Co., Cleveland, Ohio.

"National Electrical Code," National Fire Protection Association, Quincy, Mass.

"Residential Wiring Design Guide," Industry Committee on Interior Wiring Design, New York.

Richter, H. P.: "Wiring Simplified," Park Publishing, Inc., St. Paul, Minn.

Schram, P. J., J. M. Caloggero, and J. A. Tedesco: "The National Electrical Code Handbook," National Fire Protection Association, Quincy, Mass.

Appendix

This Appendix contains:

1. Commonly used abbreviations of terms found in this book.
2. Glossary of terms used in this book.
3. Tables quoted from the National Electrical Code[1] (in part from the body of the Code, others from Chap. 9 of the Code). Some of the tables shown in the Code are not repeated here because they refer to subjects outside the scope of this book. Others are omitted because they are rarely used and can be found in your copy of the Code. Portions of some tables are omitted for the same reason.
4. The Code in Chap. 9 includes specific examples illustrating the application of Code requirements. These are included in this Appendix.

[1] Every student is urged to study the National Electrical Code.

TABLE A-1, ABBREVIATIONS AND SYMBOLS

A, amp	amperes
AWG	American Wire Gauge
B&S	Brown & Sharpe (wire gauge)
c, cyc	cycles (see hertz)
C	Celsius, centigrade (temperature scale)
c.m., cmil	circular mils
cb, CB	circuit breaker
cps	cycles per second (see hertz)
EMT	electrical metallic tubing (thin-wall conduit)
F	Fahrenheit (temperature scale)
fc	footcandles
ft	foot, feet
GFI, GFCI	ground-fault circuit-interrupter
h	hours
HID	high-intensity discharge (lamps)
hp	horsepower
Hz	hertz (= cycles per second)
in	inches
k	kilo- (a thousand, watts, etc.)
kcmil	thousand circular mils
kHz	thousands of hertz
kVA	kilovolt-amperes
kW	kilowatts
kWh	kilowatthours
lm	lumens
mA	milliamperes
MCM, mcm	thousand circular mils
mega-	million (watts, volts, etc.)
micro-	one-millionth (volt, ampere, etc.)
mil	mils (thousandths of an inch)
milli-	one-thousandth (volt, ampere, etc.)
NEC	National Electrical Code
p	pole (1-p = single-pole, etc.)
ph, ϕ	phase
r/min	revolutions per minute
RI	repulsion-induction (motor)
s	seconds
UL	Underwriters Laboratories
V	volts
VA	volt-amperes
W	watts
wp, wpf	waterproof
μ (Greek *mu*)	micro- (one millionth)
ϕ (Greek *phi*)	phase

Ω (Greek *omega*)	ohms
1-p	single-pole
2-p	two-pole, double-pole
3-p	three-pole

Glossary

These nontechnical but concise definitions will assist you in better understanding the text. For some terms also defined in the NEC, the definitions may differ, but they do not conflict.

Alternating current (ac). Current which periodically reverses, having alternately positive and negative values.

Ambient temperature. Temperature of the medium (air, earth, water, etc.) surrounding conductors, devices, or utilization equipment.

Ampacity. Current in amperes that a conductor can carry continuously under the conditions of use without exceeding its temperature rating.

Ampere (A). Unit used to express the flow of electricity. One coulomb per second = one ampere. Water analogy: gallons per minute.

Approved. Acceptable to the inspector.

Arc. Unwanted flow of electricity through an insulating medium (such as air), characterized by the emission of light and heat (as opposed to controlled arcs, such as in electric-discharge lamps).

Automatic. Nonmanual; self-acting by its own mechanism when actuated by some impersonal influence such as a change in temperature or pressure.

AWG (American Wire Gauge). Standard for measurement of wire diameters in the United States.

Ballast. Auxiliary device used with electric-discharge lamps to provide starting voltage and proper operating voltage and to limit the current.

Battery. Device which stores electrical energy in the form of chemical energy.

Bonding. Effective connection of metallic parts to form a conductive path, to maintain the parts at or near the same potential, and to conduct safely the current likely to be imposed.

Branch circuit. Conductors between the last overcurrent device and the outlet(s).

British thermal unit (Btu). Unit used to express heat. The heat required to raise the temperature of one pound of water by 1°F = 1 Btu. 1 Btu = 0.29 watt.

Bus. Conductor, usually a bar of rectangular cross section, serving as the common connection for two or more smaller conductors.

BX. Trade name for Code Type AC armored cable.

Cable. Assembly of two or more conductors. Individual large-size conductors are also referred to as cables, since they are made up of several strands of smaller conductors.

Capacitance. Measure, in farads, of the opposition to voltage changes in an ac circuit, causing voltage to lag behind current; exhibited by condensers, two conductors separated by a nonconductor.

Circuit. Arrangement of conductors, devices, and utilization equipment (loads) such that current will pass through them.

Circuit breaker. Switching device, manually operable, which automatically opens a circuit at a predetermined level of overcurrent.

Conductor. Material, usually metal, having relatively low resistance, through which current will readily flow.

Continuous load. Load which remains on for 3 h or longer.

Cycle. In alternating current, the time (usually 1/60 s) during which the voltage goes from zero to + maximum, back through zero to − maximum, and back to zero again.

Demand. Kilowatts consumed during a short time period, usually 15, 30, or 60 min.

Demand factor. Ratio of the maximum demand to the total connected load.

Device. Something (such as a switch or a receptacle) designed to carry, but not consume, electrical energy.

Fault, arcing. Unwanted-high-impedance-current path causing arcing.

Fault, bolted. Unwanted-low-impedance-current path.

Fault, ground. Unwanted-current path to ground.

Fuse. Enclosure containing a soft-metal link which melts and automatically opens a circuit at a predetermined level of overcurrent.

Ground. The earth or some conducting body that serves in place of the earth; a conducting connection to earth.

Ground-fault circuit-interrupter. Device intended for personnel protection that senses abnormal current to ground and then interrupts the circuit.

Ground-fault protection. Device intended for equipment protection that senses abnormal current to ground and then interrupts the circuit.

Grounded. Connected to earth or to some conducting body that serves in place of the earth.

Grounding conductor, equipment. Conductor connecting non-current-carrying metal parts of equipment, raceways, and other enclosures to the system grounded conductor and the grounding electrode conductor at the service location.

Grounding electrode conductor. Conductor connecting the system grounded conductor and the equipment grounding conductor at the service location to the grounding electrode system.

Hertz (Hz). Unit used to express frequency. One hertz = one cycle per second.

Horsepower (hp). Unit used to express rate of work, or power. One horsepower = 746 watts.

Hot. Designating any point in the electrical system having a voltage above ground potential. Every point should be considered to be hot until you have tested it for the presence of voltage.

Impedance (Z). Measure, in ohms, of the opposition to current flow in an ac circuit. Impedance includes resistance, capacitive reactance, and inductive reactance.

Inductance. Measure, in henrys, of the opposition to current change in an ac circuit, causing current to lag behind voltage; exhibited by turns of wire with or without an iron core.

Insulation. Material having relatively high resistance, through which current will not readily flow when the material is applied within its voltage rating.

Inverter. Equipment which changes direct current to alternating current.

Leakage current. Small undesirable current which flows through insulation whenever voltage is present.

Neutral. Conductor, usually grounded, so related to the other two or more conductors of the circuit that it has the same potential to each of them and that, with balanced load and no harmonic currents present, it carries no current.

Ohm. Unit used to express electrical resistance or impedance. Ohms = volts ÷ amperes.

Outlet. Point on the wiring system at which current is taken to supply utilization equipment.

Overcurrent. Current in excess of the rated current of equipment or the ampacity of a conductor, which may result from overload, short circuit, or ground fault.

Overload. Operation of equipment in excess of normal, full-load current rating or of a conductor in excess of rated ampacity for a sufficient length of time to cause damage or dangerous overheating.

Parallel circuit. Connection of two or more devices or loads across the same conductors of the circuit, current flow through each being independent of the others.

Phase conductor. Energized conductor, usually ungrounded, and never the neutral.

Plug. Male connector attached to a flexible cord for insertion into a receptacle; also called attachment plug and attachment plug cap.

Power factor. In alternating current, the ratio of the actual or effective power in watts to the apparent power in volt-amperes, expressed as a percentage. Inductive loads cause the current to lag behind the voltage, resulting in a power factor of less than 100%.

Reactance (*X*). Measure, in ohms, of the opposition to alternating current due to capacitance (X_C) or inductance (X_L).

Receptacle. Contact device installed at an outlet for the connection of a single plug.

Rectifier. Equipment which changes alternating current to direct current.

Resistance (*R*). Measure, in ohms, of opposition to current flow in a conductor, device, or load. In direct current, volts ÷ amperes = ohms resistance. (For alternating current, see *Impedance*.)

Rheostat. Resistor which can be varied while energized.

Romex. Trade name for Code Type NM nonmetallic-sheathed cable.

Series circuit. Connection of two or more devices or loads in tandem so that the current flowing through each also flows through all the others.

Service. That portion of the system between the distribution lines of the serving agency and the premises wiring, including the main circuit breaker or fused switch.

Short circuit. Unwanted current flow between conductors.

Snap switch. Switch designed for installation in a box or for surface mounting, rated in amperes and volts. Markings may include T, indicating suitability for controlling tungsten-lamp loads; AC, indicating use on ac circuits only; ____ hp, indicating suitability for motor control; and CO/ALR, up to 20 amp, and AL-CU, 30 amp and larger, indicating suitability for connection to copper, copper-clad aluminum, or aluminum conductors.

Source. Point of origin of the circuit or system under consideration. For a wiring system the source is the serving agency's transformer, for a branch circuit it is the overcurrent device at the panelboard, and for an added outlet it is the existing outlet from which the extension is made.

Switch. Device serving as a point of control for turning something on or off.

Toggle switch. Snap switch, also often called a wall switch.

Transformer. Device with no moving parts, having two (or more) insulated windings on a laminated-steel core, used to raise or lower ac voltage by inductive coupling. Volt-amperes into the primary and volt-amperes out of the secondary are the same, less the small current necessary to magnetize the core.

Volt (V). Unit used to express the electrical force which causes current to flow. Water analogy: pounds per square inch of water pressure.

Volt-amperes. Volts × amperes. In direct current, volts × amperes = watts, but in alternating current inductance or capacitance in the circuit may introduce reactance. (See *Impedance; Power factor.*)

Voltage drop. Difference in voltage between two points in a circuit due to the intervening impedance or resistance.

Watt (W). Unit used to express electrical power. In direct current, one volt × one ampere = one watt.

Table 220-2(b). General Lighting Loads by Occupancies†

Type of occupancy	Unit load per sq ft (volt-amperes)
Armories and auditoriums	1
Banks	3½**
Barber shops and beauty parlors	3
Churches	1
Clubs	2
Court rooms	2
Dwelling units*	3
Garages—commercial (storage)	½
Hospitals	2
Hotels and motels, including apartment houses without provisions for cooking by tenants*	2
Industrial commercial (loft) buildings	2
Lodge rooms	1½
Office buildings	3½**
Restaurants	2
Schools	3
Stores	3
Warehouses (storage)	¼
In any of the above occupancies except one-family dwellings and individual dwelling units of two-family and multifamily dwellings:	
Assembly halls and auditoriums	1
Halls, corridors, closets, stairways	½
Storage spaces	¼

For SI units: one square foot = 0.093 square meter.

* All receptacle outlets of 20-amp or less rating in one-family, two-family, and multifamily dwellings and in guest rooms of hotels and motels [except those connected to the receptacle circuits specified in Sec. 220-3(b)] shall be considered as outlets for general illumination, and no additional load calculations shall be required for such outlets.

** In addition, a unit load of 1 volt-ampere per square foot shall be included for general-purpose-receptacle outlets when the actual number of general-purpose-receptacle-outlets is unknown.

TABLE 220-11 Lighting Load Feeder Demand Factors†

Type of occupancy	Portion of lighting load to which demand factor applies (volt-amperes)	Feeder demand factor, %
Dwelling units	First 3000 or less at	100
	From 3001 to 120 000 at	35
	Remainder over 120 000 at	25
Hospitals*	First 50 000 or less at	40
	Remainder over 50 000 at	20
Hotels and motels including apartment houses without provision for cooking by tenants*	First 20 000 or less at	50
	From 20 001 to 100 000 at	40
	Remainder over 100 000 at	30
Warehouses (storage)	First 12 500 or less at	100
	Remainder over 12 500 at	50
All others	Total volt-amperes	100

* The demand factors of this table shall not apply to the computed load of feeders to areas in hospitals, hotels, and motels where the entire lighting is likely to be used at one time, as in operating rooms, ballrooms, or dining rooms.

TABLE 220-18 Demand Factors for Household Electric Clothes Dryers†

Number of dryers	*Demand factor, %*	*Number of dryers*	*Demand factor, %*
1	100	10	50
2	100	11–13	45
3	100	14–19	40
4	100	20–24	35
5	80	25–29	32.5
6	70	30–34	30
7	65	35–39	27.5
8	60	40 and over	25
9	55		

TABLE 220-19 Demand Loads for Household Electric Ranges, Wall-Mounted Ovens, Counter-Mounted Cooking Units, and Other Household Cooking Appliances over $1^3/_4$-kW Rating*

(Column *A* to be used in all cases except as otherwise permitted in Note 3 on next page)

Number of appliances	Maximum demand (see notes)	Demand factors, % (see Note 3)	
	Column A (not over 12-kW rating)	Column B (less than $3\frac{1}{2}$-kW rating)	Column C ($3\frac{1}{2}$- to $8\frac{3}{4}$-kW rating)
1	8	80	80
2	11	75	65
3	14	70	55
4	17	66	50
5	20	62	45
6	21	59	43
7	22	56	40
8	23	53	36
9	24	51	35
10	25	49	34
11	26	47	32
12	27	45	32
13	28	43	32
14	29	41	32
15	30	40	32
16	31	39	28
17	32	38	28
18	33	37	28
19	34	36	28
20	35	35	28
21	36	34	26
22	37	33	26
23	38	32	26
24	39	31	26
25	40	30	26
26–30	15 plus 1 kW for each range	30	24
31–40		30	22
41–50	25 plus $\frac{3}{4}$ kW for each range	30	20
51–60		30	18
61 & over		30	16

See notes for this table on next page.

Notes to Table 220-19

Note 1. Over 12-kW through 27-kW ranges all of the same rating. For ranges, individually rated more than 12 kW but not more than 27 kW, the maximum demand in Column *A* shall be increased 5% for each additional kilowatt of rating or major fraction thereof by which the rating of individual ranges exceeds 12 kW.

Note 2. Over 12-kW through 27-kW ranges of *unequal ratings.* For ranges individually rated more than 12 kW and of different ratings but none exceeding 27 kW an average value of rating shall be computed by adding together the ratings of all ranges to obtain the total connected load (using 12 kW for any range rated less than 12 kW) and dividing by the total number of ranges; and then the maximum demand in Column *A* shall be increased 5% for each kW or major fraction thereof by which this average value exceeds 12 kW.

Note 3. Over 1¾ kW through 8¾ kW. In lieu of the method provided in Column *A*, it shall be permissible to add the name-plate ratings of all ranges rated more than 1¾ kW but not more than 8¾ kW and multiply the sum by the demand factors specified in Column *B* or *C* for the given number of appliances.

Note 4. Branch-circuit load. It shall be permissible to compute the branch-circuit load for one range in accordance with Table 220-19. The branch-circuit load for one wall-mounted oven or one counter-mounted cooking unit shall be the name-plate rating of the appliance. The branch-circuit load for a counter-mounted cooking unit and not more than two wall-mounted ovens, all supplied from a single branch circuit and located in the same room, shall be computed by adding the name-plate rating of the individual appliances and treating this total as equivalent to one range.

Note 5. This table also applies to household cooking appliances rated over 1¾ kW and used in instructional programs.

See Table 220-20 for commercial cooking equipment.

See Examples, Chap. 9.

TABLE 220-20 Feeder Demand Factors for Kitchen Equipment—Other than Dwelling Unit(s)†

Number of units of equipment	*Demand factor, %*
1	100
2	100
3	90
4	80
5	70
6 and over	65

Table 250-94
Grounding Electrode Conductor for AC Systems†

Size of Largest Service-Entrance Conductor or Equivalent Area for Parallel Conductors		Size of Grounding Electrode Conductor	
Copper	**Aluminum or Copper-Clad Aluminum**	**Copper**	***Aluminum or Copper-Clad Aluminum**
2 or smaller	0 or smaller	8	6
1 or 0	2/0 or 3/0	6	4
2/0 or 3/0	4/0 or 250 MCM	4	2
Over 3/0 thru 350 MCM	Over 250 MCM thru 500 MCM	2	0
Over 350 MCM thru 600 MCM	Over 500 MCM thru 900 MCM	0	3/0
Over 600 MCM thru 1100 MCM	Over 900 MCM thru 1750 MCM	2/0	4/0
Over 1100 MCM	Over 1750 MCM	3/0	250 MCM

Where there are no service-entrance conductors, the grounding electrode conductor size shall be determined by the equivalent size of the largest service-entrance conductor required for the load to be served.

* See installation restrictions in Section 250-92(a).

See Section 250-23(b).

Table 310-16. Ampacities of Insulated Conductors Rated 0-2000 Volts, 60° to 90°C‡

Not more than three conductors in raceway or cable or earth (directly buried), based on ambient temperature of 30°C (86°F)

Size	Temperature Rating of Conductor, See Table 310-13								Size
	60°C (140°F)	75°C (167°F)	85°C (185°F)	90°C (194°F)	60°C (140°F)	75°C (167°F)	85°C (185°F)	90°C (194°F)	
AWG MCM	TYPES †RUW, †T, †TW, †UF	TYPES †FEPW, †RH, †RHW, †RUH, †THW, †THWN, †XHHW, †USE, †ZW	TYPES V, MI	TYPES TA, TBS, SA, AVB, SIS, †FEP, †FEPB, †RHH †THHN, †XHHW*	TYPES †RUW, †T, †TW, †UF	TYPES †RH, †RHW, †RUH, †THW †THWN, †XHHW, †USE	TYPES V, MI	TYPES TA, TBS, SA, AVB, SIS, †RHH, †THHN, †XHHW*	AWG MCM
	COPPER				ALUMINUM OR COPPER-CLAD ALUMINUM				
18	....	....	....	14	....	....	....	....	....
16	....	....	18	18	....	....	....	....	....
14	20†	20†	25	25†	....	....	....	....	....
12	25†	25†	30	30†	20†	20†	25	25†	12
10	30	35†	40	40†	25	30†	30	35†	10
8	40	50	55	55	30	40	40	45	8
6	55	65	70	75	40	50	55	60	6
4	70	85	95	95	55	65	75	75	4
3	85	100	110	110	65	75	85	85	3
2	95	115	125	130	75	90	100	100	2
1	110	130	145	150	85	100	110	115	1
0	125	150	165	170	100	120	130	135	0
00	145	175	190	195	115	135	145	150	00
000	165	200	215	225	130	155	170	175	000
0000	195	230	250	260	150	180	195	205	0000
250	215	255	275	290	170	205	220	230	250
300	240	285	310	320	190	230	250	255	300
350	260	310	340	350	210	250	270	280	350
400	280	335	365	380	225	270	295	305	400
500	320	380	415	430	260	310	335	350	500
600	355	420	460	475	285	340	370	385	600
700	385	460	500	520	310	375	405	420	700
750	400	475	515	535	320	385	420	435	750
800	410	490	535	555	330	395	430	450	800
900	435	520	565	585	355	425	465	480	900
1000	455	545	590	615	375	445	485	500	1000
1250	495	590	640	665	405	485	525	545	1250
1500	520	625	680	705	435	520	565	585	1500
1750	545	650	705	735	455	545	595	615	1750
2000	560	665	725	750	470	560	610	630	2000

AMPACITY CORRECTION FACTORS

Ambient Temp. °C	For ambient temperatures other than 30°C, multiply the ampacities shown above by the appropriate factor shown below.								Ambient Temp. °F
31-40	.82	.88	.90	.91	.82	.88	.90	.91	87-104
41-45	.71	.82	.85	.87	.71	.82	.85	.87	105-113
46-50	.58	.75	.80	.82	.58	.75	.80	.82	114-122
51-60	....	.58	.67	.71	....	.58	.67	.71	123-141
61-70	....	.35	.52	.58	....	.35	.52	.58	142-158
71-80	....	....	.30	.41	....	....	.30	.41	159-176

† The overcurrent protection for conductor types marked with an obelisk (†) shall not exceed 15 amperes for 14 AWG, 20 amperes for 12 AWG, and 30 amperes for 10 AWG copper; or 15 amperes for 12 AWG and 25 amperes for 10 AWG aluminum and copper-clad aluminum after any correction factors for ambient temperature and number of conductors have been applied.

* For dry locations only. See 75°C column for wet locations.

Note: **Be sure** to read Notes following Table 310-19.

Table 310-17. Ampacities of Insulated Conductors Rated 0-2000 Volts, 60° to 90°C‡

Single conductors in free air, based on ambient temperature of 30°C (86°F)

Size	Temperature Rating of Conductor, See Table 310-13								Size
	60°C (140°F)	75°C (167°F)	85°C (185°F)	90°C (194°F)	60°C (140°F)	75°C (167°F)	85°C (185°F)	90°C (194°F)	
AWG MCM	TYPES †RUW, †T, †TW	TYPES †FEPW, †RH, †RHW, †RUH, †THW, †THWN, †XHHW, †ZW	TYPES V, MI	TYPES TA, TBS, SA, AVB, SIS, †FEP, †FEPB, †RHH †THHN, †XHHW*	TYPES †RUW, †T, †TW	TYPES †RH, †RHW, †RUH, †THW, †THWN, †XHHW	TYPES V, MI	TYPES TA, TBS, SA, AVB, SIS, †RHH, †THHN, †XHHW*	AWG MCM
	COPPER				ALUMINUM OR COPPER-CLAD ALUMINUM				
18	....	....	....	18	....	....	....	....	....
16	....	....	23	24	....	....	....	....	....
14	25†	30†	30	35†	....	....	....	....	....
12	30†	35†	40	40†	25†	30†	30	35†	12
10	40†	50†	55	55†	35†	40†	40	40†	10
8	60	70	75	80	45	55	60	60	8
6	80	95	100	105	60	75	80	80	6
4	105	125	135	140	80	100	105	110	4
3	120	145	160	165	95	115	125	130	3
2	140	170	185	190	110	135	145	150	2
1	165	195	215	220	130	155	165	175	1
0	195	230	250	260	150	180	195	205	0
00	225	265	290	300	175	210	225	235	00
000	260	310	335	350	200	240	265	275	000
0000	300	360	390	405	235	280	305	315	0000
250	340	405	440	455	265	315	345	355	250
300	375	445	485	505	290	350	380	395	300
350	420	505	550	570	330	395	430	445	350
400	455	545	595	615	355	425	465	480	400
500	515	620	675	700	405	485	525	545	500
600	575	690	750	780	455	540	595	615	600
700	630	755	825	855	500	595	650	675	700
750	655	785	855	885	515	620	675	700	750
800	680	815	885	920	535	645	700	725	800
900	730	870	950	985	580	700	760	785	900
1000	780	935	1020	1055	625	750	815	845	1000
1250	890	1065	1160	1200	710	855	930	960	1250
1500	980	1175	1275	1325	795	950	1035	1075	1500
1750	1070	1280	1395	1445	875	1050	1145	1185	1750
2000	1155	1385	1505	1560	960	1150	1250	1335	2000

AMPACITY CORRECTION FACTORS

Ambient Temp. °C	For ambient temperatures other than 30°C, multiply the ampacities shown above by the appropriate factor shown below.								Ambient Temp. °F
31-40	.82	.88	.90	.91	.82	.88	.90	.91	87-104
41-45	.71	.82	.85	.87	.71	.82	.85	.87	105-113
46-50	.58	.75	.80	.82	.58	.75	.80	.82	114-122
51-60	....	.58	.67	.71	....	.58	.67	.71	123-141
61-70	....	.35	.52	.58	....	.35	.52	.58	142-158
71-80	....	....	.30	.41	....	....	.30	.41	159-176

† The overcurrent protection for conductor types marked with an obelisk (†) shall not exceed 20 amperes for 14 AWG, 25 amperes for 12 AWG, and 40 amperes for 10 AWG copper, or 20 amperes for 12 AWG and 30 amperes for 10 AWG aluminum and copper-clad aluminum after any correction factor for ambient has been applied.

* For dry locations only. See 75°C column for wet locations.

Note: **Be sure** to read Notes following Table 310-19.

Table 310-18. Ampacities for Insulated Conductors Rated 0-2000 Volts, 110 to 250°C*

Not more than three conductors in raceway or cable based on ambient temperature of 30°C (86°F)

Size	Temperature Rating of Conductor. See Table 310-13								Size
AWG MCM	110°C (230°F)	125°C (257°F)	150°C (302°F)	200°C (392°F)	250°C (482°F)	110°C (230°F)	125°C (257°F)	200°C (392°F)	AWG MCM
	TYPES AVA, AVL	TYPES AI, AIA	TYPE Z	TYPES A, AA, FEP, FEPB, PFA	TYPES PFAH, TFE	TYPES AVA, AVL	TYPES AI, AIA	TYPES A, AA	
	COPPER				NICKEL OR NICKEL-COATED COPPER	ALUMINUM OR COPPER-CLAD ALUMINUM			
14	30	30	30	30	40	....	....	....	....
12	35	40	40	40	55	25	30	30	12
10	45	50	50	55	75	35	40	45	10
8	60	65	65	70	95	45	50	55	8
6	80	85	90	95	120	60	65	75	6
4	105	115	115	120	145	80	90	95	4
3	120	130	135	145	170	95	100	115	3
2	135	145	150	165	195	105	115	130	2
1	160	170	180	190	220	125	135	150	1
0	190	200	210	225	250	150	160	180	0
00	215	230	240	250	280	170	180	200	00
000	245	265	275	285	315	195	210	225	000
0000	275	310	325	340	370	215	245	270	0000
250	315	335	....	....	....	250	270	....	250
300	345	380	....	....	....	275	305	....	300
350	390	420	....	....	....	310	335	....	350
400	420	450	....	....	....	335	360	....	400
500	470	500	....	....	....	380	405	....	500
600	525	545	....	....	....	425	440	....	600
700	560	600	....	....	....	455	485	....	700
750	580	620	....	....	....	470	500	....	750
800	600	640	....	....	....	485	520	....	800
1000	680	730	....	....	....	560	600	....	1000
1500	785	....	....	....	....	650	....	....	1500
2000	840	....	....	....	....	705	....	....	2000
	AMPACITY CORRECTION FACTORS								
Ambient Temp. °C	For ambient temperatures other than 30°C, multiply the ampacities shown above by the appropriate factor shown below.								Ambient Temp. °F
31-40	.94	.95	.96	....	....	.94	.95	....	87-104
41-45	.90	.92	.94	....	....	.90	.92	....	105-113
46-50	.87	.89	.91	....	....	.87	.89	....	114-122
51-55	.83	.86	.89	....	....	.83	.86	....	123-131
56-60	.79	.83	.87	.91	.95	.79	.83	.91	132-141
61-70	.71	.76	.82	.87	.91	.71	.76	.87	142-158
71-75	.66	.72	.79	.86	.89	.66	.72	.86	159-167
76-80	.61	.68	.76	.84	.87	.61	.69	.84	168-176
81-90	.50	.61	.71	.80	.83	.50	.61	.80	177-194
91-100	....	.51	.65	.77	.80	....	.51	.77	195-212
101-120	....	....	.50	.69	.72	....	....	.69	213-248
121-140	....	....	.29	.59	.59	....	....	.59	249-284
141-160	....	....	....	....	.54	....	....	....	285-320
161-180	....	....	....	....	.50	....	....	....	321-356
181-200	....	....	....	....	.43	....	....	....	357-392
201-225	....	....	....	....	.30	....	....	....	393-437

Note: **Be sure** to read Notes following Table 310-19.

Table 310-19. Ampacities for Insulated Conductors Rated 0-2000 Volts, 110 to 250°C, and for Bare or Covered Conductors*

Single conductors in free air,
based on ambient temperature of 30°C (86°F)

Size	Temperature Rating of Conductor. See Table 310-13.										Size
AWG MCM	110°C (230°F) TYPES AVA, AVL	125°C (257°F) TYPES AI, AIA	150°C (302°F) TYPE Z	200°C (392°F) TYPES A, AA, FEP, FEPB, PFA	Bare or covered conductors	250°C (482°F) TYPES PFAH, TFE	110°C (230°F) TYPES AVA, AVL	125°C (257°F) TYPES AI, AIA	200°C (392°F) TYPES A, AA	Bare or covered conductors	AWG MCM
	COPPER					NICKEL OR NICKEL-COATED COPPER	ALUMINUM OR COPPER-CLAD ALUMINUM				
14	40	40	40	45	30	60	....	....	....	....	....
12	50	50	50	55	40	80	40	40	45	30	12
10	65	70	70	75	55	110	50	55	60	45	10
8	85	90	95	100	70	145	65	70	80	55	8
6	120	125	130	135	100	210	95	100	105	80	6
4	160	170	175	180	130	285	125	135	140	100	4
3	180	195	200	210	150	335	140	150	165	115	3
2	210	225	230	240	175	390	165	175	185	135	2
1	245	265	270	280	205	450	190	205	220	160	1
0	285	305	310	325	235	545	220	240	255	185	0
00	330	355	360	370	275	605	255	275	290	215	00
000	385	410	415	430	320	725	300	320	335	250	000
0000	445	475	490	510	370	850	345	370	400	290	0000
250	495	530	....	....	410	....	385	415	320	320	250
300	555	590	....	....	460	....	435	460	360	360	300
350	610	655	....	....	510	....	475	510	400	400	350
400	665	710	....	....	555	....	520	555	435	435	400
500	765	815	....	....	630	....	595	635	490	490	500
600	855	910	....	....	710	....	675	720	....	560	600
700	940	1005	....	....	780	....	745	795	....	615	700
750	980	1045	....	....	810	....	775	825	....	640	750
800	1020	1085	....	....	845	....	805	855	....	670	800
900	....	....	....	....	905	....	....	....	....	725	900
1000	1165	1240	....	....	965	....	930	990	....	770	1000
1500	1450	....	....	....	1215	....	1175	....	....	985	1500
2000	1715	....	....	....	1405	....	1425	....	....	1165	2000

AMPACITY CORRECTION FACTORS

Ambient Temp. °C	For ambient temperatures other than 30°C, multiply the ampacities shown above by the appropriate factor shown below.										Ambient Temp. °F
31-40	.94	.95	.96	....		....	.94	.95	....		87-104
41-45	.90	.92	.94	....		....	.90	.92	....		105-113
46-50	.87	.89	.91	....		....	.87	.89	....		114-122
51-55	.83	.86	.89	....		....	.83	.86	....		123-131
56-60	.79	.83	.87	.91		.95	.79	.83	.91		132-141
61-70	.71	.76	.82	.87		.91	.71	.76	.87		142-158
71-75	.66	.72	.79	.86		.89	.66	.72	.86		159-167
76-80	.61	.68	.76	.84		.87	.61	.69	.84		168-176
81-90	.50	.61	.71	.80		.83	.50	.61	.80		177-194
91-100	....	.51	.65	.77		.80	....	.51	.77		195-212
101-120	....	....	.50	.69		.72	....	....	.69		213-248
121-140	....	....	.29	.59		.59	....	....	.59		249-284
141-160	....	....	....	....		.54	....	....	....		285-320
161-180	....	....	....	....		.50	....	....	....		321-356
181-200	....	....	....	....		.43	....	....	....		357-392
201-225	....	....	....	....		.30	....	....	....		393-437

Note: **Be sure** to read Notes following Table 310-19.

Notes to Tables 310-16 through 310-19*

1. Explanation of Tables. For explanation of Type Letters, and for recognized size of conductors for the various conductor insulations, see Section 310-13. For installation requirements, see Sections 310-1 through 310-10, and the various articles of this Code. For flexible cords, see Tables 400-4 and 400-5.

3. Three-Wire, Single-Phase Dwelling Services. In dwelling units, conductors, as listed below, shall be permitted to be utilized as three-wire, single-phase, service-entrance conductors and the three-wire, single-phase feeder that carries the total current supplied by that service.†

Conductor Types and Sizes
RH-RHH-RHW-THW-THWN-THHN-XHHW

Copper	Aluminum and Copper-Clad AL	Service Rating in Amps
AWG	AWG	
4	2	100
3	1	110
2	1/0	125
1	2/0	150
1/0	3/0	175
2/0	4/0	200

4. Type MC Cable. The ampacities of Type MC cables are determined by the temperature limitation of the insulated conductors incorporated within the cable. Hence the ampacities of Type MC cable may be determined from the columns in Tables 310-16 and 310-18 applicable to the type of insulated conductors employed within the cable.

5. Bare Conductors. Where bare conductors are used with insulated conductors, their allowable ampacities shall be limited to that permitted for the insulated conductors of the same size.

6. Mineral-Insulated, Metal-Sheathed Cable. The temperature limitation on which the ampacities of mineral-insulated, metal-sheathed cable are based is determined by the insulating materials used in the end seal. Termination fittings incorporating unimpregnated, organic, insulating materials are limited to 85°C operation.

7. Type MTW Machine Tool Wire. The ampacities of Type MTW wire are specified in Table 200-B of the Electrical Standard for Metalworking Machine Tools and Plastics Machinery (NFPA 79-1980).

8. More Than Three Conductors in a Raceway or Cable. Where the number of conductors in a raceway or cable exceeds three, the ampacities given in Tables 310-16 and 310-18 shall be reduced as shown in the following table:

Number of Conductors	Percent of Values in Tables 310-16 and 310-18 as Adjusted for Ambient Temperature if Necessary
4 thru 6	80
7 thru 24	70
25 thru 42	60
43 and above	50

Where single conductors or multiconductor cables are stacked or bundled longer than 24 inches (610 mm) without maintaining spacing and are not installed in raceways, the ampacity of each conductor shall be reduced as shown in the above table.

Exception No. 1: When conductors of different systems, as provided in Section 300-3, are installed in a common raceway the derating factors shown above shall apply to the number of power and lighting (Articles 210, 215, 220, and 230) conductors only.

Exception No. 2: The derating factors

† *Note: These higher-ampacity conductors should not terminate in equipment unless the equipment has been tested at the higher temperatures that will be developed at the terminations.*

of Sections 210-22(c), 220-2(a) and 220-10(b) shall not apply when the above derating factors are also required.

Exception No. 3: For conductors installed in cable trays, the provisions of Section 318-10 shall apply.

Exception No. 4: Derating factors do not apply to conductors in nipples having a length not exceeding 24 inches (610 mm).

9. Overcurrent Protection. Where the standard ratings and settings of overcurrent devices do not correspond with the ratings and settings allowed for conductors, the next higher standard rating and setting shall be permitted.

Exception: As limited in Section 240-3.

10. Neutral Conductor.

(a) A neutral conductor which carries only the unbalanced current from other conductors, as in the case of normally balanced circuits of three or more conductors, shall not be counted when applying the provisions of Note 8.

(b) In a 3-wire circuit consisting of 2-phase wires and the neutral of a 4-wire, 3-phase wye-connected system, a common conductor carries approximately the same current as the other conductors and shall be counted when applying the provisions of Note 8.

(c) On a 4-wire, 3-phase wye circuit where the major portion of the load consists of electric-discharge lighting, data processing, or similar equipment, there are harmonic currents present in the neutral conductor and the neutral shall be considered to be a current-carrying conductor.

11. Grounding Conductor. A grounding conductor shall not be counted when applying the provisions of Note 8.

TABLE 430-148 Full-Load Currents in Amperes Single-Phase Alternating-Current Motors*

The following values of full-load currents are for motors running at usual speeds and motors with normal torque characteristics. Motors built for especially low speeds or high torques may have higher full-load currents, and multispeed motors will have full-load current varying with speed, in which case the name-plate current ratings shall be used.

To obtain full-load currents of 208- and 200-volt motors, increase corresponding 230-volt motor full-load currents by 10 and 15%, respectively.

The voltages listed are rated motor voltages. The currents listed shall be permitted for system voltage ranges of 110 to 120 and 220 to 240.

Horsepower	115 volts	230 volts
1/6	4.4	2.2
1/4	5.8	2.9
1/3	7.2	3.6
1/2	9.8	4.9
3/4	13.8	6.9
1	16	8
1 1/2	20	10
2	24	12
3	34	17
5	56	28
7 1/2	80	40
10	100	50

TABLE 430-150 Full-Load Current*
Three-Phase AC Motors‡

Horsepower	Induction type, squirrel-cage and wound rotor, amperes					Synchronous type, unity power factor, amperes†			
	115 volts	230 volts	460 volts	575 volts	2300 volts	230 volts	460 volts	575 volts	2300 volts
½	4	2	1	.8					
¾	5.6	2.8	1.4	1.1					
1	7.2	3.6	1.8	1.4					
1½	10.4	5.2	2.6	2.1					
2	13.6	6.8	3.4	2.7					
3		9.6	4.8	3.9					
5		15.2	7.6	6.1					
7½		22	11	9					
10		28	14	11					
15		42	21	17					
20		54	27	22					
25		68	34	27		53	26	21	
30		80	40	32		63	32	26	
40		104	52	41		83	41	33	
50		130	65	52		104	52	42	
60		154	77	62	16	123	61	49	12
75		192	96	77	20	155	78	62	15
100		248	124	99	26	202	101	81	20
125		312	156	125	31	253	126	101	25
150		360	180	144	37	302	151	121	30
200		480	240	192	49	400	201	161	40

For full-load currents of 208- and 200-volt motors, increase the corresponding 230-volt motor full-load current by 10 and 15%, respectively.

* These values of full-load current are for motors running at speeds usual for belted motors and motors with normal torque characteristics. Motors built for especially low speeds or high torques may require more running current, and multispeed motors will have full-load current varying with speed, in which case the name-plate current rating shall be used.

† For 90 and 80% power factor the above figures shall be multiplied by 1.1 and 1.25, respectively.

The voltages listed are rated motor voltages. The currents listed shall be permitted for system voltage ranges of 110 to 120, 220 to 240, 440 to 480, and 550 to 600 volts.

TABLE 430-151 Conversion Table of Locked-Rotor Currents for Selection of Disconnecting Means and Controllers as Determined from Horsepower and Voltage Rating†

For use only with Sections 430-110, 440-12, and 440-41.

Motor Locked-Rotor Current, Amperes*							
Single Phase		Two or Three Phase					Max. hp rating
115V	230V	115V	200V	230V	460V	575V	
58.8	29.4	24	18.8	12	6	4.8	½
82.8	41.4	33.6	19.3	16.8	8.4	6.6	¾
96	48	43.2	24.8	21.6	10.8	8.4	1
120	60	62	35.9	31.2	15.6	12.6	1½
144	72	81	46.9	40.8	20.4	16.2	2
204	102		66	58	26.8	23.4	3
336	168		105	91	45.6	36.6	5
480	240		152	132	66	54	7½
600	300		193	168	84	66	10
			290	252	126	102	15
			373	324	162	132	20
			469	408	204	162	25
			552	480	240	192	30
			718	624	312	246	40
			897	780	390	312	50
			1063	924	462	372	60
			1325	1152	576	462	75
			1711	1488	744	594	100
			2153	1872	936	750	125
			2484	2160	1080	864	150
			3312	2880	1440	1152	200

* These values of motor locked-rotor current are approximately six times the full-load current values given in Tables 430-148 and 430-150.

TABLE 430-152 Maximum Rating or Setting of Motor Branch-Circuit Short-Circuit and Ground-Fault Protective Devices‡

Type of motor	Percent of full-load current			
	Nontime-delay fuse	Dual-element (time-delay) fuse	Instantaneous trip breaker	Inverse time breaker*
Single-phase, all types				
No code letter	300	175	700	250
All AC single-phase and polyphase squirrel-cage and synchronous motors† with full-voltage, resistor or reactor starting:				
No code letter	300	175	700	250
Code letter F to V	300	175	700	250
Code letter B to E	250	175	700	200
Code letter A	150	150	700	150
All AC squirrel-cage and synchronous motors† with auto-transformer starting:				
Not more than 30 amp				
No code letter	250	175	700	200
More than 30 amp				
No code letter	200	175	700	200
Code letter F to V	250	175	700	200
Code letter B to E	200	175	700	200
Code letter A	150	150	700	150
High-reactance squirrel-cage				
Not more than 30 amp				
No code letter	250	175	700	250
More than 30 amp				
No code letter	200	175	700	200
Wound-rotor—No code letter	150	150	700	150
Direct-current (constant voltage)				
No more than 50 hp				
No code letter	150	150	250	150
More than 50 hp				
No code letter	150	150	175	150

For explanation of Code Letter Marking, see Table 430-7(b).

For certain exceptions to the values specified see Sections 430-52 through 430-54.

*The values given in the last column also cover the ratings of nonadjustable inverse-time types of circuit breakers that may be modified as in Section 430-52.

†Synchronous motors of the low-torque, low-speed type (usually 450 r/min or lower), such as are used to drive reciprocating compressors, pumps, etc., which start unloaded, do not require a fuse rating or circuit-breaker setting in excess of 200% of full-load current.

Code Chapter 9—Tables

Code Chapter 9 consists of two parts:

A. Tables not shown elsewhere in the Code. Only Tables 1, 3A, 3B, 3C, 4, 5, and 8 are of sufficient general interest to warrant repeating here. The others are of minor interest or concern projects beyond the scope of this book. However, *parts* of some of these tables have already been shown in the text of this book.

B. Examples. These show how to apply Code rules and will be presented in the pages following.

Notes to Tables*

1. Tables 3A, 3B, and 3C apply only to complete conduit or tubing systems and are not intended to apply to short sections of conduit or tubing used to protect exposed wiring from physical damage.

2. Equipment grounding conductors, when installed, shall be included when calculating conduit or tubing fill. The actual dimensions of the equipment grounding conductor (insulated or bare) shall be used in the calculation.

3. When conduit nipples having a maximum length not to exceed 24 inches (610 millimeters) are installed between boxes, cabinets, and simi-

lar enclosures, the nipple shall be filled to 60 percent of its total cross-sectional area, and Note 8 of Tables 310-16 through 310-19 does not apply to this condition.

4. For conductors not included in Chapter 9, such as compact or multiconductor cables, the actual dimensions shall be used.

5. See Table 1 for allowable percentage of conduit or tubing fill.†

Table 1 Percent of Cross Section of Conduit and Tubing for Conductors (See Table 2 for Fixture Wires)*

Number of conductors	1	2	3	4	Over 4
All conductor types except lead-covered (new or rewiring)	53	31	40	40	40
Lead-covered conductors	55	30	40	38	35

Note 1. See Tables 3A, 3B and 3C for numbers of conductors all of the same size in trade sizes of conduit ½ inch through 6 inch.

Note 2. For conductors larger than 750 MCM or for combinations of conductors of different sizes use Tables 4 through 8, Chapter 9, for dimensions of conductors, conduit and tubing.

Note 3. Where the calculated number of conductors, all of the same size, includes a decimal fraction, the next higher whole number shall be used where this decimal is 0.8 or larger.

Note 4. When bare conductors are permitted by other Sections of this Code, the dimensions for bare conductors in Table 8 of Chapter 9 shall be permitted.

Note 5. A multiconductor cable of two or more conductors shall be treated as a single-conductor cable for calculating percentage conduit fill area. For cables that have elliptical cross section, the cross-sectional area calculation shall be based on using the major diameter of the ellipse as a circle diameter.

† Table 1 is based on common conditions of proper cabling and alignment of conductors where the length of the pull and the number of bends are within reasonable limits. It should be recognized that for certain conditions a larger size conduit or a lesser conduit fill should be considered.

Table 3A Maximum Number of Conductors in Trade Sizes of Conduit or Tubing*
(Based on Table 1, Chapter 9)

Conduit Trade Size (Inches)		½	¾	1	1¼	1½	2	2½	3	3½	4	5	6
Type Letters	Conductor Size AWG, MCM												
TW, T, RUH,	14	9	15	25	44	60	99	142					
RUW,	12	7	12	19	35	47	78	111	171				
XHHW (14 thru 8)	10	5	9	15	26	36	60	85	131	176			
	8	2	4	7	12	17	28	40	62	84	108		
RHW and RHH	14	6	10	16	29	40	65	93	143	192			
(without outer	12	4	8	13	24	32	53	76	117	157			
covering),	10	4	6	11	19	26	43	61	95	127	163		
THW	8	1	3	5	10	13	22	32	49	66	85	133	
TW,	6	1	2	4	7	10	16	23	36	48	62	97	141
T,	4	1	1	3	5	7	12	17	27	36	47	73	106
THW,	3	1	1	2	4	6	10	15	23	31	40	63	91
RUH (6 thru 2),	2	1	1	2	4	5	9	13	20	27	34	54	78
RUW (6 thru 2),	1		1	1	3	4	6	9	14	19	25	39	57
FEPB (6 thru 2),	0		1	1	2	3	5	8	12	16	21	33	49
RHW and	00		1	1	1	3	5	7	10	14	18	29	41
RHH (with-	000		1	1	1	2	4	6	9	12	15	24	35
out outer	0000			1	1	1	3	5	7	10	13	20	29
covering)	250			1	1	1	2	4	6	8	10	16	23
	300			1	1	1	2	3	5	7	9	14	20
	350				1	1	1	3	4	6	8	12	18
	400				1	1	1	2	4	5	7	11	16
	500				1	1	1	1	3	4	6	9	14
	600					1	1	1	3	4	5	7	11
	700					1	1	1	2	3	4	7	10
	750					1	1	1	2	3	4	6	9

*Reprinted with permission from NFPA 70-1984, National Electrical Code®, Copyright © 1983, National Fire Protection Association, Quincy, Massachusetts 02269. This reprinted material is not the complete and official position of the NFPA on the referenced subject, which is represented only by the standard in its entirety.

TABLE 3B Maximum Number of Conductors in Trade Sizes of Conduit or Tubing*
(Based on Table 1, Chapter 9 of the Code)

Conduit Trade Size (Inches)		½	¾	1	1¼	1½	2	2½	3	3½	4	5	6
Type Letters	Conductor Size AWG, MCM												
	14	13	24	39	69	94	154						
	12	10	18	29	51	70	114	164					
THWN,	10	6	11	18	32	44	73	104	160				
	8	3	5	9	16	22	36	51	79	106	136		
THHN, FEP (14 thru 2),	6	1	4	6	11	15	26	37	57	76	98	154	
FEPB (14 thru 8),	4	1	2	4	7	9	16	22	35	47	60	94	137
PFA (14 thru 4/0)	3	1	1	3	6	8	13	19	29	39	51	80	116
PFAH (14 thru 4/0)	2	1	1	3	5	7	11	16	25	33	43	67	97
Z (14 thru 4/0)	1		1	1	3	5	8	12	18	25	32	50	72
XHHW (4 thru 500MCM)	0		1	1	3	4	7	10	15	21	27	42	61
	00		1	1	2	3	6	8	13	17	22	35	51
	000		1	1	1	3	5	7	11	14	18	29	42
	0000		1	1	1	2	4	6	9	12	15	24	35
	250			1	1	1	3	4	7	10	12	20	28
	300			1	1	1	3	4	6	8	11	17	24
	350			1	1	1	2	3	5	7	9	15	21
	400				1	1	1	3	5	6	8	13	19
	500				1	1	1	2	4	5	7	11	16
	600				1	1	1	1	3	4	5	9	13
	700					1	1	1	3	4	5	8	11
	750					1	1	1	2	3	4	7	11
XHHW	6	1	3	5	9	13	21	30	47	63	81	128	185
	600				1	1	1	1	3	4	5	9	13
	700					1	1	1	3	4	5	7	11
	750					1	1	1	2	3	4	7	10

Table 3C Maximum Number of Conductors in Trade Sizes of Conduit or Tubing* (Based on Table 1, Chapter 9)

Type Letters	Conductor Size AWG, MCM	½	¾	1	1¼	1½	2	2½	3	3½	4	5	6
	14	3	6	10	18	25	41	58	90	121	155		
	12	3	5	9	15	21	35	50	77	103	132		
RHW,	10	2	4	7	13	18	29	41	64	86	110		
	8	1	2	4	7	9	16	22	35	47	60	94	137
RHH	6	1	1	2	5	6	11	15	24	32	41	64	93
	4	1	1	1	3	5	8	12	18	24	31	50	72
(with	3	1	1	1	3	4	7	10	16	22	28	44	63
outer	2		1	1	3	4	6	9	14	19	24	38	56
covering)	1		1	1	1	3	5	7	11	14	18	29	42
	0		1	1	1	2	4	6	9	12	16	25	37
	00			1	1	1	3	5	8	11	14	22	32
	000			1	1	1	3	4	7	9	12	19	28
	0000			1	1	1	2	4	6	8	10	16	24
	250				1	1	1	3	5	6	8	13	19
	300				1	1	1	3	4	5	7	11	17
	350				1	1	1	2	4	5	6	10	15
	400				1	1	1	1	3	4	6	9	14
	500				1	1	1	1	3	4	5	8	11
	600					1	1	1	2	3	4	6	9
	700					1	1	1	1	3	3	6	8
	750						1	1	1	3	3	5	8

Tables 4 through 8, Chapter 9. Tables 4 through 8 give the nominal size of conductors and conduit or tubing for use in computing size of conduit or tubing for various combinations of conductors. The dimensions represent average conditions only, and variations will be found in dimensions of conductors and conduits of different manufacture.

Table 4 Dimensions and Percent Area of Conduit and of Tubing Areas of Conduit or Tubing for the Combinations of Wires Permitted in Table 1, Chapter 9*

Trade Size	Internal Diameter Inches	Area — Square Inches								
			Not Lead Covered			Lead Covered				
		Total 100%	2 Cond. 31%	Over 2 Cond. 40%	1 Cond. 53%	1 Cond. 55%	2 Cond. 30%	3 Cond. 40%	4 Cond. 38%	Over 4 Cond. 35%
½	.622	.30	.09	.12	.16	.17	.09	.12	.11	.11
¾	.824	.53	.16	.21	.28	.29	.16	.21	.20	.19
1	1.049	.86	.27	.34	.46	.47	.26	.34	.33	.30
1¼	1.380	1.50	.47	.60	.80	.83	.45	.60	.57	.53
1½	1.610	2.04	.63	.82	1.08	1.12	.61	.82	.78	.71
2	2.067	3.36	1.04	1.34	1.78	1.85	1.01	1.34	1.28	1.18
2½	2.469	4.79	1.48	1.92	2.54	2.63	1.44	1.92	1.82	1.68
3	3.068	7.38	2.29	2.95	3.91	4.06	2.21	2.95	2.80	2.58
3½	3.548	9.90	3.07	3.96	5.25	5.44	2.97	3.96	3.76	3.47
4	4.026	12.72	3.94	5.09	6.74	7.00	3.82	5.09	4.83	4.45
5	5.047	20.00	6.20	8.00	10.60	11.00	6.00	8.00	7.60	7.00
6	6.065	28.89	8.96	11.56	15.31	15.89	8.67	11.56	10.98	10.11

Table 5 Dimensions of Rubber-Covered and Thermoplastic-Covered Conductors‡

Size AWG MCM	Types RFH-2, RH, RHH,*** RHW,*** SF-2		Types TF, T, THW,† TW, RUH,** RUW**		Types TFN, THHN, THWN		Types**** FEP, FEPB, FEPW, TFE, PF, PFA, PFAH, PGF, PTF, Z, ZF, ZFF		Type XHHW, ZW††		Types KF-1, KF-2, KFF-1, KFF-2	
	Approx. Diam. Inches	Approx. Area Sq. In.	Approx. Diam. Inches	Approx. Area Sq. In.	Approx. Diam. Inches	Approx. Area Sq. In.	Approx. Diam. Inches	Approx. Area Sq. Inches	Approx. Diam. Inches	Approx. Area Sq. In.	Approx. Diam. Sq. In.	Approx. Area Sq. In.
Col. 1	Col. 2	Col. 3	Col. 4	Col. 5	Col. 6	Col. 7	Col. 8	Col. 9	Col. 10	Col. 11	Col. 12	Col. 13
18	.146	.0167	.106	.0088	.089	.0062	.081	.0052	...	....	.065	.0033
16	.158	.0196	.118	.0109	.100	.0079	.092	.0066	...	....	.070	.0038
14	30 mils .171	.0230	.131	.0135	.105	.0087	105 .105	.0087 .0087	...	....	.083	.0054
14	45 mils .204*	.0327*	...	....	...	....	... ...	....	...	....	...	....
14	....	....	.162†	.0206†	...	....	... ...	....	.129	.0131	...	....
12	30 mils .188	.0278	.148	.0172	.122	.0117	.121 .121	.0115 .0115	...	....	.102	.0082
12	45 mils .221*	.0384*	...	....	...	....	... ...	....	...	....	...	....
12	....	....	.179†	.0252†	...	....	... ...	....	.146	.0167	...	....
10	...... .242	.0460	.168	.0222	.153	.0184	.142 .142	.0158 .0158	...	....	.124	.0121
10	....	....	.199†	.0311†	...	....	... ...	....	.166	.0216	...	....
8	...... .328	.0845	.245	.0471	.218	.0373	.206 .186	.0333 .0272	...	....	...	....
8	....	....	.276†	.0598†	...	....	... ...	....	.241	.0456	...	....
6	.397	.1238	.323	.0819	.257	.0519	.244 .302	.0468 .0716	.282	.0625	...	....
4	.452	.1605	.372	.1087	.328	.0845	.292 .350	.0670 .0962	.328	.0845	...	....
3	.481	.1817	.401	.1263	.356	.0995	.320 .378	.0804 .1122	.356	.0995	...	....
2	.513	.2067	.433	.1473	.388	.1182	.352 .410	.0973 .1320	.388	.1182	...	....
1	.588	.2715	.508	.2027	.450	.1590	.420 ...	.1385	.450	.1590	...	....
0	.629	.3107	.549	.2367	.491	.1893	.462 ...	.1676	.491	.1893	...	....
00	.675	.3578	.595	.2781	.537	.2265	.498 ...	.1948	.537	.2265	...	....
000	.727	.4151	.647	.3288	.588	.2715	.560 ...	.2463	.588	.2715	...	....
0000	.785	.4840	.705	.3904	.646	.3278	.618 ...	.3000	.646	.3278	...	....

‡Reprinted with permission from NFPA 70-1984, National Electrical Code®, Copyright© 1983, National Fire Protection Association, Quincy, Massachusetts 02269. This reprinted material is not the complete and official position of the NFPA on the referenced subject, which is represented only by the standard in its entirety.

Table 5 (Continued)‡

Size AWG MCM	Types RFH-2, RH, RHH,*** RHW,*** SF-2		Types TF, T, THW,† TW, RUH,** RUW**		Types TFN, THHN, THWN		Types**** FEP, FEPB, FEPW, TFE, PF, PFA, PFAH, PGF, PTF, Z, ZF, ZFF		Type XHHW, ZW††	
	Approx. Diam. Inches	Approx. Area Sq. In.	Approx. Diam. Inches	Approx. Area Sq. In.	Approx. Diam. Inches	Approx. Area Sq. In.	Approx. Diam. Inches	Approx. Area Sq. Inches	Approx. Diam. Inches	Approx. Area Sq. In.
Col. 1	Col. 2	Col. 3	Col. 4	Col. 5	Col. 6	Col. 7	Col. 8	Col. 9	Col. 10	Col. 11
250	.868	.5917	.788	.4877	.716	.4026	...	....	.716	.4026
300	.933	.6837	.843	.5581	.771	.4669	...	....	.771	.4669
350	.985	.7620	.895	.6291	.822	.5307	...	....	.822	.5307
400	1.032	.8365	.942	.6969	.869	.5931	...	....	.869	.5931
500	1.119	.9834	1.029	.8316	.955	.7163	...	....	.955	.7163
600	1.233	1.1940	1.143	1.0261	1.058	.8791	...	....	1.073	.9043
700	1.304	1.3355	1.214	1.1575	1.129	1.0011	...	....	1.145	1.0297
750	1.339	1.4082	1.249	1.2252	1.163	1.0623	...	....	1.180	1.0936
800	1.372	1.4784	1.282	1.2908	1.196	1.1234	...	....	1.210	1.1499
900	1.435	1.6173	1.345	1.4208	1.259	1.2449	...	....	1.270	1.2668
1000	1.494	1.7530	1.404	1.5482	1.317	1.3623	...	....	1.330	1.3893
1250	1.676	2.2062	1.577	1.9532	...	....	...	....	1.500	1.7671
1500	1.801	2.5475	1.702	2.2751	...	....	...	....	1.620	2.0612
1750	1.916	2.8832	1.817	2.5930	...	....	...	....	1.740	2.3779
2000	2.021	3.2079	1.922	2.9013	...	....	...	....	1.840	2.6590

* The dimensions of Types RHH and RHW.
** No. 14 to No. 2.
† Dimensions of THW in sizes No. 14 to No. 8. No. 6 THW and larger is the same dimension as T.
*** Dimensions of RHH and RHW without outer covering are the same as THW No. 18 to No. 10, solid; No. 8 and larger, stranded.
**** In Columns 8 and 9 the values shown for sizes No. 1 thru 0000 are for TFE and Z only. The right-hand values in Columns 8 and 9 are for FEPB, Z, ZF, and ZFF only.
†† No. 14 to No. 2.

Table 8 Conductor Properties*

Size AWG/ MCM	Area Cir. Mils	DC Resistance at 75°C, 167°F Stranding Quantity	Stranding Diam. In.	Overall Diam. In.	Overall Area In.^{2}	Copper Uncoated ohm/MFT	Copper Coated ohm/MFT	Aluminum ohm/MFT
18	1620	1	—	0.040	0.001	7.77	8.08	12.8
18	1620	7	0.015	0.046	0.002	7.95	8.45	13.1
16	2580	1	—	0.051	0.002	4.89	5.08	8.05
16	2580	7	0.019	0.058	0.003	4.99	5.29	8.21
14	4110	1	—	0.064	0.003	3.07	3.19	5.06
14	4110	7	0.024	0.073	0.004	3.14	3.26	5.17
12	6530	1	—	0.081	0.005	1.93	2.01	3.18
12	6530	7	0.030	0.092	0.006	1.98	2.05	3.25
10	10380	1	—	0.102	0.008	1.21	1.26	2.00
10	10380	7	0.038	0.116	0.011	1.24	1.29	2.04
8	16510	1	—	0.128	0.013	0.764	0.786	1.26
8	16510	7	0.049	0.146	0.017	0.778	0.809	1.28
6	26240	7	0.061	0.184	0.027	0.491	0.510	0.808
4	41740	7	0.077	0.232	0.042	0.308	0.321	0.508
3	52620	7	0.087	0.260	0.053	0.245	0.254	0.403
2	66360	7	0.097	0.292	0.067	0.194	0.201	0.319
1	83690	19	0.066	0.332	0.087	0.154	0.160	0.253
1/0	105600	19	0.074	0.373	0.109	0.122	0.127	0.201
2/0	133100	19	0.084	0.419	0.138	0.0967	0.101	0.159
3/0	167800	19	0.094	0.470	0.173	0.0766	0.0797	0.126
4/0	211600	19	0.106	0.528	0.219	0.0608	0.0626	0.100
250	—	37	0.082	0.575	0.260	0.0515	0.0535	0.0847
300	—	37	0.090	0.630	0.312	0.0429	0.0446	0.0707
350	—	37	0.097	0.681	0.364	0.0367	0.0382	0.0605
400	—	37	0.104	0.728	0.416	0.0321	0.0331	0.0529
500	—	37	0.116	0.813	0.519	0.0258	0.0265	0.0424
600	—	61	0.992	0.893	0.626	0.0214	0.0223	0.0353
700	—	61	0.107	0.964	0.730	0.0184	0.0189	0.0303
750	—	61	0.111	0.998	0.782	0.0171	0.0176	0.0282
800	—	61	0.114	1.03	0.834	0.0161	0.0166	0.0265
900	—	61	0.122	1.09	0.940	0.0143	0.0147	0.0235
1000	—	61	0.128	1.15	1.04	0.0129	0.0132	0.0212
1250	—	91	0.117	1.29	1.30	0.0103	0.0106	0.0169
1500	—	91	0.128	1.41	1.57	0.00858	0.00883	0.0141
1750	—	127	0.117	1.52	1.83	0.00735	0.00756	0.0121
2000	—	127	0.126	1.63	2.09	0.00643	0.00662	0.0106

These resistance values are valid ONLY for the parameters as given. Using conductors having coated strands, different stranding type, and especially, other temperatures, change the resistance.

Formula for temperature change: $R_2 = R_1 [1+\alpha(T_2-20)]$ where: $\alpha_{cu} = 0.00393$, $\alpha_{AL} = 0.00403$.

Class B stranding is listed as well as solid for some sizes. Its overall diameter and area is that of its circumscribing circle. The construction information is per NEMA WC8-1976 (Rev 5-1980). The resistance is calculated per National Bureau of Standards Handbook 100, dated 1966, and Handbook 109, dated 1972.

Conductors with compact and compressed stranding have about 9 percent and 3 percent, respectively, smaller bare conductor diameters than those shown.

The IACS conductivities used: bare copper = 100%, aluminum = 61%.

Code Chapter 9—Examples*

Selection of Conductors. In the following examples, the results are generally expressed in amperes. To select conductor sizes refer to Tables 310-16 through 310-19 and the Notes that pertain to such tables.

Voltage. For uniform application of Arts. 210, 215, and 220 a nominal voltage of 120, 120/240, 240, and 208Y/120 volts shall be used in computing the ampere load on the conductor.

Fractions of an Ampere. Except where the computations result in a major fraction of an ampere (0.5 or larger), such fractions may be dropped.

Ranges. For the computation of the range loads in these examples Col. A of Table 220-19 has been used. For optional methods, see Cols. B and C of Table 220-19.

SI Units. For SI units, one square foot = 0.093 square meter; one foot = 0.3048 meter.

Example No. 1(a). One-Family Dwelling†

The dwelling has a floor area of 1500 sq. ft. exclusive of unoccupied cellar, unfinished attic, and open porches. Appliances are a 12-kW range and a 5.5 kW, 240-volt dryer. Assume range and dryer kW ratings equivalent to kVA ratings in accordance with Sections 220-18 and 220-19.

Computed Load [see Section 220-10(a)]:

General Lighting Load:

1500 sq. ft. at 3 volt-amperes per sq. ft. = 4500 volt-amperes.

Minimum Number of Branch Circuits Required [see Section 220-2(b)]:

General Lighting Load:

4500 volt-amperes ÷ 120 volts = 37.5 A: This requires three 15 A 2-wire or two 20 A 2-wire circuits

Small Appliance Load: Two 2-wire 20 A circuits [see Section 220-3(b)]

Laundry Load: One 2-wire 20 A circuit [see Section 220-3(c)]

Minimum Size Feeder Required [see Section 220-10(a)]:

General Lighting	4500 volt-amperes
Small Appliance Load	3000 volt-amperes
Laundry	1500 volt-amperes
Total General Light & Small Appliance	9000 volt-amperes
3000 volt-amperes at 100%	3000 volt-amperes
9000 − 3000 = 6000 volt-amperes at 35%	2100 volt-amperes
Net General Lighting & Small Appliance Load	5100 volt-amperes
Range Load (see Table 220-19)	8000 volt-amperes
Dryer Load (see Table 220-18)	5500 volt-amperes
Total Load	18,600 volt-amperes

For 120/240-volt 3-wire single-phase service or feeder, 18,600 volt-amperes ÷ 240 volts = 77.5 A.

Net computed load exceeds 10 kVA. Service conductors shall be 100 amperes [see Section 230-42(b)(2)].

Neutral for Feeder and Service

Lighting and Small Appliance Load	5100 volt-amperes
Range Load 8000 volt-amperes at 70%	5600 volt-amperes
Dryer Load 5500 volt-amperes at 70%	3850 volt-amperes
Total	14,550 volt-amperes

14,550 VA ÷ 240 V = 60.6 amperes

Example No. 1(b). One-Family Dwelling†

Same conditions as Example No. 1(a), plus addition of one 6-ampere 230-volt room air-conditioning unit and one 12-ampere 115-volt room air-conditioning unit*, one 8-ampere 115-volt rated disposal and one 10-ampere 120-volt rated dishwasher*. See Article 430 for general motors and Article 440, Part G for air-conditioning equipment. Motors have nameplate ratings of 115 V and 230 V for use on 120 V and 240 V nominal voltage systems.

From previous Example No. 1(a), feeder current is 78 amperes (3-wire 240 volts).

	Line A	Neutral	Line B
Amperes from Example No. 1(a)	78	61	78
One 230 V air conditioner	6	—	6
One 115 V air conditioner and 120 V dishwasher	12	12	10
One 115 V disposal	—	8	8
25% of largest motor (Section 430-24)	3	3	2
Amperes per line	99	84	104

* For feeder neutral, use largest of the two appliances for unbalance.

Example No. 2(a). Optional Calculation for One-Family Dwelling Heating Larger than Air-Conditioning†

(See Section 220-30.)

Dwelling has a floor area of 1500 sq. ft. exclusive of unoccupied cellar, unfinished attic, and open porches. It has a 12-kW range, a 2.5-kW water heater, a 1.2-kW dishwasher, 9 kW of electric space heating installed in five rooms, a 5-kW clothes dryer, and a 6-ampere 230-volt room air-conditioning unit. Assume range, water heater, dishwasher, space heating, and clothes dryer kW ratings equivalent to kVA.

Air conditioner kVA is 6 × 230 ÷ 1000 = 1.38 kVA

1.38 kVA is less than the connected load of 9 kVA of space heating; therefore, the air conditioner load need not be included in the service calculation (see Section 220-21).

1500 sq. ft. at 3 volt-amperes	4.5 kVA
Two 20-amp. appliance outlet circuits at 1500 volt-amperes each	3.0 kVA
Laundry circuit	1.5 kVA
Range (at nameplate rating)	12.0 kVA
Water heater	2.5 kVA
Dishwasher	1.2 kVA
Space heating	9.0 kVA
Clothes dryer	5.0 kVA
	38.7 kVA

First 10 kVA at 100% = 10.00 kVA
Remainder at 40% (28.7 kVA × .4) = 11.48 kVA

Calculated load for service size 21.48 kVA = 21,480 volt-amperes
21,480 VA ÷ 240 volts = 89.5 amperes

Therefore, this dwelling may be served by a 100-ampere service.

Feeder Neutral Load, per Section 220-22:

1500 sq. ft. at 3 volt-amperes	4500 volt-amperes
Three 20-amp. circuits at 1500 volt-amperes	4500 volt-amperes
Total	9000 volt-amperes
3000 volt-amperes at 100%	3000 volt-amperes

[*Example 2(a) continued next page*]

Example No. 2(a) (Continued)

9000 VA−3000 VA = 6000 volt-amperes at 35%..........	2100 volt-amperes
	5100 volt-amperes
Range—8 kVA at 70%	5600 volt-amperes
Clothes dryer—5 kVA at 70%	3500 volt-amperes
Dishwasher	1200 volt-amperes
Total	15,400 volt-amperes

15,400 VA ÷ 240 volts = 64.2 amperes

Example No. 2(b). Optional Calculation for One-Family Dwelling Air Conditioning Larger than Heating†

(See Section 220-30.)

Dwelling has a floor area of 1500 sq. ft. exclusive of unoccupied cellar, unfinished attic, and open porches. It has two 20-ampere small appliance circuits, one 20-ampere laundry circuit, two 4-kW wall-mounted ovens, one 5.1-kW counter-mounted cooking unit, a 4.5-kW water heater, a 1.2-kW dishwasher, a 5-kW combination clothes washer and dryer, six 7-ampere 230-volt room air-conditioning units, and a 1.5-kW permanently installed bathroom space heater. Assume wall-mounted ovens, counter-mounted cooking unit, water heater, dishwasher and combination clothes washer and dryer kW ratings equivalent to kVA.

Air Conditioning kVA Calculation:

Total amperes 6 × 7 = 42.00 amperes

42 × 240 ÷ 1000 = 10.08 kVA of air-conditioned load assume P.F. = 1.0

Load Included at 100%:

Air conditioning	10.08 kVA
Space heater (omit, see Section 220-21)	

Other Load:

	kVA
1500 sq. ft. at 3 volt-amperes	4.5
Two 20-amp. small appliance circuits at 1500 volt-amperes	3.0
Laundry circuit	1.5
Two ovens	8.0
One cooking unit	5.1
Water heater	4.5
Dishwasher	1.2
Washer/dryer	5.0
Total other load	32.8

1st 10 kVA at 100%	10.0 kVA
Remainder at 40% (22.8 kVA × .4)	9.12 kVA
Total calculated load	29.2 kVA = 29,200 volt-amperes

29,200 VA ÷ 240 V = 122 amperes (service rating)

Feeder Neutral Load, per Section 220-22:

(It is assumed that the two 4 kVA wall-mounted ovens are supplied by one branch circuit, the 5.1 kVA counter-mounted cooking unit by a separate circuit.)

1500 sq. ft. at 3 volt-amperes	4500 volt-amperes
Three 20-amp. circuits at 1500 volt-amperes	4500 volt-amperes
Total	9000 volt-amperes
3000 volt-amperes at 100%	3000 volt-amperes
9000 VA−3000 VA = 6000 volt-amperes at 35%	2100 volt-amperes
	5100 volt-amperes

Two 4 kVA ovens plus one 5.1 kVA cooking unit totals 13.1 kVA	
Table 220-19 permits 55% demand factor	
13.1 kVA × .55 = 7.2 kVA feeder capacity	
7200 VA × 70% for neutral load	5040 volt-amperes
Clothes washer/dryer — 5 kVA × 70% for neutral load	3500 volt-amperes
Dishwasher	1200 volt-amperes
Total	14,840 volt-amperes

14,840 VA ÷ 240 V = 61.83, use 62 amperes

Example No. 3. Store Building†

A store 50 ft. by 60 ft., or 3000 sq. ft., has 30 ft. of show window. There are a total of 80 duplex receptacles. The service is 120/240-volt, single-phase (3-wire service).

Computed Load (Section 220-10):

* General Lighting Load:	
3000 sq. ft. at 3 volt-amperes per sq. ft. × 1.25	11,250 volt-amperes
Show Window Lighting Load:	
30 ft. at 200 volt-amperes per foot	6000 volt-amperes
Receptacle Load (Section 220-13)	
80 receptacles at 180 VA = 14,400 VA	
10,000 VA at 100%	10,000 volt-amperes
(14,400—10,000) VA at 50%	2200 volt-amperes
Outside sign circuit 1200 volt-amperes [Section 600-6(c)].	1,200 volt-amperes
Total	30,650 volt-amperes

Minimum Number of Branch Circuits Required:

General Lighting Load:

11,250 volt-amperes ÷ 240 volts = 47 amperes for 3-wire, 120/240.

The lighting load may be served by 2-wire or 3-wire 15- or 20-ampere circuits with combined capacity equal to 47 amperes or greater for 3-wire circuits or 94 amperes or greater for 2-wire circuits.

Show Window:

6000 volt-amperes ÷ 240 volts = 25 amperes for 3-wire, 120/240.

The show window lighting may be served by 2-wire or 3-wire circuits with a capacity equal to 25 amperes or greater for 3-wire circuits or 50 amperes or greater for 2-wire circuits.

[*Example 3 continued next page*]

Example No. 3 (Continued)

Receptacles required by Section 210-62 are assumed to be included in the receptacle load above if these receptacles do not supply the show window lighting load.

Receptacle Load: 14,400 volt-amperes ÷ 240 volts = 60 amperes for 3-wire, 120/240.

The receptacle load may be served by 2-wire or 3-wire circuits with a capacity equal to 60 amperes or greater for 3-wire circuits or 120 amperes or greater for 2-wire circuits.

Minimum Size Feeders (or Service Conductors) Required (Section 215-3):

For 120/240-volt, 3-wire system:

30,650 volt-amperes ÷ 240 volts = 128 amperes

* The above examples assume that the entire lighting load is continuous. The general lighting load is increased by 25 percent in accordance with Section 220-2. No branch circuit may serve a continous lighting load greater than 80 percent of its rating.

Example No. 4(a). Multifamily Dwelling†

Multifamily dwelling having 40 dwelling units.

Meters in two banks of 20 each and individual subfeeders to each dwelling unit.

One-half of the dwelling units are equipped with electric ranges not exceeding 12 kW each. Assume range kW rating equivalent to kVA rating in accordance with Section 220-19. Other half of ranges are gas ranges.

Area of each dwelling unit is 840 sq. ft.

Laundry facilities on premises available to all tenants. Add no circuit to individual dwelling unit. Add 1500 volt-amperes for each laundry circuit to house load and add to the example as a "house load."

Computed Load for Each Dwelling Unit (Article 220):

General Lighting Load:

840 sq. ft. at 3 volt-amperes per sq. ft. .... 2520 volt-amperes

Special Appliance Load:

Electric Range (Section 220-19) .... 8000 volt-amperes

Minimum Number of Branch Circuits Required for Each Dwelling Unit (Section 220-3):

General Lighting Load: 2520 volt-amperes ÷ 120 volts = 21 amperes or two 15-ampere, 2-wire circuits; or two 20-ampere, 2-wire circuits.

Small Appliance Load: Two 2-wire circuits of No. 12 wire. [See Section 220-3(b).]

Range Circuit: 8000 volt-amperes ÷ 240 = 33 amperes or a circuit of two No. 8 and one No. 10 as permitted by Section 220-22. (See Section 210-19.)

Minimum Size Subfeeder Required for Each Dwelling Unit (Section 215-2):

Computed Load (Article 220):

General Lighting Load	2520 volt-amperes
Small Appliance Load, two 20-ampere circuits	3000 volt-amperes
Total Computed Load (without ranges)	5520 volt-amperes

Application of Demand Factor:

3000 volt-amperes at 100%	3000 volt-amperes
2520 volt-amperes at 35%	882 volt-amperes
Net Computed Load (without ranges)	3882 volt-amperes
Range Load	8000 volt-amperes
Net Computed Load (with ranges)	11,882 volt-amperes

†Reprinted with permission from NFPA 70-1984, National Electrical Code®, Copyright© 1983, National Fire Protection Association, Quincy, Massachusetts 02269. This reprinted material is not the complete and official position of the NFPA on the referenced subject, which is represented only by the standard in its entirety.

Size of Each Subfeeder (see Section 215-2).

For 120/240-volt, 3-wire system (without ranges):

Net Computed Load, 3882 volt-amperes ÷ 240 volts = 16.2 amperes.

For 120/240-volt, 3-wire system (with ranges):

Net Computed Load, 11,882 volt-amperes ÷ 240 volts = 49.5 amperes.

Subfeeder Neutral:

Lighting and Small Appliance Load	3882 volt-amperes
Range Load, 8000 volt-amperes at 70% (see Section 220-22)	5600 volt-amperes
(Not included for apartments without electric range)	
Net Computed Load (neutral)	9482 volt-amperes

9482 volt-amperes ÷ 240 volts = 39.5 amperes

Minimum Size Feeders Required from Service Equipment to Meter Bank (For 20 Dwelling Units — 10 with Ranges):

Total Computed Load:

Lighting and Small Appliance Load, 20 × 5520 volt-amperes	110,400 volt-amperes
Application of Demand Factor:	
3000 volt-amperes at 100%	3000 volt-amperes
107,400 volt-amperes at 35%	37,950 volt-amperes
Net Computed Lighting and Small Appliance Load	40,590 volt-amperes
Range Load, 10 ranges (less than 12 kVA, Col. A, Table 220-19)	25,000 volt-amperes
Net Computed Load (with ranges)	65,590 volt-amperes

For 120/240-volt, 3-wire system:

Net Computed Load, 65,590 volt-amperes ÷ 240 volt = 273 amperes.

Feeder Neutral:

Lighting and Small Appliance Load	40,590 volt-amperes
Range Load: 25,000 volt-amperes at 70% (see Section 220-22)	17,500 volt-amperes
Computed Load (neutral)	58,090 volt-amperes

58,090 volt-amperes ÷ 240 volts = 242 amperes.

Further Demand Factor (Section 220-22):

200 amperes at 100%	200 amperes
42 amperes at 70%	29 amperes
Net Computed Load (neutral)	229 amperes

Minimum Size Main Feeder (or Service Conductors) Required (less house load). (For 40 Dwelling Units — 20 with Ranges):

Total Computed Load:

Lighting and Small Appliance Load, 40 × 5520 volt-amperes	220,800 volt-amperes
Application of Demand Factor:	
3000 volt-amperes at 100%	3000 volt-amperes
117,000 volt-amperes at 35%	40,950 volt-amperes
100,800 volt-amperes at 25%	25,200 volt-amperes
Net Computed Lighting and Small Appliance Load	69,150 volt-amperes
Range Load, 20 ranges (less than 12 kVA, Col. A, Table 220-19)	35,000 volt-amperes
Net Computed Load	104,150 volt-amperes

For 120/240-volt, 3-wire system:

Net Computed Load, 104,150 volt-amperes ÷ 240 volts = 434 amperes.

[*Example 4(a) continued next page*]

Example No. 4(a) (Continued)

Feeder Neutral:

Lighting and Small Appliance Load	69,150 volt-amperes
Range Load, 35,000 volt-amperes at 70% (see Section 220-22)	24,500 volt-amperes
Computed Load (neutral)	93,650 volt-amperes

93,650 volt-amperes ÷ 240 volts = 390 amperes.

Further Demand Factor (see Section 220-22):

200 amperes at 100%	200 amperes
190 amperes at 70%	133 amperes
Net Computed Load (neutral)	333 amperes

See Tables 310-16 through 310-19, Notes 8 and 10.

Example No. 4(b). Optional Calculation for Multifamily Dwelling†

Multifamily dwelling equipped with electric cooking and space heating or air conditioning and having 40 dwelling units.

Meters in two banks of 20 each plus house metering and individual subfeeders to each dwelling unit.

Each dwelling unit is equipped with an electric range of 8-kW nameplate rating, four 1.5 kW separately controlled 240-volt electric space heaters, and a 2.5-kW 240-volt electric water heater. Assume range, space heater, and water heater kW ratings equivalent to kVA.

A common laundry facility is available to all tenants [Section 210-52(e), Exception No. 1].

Area of each dwelling unit is 840 square feet.

Computed Load for Each Dwelling Unit (Article 220):

General Lighting Load:

840 sq. ft. at 3 volt-amperes per sq. ft.	2520 volt-amperes
Electric Range	8000 volt-amperes
Electric Heat 6 kVA (or air conditioning if larger)	6000 volt-amperes
Electric Water Heater	2500 volt-amperes

Minimum Number of Branch Circuits Required for Each Dwelling Unit:

General Lighting Load 2520 volt-amperes ÷ 120 volts = 21 amperes or two 15-ampere 2-wire circuits or two 20-ampere 2-wire circuits.

Small Appliance Loads: Two 2-wire circuits of No. 12 [see Section 220-3(b)].

Range circuit 8000 volt-amperes × 80% ÷ 240 volts = 27 amperes on a circuit of three No. 10 AWG as permitted in Column C of Table 220-19.

Space Heating 6000 volt-amperes ÷ 240 volts = 25 amperes

No. of circuits (see Section 220-3).

Minimum Size Subfeeder Required for Each Dwelling Unit (Section 215-2):

Computed Load (Article 220):

General Lighting Load	2520 volt-amperes
Small Appliance Load, two 20-ampere circuits	3000 volt-amperes
Total Computed Load (without range and space heating)	5520 volt-amperes

Application of Demand Factor:

3000 volt-amperes at 100%	3000 volt-amperes
2520 volt-amperes at 35%	882 volt-amperes
Net Computed Load (without range and space heating)	3882 volt-amperes
Range Load	6400 volt-amperes
Space Heating (Section 220-15)	6000 volt-amperes
Water Heater	2500 volt-amperes
Net Computed Load for individual dwelling unit	18,782 volt-amperes

For 120/240-volt 3-wire system
Net Computed Load 18,782 volt-amperes ÷ 240 volts = 78 amperes

Subfeeder Neutral (Section 220-22)

Lighting and Small Appliance Load	3882 volt-amperes
Range Load 6400 volt-amperes at 70% (see Section 220-22)	4480 volt-amperes
Space and Water Heating (no neutral) 240 volt	0 volt-amperes
Net Computed Load (neutral)	8362 volt-amperes

8362 volt-amperes ÷ 240 volts = 35 amperes

Minimum Size Feeder Required from Service Equipment to Meter Bank for 20 Dwelling Units:

Total Computed Load:

Lighting and Small Appliance Load 20 × 5520	110,400 volt-amperes
Water and Space Heating Load 20 × 8500	170,000 volt-amperes
Range Load 20 × 8000 volt-amperes	160,000 volt-amperes
Net Computed Load (20 dwelling units)	440,400 volt-amperes

Net Computed Load Using Optional Calculation (Table 220-32)
440,400 volt-amperes × .38 .... 167,352 volt-amperes
167,352 volt-amperes ÷ 240 volts = 697 amperes

Minimum Size Main Feeder Required (less house load) for 40 Dwelling Units:

Total Computed Load:

Lighting and Small Appliance Load 40 × 5520	220,800 volt-amperes
Water and Space Heating 40 × 8500	340,000 volt-amperes
Range Load 40 × 8000 volt-amperes	320,000 volt-amperes
Net Computed Load (40 dwelling units)	880,800 volt-amperes

Net Computed Load Using Optional Calculation (Table 220-32)
880,800 volt-amperes × .28 .... 246,624 volt-amperes
246,624 volt-amperes ÷ 240 volts = 1028 amperes

Feeder Neutral Load for Feeder from Service Equipment to Meter Bank for 20 Dwelling Units:

Lighting and Small Appliance Load 20 × 5520 volt-amperes	110,400 volt-amperes
First 3000 volt-amperes at 100%	3000 volt-amperes
107,400 volt-amperes at 35%	37,500 volt-amperes
Subtotal	40,500 volt-amperes
20 Ranges = 35,000 volt-amperes at 70% (See Table 220-19 and Section 220-22.)	24,500 volt-amperes
Total	65,000 volt-amperes

65,000 volt-amperes ÷ 240 volts = 271 amperes

Further Demand Factor (Section 220-22)

First 200 amperes at 100%	200 amperes
Balance: 71 amperes at 70%	50 amperes
Total	250 amperes

[Example 4(b) continued next page]

Feeder Neutral Load of Main Feeder (less house load) for 40 Dwelling Units:

Lighting and Small Appliance Load

40 × 5520 volt-amperes	220,800 volt-amperes
First 3000 volt-amperes at 100%	3000 volt-amperes
120,000 volt-amperes—3000 volt-amperes = 117,000 volt-amperes at 35%	40,950 volt-amperes
220,800 volt-amperes—120,000 volt-amperes = 100,800 volt-amperes at 25%	25,200 volt-amperes
Net Computed Lighting and Small Appliance Load	69,150 volt-amperes
40 Ranges = 55,000 volt-amperes at 70% .... (See Table 220-19 and Section 220-22)	38,500 volt-amperes
Total	107,650 volt-amperes

107,650 volt-amperes ÷ 240 volts = 449 amperes

Further Demand Factor (Section 220-22)

First 200 amperes at 100%	200 amperes
Balance: 249 amperes at 70%	174 amperes
Total	374 amperes

Example No. 5(a). Multifamily Dwelling Served at 208Y/120 Volts, Three Phase†

All conditions and calculations the same as for Multifamily Dwelling [Example No. 4(a)] served at 120/240 volts, single phase except as follows: Service to each dwelling unit shall be two phase legs and neutral.

Minimum Number of Branch Circuits Required for Each Dwelling Unit (Section 220-3):

Range Circuit: 8000 volt-amperes ÷ 208 volts = 38 amperes or a circuit of two No. 8 and one No. 10 as permitted by Section 220-22.

Minimum Size Subfeeder Required for Each Dwelling Unit (Section 215-2):

For 120/208-volt, 3-wire system (without ranges)
Net Computed Load: 3882 volt-amperes ÷ 208 volts = 18.7 amperes
For 120/208-volt, 3-wire system (with ranges)
Net Computed Load: 11,882 volt-amperes ÷ 208 volts = 57.1 amperes
Subfeeder neutral: 9482 volt-amperes ÷ 208 volts = 45.6 amperes

Minimum Size Feeders Required from Service Equipment to Meter Bank (For 20 Dwelling Units—10 with Ranges):

For 208Y/120-volt, 3-phase, 4-wire system
Ranges: Maximum number between any two phase legs = 4
Twice 4 = 8. Table 220-19 Demand = 23,000 volt-amperes.
Per phase demand: 23,000 volt-amperes ÷ 2 = 11,500 volt-amperes
Equivalent 3-phase load = 34,500 volt-amperes
Net Computed Load (total): 40,590 volt-amperes + 34,500 volt-amperes = 75,090 volt-amperes
75,090 volt-amperes ÷ (208)(1.732) = 208.4 amperes

Feeder Neutral Size
40,590 volt-amperes + 34,500 volt-amperes at 70% = 64,700 volt-amperes

Net Computed Neutral Load:
64,700 volt-amperes ÷ (208)(1.732) = 179.6 amperes

Minimum Size Main Feeder (less house load) (For 40 Dwelling Units—20 with Ranges):

For 208Y/120-volt, 3-phase, 4-wire system

Ranges: Maximum number between any two phase legs = 7

Twice 7 = 14. Table 220-19 Demand = 29,000 volt-amperes.

Per phase demand: 29,000 ÷ 2 = 14,500 volt-amperes

Equivalent 3-phase load = 43,500 volt-amperes

Net Computed Load (total): 69,150 volt-amperes + 43,500 volt-amperes = 112,650 volt-amperes

112,650 volt-amperes ÷ (208)(1.732) = 312.7 amperes

Main Feeder Neutral Size

69,150 volt-amperes + 43,500 volt-amperes at 70% = 99,600 volt-amperes

99,600 volt-amperes ÷ (208)(1.732) = 276.5 amperes

Further Demand Factor (Section 220-22)

200 amperes at 100% .... 200.0 amperes

76.5 amperes at 70% .... 53.6 amperes

Net Computed Load .... 253.6 amperes

Example No. 5(b). Optional Calculation for Multifamily Dwelling Served at 208Y/120 Volts, Three Phase†

All conditions and calculations the same as for Optional Calculation for Multifamily Dwelling [Example No. 4(b)] served at 120/240 volt, single phase except as follows:

Service to each dwelling unit shall be two phase legs and neutral.

Minimum Number of Branch Circuits Required for Each Dwelling Unit (Section 220-3):

Range Circuit: 8000 volt-amperes at 80% ÷ 208 volts = 30.7 amperes or a circuit of two No. 8 and one No. 10 as permitted by Section 220-22.

Space Heating: 6000 volt-amperes ÷ 208 volts = 28.8 amperes. Two 20-ampere, 2-pole circuits required, No. 12.

Minimum Size Subfeeder Required for Each Dwelling Unit.

Computed Load (120/208-volt, 3-wire circuit)

Net Computed Load: 18,782 volt-amperes ÷ 208 volts = 90.3 amperes

Net Computed Load (neutral): 3882 volt-amperes + 6000 volt-amperes + 2500 volt-amperes + 6400 volt-amperes at 70% = 16,862 volt-amperes.

16,862 volt-amperes ÷ 208 volts = 81.1 amperes

Minimum Size Feeder Required for Service Equipment to Meter Bank (For 20 dwelling units):

Net Computed Load: 167,352 volt-amperes ÷ (208)(1.732) = 464.9 amperes

Feeder Neutral Load:

65,000 volt-amperes ÷ (208)(1.732) = 180.4 amperes

Minimum Size Main Feeder Required (less house load) (for 40 dwelling units):

Net Computed Load: 246,624 volt-amperes ÷ (208)(1.732) = 685.1 amperes

Main Feeder Neutral Load:

107,650 volt-amperes ÷ (208)(1.732) = 298.8 amperes

Further Demand Factor (Section 220-22)

200 amperes at 100% .... 200.0 amperes

98.8 amperes at 70% .... 69.2 amperes

Net Computed Load .... 269.2 amperes

Example No. 6. Maximum Demand for Range Loads†

Table 220-19, Column A applies to ranges not over 12 kW. The application of Note 1 to ranges over 12 kW (and not over 27 kW) is illustrated in the following examples:

A. Ranges all of same rating.

Assume 24 ranges each rated 16 kW.

From Column A the maximum demand for 24 ranges of 12 kW rating is 39 kW.

16 kW exceeds 12 kW by 4.
5% × 4 = 20% (5% increase for each kW in excess of 12).
39 kW × 20% = 7.8 kW increase.
39 + 7.8 = 46.8 kW: value to be used in selection of feeders.

B. Ranges of unequal rating.

Assume 5 ranges each rated 11 kW.
2 ranges each rated 12 kW.
20 ranges each rated 13.5 kW.
3 ranges each rated 18 kW

5 × 12 = 60 Use 12 kW for range rated less than 12.
2 × 12 = 24
20 × 13.5 = 270
3 × 18 = 54

30 408 kW

408 ÷ 30 = 13.6 kW (average to be used for computation)

From Column A the demand for 30 ranges of 12 kW rating is 15 + 30 = 45 kW.

13.6 exceeds 12 by 1.6 (use 2).
5% × 2 = 10% (5% increase for each kW in excess of 12).
45 kW × 10% = 4.5 kW increase.
45 + 4.5 = 49.5 kW = value to be used in selection of feeders.

Example No. 8. Motors, Conductors, Overload, and Short-Circuit and Ground-Fault Protection†

(See Sections 430-6, 430-7, 430-22, 430-24, 430-32; 430-34, 430-52, 430-62, and Tables 430-150, and 430-152.)

Determine the conductor size, the motor overload protection, the branch-circuit short-circuit and ground-fault protection, and the feeder protection for one 25-h.p. squirrel-cage induction motor (full-voltage starting, service factor 1.15, Code letter F), and two 30-h.p. wound-rotor induction motors (40°C rise), on a 460-volt, 3-phase, 60-Hertz supply.

Conductor Loads

The full-load current of the 25-h.p. motor is 34 amperes (Table 430-150). A full-load current of 34 amperes × 1.25 = 42.5 amperes (Section 430-22). The full-load current of the 30-h.p. motor is 40 amperes (Table 430-150). A full-load current of 40 amperes × 1.25 = 50 amperes (Section 430-22).

The feeder ampacity will be 125 percent of 40 plus 40 plus 34, or 124 amperes (Section 430-24).

Overload and Short-Circuit and Ground-Fault Protection

Overload. Where protected by a separate overload device, the 25-horsepower motor, with full-load current of 34 amperes, must have overload protection of not over 42.5 amperes [Sections 430-6(a) and 430-32(a)(1)]. Where protected by a separate overload device, the 30-horsepower motor, with full-load current of 40 amperes, must have overload protection of not over 50 amperes [Sections 430-6(a) and 430-32(a)(1)]. If the overload protection is not sufficient to start the motor or to carry the load, it may be increased according to Section 430-34. For a motor marked "thermally protected," overload protection is provided by the thermal protector [See Sections 430-7(a)(12) and 430-32(a)(2)].

Branch-Circuit Short-Circuit and Ground-Fault. The branch circuit of the 25-horsepower motor must have branch-circuit short-circuit and ground-fault protection of not over 300 percent for a nontime-delay fuse (Table 430-152) or $3.00 \times 34 = 102$ amperes. The next smaller standard size fuse is 100 amperes. The fuse size may be increased to 110 or 125 amperes (Section 430-52, Exception No. 1) if the 100-ampere fuse is not sufficient for the starting current of the motor.

For the 30-horsepower motor, the branch-circuit short-circuit and ground-fault protection is 150 percent (Table 430-152) or $1.50 \times 40 = 60$ amperes. Where the maximum value of branch-circuit short-circuit and ground-fault protection is not sufficient to start the motor, the value for a nontime-delay fuse may be increased to 400 percent [Section 430-52, Exception a.].

Feeder Circuit. The maximum rating of the feeder short-circuit and ground-fault protection device is based on the sum of the largest branch-circuit protective device (110-ampere fuse) plus the sum of the full-load currents of the other motors or 110 plus 40 plus 40 = 190 amperes. The nearest standard fuse which does not exceed this value is 175 amperes [Section 430-62(a)].

Index